The Human
Impact

ON THE NATURAL
ENVIRONMENT

Fifth Edition

Andrew Goudie

The Human Impact

ON THE NATURAL ENVIRONMENT

Fifth Edition

The MIT Press
Cambridge, Massachusetts

This MIT Press edition, 2000

Printed and bound in Great Britain

Library of Congress Cataloging-in-Publication Data

Goudie, Andrew.
    The human impact on the natural environment / Andrew Goudie. — 5th ed.
      p. cm.
    Includes bibliographical references and index.
    ISBN 0–262–07202–5 (hc : alk. paper) — ISBN 0–262–57138–2 (pb : alk. paper)
    1. Nature—Effect of human beings on. I. Title.

  GF75.G68   2000
  304.2—dc21   99–025279

Contents

Preface to the Fifth Edition   ix
Acknowledgements   x

1 Introduction   1

The development of ideas   1
The development of human population and stages
    of cultural development   12
Hunting and gathering   16
Humans as cultivators, keepers and metal workers   20
Modern industrial and urban civilizations   31

2 The Human Impact on Vegetation   37

Introduction   37
The use of fire   38
Fires: natural and anthropogenic   41
The temperatures attained in fires   44
Some consequences of fire suppression   45
Some effects of fire on vegetation   48
The role of grazing   49
Deforestation   52
Secondary rain forest   61
The human role in the creation and maintenance
    of savanna   62
The spread of desert vegetation on desert margins   67
The maquis of the Mediterranean lands   76
The prairie problem   78
Post-glacial vegetational change in Britain and
    Europe   80
Introduction, invasion and explosion   83
Air pollution and its effects on plants   91
Forest decline   94
Miscellaneous causes of plant decline   99
The change in genetic and species diversity   103

3   Human Influence on Animals   **106**

Introduction   106
Domestication of animals   106
Dispersal and invasions of animals   107
Human influence on the expansion of animal
    populations   115
Causes of animal contractions and decline: pollution   121
Habitat change and animal decline   130
Other causes of animal decline   137
Animal extinctions in prehistoric and historic times   142
Modern-day extinctions   150

4   The Human Impact on the Soil   **160**

Introduction   160
Salinity: natural sources   161
Human agency and increased salinity   163
The spread of salinity   168
Consequences of salinity   171
Reclamation of salt-affected lands   173
Lateritization   173
Accelerated podzolization and acidification   175
Soil structure alteration   177
Soil drainage and its impact   180
Soil fertilization   183
Fires and soil   185
Soil erosion: general considerations   185
Soil erosion associated with deforestation and
    agriculture   188
Soil erosion produced by fire   196
Soil erosion associated with construction and
    urbanization   197
Attempts at soil conservation   199

5   The Human Impact on the Waters   **203**

Introduction   203
Deliberate modification of rivers   203
Urbanization and its effects on river flow   219
Deforestation and its effects on river flow   222
The human impact on lake levels   226
Changes in groundwater conditions   229
Water pollution   235
Chemical pollution by agriculture and other
    activities   239
Deforestation and its effects on water quality   251
Thermal pollution   254
Pollution with suspended sediments   257
Marine pollution   257

6 Human Agency in Geomorphology 261

Introduction 261
Landforms produced by excavation 263
Landforms produced by construction and dumping 269
Accelerated sedimentation 275
Ground subsidence 280
Arroyo trenching, gullies and peat haggs 288
Accelerated weathering and the tufa decline 293
Accelerated mass movements 294
Deliberate modification of channels 299
Non-deliberate river-channel changes 300
Reactivation and stabilization of sand dunes 307
Accelerated coastal erosion 311
Changing rates of salt marsh accretion 322
The human impact on seismicity and volcanoes 323

7 The Human Impact on Climate and the Atmosphere 328

World climates 328
The greenhouse gases – carbon dioxide 333
Other greenhouse gases 336
Aerosols 340
Vegetation and albedo change 343
Forests, irrigation and climate 345
The possible effects of water diversion schemes 347
Lakes 347
Urban climates 348
Urban air pollution 355
Air pollution: some further effects 362
Stratospheric ozone depletion 371
Deliberate climate modification 374
Conclusion 377

8 The Future 379

Introduction 379
The climatic future 381
The patterning of change 385
Potential consequences of global warming 389
Changes in tropical cyclones 390
The cryosphere 392
The biosphere 396
The hydrosphere 400
The aeolian environment 403
Coastal environments 404
Future uncertainty 415
Future responses 416

9   Conclusion   **419**

The power of non-industrial and pre-industrial
    civilizations   419
The proliferation of impacts   420
Changes are reversible   425
The susceptibility to change   428
Human influence or nature?   431
Into the unknown   433

Guide to Reading   434
References   438
Index   490

Preface to the Fifth Edition

In this edition I have retained the structure of the four previous editions. However, I have updated substantial portions of the text to take account of the ever-increasing amount of material appearing on this theme. I have also revised and expanded the guides to further reading, augmented and updated the references, and included many new tables and figures.

A.S.G.

Acknowledgements

The publisher and author are grateful for permission to use figures and photographs from the following publications: Atkinson, B. W., 'Thunder in south-east England', 'Total thunder rain in south-east England' and 'number of days with thunder overhead in south-east England', from *Transactions of the Institute of British Geographers*, 44, 1968; Birkeland, P. W. and Larson, E. E., 'Correlation between quantity of waste water pumped into a deep well and the number of earthquakes near Denver, Colorado', from *Putnam's geology* (Oxford University Press, New York, reprinted by permission of Oxford University Press, 1978); Bockh, A., 'Variations in the level of Lake Valencia, Venezuela to 1968', from M. T. Farvar, J. P. Milton, eds, *The careless technology* (Tom Stacey, London, 1973); Brampton, A. H., 'A selection of "hard engineering" structures designed to afford coastal protection', from A. H. Brampton, 'Cliff conservation and protection: methods and Practices to Resolve Conflicts', in J. Hooke, ed., *Coastal defence and earth science conservation* (Geological Society Publishing House, 1998); Brimblecombe, P., 'Thunder in south-east England' and 'Thunderstorms per year in London', from 'London air pollution 1500–1900', *Atmospheric environment*, 11, 1977; Brookes, A., 'Principal types of adjustment in straightened river channels', from 'The distribution and management of channelized streams in Denmark', *Regulated rivers*, 1, 1987; Brown, A. A. and Davis, K. P., 'The reduction in area burned per area protected for the USA between 1926 and 1969 as a result of fire-suppression policies', from *Forest fire control and its use*, 2nd edn (McGraw-Hill, New York, 1973, reproduced with permission of The McGraw-Hill Companies); Budyko, M. I. and Izrael, Y. A., 'Deviations in annual precipitation means (cm) for two past warm phases', from *Anthropogenic climatic change* (Copyright © 1991 The Arizona Board of Regents. Reprinted by permission of the University of Arizona Press); Budyko, M. I. and Izrael, Y. A., 'Relative changes in mean latitudinal precipitation on the continents of the Northern hemisphere with 1°C higher mean surface air temperature', from *Anthropogenic climatic change* (Copyright © 1991 The Arizona Board of Regents. Reprinted by permission of the University of Arizona Press); Bussing, C., 'Cattle on feed lots in Colorado 1917–1971', from D. D. McPhail, ed., *The high plains: problem of semiarid environments* (Colorado State University, Fort Clark, R. R., 'Flint mines at Grimes Graves, Norfolk', from *Grimes Graves* (Her Majesty's Stationery Office, London, 1963); Cochrane, R., 'The changing state of the vegetation cover in New Zealand', from A. G. Anderson, ed., *New Zealand in maps* (Hodder & Stoughton Educational, London, 1977); Cohen and Rushton, 'The effect of air quality on plant growth in Leeds, England, in 1913', data from R. Barrass, *Biology, food and people* (Hodder & Stoughton Educational, London, 1974); Cole, S., 'A neolithic chert axe-blade from Denmark, of the type which has been shown to be effective at cutting forest in experimental studies', from *The Neolithic Revolution*, 5th edn (The British Museum [Natural History], London, 1970); Cooke, R. U., and Reeves, R. W., 'A model for the formation of arroyos (gullies) in the south-western USA', from *Arroyos and environmental change in the American south-west* (Clarendon Press, Oxford, 1976, reprinted with permission of Oxford Univeristy Press); Council on Environmental Quality, 'Changes in water pollution in the Great Lakes of North America', from 17th (1986) and 22nd (1992) Annual Reports; Currey, D. T., 'The effect of an irrigation channel in causing a local rise in the water table, through the development of a pressure barrier which influences the movement of sub-surface drainage', from J. R. Hails, ed., *Applied geomorphology* (Elsevier Publishing, Amsterdam, 1977); Currey, D. T., 'Piezometer record to show ground-water level changes in the Murray Valley, Victoria, Australia from 1962 to 1974 as a result of the extension of irrigation', from J. R. Hails, ed., *Applied geomorphology* (Elsevier Publishing, Amsterdam, 1977); Darby, H. C., 'The changing distribution of forest in Central Europe between (a) AD 900 and (b) AD 1900', from W. L. Thomas, ed., *Man's role in changing the face of the Earth* (University of Chicago Press, Chicago 1956, by permission of the University of Chicago Press © The University of Chicago 1956); Dolan, R., Godfrey, P. J. and Odum, W. E., 'Cross-sections of two barrier islands in North Carolina, USA', from *Man's impact on the barrier islands of North Carolina American Scientist*, 61, 1973; Doughty, R. W., 'The spread of the English house sparrow in the New World', from *The English sparrow in the American landscape: a paradox in nineteenth century wildlife conservation*, Research paper 19, School of Geography, University of Oxford, 1978; Edmonson, W. T., 'Changes in the state of Lake Washington, USA, associated with levels of untreated sewage from 1933 to

1973', from W. W. Murdoch, ed., *Environment* (Sinauer Associates, Sunderland, 1975); Ehrlich, P. R., Ehrlich, A. H., and Holdren, J. P., 'The growth of human numbers for the past half million years', from *Ecoscience: population, resources, environment* (W. H. Freeman, San Francisco, 1977); Elton, C. S., 'The spread of the Japanese beetle, *Popilla japonica*, in the eastern USA', from *The ecology of invasion by plants and animals* (Methuen, London, 1958); Frenkel, R. E., 'Estimation of the establishment of alien plant species in California, USA since 1750', from *Ruderal vegetation along some California roadsides*, UC Publications in Geography, no. 20 © 1970; Gameson, A. L. H., and Wheeler, A., 'The average dissolved oxygen content of the River Thames at half-tide in the July–September quarter since 1890', from J. Cairns, K. L. Dickson and E. E. Herricks, eds, *Recovery and restoration of damaged ecosystems* (University Press of Virginia, Charlottesville, 1977. Reprinted with permission of the University Press of Virginia); Gilbert, 'Air pollution in north-east England and its impact upon growth area for lichens', from R. Barrass, *Biology, food and people* (Hodder & Stoughton Educational, London, 1974); Gorman, M., 'A set of general design rules for nature reserves based on theories of island biogeography', from *Island ecology* (Chapman and Hall Publishers Ltd, London, 1979, copyright with kind permission of Kluwer Academic Publishers); Gorman, M., 'Some relationships between the size of "islands" and numbers of species', from *Island ecology* (Chapman and Hall Publishers Ltd, London, 1979, copyright with kind permission of Kluwer Academic Publishers); Goudie, A. S. 'A schematic representation of some of the possible influences causing climatic change', from *Environmental change*, 3rd edn (Clarendon Press, Oxford, 1992, reprinted by permission of Oxford University Press); Goudie, A. S., 'The concentration of dust storms in the USA in 1939', from 'Dust storms in space and time', *Progress in physical geography*, 7, 1983; Goudie, A. S., and Wilkinson, J. C., 'The Ghyben–Herzberg relationship between fresh and saline ground water and the effect of excessive pumping from the well', from *The warm desert environment* (Cambridge University Press, Cambridge, 1977); Green, F. H. W. 'Annual total area of field drainage by the tile drains in England and Wales', from *Recent changes in land use and treatment, Geographical Journal*, 142, 1996; Green, F. H. W., 'Percentages of drained agricultural land in Europe', from *Field drainage in Europe, Geographical Journal*, 144, 1978; Haagen-Smit, 'Possible reactions involving primary and secondary pollutants', from R. A. Bryson and J. E. Kutzbach, eds, *Air pollution* (Commission on College Geography Resource Paper 2. Washington DC: Association of American Geographers); Haggett, P., 'Land rotation and population density', from *Geography: a modern synthesis*, 3rd edn (Prentice-Hall, London,

1979); Haigh, M. J., 'Some shapes produced by shale tipping', from *Evolution of slopes on artificial landforms – Blaenavon, UK*, Research Paper 183, Department of Geography, University of Chicago; Hollis, G. E., 'Some hydrological consequences of urbanization', from 'The effects of urbanization on floods of different recurrence intervals', *Water resources research*, 11, 1975; Hughes, R. J., Sullivan, M. E., & York, D., 'Rates of erosion in Papus New Guinea in the Holocene derived from rates of sedimentation in Kuk Swamp', from *Human-induced erosion in a highlands catchment in Papua New Guinea: the prehistoric and contemporary records, Zeitschrift fur Geomorphologie, Supplementband*, 83, 1991; Johansen, 'The spread of contour-strip soil conservation methods in Wisconsin, USA, between 1939 and 1967', from S. W. Trimble and S. W. Lund, eds, *Soil conservation and the reduction of erosion and sediment in the Coon Creek Basin, Wisconsin, US Geological Survey Professional Paper*, 1234, 1982; Judd, W. R., 'Relationships between reservoir levels and earthquake frequencies', from *Seismic effects of reservoir impounding, Engineering geology*, 8, 1974; Keller, E. A., 'Comparison of the natural channel morphology and hydrology with that of a channelized stream', from D. R. Coates, ed., *Geomorphology and engineering* (Hutchinson and Ross, 1976); Komar, P. D., 'Examples of the effects of shoreline installations on beach and shoreline morphology', from *Beach processes and sedimentation* (Prentice-Hall, Englewood Cliffs, 1976); Laporte, L. F., 'Maximum temperatures for the spawning and growth of fish', from *Encounter with the Earth* (copyright © by permission of Harper and Row Publishers, Inc., 1975); MacCrimmon, 'The original area of distribution of the brown trout and areas where it has been artificially naturalized', from J. Illies, ed., *Introduction to zoogeography* (Macmillan, London, 1974); Manshard, W., 'The irrigated areas in Sind (Pakistan) along the Indus Valley', from *Tropical agriculture* (Addison-Wesley-Longman, 1974, © Bibliographisches Institut AG, Mannheim); Meade, R. H. and Trimble, S. W., 'The decline in suspended sediment discharge to the eastern seaboard of the USA between 1910 and 1970 as a result of soil conservation measures . . .', from *Changes in sediment loads in rivers of the Atlantic drainage of the United States since 1900, Publication of the International Association of Hydrological Science*, 113, 1974; Mellanby, K., 'The increase of lichen cover on trees outside the city of Belfast, Northern Ireland (after Fenton)', from *Pesticides and pollution* (Fontana, London, 1967); Murton, R. K. 'The changing range of the little ringed plover: the increase in the number summering in Britain', from *Man and birds* (HarperCollins Publishers, London, 1971); Nature Conservancy Council, 'The changing range of the little ringed plover, related to habitat change, especially as a result of the increasing number of gravel pits', from *Nature conservation and agriculture* (Nature Conservancy Council, Her Majesty's Stationery Office, London, 1977); Nature Conservancy Council, 'Examples of habitat loss in England since the mid-eighteenth century' (Nature Conservancy Council: Her Majesty's Stationery Office, London, 1997); Nature Conservancy Council, 'The loss of lowland chalk grasslands in Dorset, southern England', from *Nature conservation in Great Britain* (Nature Conservancy Council, Shrewsbury, 1984); Nature Conservancy Council, 'Losses of lowland heath in southern England', from *Nature conservation in Great Britain* (Nature Conservancy Council, Shrewsbury, 1984); Nature Conservancy Council, 'Reduction in the range of the silver spotted skipper butterfly (Hesperia comma)', from *Nature conservation and agriculture* (Nature Conservancy Council, Her Majesty's Stationery Office, London, 1977); Nature Conservancy Council, 'Reduction in the range of species related to habitat loss', from *Nature conservation and agriculture* (Nature Conservancy Council: Her Majesty's Stationery Office, London, 1977); Nature Conservancy Council, 'Reduction in the range of species related to habitat loss associated with drainage activities' (Nature Conservancy Council: Her Majesty's Stationery Office, London, 1997); Nature Conservancy Council, 'Summary of recent habitat and species changes in Britain', from *Nature conservation in Great Britain* (Nature Conservancy Council, Shrewsbury, 1984); Pereira, H. C., 'The increase of water yield after clear-felling a forest: a unique confirmation from the Coweeta catchment in North Carolina', from *Land use and water resources in temperate and tropical climates* (Cambridge University Press, Cambridge, 1973); Rapp, A., Le Houerou, H. N. and Lundholm, B., 'Desert encroachment in the northern Sudan 1958–75, as represented by the position of the boundary between sub-desert scrub and grassland in the desert', from *Ecological Bulletin*, 24, 1976; Rapp, A., 'Relation between spacing of wells and over-grazing', from *A review of desertization in Africa – water, vegetation and man* (Secretariat for International Ecology, Stockholm, Report no. 1, 1974); Richards, K. S. and Wood, R., 'Trends in 25 month moving averages of runoff as a percentage of 25 month moving averages of rainfall for the River Tame, Lee Marston, England 1957–74', from K. J. Gregory, ed., *River channel changes* (copyright © 1977 John Wiley & Sons, Inc., reprinted by permission of John Wiley & Sons, Inc.); Roberts, Neil, 'The human colonization of Ice-Age earth', from *The Holocene: an environmental history* (Blackwell Publishers, Oxford, 1989); Shiklomanov, A. I., 'Changes in annual runoff in the CIS due to human activity during 1936–2000', from J. C. Rodda, ed., *Facets of hydrology II* (copyright © 1985 John Wiley & Sons, Inc., reprinted by permission of John Wiley & Sons Inc.); Shiklomanov, A. I., 'Some major schemes proposed for large-scale inter-basin water transfers', from J. C. Rodda, ed., *Facets of hydrology II* (copyright © 1985 John Wiley & Sons, Inc., reprinted by permission of John Wiley & Sons Inc.); Simpson, 'Evidence of salt-water intrusion exemplified by chemical analyses of a line of well-waters from the Salinas valley, California', from D. K. Todd, ed., *Groundwater hydrology* (copyright © 1959 John Wiley & Sons, Inc., reprinted by permission of John Wiley & Sons Inc.); Spate, O. H. K. and Learmonth, A. T. A., 'The Madurai–Ramanthapuram tank country in south India', from *India and Pakistan* (Methuen, London, 1967); Spencer, J. E. and Thomas, W. L., 'The spread of plough agriculture in the Old World', from *Introducing cultural geography*, 2nd edn (copyright © 1978 John Wiley & Sons, Inc., reprinted by permission of John Wiley & Sons, Inc.); Spencer, J. E. and Thomas, W. L., 'The early development of transport', from *Introducing cultural geography*, 2nd edn (copyright © 1978 John Wiley & Sons, Inc., reprinted by permission of John Wiley & Sons, Inc.); Spencer, J. E., and Thomas, W. L., 'The diffusion of mining and smelting in the Old World', from *Introducing cultural geography*, 2nd edn (copyright © 1978 John Wiley & Sons, Inc., reprinted by permission of John Wiley & Sons, Inc.); Strandberg, C. H. 'Biological concentration occurs when relatively indestructible substances (DDT for example) are ingested by lesser organisms at the base of the food pyramid', from G. H. Smith, ed., *Conservation of natural resources*, 4th edn (copyright © 1971 John Wiley & Sons, Inc., reprinted by permission of John Wiley & Sons, Inc.); Taylor, C., 'Plan of Neolithic plough marks: some of the first tangible evidence of the first agricultural revolution in the British Isles', from *Fields in the English landscape* (Dent & Sons Ltd, London 1975, © The Orion Publishing Group Ltd); Trimble, S. W., 'Changes in the evolution of fluvial landscapes in the Piedmont of Georgia, USA in response to land-use change between 1700 and 1970', from *Man-induced soil erosion on the southern Piedmont* (Soil Conservation Society of America, © Soil Conservation Society of America, 1974); US Geological Survey, 'Historical sediment and water discharge trends for the Colorado River', ed. H. E. Schwarz, J. Emel, W. J. Dickens, P. Rogers and J. Thompson, 'Water quality and flows', from B. L. Turner, W. C. Clark, R. W. Kates, J. F. Richards, J. T. Matthews and W. B. Meyer, eds, *The earth as transformed by human action* (Cambridge University Press, Cambridge, 1991); Wallen, 'Annual mean concentration of sulphate in precipitation in Europe', from M. W. Holdgate, M. Kassas and G. F. White, eds, *The world environment 1972– 1982* (Tycooly, Dublin, 1982); Wallwork, K. L., 'Subsidence in the salt area of mid-Cheshire, England, in 1954', from *Subsidence in the mid-Cheshire industrial area*, *Geographical Journal*, 122, 1956; Watson, A., 'The distribution of pits and ponds in a portion of north-western England', from 'The origin and distribution of closed depressions in south-west Lancashire and north-west Cheshire', unpublished BA dissertation, University of Oxford; Wilkinson, W. B., and Brassington, F. C., 'Ground-water levels in the London area', from R. A. Downing and W. B. Wilkinson, eds, *Applied groundwater hydrology – a British perspective* (Clarendon Press, Oxford, 1991, by permission of Oxford University Press); Wooster, W. S., 'Decrease in the salinity of the Great Bitter Lake, Egypt, resulting from the intrusion of fresher water by way of the Suez Canal', from 'The ocean and man', *Scientific American*, 221, 1969; Ziswiler, 'The former and present distribution of the bison in North America', from J. Illies, ed., *Introduction of zoogeography* (Macmillan, London, 1974); Ziswiler, 'Time series of animal extinctions since the seventeenth century, in relation to the human population increase', figure (b), from *The earth and human affairs* (National Academy of Sciences, 1972).

Above all, I am most grateful to Jan Magee for preparing the manuscript, to Ailsa Allen and Neil McKintosh for preparing the figures, and to Martin Barfoot for preparing some of the plates.

Introduction

1

The development of ideas

In the history of Western thought, men have persistently asked three questions concerning the habitable earth and their relationship to it. Is the earth, which is obviously a fit environment for man and other organic life, a purposefully made creation? Have its climates, its relief, the configuration of continents influenced the moral and social nature of individuals, and have they had an influence in moulding the character and nature of human culture? In his long tenure of the earth, in what manner has man changed it from its hypothetical pristine condition?

(C. J. Glacken, 1967, p. vii)

Glacken's third question forms the focus of this book. Since, he wrote, its importance has been increasingly recognized, for as Yi-fu Tuan (1971) put it: 'The fact of diminishing nature and of human ubiquity is now obvious.'

This last question received relatively little systematic attention until the eighteenth century, and overall it has been less central in the history of geographical thought than the second – the one which asks whether environment has an influence on people (Glacken, 1967). Indeed, in the late seventeenth century, when the Creation was deemed to have occurred in 4004 BC, a grand conception of nature as a divinely created order for the well-being of all life was formulated (Glacken, 1956). In other words, it was the first question to which people addressed themselves. Men and women were seen to be caretakers, stewards of God, and their task was thought to improve the primeval aspect of the earth through tillage and in other ways. Many thinkers at this time, such as John Ray, author of *The wisdom of God manifested in the works of the Creation* (1691), believed people were living in a world – which they could change and improve – where the order and beauty of nature, manifestations of the Creator's skill and wisdom, could be seen everywhere. Indeed, the biblical notion expressed in the book of Genesis, which set humans above and against nature, entitled to rule like absolute despots (Passmore, 1974: 9), was formulated into a philosophy of science and progress by men like Bacon, Galileo and Descartes (Swift, 1974). Bacon foresaw the need to combine knowledge with invention to give humankind control of nature. Also, for a time in the nineteenth century, biological interpretations fortified people's belief in the right to deal with nature however they chose. Darwin's theory of natural selection was transformed by Herbert Spencer into the doctrine of 'the survival of the fittest'.

Count Buffon can be regarded as the first Western scientist to be concerned directly and intimately with the human impact on the natural environment (Glacken, 1963). He contrasts the appearance of inhabited and uninhabited lands: the anciently inhabited countries have few woods, lakes or marshes, but they have many heaths and scrub; their mountains are bare, and their soils are less fertile because they lack the organic matter which woods, felled in inhabited countries, supply, and the herbs are browsed. Buffon was also much involved with the domestication of plants and animals – one of the major transformations in nature brought about by human actions.

Studies of the torrents of the French and Austrian Alps, undertaken in the late eighteenth and early nineteenth centuries, deepened immeasurably the realization of human capacity to change the environment. Fabre and Surell studied the flooding, siltation, erosion and division of watercourses brought about by deforestation in the Alps. Similarly, de Saussure showed that Alpine lakes had suffered a lowering of water levels in recent times because of deforestation. In Venezuela, in the valley of Aragua, von Humboldt concluded that the lake level of Lake Valencia in 1800 (the year of his visit) was lower than it had been in previous times, and that deforestation, the clearing of plains and the cultivation of indigo were among the causes of the gradual drying up of the basin.

Comparable observations were made by the French rural economist Boussingault (1845). He returned to Valencia some 25 years after Humboldt and noted that the lake was actually rising. He described this reversal to political and social upheavals following the granting of independence to the colonies of the erstwhile Spanish Empire. The freeing of slaves had led to a decline in agriculture, a reduction in the application of irrigation water and the re-establishment of forest.

Boussingault also reported some pertinent hydrological observations that had been made on Ascension Island:

In the Island of Ascension there was an excellent spring situated at the foot of a mountain originally covered with wood; the spring became scanty and dried up after the trees which covered the mountain had been felled. The loss of the spring was rightly ascribed to the cutting down of the timber. The mountain was therefore planted anew. A few

Plate 1.1
Mary Somerville (1780–1872)

years afterwards the spring reappeared by degrees, and by and by
followed with its former abundance.

(p. 685)

Charles Lyell, in his *Principles of geology*, one of the most
influential of all scientific works, referred to the human impact
and recognized that tree-felling and drainage of lakes and marshes
tended 'greatly to vary the state of the habitable surface'. Overall,
however, he believed that the forces exerted by people were
insignificant in comparison with those exerted by nature:

If all the nations of the earth should attempt to quarry away the lava
which flowed from one eruption of the Icelandic volcanoes in 1783,
and the two following years, and should attempt to consign it to the
deepest abysses of the ocean they might toil for thousands of years
before their task was accomplished. Yet the matter borne down by
the Ganges and Burrampooter, in a single year, probably very much
exceeds, in weight and volume, the mass of Icelandic lava produced
by that great eruption.

(Lyell, 1835: 197)

Lyell somewhat modified his views in later editions of the *Principles* (see e.g. Lyell, 1875), largely as a result of his experiences
in the United States, where recent deforestation in Georgia and
Alabama had produced numerous ravines of impressive size.

One of the most important physical geographers to show
concern with our theme was Mary Somerville (1858) (plate 1.1),
who clearly appreciated the unexpected results that occurred
as man 'dextrously avails himself of the powers of nature to
subdue nature':

Man's necessities and enjoyments have been the cause of great changes
in the animal creation, and his destructive propensity of still greater.
Animals are intended for our use, and field-sports are advantageous by

Plate 1.2
George Perkins Marsh in 1861.
Marsh was the author of *Man
and nature* – probably the
most important landmark in
the history of the study of the
role of humans in changing
the face of the earth

encouraging a daring and active spirit in young men; but the utter
destruction of some races in order to protect those destined for his
pleasure, is too selfish, and cruelty is unpardonable: but the ignorant
are often cruel. A farmer sees the rook pecking a little of his grain, or
digging at the roots of the springing corn, and poisons all his neigh-
bourhood. A few years after he is surprised to find his crop destroyed
by grubs. The works of the Creator are nicely balanced, and man
cannot infringe his Laws with impunity.

(Somerville, 1858: 493)

This is in effect a statement of one of the basic laws of ecology:
that everything is connected to everything else and that one
cannot change just one thing in nature.

   Considerable interest in conservation, climatic change and extinctions arose among European colonialists who witnessed some
of the consequences of Western-style economic development in
tropical lands (Grove, 1997). However, the extent of human
influence on the environment was not explored in detail and on
the basis of sound data until George Perkins Marsh (plate 1.2)
published *Man and nature* (1864), in which he dealt with human
influence on the woods, the waters and the sands. The following
extract illustrates the breadth of his interests and the ramifying
connections he identified between human actions and environmental changes:

Vast forests have disappeared from mountain spurs and ridges; the
vegetable earth accumulated beneath the trees by the decay of leaves
and fallen trunks, the soil of the alpine pastures which skirted and in-
dented the woods, and the mould of the upland fields, are washed away;
meadows, once fertilized by irrigation, are waste and unproductive,

because the cisterns and reservoirs that supplied the ancient canals are broken, or the springs that fed them dried up; rivers famous in history and song have shrunk to humble brooklets; the willows that ornamented and protected the banks of lesser watercourses are gone, and the rivulets have ceased to exist as perennial currents, because the little water that finds its way into their old channels is evaporated by the droughts of summer, or absorbed by the parched earth, before it reaches the lowlands; the beds of the brooks have widened into broad expanses of pebbles and gravel, over which, though in the hot season passed dryshod, in winter sealike torrents thunder; the entrances of navigable streams are obstructed by sandbars, and harbours, once marts of an extensive commerce, are shoaled by the deposits of the rivers at whose mouths they lie; the elevation of the beds of estuaries, and the consequently diminished velocity of the streams which flow into them, have converted thousands of leagues of shallow sea and fertile lowland into unproductive and miasmatic morasses.

(Marsh, 1965: 9)

More than a third of the book is concerned with 'the woods'; Marsh does not touch upon important themes like the modifications of mid-latitude grasslands, and he is much concerned with Western civilization. Nevertheless, employing an eloquent style and copious footnotes, Marsh, the versatile Vermonter, stands as a landmark in the study of environment (Thomas, 1956).

Marsh, however, was not totally pessimistic about the future role of humankind or entirely unimpressed by positive human achievements (1965: 43–4):

New forests have been planted; inundations of flowing streams restrained by heavy walls of masonry and other constructions; torrents compelled to aid, by depositing the slime with which they are charged, in filling up lowlands, and raising the level of morasses which their own overflows had created; ground submerged by the encroachment of the ocean, or exposed to be covered by its tides, has been rescued from its dominion by diking; swamps and even lakes have been drained, and their beds brought within the domain of agricultural industry; drifting coast dunes have been checked and made productive by plantation; sea and inland waters have been repeopled with fish, and even the sands of the Sahara have been fertilized by artesian fountains. These achievements are far more glorious than the proudest triumphs of war . . .

Reclus (1873), one of the most prominent French geographers of his generation, and an important influence in the United States, also recognized that the 'action of man may embellish the earth, but it may also disfigure it; according to the manner and social condition of any nation, it contributes either to the degradation or glorification of nature' (p. 522). He warned rather darkly (p. 523) that 'in a spot where the country is disfigured, and where all the grace of poetry has disappeared from the landscape, imagination dies out, and the mind is impoverished; a spirit of routine and servility takes possession of the soul, and leads it on to torpor and death.' Reclus (1871) also displayed a concern with the relationship between forests, torrents (p. 335) and sedimentation (p. 338):

It is, however, the improvidence of the inhabitants, and not so much the geological constitution of the soil, which is the principal cause of the devastating action of the streams. In the mountains of Dauphiny and Provence, during the course of centuries, the trees have been cut down by greedy speculators, and by senseless farmers who wished to add some little strips of land to the fields in the valleys and to the pastures on the summits; but when they destroyed the forest they also destroyed the very land it stood on.

Although the torrents lower the mounts, on the other hand they elevate the plains; but their deposits not being pulverised into clays and sands, are often the means of bringing another disaster on the inhabitants, who find their fertile land covered beneath enormous masses of rocks and pebbles.

Reclus obtained many of his ideas from Marsh, and another important contributor, Woeikof (1842–1914) of St Petersburg, obtained some of his ideas from Reclus. Woeikof believed that the earth's surface was composed of mobile bodies (*corps meubles*), such as soil, subsoil and gravels, and that humans influenced them through an intermediary, vegetation. He refers to the effects of fire, the climatic effects of forest removal, the destruction of chernozems (black soils) by ploughing, gullying, the siltation of streams and the barrenness of karst.

In 1904 Friedrich coined the term 'Raubwirtschaft', which can be translated as economic plunder, robber economy or, more simply, devastation. This concept has been extremely influential but is open to criticism. He believed that destructive exploitation of resources leads of necessity to foresight and to improvements, and that after an initial phase of ruthless exploitation and resulting deprivation human measures would, as in the old countries of Europe, result in conservation and improvement:

I can conceive the Raubwirtschaft of our times as a depletion of land, wild plants, and animals which will disappear quickly as the culture of northwest Europe is substituted for that of colonial and primitive peoples. Such a substitution will bring an intensive use of earth resources, which aims at a firmer and firmer establishment of man on the earth.

(Friedrich, cited in Whitaker, 1940: 157)

This idea was opposed by Sauer (1938) and Whitaker (1940), the latter pointing out that some soil erosion could well be irreversible (p. 157):

It is surely impossible for anyone who is familiar with the eroded loessial lands of northwestern Mississippi, or the burned and scarred rock hills of north central Ontario, to accept so complacently the damage to resources involved in the process of colonization, or to be so certain that resource depletion is but the forerunner of conservation.

None the less Friedrich's concept of robber economy was adopted and modified by the great French geographer, Jean Brunhes, in his *Human geography* (1920). He recognized the interrelationships involved in anthropogenic environmental change (p. 332): 'Devastation always brings about, not a catastrophe, but a series of catastrophes, for in nature things are dependent one upon the other.' Moreover, Brunhes acknowledged that the 'essential facts' of human geography included 'Facts of Plant and Animal Conquest' and 'Facts of Destructive Exploitation'. At much the same time other significant studies were made of the same theme. Shaler of Harvard (*Man and the earth*, 1912) was very much concerned with the destruction of mineral resources (a topic largely neglected by Marsh).

In spite of this work, however, the subject of the human role in changing the face of the earth received relatively little attention from geographers in the nineteenth century and the early decades of the twentieth. One of the main features of the 'classical' view of geography put forward by men like Ratzel (Wrigley, 1965: 5) was the investigation of the effects of the physical environment on the functioning and development of societies, and environmentalism persisted as a potent force, notably in American geography, through the works of Davis, Brigham, Semple and Huntington.

Very definite shifts of emphasis appeared, however, in the 1920s and 1930s when, under the influence of Harlan Barrows and Carl Sauer, there was an eager search for an alternative to so-called 'naive environmental determinism'. Sauer (plate 1.3) led an effective campaign against destructive exploitation (Speth, 1977), reintroduced Marsh to a wide public, recognized the ecological virtues of some so-called primitive peoples, concerned himself with the great theme of domestication, concentrated on the landscape changes that resulted from human action, and gave clear and far-sighted warnings about the need for conservation (Sauer, 1938: 494):

We have accustomed ourselves to think of ever expanding productive capacity, of ever fresh spaces of the world to be filled with people, of ever new discoveries of kinds and sources of raw materials, of continuous technical progress operating indefinitely to solve problems of supply. We have lived so long in what we have regarded as an expanding world, that we reject in our contemporary theories of economics and of population the realities which contradict such views. Yet our

Plate 1.3
Carl Ortwin Sauer (1889–1975),
a cultural and historical
geographer, whose studies
examined such central themes
as domestication, the effects
of resource exploitation,
conservation and the evolution
of cultural landscapes

modern expansion has been effected in large measure at the cost of an actual and permanent impoverishment of the world.

Likewise, in the former USSR, in spite of the fact that in *Das Kapital* Marx took an environmental line (Matley, 1966), the era of Stalin produced a strong attack on environmentalist views. Grandiose 'practical' schemes were developed for the alteration of the natural environment, such as the 1948 'Stalin Plan for the Transformation of Nature'. These views in turn, however, were thought by some geographers to be excessive. Indeed, much Soviet geography was concerned with the sometimes undesirable transformation of the natural environment by human actions (see Komarov, 1978).

Gerasimov et al. (1971) point to some of the effects produced by the exploitation of natural resources in the erstwhile USSR, including erosion and pollution, and Gerasimov (1976: 77) has placed the human impact on nature in its historical context:

By taking energy and matter from the environment and returning them in converted – industrial, domestic and other – forms society interferes with the dynamically balanced cycles of natural processes. However, as a result of its long evolution, nature has acquired an ability to restore disrupted natural processes . . . Thus the natural environment taken as a whole was able, up to a point, to withstand anthropogenic disturbances, although there were also local irreversible changes. Since the industrial revolution, the general intensity of human impact on the environment has exceeded its potential for restoration in many large areas of the earth's surface, leading to irreversible changes not only on a local but also on a regional scale.

He also indicated that some systems had been more disrupted than others (1976: 79):

Almost all European capitalist countries find themselves 'pressed' between the two jaws of a vice – the complex historical inheritance of a considerably disrupted natural environment and the increasing negative effects on the environment of industrialization and urbanization.

Gerasimov wrote in the same work (p. 78) of a subdiscipline of *'constructive geography'* ('a science of the planned transformation and management of the natural environment for the sake of the future of mankind'). This concept has been criticized for being insufficiently geographical and for lacking a distinctive object of study by Mil'kov and his co-workers (Nesterov, 1974), who propose an alternative subdiscipline, *'anthropogenic landscape science'*, concerned with the human construction of entire landscapes. Isachenko (1974, 1975) in turn attacks the Mil'kov school with vigour and urges the need to see natural and human influences working together in the landscape:

An oasis created in desert remains part of the desert just as a clearing and plot of cropped land in the tayga belongs to the tayga and will continue to belong to it as long as zonal and azonal external conditions will be preserved or until man acquires the ability to control incoming solar radiation, the circulation of air masses and tectonic processes.

(1974: 474)

To the non-Russian geographer the wonder is not so much that their Russian colleagues held such firmly divergent viewpoints, but that Gerasimov, Mil'kov and Isachenko were all so greatly concerned with the human impact.

The theme of the human impact on the environment has, however, been central to some Western historical geographers studying the evolution of the cultural landscape. The clearing of woodland (Darby, 1956; Williams, 1989), the domestication process (Sauer, 1952), the draining of marshlands (Williams, 1970), the introduction of alien plants and animals (McKnight, 1959) and the reclamation of heathland are among some of the recurrent themes of a fine tradition of historical geography.

In 1956 some of these themes were explored in detail in a major symposium volume, *Man's role in changing the face of the earth* (Thomas, 1956). Its impact is difficult to assess, for although it contained many brilliant individual contributions it did not really herald a new generation of environmental research. Kates et al. (1991: 4) write of it:

Man's role seems at least to have anticipated the ecological movement of the 1960s, although direct links between the two have not been demonstrated. Its dispassionate, academic approach was certainly foreign to the style of the movement . . . Rather, *Man's role* appears to have exerted a much more subtle, and perhaps more lasting, influence as a reflective, broad-ranging and multidimensional work.

In the last 30 years many Anglo-Saxon geographers have largely retreated from examining relationships between human agency and the physical environment. Many have focused on spatial and distributional analysis and has found little space for consideration of the natural environment (see e.g. Abler et al., 1971), while others have developed a very strong social, cultural and humanistic approach in which the natural environment appears to enjoy no place. None the less, this retreat has not been total, though it is salutary to reflect that at a time of increasing anxiety about environmental issues at public and political levels, one of the main traditional concerns of geography, human–land relationships, has been effectively carried out by other disciplines under the guise of ecology, environmental studies and the like.

There have, however, been notable exceptions to this tendency, and many geographers have contributed to, and been affected by, the phenomenon which is often called the environmental revolution or the ecological movement. The subject of the human impact on the environment, dealing as it does with such matters as environmental degradation, pollution and desertification, has close links with these developments, and is once again a theme in many textbooks and research monographs in geography (see e.g. Manners and Mikesell, 1974; Wagner, 1974; Cooke and Reeves, 1976; Gregory and Walling, 1979; Simmons, 1979; Tivy and O'Hare, 1981; Turner et al., 1990; Bell and Walker, 1992; Middleton, 1999; Meyer, 1996; Mannion, 1997).

It is also a theme that has been taken up more generally in recent years as the concepts of global change or global environmental change have developed. These phrases are much used, but seldom rigorously defined. Wide use of the term 'global change' seems to have emerged in the 1970s, but in that period was used principally (though by no means invariably) to refer to changes in international social, economic and political systems (Price, 1989). It included such issues as proliferation of nuclear weapons, population growth, inflation and matters relating to international insecurity and decreases in the quality of life. As Pernetta (1995: 1) pointed out, the first use of the term was essentially anthropocentric.

From the early 1980s the concept of global change took on another meaning which was more geocentric in focus. As Price (1989, p. 20) remarked, 'the concept has effectively been captured by the physical and biological sciences'.

The geocentric meaning of global change can be seen in the development of the International Geosphere–Biosphere Programme: A Study of Global Change (IGBP). This was established in 1986 by the International Council of Scientific Unions (ICSU), 'to describe and understand the interactive physical, chemical and biological processes that regulate the total Earth system, the unique environment that it provides for life, the changes that are occurring in this system, and the manner in which they are influenced by human activities'.

The term 'global environmental change' has in many senses come to be used synonymously with the more geocentric use of 'global change'. Its validity and wide currency were recognized when *Global environmental change* was established in 1990 as

an international journal that addresses the human ecological and public policy dimensions of the environmental processes which are threatening the sustainability of life on Earth. Topics include, but are not limited to, deforestation, desertification, soil degradation, species extinction, sea level rise, acid precipitation, destruction of the ozone layer, atmospheric warming/cooling, nuclear winter, the emergence of new technological hazards, and the worsening effects of natural disasters.

Likewise, the UK Research Councils established a UK Global Environmental Research Office, and recognized two scales of issues (*The globe*, March 1991, 1: 1): '(a) those that are truly global (e.g. ozone depletion, climate modification) and (b) local and regional problems which occur on a world-wide scale (e.g. acid rain)'. They recognized that global environmental change 'has tended to become synonymous with understanding the complex interacting processes influencing the Earth's environment and the natural and man-induced factors, that affect them'.

The term 'global change' is now generally regarded as being synonymous with 'global environmental change' and we may take as a definition one that has been adopted in the USA (*Our changing planet, the FY95 US Global Change Research Program*, a report of the United States National Science and Technology Council, 1994, p. 2; quoted in Munn, 1996):

Global change is a term intended to encompass the full range of global issues and interactions concerning natural and human-induced changes in the Earth's environment. The United States Global Change Research Act of 1990 defines *global change* as changes in the global environment (including alterations in climate, land productivity, oceans or other water resources, atmospheric chemistry and ecological systems) that may alter the capacity of the Earth to sustain life.

This definition is very similar to that used by the European Commission's European Network for Research in Global Change:

Global Change research seeks to understand the integrated Earth system in order to: (i) identify, explain and predict natural and anthropogenic changes in the global environment; (ii) assess the potential regional and local impacts of those changes on natural and human systems; and (iii) provide a scientific basis for the development of appropriate technical, economic and societal mitigation/adaptation strategies.

The development of human population and stages of cultural development

During the early part of the Ice Age, some three million years ago, human life probably first appeared on earth. The oldest remains have been found either in sediments from the Rift Valleys of East Africa, notably in Tanzania, Kenya and Ethiopia, or in cave breccias in South Africa. Since that time the human population has spread over virtually the entire land surface of the planet (figure 1.1). *Homo* may have reached Asia by around two million years ago (Larick and Ciochon, 1996). Table 1.1 gives data on recent views of the dates for the arrival of humans in selected areas. Some of these dates are controversial, and this is especially true of Australia, where they range from *c.*40,000 years to as much as 150,000 years (Kirkpatrick, 1994: 28–30).

There are at least three interpretations of global population trends over the last three million years (Whitmore et al., 1990). The first, described as the 'arithmetic–exponential' view, sees the history of the global population as a two-stage phenomenon: the first stage is one of slow growth, while the second stage, related

Figure 1.1
The human colonization of Ice-Age earth (after Roberts, 1989, figure 3.7)

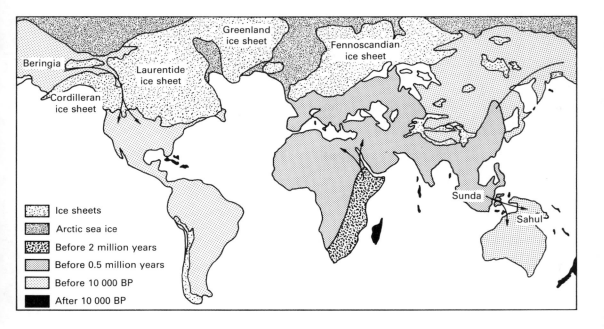

Ice sheets
Arctic sea ice
Before 2 million years
Before 0.5 million years
Before 10 000 BP
After 10 000 BP

Table 1.1 Dates of human arrivals

Area	Source	Date (years BP)
Africa	Klein (1983)	2,700,000–2,900,000
China	Huang et al. (1995)	1,900,000
Java	Swisher et al. (1994)	1,800,000
Europe	Champion et al. (1984)	c.1,600,000 but most post-350,000
Britain	Green (1981)	c.200,000
Japan	Ikawa-Smith (1982)	c.50,000
New Guinea	Bulmer (1982)	c.50,000
Australia	Allen et al. (1977)	c.40,000
North America	Irving (1985)	15,000–40,000
South America	Guidon and Delibrias (1986)	32,000
Peru	Keefer et al. (1998)	12,500–12,700
Ireland	Edwards (1985)	9,000
Caribbean	Morgan and Woods (1986)	4,500
Polynesia	Kirch (1982)	2,000
Madagascar	Battistinti and Verin (1972)	c.AD 500
New Zealand	Green (1975)	AD 700–800

to the industrial revolution, displays a staggering acceleration in growth rates. The second view, described as 'logarithmic–logistic', sees the last million or so years in terms of three revolutions – the tool, agricultural and industrial revolutions. In this view, humans have increased the carrying capacity of the earth at least three times. There is also a third view, described as 'arithmetic–logistic', which sees the global population history over the last 12,000 years as a set of three cycles: the 'primary cycle', the 'medieval cycle' and the 'modernization cycle': these three alternative models are presented graphically in figure 1.2.

Estimates of population levels in the early stages of human development are difficult to make with any degree of certainty (figure 1.3). Before the agricultural 'revolution' some 10,000 years ago, human groups lived by hunting and gathering in parts of the world where this was possible. At that time the world population *may* have been of the order of five million people (Ehrlich et al., 1977: 182) and large areas would only recently have witnessed human migration. The Americas and Australia, for example, were virtually uninhabited until about 30,000 years ago (or possibly even later).

The agricultural revolution probably enabled an expansion of the total human population to about 200 million by the time of Christ, and to 500 million by AD 1650. It is since that time, helped by the medical and industrial revolutions and by developments in agriculture and the colonization of new lands, that human population has exploded, reaching about 1,000 million by AD 1850, 2,000 million by AD 1930 and 4,000 million by AD 1975. The figure had reached 5,732 million by 1995. Victory over malaria, smallpox, cholera and other diseases has been

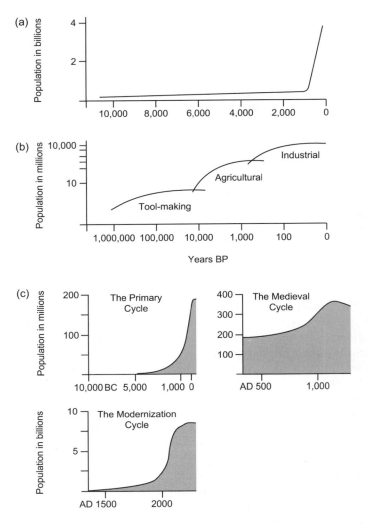

Figure 1.2
Three interpretations of
global population trends over
the millennia: (a) the
arithmetic–exponential;
(b) the logarithmic–logistic;
(c) the arithmetic–logistic
(after Whitmore et al., 1990,
figure 2.1)

responsible for marked decreases in death rates throughout the
non-industrial world, but death-rate control has not in general
been matched by birth control. Thus the annual population
growth rate in the late 1980s in South Asia was 2.64 per cent,
in Africa 2.66 per cent and in Latin America (where population
increased sixfold between 1850 and 1950) 2.73 per cent. The
global annual growth in population over the 1990s has been
just under 90 million people.

The history of the human impact, however, has not been a
simple process of increasing change in response to linear population growth over time, for in specific places at specific times
there have been periods of reversal in population growth and
ecological change as cultures collapsed, wars occurred, disease
struck and habitats were abandoned. Denevan (1992), for

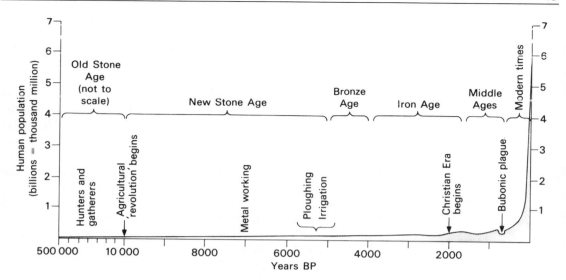

Figure 1.3
The growth of human numbers for the past half million years (after Ehrlich et al., 1977, figure 5.2)

example, has pointed to the decline of Native American populations in the New World following European entry into the Americas. This created what was 'probably the greatest demographic disaster ever'. The overall population of the western hemisphere in 1750 was perhaps less than a third of what it may have been in 1492, and the ecological consequences were legion.

Clearly, this growth of the human population of the earth is in itself likely to be a highly important cause of the transformation of nature. Of no lesser importance, however, has been the growth and development of culture and technology. Sears (1957: 51) has put the power of humankind into the context of other species:

Man's unique power to manipulate things and accumulate experience presently enabled him to break through the barriers of temperature, aridity, space, seas and mountains that have always restricted other species to specific habitats within a limited range. With the cultural devices of fire, clothing, shelter, and tools he was able to do what no other organism could do without changing its original character. Cultural change was, for the first time, substituted for biological evolution as a means of adapting an organism to new habitats in a widening range that eventually came to include the whole earth.

The evolving impact of humans on the environment has often been expressed in terms of a single equation:

$$I = P\,A\,T$$

where I is the amount of pressure or impact that humans apply on the environment, P is the number of people, A is the affluence (or the demand on resources per person), and T is a technological factor (the power that humans can exert through technological

change). *P*, *A* and *T* have been seen by some as 'the three horse-
men of the environmental apocalypse' (Meyer, 1996, p. 24).
There may be considerable truth in the equation and in that
sentiment; but, as Meyer points out, the formula cannot be
applied in too mechanistic a way. The 'cornucopia view', indeed,
sees population not as the ultimate depleter of resources but as
itself the ultimate resource capable of causing change for the
better (see, for example, Simon, 1981 and 1996). There are
cases where strong population growth has appeared to lead to a
reduction in environmental degradation (Tiffen et al., 1994).
Likewise, there is debate about whether it is poverty or affluence that creates deterioration in the environment. Many poor
countries have severe environmental problems and do not have
the resources to clear them up, whereas affluent countries do.
Conversely, it can be argued that affluent countries have plun-
dered and fouled less fortunate countries, and that it would be
environmentally catastrophic if all countries used resources at
the rate that the rich countries do. Similarly, it would be naïve
to see all technologies as malign, or indeed benign. Technology
can be a factor either of mitigation and improvement or of
damage. Sometimes it is the problem (as when ozone depletion
has been caused by a new technology – the use of chlorofluorocarbons) and sometimes it can be the solution (as when
renewable energy sources replace the burning of polluting lignite
in power stations).

In addition to the three factors of population, affluence and
technology, environmental changes also depend on variations in
the way in which different societies are organized and in their
economic and social structures (see Meyer, 1996: 39–49 for an
elaboration of this theme). For example, the way in which land
is owned is a crucial issue.

The controls of environmental changes caused by the human
impact are thus complex and in many cases contentious, but all
the factors discussed play a role of some sort, at some places,
and at some times.

We now turn to a consideration of the major cultural and
technical developments that have taken place during the past
two to three million years. Three main phases will form the
basis of this analysis: the phase of hunting and gathering; the
phase of plant cultivation, animal keeping and metal working;
and the phase of modern urban and industrial society. These
developments are treated in much greater depth by Ponting
(1991) and Simmons (1996).

Hunting and gathering

The definition of 'human' is something of a problem, not least
because, as is the case with all existing organisms, new forms

tend to emerge by perceptible degrees from antecedent ones. Moreover, the fossil evidence is scarce, fragmentary and can rarely be dated with precision. Although it is probably justifiable to separate the hominids from the great apes on the basis of their assumption of an upright posture, it is much less justifiable or indeed possible to distinguish on purely zoological grounds between those hominids that remained pre-human and those that attained human status. To qualify as a human, a hominid must demonstrate cultural development: the systematic manufacture of implements as an aid to manipulating the environment.

The oldest records of human activity and technology, pebble tools (crude stone tools which consist of a pebble with one end chipped into a rough cutting edge), have been found with human bone remains in various parts of Africa. For example, at Lake Turkana in northern Kenya, and the Omo Valley in southern Ethiopia, a tool-bearing bed of volcanic material called tuff has been dated by isotopic means at about 2.6 million years old and another from Gona in the north-east of Ethiopia at about 2.5 million years old (Semaw et al., 1997), while another bed at the Olduvai Gorge in Tanzania, made famous by the researches of the Leakey family, has been dated by similar means at 1.75 million years. Indeed, these very early tools are generally termed 'Oldowan'.

As the Stone Age progressed the tools became more sophisticated, more varied and more effective. Greater exploitation of plant and animal resources became feasible. It is possible to observe a clear progression during the Palaeolithic Age in the technology of working stone (figure 1.4), although it needs to be stressed that in reality the stages were not synchronous in different areas nor among different human groups in the same area. In addition, particular industries are often seen to combine techniques from more than one stage of development.

The lithic stone technology in the Old Stone Age in Europe has been divided into five main modes. Mode 1 (figure 1.4a) consists of chopper tools of Lower Palaeolithic type; mode 2 (figure 1.4b and c) comprises bifacially flaked hand axes of the Lower Palaeolithic; mode 3 (figure 1.4d), Middle Palaeolithic flake tools from prepared cores; mode 4 (figure 1.4e), Upper Palaeolithic punch-struck blades with steep retouch; and mode 5 (figure 1.4f), microlithic components of composite artefacts from the Mesolithic Age.

Stone may not, however, have been the only material used by early civilizations. Sticks and animal bones, which are less likely to be preserved than stone, are among the first objects that may have been used as implements, although the sophisticated utilization of antler and bone as materials, for weapons and implements appears to have developed surprisingly late in prehistory. There is certainly a great deal of evidence for the use of wood

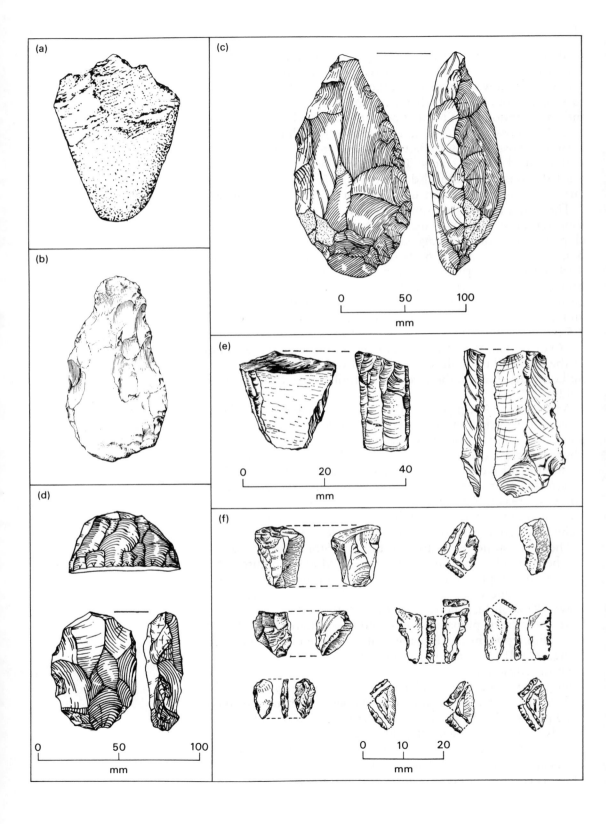

(a)

(b)

(c)

0   50   100

mm

(d)

0   50   100

mm

(e)

0   20   40

mm

(f)

0   10   20

mm

throughout the Palaeolithic Age, for building ladders, fire, pigment (charcoal), the drying of wood and digging sticks. Tyldesley and Bahn (1983: 59) go so far as to suggest that 'The Palaeolithic might more accurately be termed the "Palaeoxylic" or "Old Wood Age".'

The building of shelters and the use of clothing became permanent features of human life as the Palaeolithic period progressed, permitting habitation in areas where the climate was otherwise not congenial. European sites from the Mousterian of the Middle Palaeolithic have revealed the presence of purposefully made dwellings as well as caves, and by the Upper Palaeolithic more complex shelters were in use, allowing people to live in the tundra lands of Central Europe and Russia.

Another feature of early society which seems to have distinguished humans from the surviving non-human primates was their seemingly omnivorous diet. Biological materials recovered from settlements in many different parts of the world indicate that in the Palaeolithic Age humans secured a wide range of animal meats, whereas the great apes, though not averse to an occasional taste of animal food, are predominantly vegetarian. One consequence of enlarging the range of their diet was that, in the long run, humans were able to explore a much wider range of environments (Clark, 1977b: 19).

Moreover, at an early stage humans discovered the use of fire, and as Sparks and West (1972) have argued, 'the regularity with which hearths are found associated with the Middle and Upper Palaeolithic sites leaves little doubt that Neanderthal Man and his successors were capable of fire production.' Fire, as we shall see in chapter 2, is a major agent by which humans have influenced their environment. It has been found in association with Peking Man in deposits at Choukoutien dating back to perhaps 600,000–700,000 years, and may have been employed even earlier in East Africa, where Gowlett et al. (1981) have claimed to find evidence for deliberate manipulation of fire from over 1.4 million years ago. As Pyne (1982: 3) has written:

It is among man's oldest tools, the first product of the natural world he learned to domesticate. Unlike floods, hurricanes or windstorms, fire can be initiated by man; it can be combated hand to hand, dissipated, buried, or 'herded' in ways unthinkable for floods or tornadoes.

Figure 1.4 (*opposite*)
Developments in stone tool technology:
(a) a primitive pebble tool from early Pleistocene strata – mode 1
(b) a hand axe of Abbeville type from a younger level – mode 2
(c) another Lower Palaeolithic hand axe from Rajasthan, India – mode 2
(d) a Middle Palaeolithic keeled scraper and core from Rajasthan – mode 3
(e) Upper Palaeolithic burin and blade from Rajasthan – mode 4
(f) Mesolithic artefacts from Rajasthan – mode 5

He goes on to stress the implications that fire had for subsequent human cultural evolution (p. 4):

It was fire as much as social organisation and stone tools that enabled early big game hunters to encircle the globe and to begin the extermination of selected species. It was fire that assisted hunting and gathering societies to harvest insects, small game and edible plants; that encouraged the spread of agriculture outside the flood plains by allowing for rapid landclearing, ready fertilization, the selection of food grains, the primitive herding of grazing animals that led to domestication, and the expansion of pasture and grasslands against climate gradients; and that, housed in machinery, powered the prime movers of the industrial revolution.

Another major difference that set humankind above the beasts was the development of communicative skills such as speech. Until hominids had developed words as symbols, the possibility of transmitting, and so accumulating, culture hardly existed. Animals can express and communicate emotions, never designate or describe objects.

Overall, compared with later stages of cultural development, early hunters and gatherers had neither the numbers nor the technological skills to exert a very substantial effect on the environment. Wymer (1982) believes that the first clear evidence of human influence on vegetation may be found at Hoxne in eastern England, where pollen analysis indicates an increase in grassland at the same time as Acheulian societies occupied the edge of a small lake. Specks of charcoal have been found in the lake muds and the inference is that hunters had set light to the forest, intentionally or accidentally, on such a scale that the local vegetational succession was disrupted.

Besides the effects of fire, early cultures may have caused some diffusion of seeds and nuts, and through hunting activities (see chapter 3) may have had some effects on animal extinctions (the so-called 'Pleistocene overkill'). Locally some eutrophication may have occurred, and around some archaeological sites phosphate and nitrate levels may be sufficiently raised to make them an indicator of habitation to archaeologists today (Holdgate, 1979). Some hunter and gatherer groups may have had conservationist tendencies, but as Simmons (1993: 8) has remarked, 'any picture of hunter-gatherers simply as responsive children of nature living solely off a provident usufruct is part of a myth of a Golden Age.'

Humans as cultivators, keepers and metal workers

It is possible to identify some key stages of economic development that have taken place since the end of the Pleistocene

Table 1.2 Five stages of economic development

Economic stage	Dates and characteristics
Hunting–gathering and early agriculture	Domestication first fully established in south-western Asia around 7500 BC; hunter–gatherers persisted in diminishing numbers until today. Hunter–gatherers generally manipulate the environment less than later cultures, and adapt closely to environmental conditions.
Riverine civilizations	Great irrigation-based economies lasting from c.4000 BC to 1st century AD in places such as the Nile Valley and Mesopotamia. Technology developed to attempt to free civilizations from some of the constraints of a dry season.
Agricultural empires	From 500 BC to around 1800 AD a number of city-dominated empires existed, often affecting large areas of the globe. Technology (e.g. terracing and selective breeding) developed to help overcome environmental barriers to increased production.
The Atlantic–industrial era	From c.1800 AD to today a belt of cities from Chicago to Beirut, and around the Asian shores to Tokyo, form an economic core area based primarily on fossil fuel use. Societies have increasingly divorced themselves from the natural environment, through air conditioning for example. These societies have also had major impacts on the environment.
The Pacific–global era	Since the 1960s there has been a shifting emphasis to the Pacific Basin as the primary focus of the global economy, accompanied by globalization of communications and the growth of multinational corporations.

Source: Adapted from Simmons (1993), pp. 2–3.

(table 1.2). First, around the beginning of the Holocene, about 10,000 years ago, humans started in various parts of the world to domesticate rather than to gather food plants and to keep, rather than just hunt, animals. This phase of human cultural development is well reviewed in Roberts (1998). By taking up farming and domesticating food plants, they reduced enormously the space required for sustaining each individual by a factor of the order of 500 at least (Sears, 1957: 54). As a consequence we see shortly thereafter, notably in the Middle East, the establishment of the first major settlements – towns. So long as man had

to subsist on the game animals, birds and fish he could catch and trap, the insects and eggs he could collect and the foliage, roots, fruits and seeds he could gather, he was limited in the kind of social life he could develop; as a rule he could only live in small groups, which gave small scope for specialization and the subdivision of labour, and in the course of a year he would have to move over extensive tracts of country, shifting his habitation so that he could tap the natural resources of successive areas. It is hardly to be wondered at that among communities whose energies were almost entirely absorbed by the mere business of keeping alive, technology remained at a low ebb.

(Clark, 1962: 76)

Although it is now recognized that some subsistence hunters and gatherers had considerable leisure, there is no doubt that through the controlled breeding of animals and plants humans were able, by hard work, to develop a more reliable and readily expandable source of food and thereby create a solid and secure basis for cultural advance, an advance which included civilization and the 'urban revolution' of Childe (1936) and others. Indeed, Isaac (1970) has termed domestication 'the single most important intervention man had made in his environment'; while Clark (1977b) has remarked that 'the milk-stool and the mattock were forerunners of the conveyor-belt and the punch-card'; and Harris (1996) has termed the transition from foraging to farming 'the most fateful change in the human career'.

A distinction can be drawn between cultivation and domestication. Whereas cultivation involves deliberate sowing or other management, and entails plants which do not necessarily differ genetically from wild populations of the same species, domestication results in genetic change brought about through conscious or unconscious human selection. This creates plants that differ morphologically from their wild relatives and which may be dependent on humans for their survival. Domesticated plants are thus necessarily cultivated plants, but cultivated plants may or may not be domesticated. For example, the first plantations of *Hevea* rubber and quinine in the Far East were established from seed which had been collected from the wild in South America. Thus at this stage in their history these crops were cultivated but not yet domesticated.

The whole question of the origin of agriculture remains controversial (Bender, 1975; Harris, 1996). Some early workers saw agriculture as a divine gift to humankind, while others thought that animals were domesticated for religious reasons. They argued that it would have been improbable that humans could have predicted the usefulness of domestic cattle before they were actually domesticated. Wild cattle are large and fierce beasts, and no one could have foreseen their utility for labour or milk until they were tamed – tamed for ritual sacrifice in connection with lunar goddess cults (the great curved horns being the reason for the association). Another major theory was that domestication was produced by crowding, possibly brought on by a combination of climatic deterioration (postglacial progressive desiccation) and population growth. Such pressure, according to Childe, forced communities to intensify their methods of food production. Current palaeo-climatological research tends not to support his precise interpretation, but that is not to say that other severe climatic changes could not have played a role (Sherratt, 1997).

Sauer (1952), a geographer, believed that the domestication of plants was initiated in South-East Asia by fishing folk, who found that lacustrine and riverine resources would underwrite a stable economy and a sedentary or semi-sedentary lifestyle. He surmised, in an almost totally theoretical model, that the initial domesticates would be multi-purpose plants set around small fishing villages to provide such items as starch foods, substances for toughening nets and lines and making them water-resistant, and drugs and poisons. He suggested that 'food production was one and perhaps not the most important reason for bringing plants under cultivation.'

Yet another model, advanced by Jacobs (1969), turned certain more traditional models upside down. Instead of following the classic pattern whereby farming leads to village which leads to town which leads to civilization, she proposed that one could be a hunter-gatherer and live in a town or city, and that agriculture originated in and around such cities rather than in the countryside. Her argument suggests that even in primitive hunter-gatherer societies particularly valuable commodities such as fine stones, pigments and shells could create and sustain a trading centre which could become large and stable. Food would be exchanged for goods, but natural produce brought any distance would have to be durable, so meat, for example, would be transported on the hoof; but not all the animals would be consumed immediately – some would be herded together and might breed. This might be the start of domestication.

The process of domestication and cultivation was also once considered a revolutionary system of land procurement that had evolved in one or two hearths and diffused over the face of the earth, replacing the older hunter-gathering systems by stimulus diffusion. It was felt that the deliberate rearing of plants and animals for food was a discovery or invention so radical and complex that it could have developed only once (or possibly twice) – the so-called 'Eureka model'.

Some of the first domesticated plants occur at approximately the same time in widely separated areas (table 1.3). This might be construed as suggesting that developments in one area triggered experiments with local plant materials in others. However, if plant domestication were proved to have begun at different times in different parts of the world, this could also be used to argue that diffusion took place from one single early centre. The balance of botanical and archaeological evidence seems to suggest that humans started experimenting with domestication and cultivation of different plants at different times in different parts of the world.

The current belief (Harlan, 1976) is that agriculture evolved through an extension and intensification of long-standing practices, it evolved over wide areas in different parts of the world (figure 1.5), and the innovative pattern was complex and diffuse. Harlan (1975b: 57) has gone so far as to write:

Table 1.3 Dates which indicate that there may have been some synchroneity of plant domestication in different centres

Centre	Dates (000 years BP)	Plant
Mesoamerica	10.7–9.8 9.0	Squash-pumpkin Bottle gourd
Near East	10.0–9.3	Emmer wheat Two-rowed barley
	9.8–9.6	Einkorn wheat Pea Lentil
Far East	8.0–7.0	Broomcorn millet Rice Gourd Water chestnut
Andes	9.4–8.0	Chile pepper Common bean Ullucu White potato

Figure 1.5
Major areas of domestication of plants identified by various workers. (1) The prime centres in which a number of plants were domesticated and which then diffused outwards into neighbouring regions. (2) Broader regions in which plant domestication occurred widely and which may have received their first domesticated plants from the prime centres

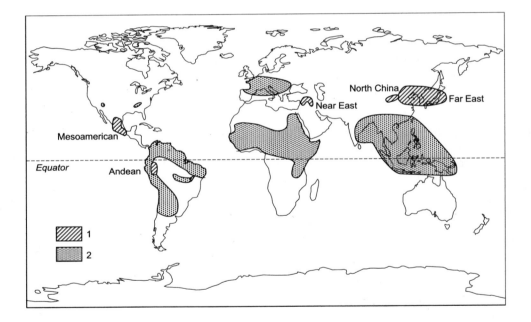

I am inclined to develop a no-model which leaves room for whole arrays of motives, actions, practices and evolutionary processes. What applies in Southeast Asia may not apply in Southwest Asia . . . A search for a single overriding cause for human behaviour is likely to be frustrating and fruitless.

Certainly the pre-eminence once accorded the Near East is now lessened (Simoons, 1974), while Jarman (1977) has stressed

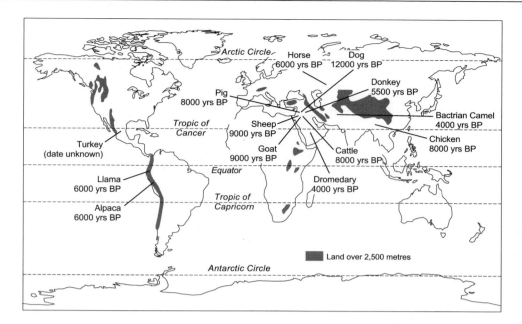

Figure 1.6
The places of origin, with approximate dates, for the most common domesticated animals

the continuity of human–animal relationships and cast doubt on any clear-cut wild-domestic dichotomy. Indeed, Jarman proposes six stages in the process of human exploitation of animals:

1 *Random predation*, during which the human group makes no effort to control or benefit from the regularity of the animal behaviour pattern, simply exploiting the animals when they happen to be available.
2 *Controlled predation*, which would not necessarily amount to a year-round association of humans and their prey. The control of animal movement in game drives and corralling are examples.
3 *Herd following*, in which the human group, or part of it, echoes the animal movements, maintaining a degree of contact, so that a given human population tends to become associated with a given animal population.
5 *Close herding*, in which a high degree of control of herds is maintained throughout the year.
6 *Factory farming*, whereby animals are kept immobile throughout their lives and maintained in a wholly artificial environment.

The locations and dates for domestication of some important domestic animals are shown in figure 1.6.

One highly important development in agriculture, because of its rapid and early effects on environment, was irrigation and the adoption of riverine agriculture. This came rather later than

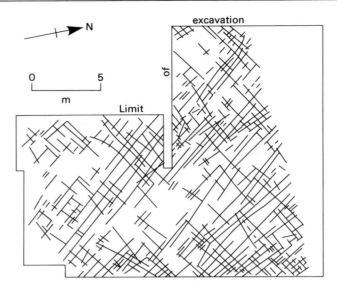

N

excavation

0   5

m

Limit

of

Figure 1.7
Plan of Neolithic plough
marks: some of the first
tangible evidence of the first
agricultural revolution in the
British Isles (after Taylor, 1975,
figure 1a)

domestication. Among the earliest evidence of artificial irrigation
is the mace-head of the Egyptian Scorpion King which shows
one of the last pre-dynastic kings ceremonially cutting an irrigation ditch around 5,050 years ago (Butzer, 1976), although it is
possible that irrigation in Iraq started even earlier.

A major difference has existed in the development of agriculture in the Old and New Worlds: in the New World there were
few counterparts to the range of domesticated animals which
were an integral part of Old World systems (Sherratt, 1981). A
further critical difference was that in the Old World the secondary applications of domesticated animals were explored. The
plough was particularly important in this process – the first
application of animal power to the mechanization of agriculture. Closely connected to this was the use of the cart, which
both permitted more intensive farming and enabled the trans-
portation of its products. Furthermore, the development of tex-
tiles from animal fibres afforded, for the first time, a commodity
which could be produced for exchange in areas where arable
farming was not the optimal form of land use. Finally, the use
of animal milk provided a means whereby large herds could use
marginal or exhausted land, encouraging the development of
the pastoral sector with transhumance or nomadism.

This secondary utilization of animals therefore had radical
effects, and the change took place over quite a short period. The
plough was invented some 5,000 years ago, and was used in
Mesopotamia, Assyria and Egypt. The remains of plough marks
have also been found beneath a burial mound at South Street,
Avebury, in England, dated at around 3000 BC (see figure 1.7),
and ever since that time have been a dominant feature of the

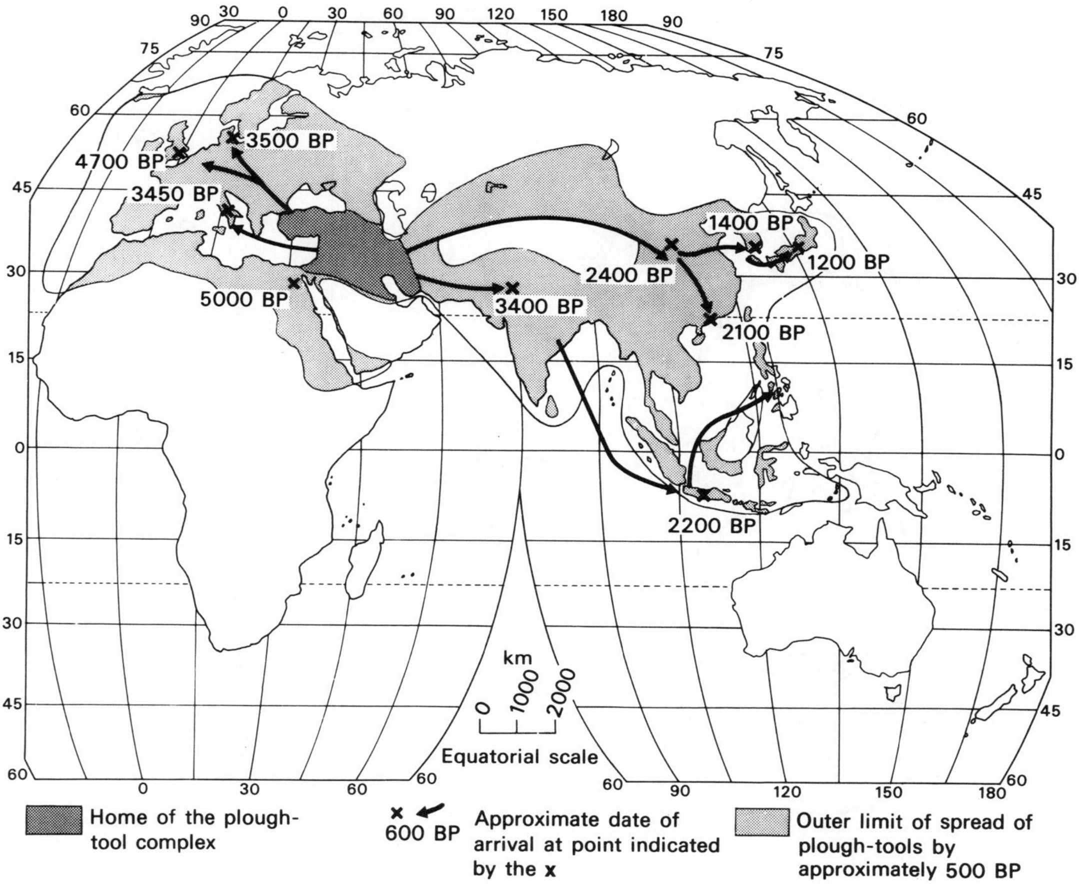
3500 BP
4700 BP
3450 BP
5000 BP
3400 BP
2400 BP
1400 BP
1200 BP
2100 BP
2200 BP
km
0 1000 2000
Equatorial scale
Home of the plough-tool complex
600 BP Approximate date of arrival at point indicated by the x
Outer limit of spread of plough-tools by approximately 500 BP

Figure 1.8
The spread of plough agriculture in the Old World (after Spencer and Thomas, 1978, figure 4.2)

English landscape (Taylor, 1975). The wheeled cart was first produced in the Near East in the fourth millennium BC, and rapidly spread from there to both Europe and India during the course of the third millennium. The diffusion of the plough is shown in figure 1.8 and that of the cart in figure 1.9.

The development of other means of transport preceded the wheel. Sledge-runners found in Scandinavian bogs have been dated to the Mesolithic period (Cole, 1970: 42), while by the Neolithic era humans had developed boats, floats and rafts that were able to cross to Mediterranean islands and sail the Irish Sea. Dugout canoes could hardly have been common before polished stone axes and adzes came into general use during Neolithic times, although some paddle and canoe remains are recorded from Mesolithic sites in northern Europe. The middens of the hunter-fishers of the Danish Neolithic contain bones of deep-sea fish such as cod, showing that these people certainly had seaworthy craft with which to exploit ocean resources.

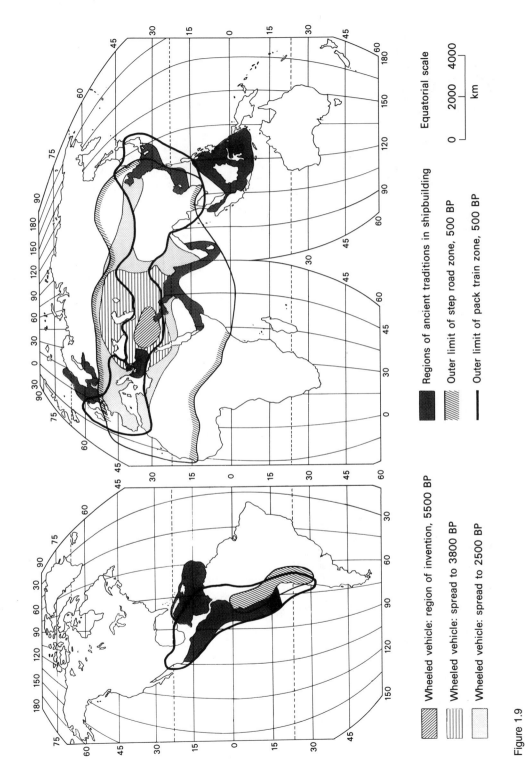

Figure 1.9
The early development of transport (after Spencer and Thomas, 1978, figure 4.5)

Wheeled vehicle: region of invention, 5500 BP

Wheeled vehicle: spread to 3800 BP

Wheeled vehicle: spread to 2500 BP

Regions of ancient traditions in shipbuilding

Outer limit of step road zone, 500 BP

Outer limit of pack train zone, 500 BP

Equatorial scale

0   2000   4000

km

Plate 1.4
The domestication of
plants such as wheat
involved the deliberate
selection of particular
genetic characteristics
thought to be advantageous

Both the domestication of animals and the cultivation of plants
have been among the most significant causes of the human impact
(see Mannion, 1995). Pastoralists have had many major effects
– for example, on soil erosion – though Passmore (1974: 12)
believes that nomadic pastoralists are probably more conscious
than agriculturalists that they share the earth with other living
things. Agriculturalists, on the other hand, deliberately transform nature in a sense which nomadic pastoralists do not. Their
main role has been to simplify the world's ecosystems. Thus in
the prairies of North America, by ploughing and seeding the
grasslands, farmers have eliminated 100 species of native prairie
herbs and grasses, replacing them with pure stands of wheat,
corn or alfalfa. This simplification may reduce stability in the
ecosystem (but see chapter 9, section on 'The susceptibility
to change'). Indeed, on a world basis (see Harlan, 1976) such
simplification is evident. Whereas people once enjoyed a highly
varied diet, and have used for food several thousand species of
plants and several hundred species of animals, with domestication their sources are greatly reduced. For example, today the
four crops (wheat, rice, maize and potatoes) at the head of the
list of food supplies contribute more tonnage to the world total
than the next twenty-six crops combined. Simmonds (1976) provides an excellent account of the history of most of the major
crops produced by human society.

It is becoming increasingly clear that Neolithic and Bronze
Age peoples were able to achieve very notable changes in the
soils, plants and animals of large areas of Europe and the Near
East (see e.g. Dimbleby, 1974).

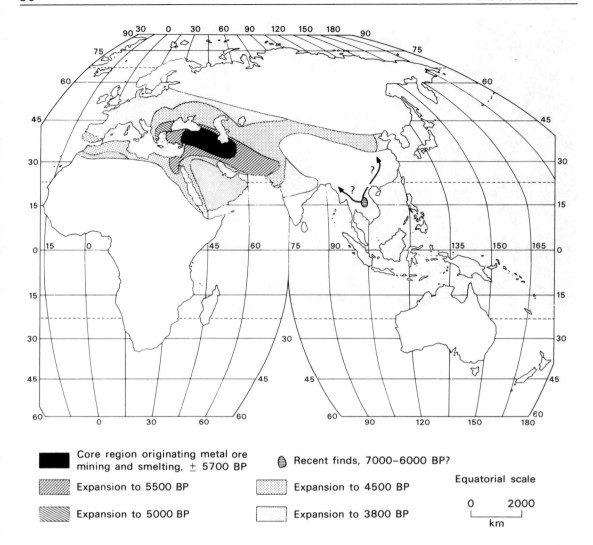

Core region originating metal ore
mining and smelting, ± 5700 BP

Recent finds, 7000–6000 BP?

Expansion to 5500 BP

Expansion to 4500 BP

Expansion to 5000 BP

Expansion to 3800 BP

Equatorial scale

0   2000
|—————————|
        km

Figure 1.10
The diffusion of mining and
smelting in the Old World
(after Spencer and Thomas,
1978, figure 4.4)

One further development in human cultural and technological life which was to increase human power was the mining of
ores and the smelting of metals. This may have started around
6,000 years ago, possibly in what is today Anatolia (Turkey)
and north-west Iran, with the smelting of copper-oxide ores
into metallic copper. Recent finds of copper and bronze implements in north-east Thailand, however, raise questions about
the dating and the place of origin, since tentative dates of
between 7,000 and 6,000 years ago have been given to the
Thailand finds. The spread of metal working into other areas
was rapid, particularly in the second half of the fifth millennium
BC (Muhly, 1997) (figure 1.10), and by 2500 BC bronze products were in use from Britain in the west to northern China in

Table 1.4 Environmental impacts of minerals extraction

Activity	Potential impacts
Excavation and ore removal	• Destruction of plant and animal habitat, human settlements, and other surface features (surface mining) • Land subsidence (underground mining) • Increased erosion: silting of lakes and streams • Waste generation (overburden) • Acid drainage (if ore or overburden contain sulphur compounds) and metal contamination of lakes, streams and groundwater
Ore concentration	• Waste generation (tailings) • Organic chemical contamination (tailings often contain residues of chemicals used in concentrators) • Acid drainage (if ore contains sulphur compounds) and metal contamination of lakes, streams, and groundwater
Smelting/refining	• Air pollution (substances emitted can include sulphur dioxide, arsenic, lead, cadmium and other toxic substances) • Waste generation (slag) • Impacts of producing energy (most of the energy used in extracting minerals goes into smelting and refining)

Source: Young, 1992, table 5.

the east. The smelting of iron ores may date back to as late as 1500 BC (Spencer and Thomas, 1978). Metal working required enormous amounts of wood, and Sir William Flinders Petrie in his investigation of the third millennium BC copper industry at Wadi Nasb in western Sinai found a bed of wood ashes 30 metres long, 15 metres wide and 0.5 metres deep.

In recent decades fossil-fuelled machinery has allowed mining activity to expand to such a degree that in terms of the amount of material moved its effects are reputed to rival the natural processes of erosion. Taking overburden into account, the total amount of material moved by the mining industry globally is probably at least 28 billion tonnes – about 1.7 times the estimated amount of sediment carried each year by the world's rivers (Young, 1992). The environmental impacts of mineral extraction are diverse but extensive, and relate not only to the processes of excavation and removal, but also to the processes of mineral concentration, smelting and refining (table 1.4).

Modern industrial and urban civilizations

In ancient times, certain cities evolved which had considerable human populations. It has been estimated that Nineveh may have had a population of 700,000, that Augustan Rome may have had a population of around one million, and that Carthage, at its fall in 146 BC, had 700,000 inhabitants (Thirgood, 1981). Such cities would have exercised a considerable influence on

their environs, but this influence was never as extensive as that
of the last few centuries; for the modern era, especially since the
late seventeenth century, has witnessed the transformation of,
or revolution in, culture and technology – the development
of major industries (Pawson, 1978). This, like domestication,
has reduced the space required to sustain each individual and
has increased the intensity with which resources are utilized.
Modern science and modern medicine have compounded these
effects, leading to accelerating population increase even in non-
industrial societies. Urbanization has gone on apace, and it is
now recognized that large cities have their own environmental
problems (Cooke et al., 1982) and a multitude of environmental effects (Douglas, 1983). If present trends continue, many
cities in the less developed countries will become inconceivably
large and crowded. For instance, it is projected that by the year
2000 Mexico City will have more than 30 million people
(roughly three times the present population of the New York
metropolitan area), and Calcutta, Greater Bombay, Greater
Cairo, Jakarta and Seoul are each expected to be in the 15–20
million range by that date. In all, around 400 cities will have
passed the million mark (Council on Environmental Quality
and the Department of State, 1982: 12), and UN estimates indicate that by the end of the century over 3,000 million people
will live in cities, compared with around 1,400 million people
in 1970. Soon after the year 2000 there will be more urban
dwellers than rural dwellers worldwide. The distribution of the
world's total, rural and urban populations by continent in 1990
is shown in table 1.5.

The perfecting of seagoing ships in the sixteenth and seventeenth centuries was part of this industrial and economic

Table 1.5 The distribution of the world's total, rural and urban populations in 1990

Region	Total population (millions)	Total population (%)	Rural population (%)	Urban population (%)	In 'million cities' (%)
World	5,285	100	100	100	100
Africa	633	12.0	14.4	8.8	7.5
Asia	3,186	60.3	72.2	44.5	45.6
Europe	722	13.7	6.7	22.8	17.9
Latin America and the Caribbean	440	8.3	4.2	13.8	14.7
North America	278	5.3	2.3	9.2	13.1
Oceania	26	0.5	0.3	0.8	1.3

Source: United Nations Centre for Human Settlements, *An urbanizing world: global report on human settlements, 1996* (Oxford: Oxford University Press), Table 1.3.

transformation, and this was the time when mainly self-contained but developing regions of the world coalesced so that the ecumene became to all intents and purposes continuous. The invention of the steam engine in the late eighteenth century, and the internal combustion engine in the late nineteenth century, massively increased human access to energy and lessened dependence on animals, wind and water.

Modern science, technology and industry have also been applied to agriculture, and in recent decades some spectacular progress has been made through, for example, the use of fertilizers and the selective breeding of plants and animals.

To conclude, we can recognize certain trends in human manipulation of the environment which have taken place in the modern era. The first of these is that the ways in which humans are affecting the environment are proliferating. For example, nearly all the powerful pesticides post-date the Second World War, and the same applies to the construction of nuclear reactors. Secondly, environmental issues that were once locally confined have become regional or even global problems. An instance of this is the way in which substances such as DDT, lead and sulphates are found at the poles, far removed from the industrial societies that produced them. Thirdly, the complexity, magnitude and frequency of impacts are probably increasing; for instance, a massive modern dam like that at Aswan in Egypt has a very different impact from a small Roman dam. Finally, compounding the effects of rapidly expanding populations is a general increase in per capita consumption and environmental impact (table 1.6). Energy resources are being developed at an ever-increasing rate, giving humans enormous power to transform the environment. One index of this is world commercial energy consumption, which trebled in size between the 1950s and 1980. Figure 1.11 shows worldwide energy consumption since 1860 on a per capita basis.

Table 1.6 Some indicators of change in the global economy from 1950 to 1995

World indicator	1950	1995	Change (xn)
Grain production (million tonnes)	631	80	2.7
Soybean production (million tonnes)	17	125	7.4
Meat production (million tonnes)	44	192	4.4
Fish catch (million tonnes)	21	109[a]	5.2
Irrigated area (million ha)	94	248[b]	2.6
Fertilizer use (million tonnes)	14	122	8.7
Oil production (million tonnes)	518	3,031	5.9
Natural gas production (million tonnes of oil equivalent)	187	2,114	11.3
Car production (million)	8	36	4.5
Bike production (million)	11	114	10.4
Human population (million)	2,555	5,732	2.2

[a] 1994 figure.
[b] 1993 figure.
Source: data processed by author from Brown et al., 1996.

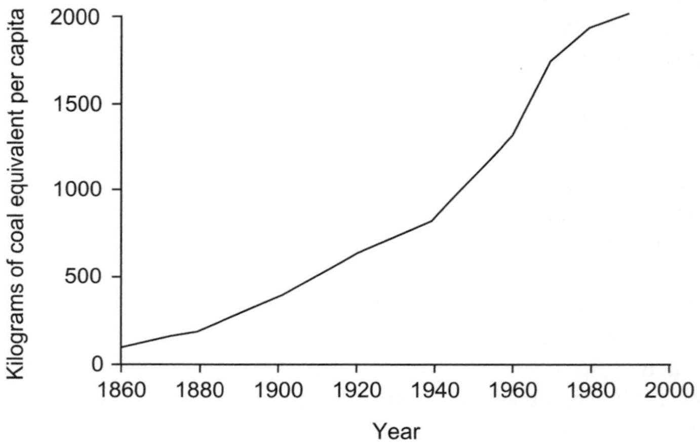

Figure 1.11
World per capita energy consumption since 1860, based on data from the United Nations

Modern technologies have immense power output. A pioneer steam engine in AD 1800 might have rated at 8–16 kilowatts. Modern railway diesels top 3.5 megawatts, and a large aero engine 60 megawatts. Figure 1.12 shows how the human impact on six 'component indicators of the biosphere' has increased over time. This graph is based on work by Kates et al. (1990). For each component indicator they defined the total net change induced by humans to be 0 per cent for 10,000 years ago (before the present = BP) and 100 per cent for 1985. They then estimated the dates by which each component had reached successive quartiles (that is, 25, 50 and 75 per cent) of its total change at 1985. They believe that about half of the components have changed more in the single generation since 1950 than in the whole of human history before that date.

Likewise, we can see stages in the pollution history of the earth. Mieck (1990), for instance, has identified a sequence of

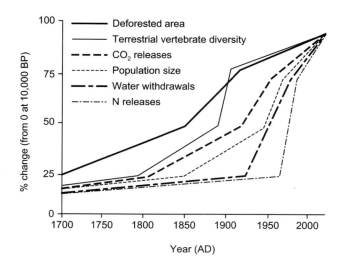

Figure 1.12
Percentage change (from assumed zero human impact at 10,000 BP) of selected human impacts on the environment

changes in the nature and causes of pollution: *pollution microbienne* or *pollution bacterielle*, caused by bacterial living and developing in decaying and putrefying materials and stagnant water associated with settlements of growing size; *pollution artisanale*, associated with small-scale craft industries such as tanneries, potteries and other workshops carrying out various rather disagreeable tasks, including soap manufacture, bone burning and glue making; *pollution industrielle*, involving large-scale and pervasive pollution over major centres of industrial activity, particularly from the early nineteenth century in areas like the Ruhr or the English 'Black Country'; *pollution fondamentale*, in which whole regions are affected by pollution, as with the desiccation and subsequent salination of the Aral Sea area; *pollution foncière*, in which vast quantities of chemicals are deliberately applied to the land as fertilizers and biocides; and finally, *pollution accidentale*, in which major accidents can cause pollution which is neither foreseen nor calculable (e.g. the Chernobyl disaster).

Above all, as a result of the escalating trajectory of environmental transformation it is now possible to talk about *global* environmental change. There are two components to this (Turner et al., 1990): systemic global change and cumulative global change. In the systemic meaning, 'global' refers to the spatial scale of operation and comprises such issues as global changes in climate brought about by atmospheric pollution. This is a topic discussed at length in chapters 7 and 8. In the cumulative meaning, 'global' refers to the areal or substantive accumulation of localized change, and a change is seen to be 'global' if it occurs on a worldwide scale, or represents a significant fraction of the total environmental phenomenon or global resource. The two types of change are closely interwined. For example, the

Table 1.7 Types of global environmental change

Type	Characteristic	Examples
Systemic	Direct impact on globally functioning system	(a) Industrial and land use emissions of 'greenhouse' gases (b) Industrial and consumer emissions of ozone-depleting gases (c) Land cover changes in albedo
Cumulative	Impact through worldwide distribution of change Impact through magnitude of change (share of global resource)	(a) Groundwater pollution and depletion (b) Species depletion/genetic alteration (biodiversity) (a) Deforestation (b) Industrial toxic pollutants (c) Soil depletion on prime agricultural lands

Source: from B. L. Turner et al., 1990, table 1.

burning of vegetation can lead to systemic change through such mechanisms as carbon dioxide release and albedo change, and to cumulative change through its impact on soil erosion and biotic diversity (table 1.7).

We can conclude this introductory chapter by quoting from Kates et al. (1991: 1):

Most of the change of the past 300 years has been at the hands of humankind, intentionally or otherwise. Our ever-growing role in this continuing metamorphosis has itself essentially changed. Transformation has escalated through time, and in some instances the scales of change have shifted from the locale and region to the earth as a whole. Whereas humankind once acted primarily upon the visible 'faces' or 'states' of the earth, such as forest cover, we are now also altering the fundamental flows of chemicals and energy that sustain life on the only inhabited planet we know.

The Human Impact on Vegetation

2

Introduction

In any consideration of the human impact on the environment it is probably appropriate to start with vegetation, for human-kind has possibly had a greater influence on plant life than on any of the other components of the environment. Through the many changes humans have brought about in land use and land cover they have modified soils (Meyer and Turner 1994) (see chapter 4), influenced climates (see chapter 7), affected geomorphic processes (see chapter 6), and changed the quality (see chapter 5) and quantity of some natural waters. Indeed, the nature of whole landscapes has been transformed by human-induced vegetation change (figure 2.1). Hannah et al. (1994) attempt to provide a map and inventory of human disturbance of world ecosystems, but many of their criteria for recognizing undisturbed areas are naïve. Large tracts of central Australia are classed as undisturbed, when plainly they are not (Fitzpatrick, 1994). Following Hamel and Dansereau (1949, cited by Frenkel, 1970) we can recognize five principal degrees of interference – each one increasingly remote from pristine conditions. These are:

1 *Natural habitats*: those that develop in the absence of human activities.
2 *Degraded habitats*: those produced by sporadic, yet incomplete disturbances; for example, the cutting of a forest, burning or the non-intensive grazing of natural grassland.
3 *Ruderal habitats*: where disturbance is sustained but where there is no intentional substitution of vegetation. Roadsides are an example of a ruderal habitat.
4 *Cultivated habitats*: when constant disturbance is accompanied by the intentional introduction of plants.
5 *Artificial habitats*: which are developed when humans modify the ambient climate and soil, as in greenhouse cultivation.

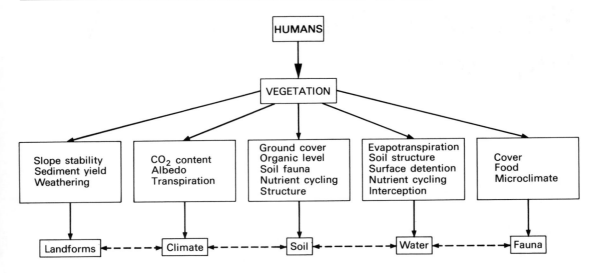

Figure 2.1
Some ramifications of human-
induced vegetation change

An alternative model for classifying the extent of human influence on vegetation is provided by Westhoff (1983), who adopts
a four-part scheme:

1   *Natural*: a landscape or ecosystem not influenced by human
    activity.
2   *Subnatural*: a landscape or ecosystem partly influenced by
    humans, but still belonging to the same (structural) forma-
    tion type as the natural system from which it derives (for
    example, a wood remaining a wood).
3   *Semi-natural*: a landscape or ecosystem in which flora and
    fauna are largely spontaneous, but the vegetation structure is
    altered so that it belongs to another formation type (for ex-
    ample, a pasture, moorland or heath deriving from a wood).
4   *Cultural*: a landscape or ecosystem in which flora and fauna
    have been essentially affected by human agency in such a
    way that the dominant species may have been replaced by
    other species (for example, arable land).

In this chapter we shall be concerned mainly with degraded and
ruderal habitats, or subnatural and semi-natural habitats; but first
we need to consider some of the processes that human societies
employ, notably fire, grazing and the physical removal of forest.

The use of fire

Humans are known to have used fire since Palaeolithic times (see
pp. 19–20). As Sauer, one of the great proponents of the role of
fire in environmental change, has put it (1969: 10–11):

Plate 2.1
Fires are an important means
by which humans have
transformed their environment.
Some major biome types,
including the grasslands of the
Highveld areas of Swaziland,
shown here, owe much of their
character to burning

Through all ages the use of fire has perhaps been the most important
skill to which man has applied his mind. Fire gave to man, a diurnal
creature, security by night from other predators . . . The fireside was the
beginning of social living, the place of communication and reflection.

People have utilized fire for a great variety of reasons (Bartlett,
1956; Stewart, 1956; Wertine, 1973): to clear forest for agriculture (plate 2.1); to improve grazing land for domestic animals
or to attract game; to deprive game of cover; to drive game from
cover in hunting; to kill or drive away predatory animals, ticks,
mosquitoes and other pests; to repel the attacks of enemies, or
to burn them out of their refuges; for cooking; to expedite travel;
to burn the dead and raise ornamental scars on the living; to
provide light; to transmit messages via smoke signalling; to break
up stone for tool making; to protect settlements or encampments
from great fires by controlled burning; to satisfy the sheer love
of fires as spectacles; to make pottery; to smelt ores; to harden
spears; to provide warmth; to make charcoal; and to assist in
the collection of insects such as crickets for eating. Given this
remarkable utility, it would be surprising if fire had not been
turned to account. Indeed, it is still much used, especially by
pastoralists such as the cattle keepers of Africa (Lemon, 1968),
and by the practitioners of shifting agriculture. For example,
the Malaysian and Indonesian *ladang* and the *milpa* system of
the Maya in Latin America involved the preparation of land for
planting by felling or deadening forest, letting the debris dry in
the hot season, and burning it before the commencement of the
rainy season. With the first rains, holes were dibbled in the soft
ash-covered earth with a planting stick. This system was suited
to areas of low population density with sufficiently extensive

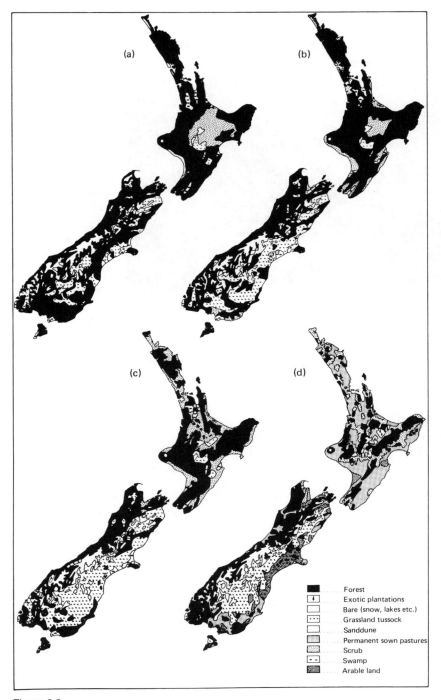

Figure 2.2
The changing state of the vegetation cover in New Zealand (from Cochrane, 1977):
(a) early Polynesian vegetation *c.*AD 700
(b) Pre-Classical Maori vegetation, *c.*AD 1200
(c) Pre-European vegetation, *c.*AD 1800
(d) present-day vegetation

forest to enable long intervals of 'forest fallow' between burnings. Land which was burned too frequently became overgrown with perennial grasses, which tended to make it difficult to farm with primitive tools. Land cultivated for too long rapidly suffered a deterioration in fertility, while land recently burned was temporarily rich in nutrients (see figure 9.1, p. 421).

The use of fire, however, has not been restricted to primitive peoples in the tropics. Remains of charcoal are found in Neolithic soil profiles in highland Britain; large parts of North America appear to have suffered fires at regular intervals prior to European settlement; and in the case of South America the 'great number of fires' observed by Magellan during the historical passage of the Strait that bears his name resulted in the toponym, 'Tierra del Fuego'. Indeed, says Sternberg (1968: 718): 'For thousands of years, man has been putting the New World to the torch, and making it a "land of fire".' Given that the Native American population may have been greater than once thought, their impact, even in Amazonia, may have been appreciable (Denevan, 1992).

Fire was also central to the way of life of the Australian Aboriginals, including those of Tasmania, and the carrying of fire sticks was a common phenomenon. As Blainey (1975: 76) has put it, 'Perhaps never in the history of mankind was there a people who could answer with such unanimity the question: "have you got a light, mate?" There can have been few if any races who for so long were able to practise the delights of incendiarism.' That is not to say that Aboriginal burning necessarily caused wholesale modification of Australian vegetation in pre-European times. That is a hotly debated topic (Kohen, 1995).

In neighbouring New Zealand, Polynesians carried out extensive firing of vegetation in pre-European settlement times, and hunters used fire to facilitate travel and to frighten and trap a major food source – the flightless moa (Cochrane, 1977). The changes in vegetation that resulted were substantial (figure 2.2). The forest cover was reduced from about 79 per cent to 53 per cent, and fires were especially effective in the drier forests of central and eastern South Island in the rain shadow of the Southern Alps. The fires continued over a period of about a thousand years up to the period of European settlement (Mark and McSweeney, 1990).

Fires: natural and anthropogenic

Although people have used fire for all the reasons that have been mentioned, before one can assess the role of this facet of the human impact one must ascertain how important fires started by human action are in comparison with those caused naturally, especially by lightning, which on average strikes the land surface of the globe 100,000 times each day (Yi-fu Tuan, 1971). Some natural fires may result from spontaneous combustion (Vogl, 1974), for in certain ecosystems heavy vegetal accumulations may become compacted, rotted and fermented, thus generating heat. Other natural fires can result from sparks produced by falling boulders and by landslides (Booysen and Tainton, 1984). Some of the available data on the causes of fire are summarized in table 2.1. In the forest lands of the western United States about half the fires are caused by lightning. Lightning starts over half the fires in the pine savanna of Belize, Central America, and about 8 per cent of the fires in the bush of Australia, while in the south of France, nearly all the fires are caused by people.

One method of gauging the long-term frequency of fires (Clark et al., 1997) is to look at tree rings and lake sediments, for these

Table 2.1   Fire frequencies and their causes for selected areas

1 Western United States

State	Total average yearly number of fires	% caused by lightning
Arizona	1,486	84
California	3,608	26
Colorado	413	36
Idaho	1,458	69
Montana	852	71
Nevada	86	34
New Mexico	614	79
Oregon	1,860	52
South Dakota	173	62
Utah	236	35
Washington	1,807	28
Wyoming	157	62
TOTAL	12,750	49

Source: Brown and Davis, 1973.

2 Montane savanna, Belize

Year	Total number of fires	% caused by lightning
1963	7	43
1964	11	55
1965	7	43
1966	2	100
1967	5	80
1968	8	100
1969	3	67
1970	7	71
1971	4	25
1972	12	42
TOTAL	66	59

Source: Kellman, 1975.

Table 2.1 *(cont'd)*

3 Bush fires in Australia

Cause	% of total
Deliberate burning off	35
Cigarettes and matches	7
Camp fires	6
Lightning	8
Trains	6
Tractors	5
Motor vehicles	6
Domestic	9
Miscellaneous	18

Source: Luke, 1962.

4 Mediterranean France (Bouches du Rhône), 1962

Cause	% of total
Children playing with fire	39
Cigarettes	17
Agricultural accidents	12.5
Railways	11.8
Rubbish tips	6.8
Spontaneous fires	6.4
Others	6.0

Source: Nicod, cited by Wright and Wanstall, 1977.

5 Fires in the fynbos heathlands of the Southern Cape, South Africa

Cause	% of total
Lightning	27.4
Trains	23.8
Unknown	15.5
Prescribed	11.9
Camp fires	10.7
Arson	7.1
Smokers	2.4
Money hunters	1.2

Source: Booysen and Tainton, 1984.

are affected by them. Studies in the United States indicate that over wide areas fires occurred with sufficient frequency to have effects on annual tree rings and lake cores every 7 to 80 years in pre-European times. The frequency of fires in different environments tends to show some variation. Rotation periods may be in excess of a century for tundra, 60 years for boreal pine forest, 100 years for spruce-dominated ecosystems, 5 to 15 years for savanna, 10 to 15 years for chaparral, and less than 5 years for semi-arid grasslands (Wein and Maclean, 1983).

Table 2.2 Temperatures attained in fires (maxima)

Vegetation	Source	Temperature (°C)
Minnesota jack pine	Ahlgren (1974)	800
Chaparral scrub	Ahlgren (1974)	538
British heath	Whittaker (1961)	840
Senegal savanna	Daubenmire (1968)	715
Japanese grassland	Daubenmire (1968)	887
Chaparral scrub	Mooney and Parsons (1973)	1,100
Nigerian savanna	Hopkins (1965)	>538–640
Sudanese savanna	Hopkins (1965)	850

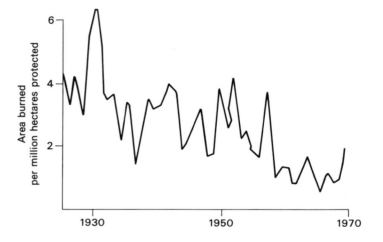

Figure 2.3
The reduction in area burned per million hectares protected for the USA between 1926 and 1969 as a result of fire-suppression policies (after Brown and Davis, 1973, figure 2.1)

The temperatures attained in fires

The effects which fires have on the environment depend very much on their size, duration and intensity. Some fires are relatively quick and cool, and destroy only ground vegetation. Other fires, crown fires, affect whole forests up to crown level and generate very high temperatures (table 2.2). In general, forest fires are hotter than grassland fires. Perhaps more significantly in terms of forest management, fires which occur with great frequency do not attain too high a temperature because there is inadequate inflammable material to feed them. However, when humans deliberately suppress fire, as has frequently been normal policy in forest areas (see figure 2.3), large quantities of inflammable materials accumulate, so that when a fire does break out it is of the hot, crown type. Such fires can be ecologically disastrous and there is now much debate about the wisdom of fire suppression given that, in many forests, fires under so-called 'natural' conditions appear, as we have seen, to have been a relatively frequent and regular phenomenon.

Some consequences of fire suppression

Given that fire has long been a feature of many ecosystems, and irrespective of whether the fires were or were not caused by people, it is clear that any deliberate policy of fire suppression will have important consequences for vegetation.

Fire suppression, as has already been suggested, can magnify the adverse effects of fire. The position has been well stated by Sauer (1969: 14):

> The great fires we have come to fear are effects of our civilization. These are the crown fires of great depths and heat, notorious aftermaths of the pyres of slash left by lumbering. We also increase fire hazard by the very giving of fire protection which permits the indefinite accumulation of inflammable litter. Under the natural and primitive order, such holocausts, that leave a barren waste, even to the destruction of the organic soil, were not common.

Recent studies have indicated that rigid fire-protection policies have often had undesirable results and as a consequence many foresters stress the need for 'environmental restoration burning' (Vankat, 1977). So, for example, in the coniferous forests of the middle upper elevations in the Sierra Nevada mountains of California fire protection since 1890 has made the stands denser, shadier and less park-like, with sequoia seedlings decreasing in number. Likewise, at lower elevations the Mediterranean semi-arid shrubland called 'chaparral' has had its character changed. The vegetation has increased in density, the amount of combustible fuels has risen, fire-intolerant species have encroached and vegetation diversity has decreased, resulting in a monotony of old-age stands, instead of a mosaic of different successional stages. Unfavourable consequences of fire suppression have also been noted in Alaska (Oberle, 1969). It has been found that when fire is excluded from many lowland sites an insulating carpet of moss tends to accumulate and raise the permafrost level. Permafrost close to the surface encourages the growth of black spruce, a low-growing species with little timber or food value. In the Kruger National Park, in South Africa, fires have occurred less frequently after the establishment of the game reserve, when it became uninhabited by natives and hunters. As a result, bush encroachment has taken place in areas that were formerly grassland, and the carrying capacity for grazing animals has declined. Controlled burning has been reinstituted as a necessary game-management operation.

However, following the severe fires that ravaged America's Yellowstone National Park in 1988 (plate 2.2), there has been considerable debate as to whether the inferno, the worst since the park was established in the 1870s, was the result of a policy

Plate 2.2
Fires are a highly important ecological process in areas like the Yellowstone National Park, USA. Fire suppression can result ultimately in forest devastation as increasingly large amounts of combustible matter accumulate through time

of fire suppression. Without such a policy the forest would burn at intervals of 10–20 years because of lightning strikes. Could it be that the suppression of fires over long periods of, say, 100 years or more, allegedly to protect and preserve the forest, led to the build-up of abnormal amounts of combustible fuel in the form of trees and shrubs in the understory? Should a programme of prescribed burning be carried out to reduce the amount of available fuel?

Fire suppression policies at Yellowstone did indeed lead to a critical build-up in flammable material. However, other factors must also be examined in explaining the severity of the fire. One of these was the fact that the last comparable fire had been in the 1700s, so that the Yellowstone forests had had nearly 300 years in which to become increasingly flammable. In other words, because of the way vegetation develops through time (a process called succession), very large fires may occur every 200–300 years as part of the natural order of things (figure 2.4). Another crucial factor was that the weather conditions in the summer of 1988 were abnormally dry, bringing a great danger of fire. Romme and Despain (1989: 28) remark in conclusion to their study of the Yellowstone fires:

It seems that unusually dry, hot and windy weather conditions in July and August of 1988 coincided with multiple ignition in a forest that was at its most flammable stage of succession. Yet it is unlikely that past suppression efforts were a major factor in exacerbating the Yellowstone fire. If fires occur naturally at intervals ranging from 200 to 400 years, then 30 or 40 years of effective suppression is simply not enough for excessive quantities of fuel to build up; major attempts at suppression in the Yellowstone forests may have merely delayed the inevitable.

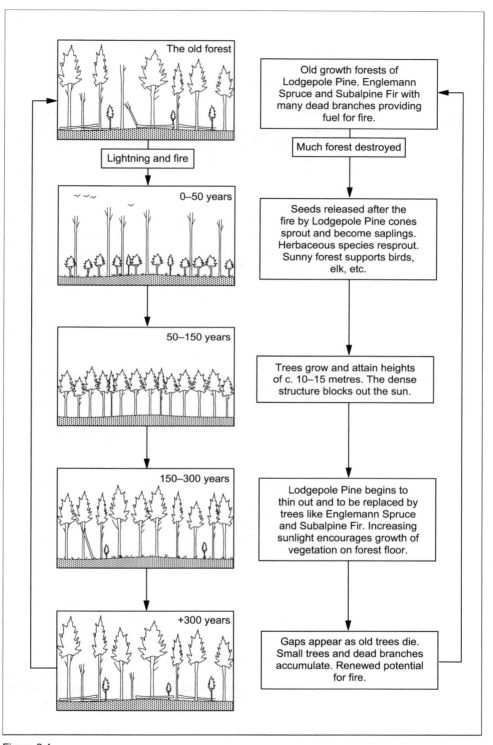

Figure 2.4
Ecological succession in response to fire in Yellowstone National Park (after Romme and Despain, 1989, pp. 24–5, heavily modified)

Some effects of fire on vegetation

There is evidence that fire has played an important role in the formation of various major types of vegetation. This applies, for instance, to some savannas, mid-latitude grasslands and shrublands (like the garrigue and maquis of the Mediterranean lands and the chaparral of the south-west United States). Before examining these, however, it is worth looking at some of the general consequences of burning (Crutzen and Goldammer, 1993).

Fire may assist in seed germination. For example, the abundant germination of dormant seeds on recently burned chaparral sites has been reported by many investigators, and it seems that some seeds of chaparral species require scarification by fire. The better germination of those not requiring scarification may be related to the removal by fire of competition, litter and some substances in the soils which are toxic to plants (Hanes, 1971). Fire alters seedbeds. If litter and humus removal are substantial, large areas of rich ash, bare soil or thin humus may be created. Some trees, such as Douglas pine and the giant sequoia, benefit from such seedbeds (Heinselman and Wright, 1973). Fire sometimes triggers the release of seeds (as with the Jack pine, *Pinus bankdiana*) and seems to stimulate the vegetative reproduction of many woody and herbaceous species. Fire can control forest insects, parasites and fungi – a process termed 'sanitization'. It also seems to stimulate the flowering and fruiting of many shrubs and herbs, and to modify the physicochemical environment of the plants. Mineral elements are released both as ash and through increased decomposition rates of organic layers. Above all, areas subject to fire often show greater species diversity, which is a factor that tends to favour stability.

One can conclude by quoting at length from Pyne (1982: 3), who provides a detailed and scholarly analysis of the history of cultural fires in America:

Hardly any plant community in the temperate zone has escaped fire's selective action, and, thanks to the radiation of *Homo sapiens* throughout the world, fire has been introduced to nearly every landscape on earth. Many biotas have consequently so adapted themselves to fire that, as with biotas frequented by floods and hurricanes, adaptation has become symbiosis. Such ecosystems do not merely tolerate fire, but often encourage it and even require it. In many environments fire is the most effective form of decomposition, the dominant selective force for determining the relative distribution of certain species, and the means for effective nutrient recycling and even the recycling of whole communities.

Plate 2.3
Grazing by domestic animals
has many environmental
consequences and assists in
the maintenance of grasslands
such as the Pampas at Entre
Rios in Argentina. Excessive
grazing can compact the soil
and contribute to both wind
and water erosion

The role of grazing

Many of the world's grasslands have long been grazed by wild
animals, like the bison of North America or the large game of
East Africa, but the introduction of pastoral economies also
affects their nature and productivity (plate 2.3) (Coupland, 1979).

Light grazing may increase the productivity of wild pastures
(Warren and Maizels, 1976). Nibbling, for example, can encourage the vigour and growth of plants, and in some species,
such as the valuable African grass *Themeda triandra*, the removal of coarse, dead stems permits succulent sprouts to shoot.
Likewise, the seeds of some plant species are spread efficiently
by being carried in cattle guts, and then placed in favourable
seedbeds of dung or trampled into the soil surface. Moreover,
the passage of herbage through the gut and out as faeces modifies the nitrogen cycle, so that grazed pastures tend to be richer
in nitrogen than ungrazed ones. Also, like fire, grazing can increase species diversity by opening out the community and creating more niches.

On the other hand, heavy grazing may be detrimental. Excessive trampling when conditions are dry will reduce the size of
soil aggregates and break up plant litter to a point where they
are subject to removal by aeolian deflational processes. Trampling, by puddling the soil surface, can accelerate soil deterioration and erosion as infiltration capacity is reduced. Heavy grazing can kill plants or lead to a marked reduction in their level of
photosynthesis. In addition, when relieved of competition from
palatable plants or plants liable to trampling damage, resistant
and usually unpalatable species expand their cover. Thus in the

western United States poisonous burroweed (*Haploplappius* spp.) has become dangerously common, and many woody species have intruded. These include the mesquite (*Prosopis juliflora*), the big sagebrush (*Artemisia tridentata*) (Vale, 1974), the one-seed juniper (*Juniperus monosperma*) (Harris, 1966), and the Pinyon pine (Blackburn and Tueller, 1970).

Grover and Musick (1990) see shrubland encroachment by creosote bush (*Larrea tridentata*) and mesquite as part and parcel of desertification, and indicate that in southern New Mexico the area dominated by them has increased several fold over the last century. This has been as a result of a corresponding decrease in the area's coverage of productive grasslands. They attribute both tendencies to livestock overgrazing at the end of the nineteenth century, but point out that this was compounded by a phase of rainfall regimes that were unfavourable for perennial grass growth.

Some of the plants that have invaded grasslands in California are aliens, and these have been studied in detail by Burcham (1970). It is clear from his long-term analysis of their progress that the aliens have moved in four distinct stages:

Stage 1 (1845 →)
   Wild oats (*Aventa fatua* and *A. barbata*) and black mustard (*Brassica nigra*).
Stage 2 (1855 →)
   Filarees (*Erodium* spp.), wild barleys (*Hordeum* spp.), nitgrass (*Gastridium ventricosum*) and native annuals.
Stage 3 (1870 →)
   Mouse barley (*Hordeum leporinum*), red brome (*Bromus rubens*), silver hairgrass (*Aira caryophyllea*), Chile tarweed (*Madia sativa*) and star thistles (*Centaurea*).
Stage 4 (1900 →)
   Medusa-head (*Taeniatherum asperum*), barb goatgrass (*Aegilops triuncialis*), dogtail grass (*Cynosaurus echinatus*) and annual falso-brome (*Brachypodium distachyon*).

Each stage involves progressively less desirable species and indicates increased intensities of grazing use. In general, Burcham demonstrates that perennials have tended to be replaced through time by greater numbers of annuals, which have the ability to germinate quickly, grow rapidly and produce large quantities of viable seed.

In Australia (Moore, 1959) the widespread adoption of sheep grazing led to significant changes in the nature of grasslands over extensive areas. In particular, the introduction of sheep led to the removal of kangaroo grass (*Themeda australis*) – a predominantly summer-growing species – and its replacement by essentially winter-growing species such as *Danthonia* and *Stipa*. Also in Australia, not least in the areas of tropical savanna in the north, large herds of introduced feral animals (e.g. *Bos taurus*, *Equus caballus*, *Camelus dromedarius*, *Bos banteng* and *Cervus unicolor*) have resulted in overgrazing and alteration of native habitats. As they appear to lack significant control by predators and pathogens, their densities, and thus their effects, became very high (Freeland, 1990).

Similarly, in Britain many plants are avoided by grazing animals because they are distasteful, hairy, prickly or even poisonous (Tivy, 1971). The persistence and continued spread of bracken (*Pteridium aquilinium*) on heavily grazed rough pasture in Scotland is aided by the fact that it is slightly poisonous, especially to young stock. The success of bracken is furthered by its reaction to burning, for with its extensive system of underground stems (rhizomes) it tends to be little damaged by fire. The survival and prevalence of shrubs such as elder (*Sambucus nigra*), gorse (*Ulex* spp.), broom (*Sarothamus scoparius*), and the common weeds ragwort (*Senecio jacobaea*) and creeping thistle (*Cirsium arvense*), in the face of grazing, can be attributed to their lack of palatability.

The role of grazing in causing marked deterioration of habitat has been the subject of further discussion in the context of upland Britain. In particular, Darling (1956) stressed that while trees bring up nutrients from rocks and keep minerals in circulation, pastoralism means the export of calcium phosphate and nitrogenous organic matter. The vegetation gradually deteriorates, the calcicoles disappear and the herbage becomes deficient in both minerals and protein. Progressively more xerophytic plants come in: *Nardus stricta*, *Molinia caerulaea*, *Erica tetralix* and then *Scripus caespitosa*. However, this is not a view that now receives universal support (Mather, 1983). Studies in several parts of upland Britain have shown that mineral inputs from precipitation are very much greater than the nutrient losses in wool and sheep carcasses. During the 1970s the idea of upland deterioration was gradually undermined.

In general terms, it is clear that in many parts of the world the grass family is well equipped to withstand grazing. Many plants have their growing points located on the apex of leaves and shoots, but grasses reproduce the bulk of fresh tissue at the base of their leaves. This part is least likely to be damaged by grazing and allows regrowth to continue at the same time as material is being removed.

Communities severely affected by the treading of animals (and indeed people) tend to have certain distinctive characteristics. These include diminutiveness (since the smaller the plant is, the more protection it will get from soil surface irregularities); strong ramification (the plant stems and leaves spread close to the ground); small leaves (which are less easily damaged by treading); tissue firmness (cell-wall strength and thickness to limit mechanical damage); a bending ability; strong vegetative increase and dispersal (for example, by stolons); small hard seeds which can be easily dispersed; and the production of a large number of seeds per plant (which is particularly important because the mortality of seedlings is high under treading and trampling conditions).

Deforestation

Controversy surrounds the meaning of the word 'deforestation' and this causes problems when it comes to assessing rates of change and causes of the phenomenon (Williams, 1994). It is best defined (Grainger, 1992) as 'the temporary or permanent clearance of forest for agriculture or other purposes'. According to this definition, if clearance does not take place then deforestation does not occur. Thus, much logging in the tropics, which is selective in that only a certain proportion of trees and only certain species are removed, does not involve clear-felling and so cannot be said to constitute deforestation. However, how clear is clear? Many shifting cultivators in the humid tropics leave a small proportion of trees standing (perhaps because they have special utility). At what point does the proportion of trees left standing permit one to say that deforestation has taken place?

The deliberate removal of forest, whether by fire or cutting, is one of the most long-standing and significant ways in which humans have modified the environment. Pollen analysis shows that temperate forests were removed in Mesolithic and Neolithic times and at an accelerating rate thereafter. Since pre-agricultural times forests have declined approximately one-fifth, from five to four billion hectares. The highest loss has occurred in temperate forests (32–35 per cent), followed by subtropical woody savannas and deciduous forests (24–25 per cent). The lowest losses have been in tropical evergreen forests (4–6 per cent) because many of them have for much of their history been inaccessible and sparsely populated (World Resources Institute, 1992: 107).

In Britain, Birks (1988) has analysed dated pollen sequences to ascertain when and where tree pollen values in sediments from upland areas drop to 50 per cent of their Holocene maximum percentages, and takes this as a working definition of 'deforestation'. From this work he identified four phases:

1 *3700–3900 years before the present (BP):* north-west Scotland and the eastern Isle of Skye;
2 *2100–2600 years BP (the pre-Roman Iron Age):* Wales, England (except the Lake District), northern Skye, and northern Sutherland;
3 *1400–1700 years BP (post-Roman):* Lake District, southern Skye, Galloway and Knapdale–Ardnamurchan;
4 *300–400 years BP:* the Grampians and the Cairngorms.

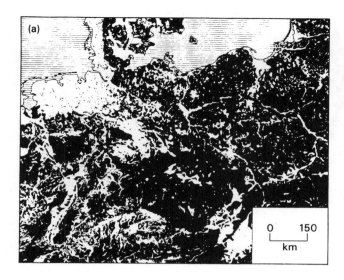

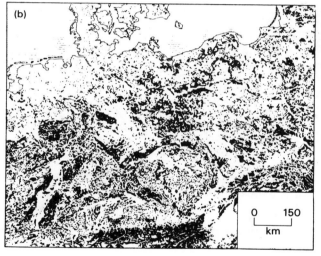

Figure 2.5
The changing distribution of
forest in Central Europe
between (a) AD 900 and
(b) AD 1900 (reprinted from
Darby, 1956, pp. 202, 203, in
*Man's role in changing the
face of the Earth*, ed. W. L.
Thomas, by permission of the
University of Chicago Press
© The University of Chicago
1956)

Sometimes forests are cleared to allow for agriculture; sometimes to provide fuel for domestic purposes, or to provide charcoal or wood for construction; sometimes to fuel locomotives,
or to smoke fish; and sometimes to smelt metals. The Phoenicians
were exporting cedars as early as 4,600 years ago (Mikesell,
1969) both to the Pharaohs and to Mesopotamia. Attica in
Greece was laid bare by the fifth century BC; classical writers
allude to the effects of fire, cutting and the destructive nibble of
the goat. The great phase of deforestation in central and western Europe, described by Darby (1956: 194) as 'the great heroic
period of reclamation', occurred properly from AD 1050 onwards
for about 200 years (figure 2.5). In particular, the Germans

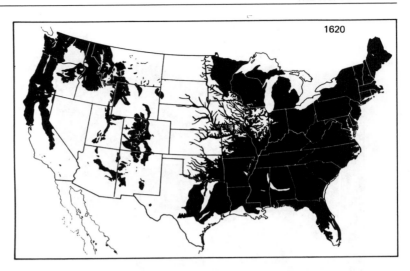

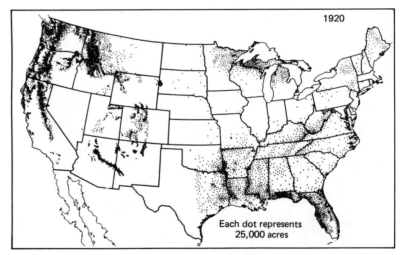

Figure 2.6
The distribution of American
natural forest in 1620 and
1920 (modified from Williams,
1989)

moved eastward: 'What the new west meant to young America
in the nineteenth century, the new east meant to Germany in
the Middle Ages' (Darby, 1956: 196). The landscape of Europe
was transformed, just as that of North America, Australia, New
Zealand and South Africa was to be as a result of the European
expansions, especially in the nineteenth century.

Temperate North America underwent particularly brutal
deforestation (Williams, 1989), and lost more woodland in 200
years than Europe did in 2,000. The first colonialists arriving in
the *Mayflower* found a continent that was wooded from the
Atlantic seaboard as far as the Mississippi river (figure 2.6). The
forest originally occupied some 170 million hectares. Today only
about 10 million hectares remain.

Table 2.3 Estimates of deforestation since 1900 (millions of hectares)

Country	Orthodox estimate of forest area lost	Forest area lost according to World Conservation Monitoring Centre	Forest area lost according to Fairhead and Leach (1998, table 9.1)
Côte d'Ivoire	13	20.2	4.3–5.3
Liberia	4–4.5	5.5	1.3
Ghana	7	12.9	3.9
Benin	0.7	12.9	3.9
Togo	0	1.7	0
Sierra Leone	0.8–5	6.7	c.0
Total	25.3–30.2	48.6	9.5–10.5

Fears have often been expressed that the mountains of High Asia (for example in Nepal) have been suffering from a wave of deforestation that has led to a whole suite of environmental consequences, which include accelerated landsliding, flooding in the Ganges plain and sedimentation in the deltaic areas of Bengal. Ives and Messerli (1989) doubt that this alarmist viewpoint is soundly based and argue (p. 67) that 'the popular claims about catastrophic post-1950 deforestation of the Middle Mountain belt and area of the high mountains of the Himalayas are much exaggerated, if not inaccurate.'

Likewise, some workers in West Africa have interpreted 'islands' of dense forest in the savanna as the relics of a once more extensive forest cover that was being rapidly degraded by human pressure. More recent research (Fairhead and Leach, 1996) suggests that this is far from the truth and that villagers have in recent decades been extending rather than reducing forest islands. There are also some areas of Kenya in East Africa where, far from recent population pressures promoting vegetation removal and land degradation, there has been an increase in woodland since the 1930s. Considerations such as these have led Fairhead and Leach (1998) to suggest that in a selection of six countries in West Africa the amount of twentieth-century deforestation 'is probably only one-third of that suggested by the estimates in international circulation' (p. 183). Their data are presented in table 2.3.

With regard to the equatorial rain forests, the researches of Flenley and others (Flenley, 1979) have indicated that forest clearance for agriculture has been going on since at least 3000 BP in Africa, 7000 BP in South and Central America, and possibly since 9000 BP or earlier in India and New Guinea.

The causes of the current spasm of tropical deforestation (plate 2.4) are complex and multifarious and are summarized by Grainger (1992) and in table 2.4. Grainger also provides an excellent review of the problems of measuring and defining loss of tropical forest area. Indeed, there are very considerable difficulties

Plate 2.4
One of the most serious
environmental problems in
tropical areas is the removal of
the rain forest. The felling of
trees in Amazonia on slopes
as steep as these will cause
accelerated erosion and loss
of soil nutrients, and may
promote lateritization

in estimating rates of deforestation, in part because different
groups of workers use different definitions of what constitutes a
forest (Allen and Barnes, 1985) and what distinguishes rain forest
from other types of forest. There is thus some variability in views
as to the present rate of rain-forest removal. Recent estimates by
the UN Food and Agriculture Organization (Lanly et al., 1991)
show that the total annual deforestation in 1990 for 62 countries (representing some 78 per cent of the tropical forest area of
the world) was 16.8 million hectares, a figure significantly higher
than the one obtained for these same countries for the period
1976–80 (9.2 million hectares per year). Myers (1992) suggests
that there has been an 89 per cent increase in the tropical deforestation rate during the 1980s (compared with the FAO estimate
of a 59 per cent increase). He believes that the annual rate of
loss in 1991 amounted to about 2 per cent of the total forest
expanse.

There is, however, a considerable variation in the rate of forest
regression in different areas (figure 2.7). Some areas are under
relatively modest threat (e.g. western Amazonia, the forests of
Guyana, Surinam and French Guyana, and much of the Zaire
basin in Central Africa (Myers, 1983, 1984). Some other areas
are being exploited so fast that minimal areas will be left in the

Table 2.4 The causes of deforestation

A. Immediate causes – land use

---

1 Shifting agriculture
   (a) Traditional long-rotation shifting cultivation
   (b) Short-rotation shifting cultivation
   (c) Encroaching cultivation
   (d) Pastoralism

2 Permanent agriculture
   (a) Permanent staple crop cultivation
   (b) Fish farming
   (c) Government sponsored resettlement schemes
   (d) Cattle ranching
   (e) Tree crop and other cash crop plantations

3 Mining

4 Hydro-electric schemes

5 Cultivation of illegal narcotics

B. Underlying causes

---

1 Socio-economic mechanisms
   (a) Population growth
   (b) Economic development

2 Physical factors
   (a) Distribution of forests
   (b) Proximity of rivers
   (c) Proximity of roads
   (d) Distance from urban centres
   (e) Topography
   (f) Soil fertility

3 Government policies
   (a) Agriculture policies
   (b) Forestry policies
   (c) Other policies

Source: from Grainger, 1992.

not too distant future; these include the Philippines, peninsular Malaya, Thailand, Australia, Indonesia, Vietnam, Bangladesh, Sri Lanka, Central America, Madagascar, West Africa and eastern Amazonia. Myers (1992) refers to particular 'hot spots' where the rates of deforestation are especially threatening, and presents data for certain locations where the percentage loss of forest is more than three times the global figure of 2 per cent: southern Mexico (10 per cent), Madagascar (10 per cent), northern and eastern Thailand (9.6 per cent), Vietnam (6.6 per cent) and the Philippines (6.7 per cent).

The rapid loss of rain forest is potentially extremely serious, because as Poore (1976: 138) has stated, these forests are a source-book of potential foods, drinks, medicines, contraceptives, abortifacients, gums, resins, scents, colorants, specific pesticides,

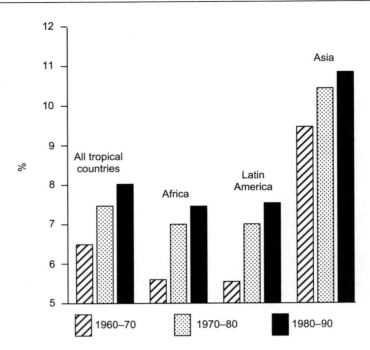

Figure 2.7
Estimated rate of tropical
deforestation, 1960–1990,
showing the particularly high
rates in Asia (modified from
World resources, 1996–97)

and so on, of which we have scarcely turned the pages. Their
removal may contribute to crucial global environmental concerns
(e.g. climatic change and loss of biodiversity), besides causing
regional and local problems, including lateritization, increased
rates of erosion and accelerated mass movements. The great
range of potential impacts of tropical deforestation is summarized in table 2.5.

Some traditional societies have developed means of exploiting
the rain-forest environment which tend to minimize the problems
posed by soil fertility deterioration, soil erosion and vegetation
degradation. Such a system is known as shifting agriculture
(swidden). As Geertz (1963: 16) has remarked:

In ecological terms, the most distinctive positive characteristic of
swidden agriculture . . . is that it is integrated into and, when genuinely
adaptive, maintains the general structure of the pre-existing natural
ecosystem into which it is projected, rather than creating and sustaining
one organised along novel lines and displaying novel dynamics.

The tropical rain forest and the swidden plots have certain
common characteristics. Both are closed cover systems, in part
because in swidden some trees are left standing, in part because
some tree crops (such as banana, papaya, areca, etc.) are planted,
but also because food plants are not planted in an open field,
crop-row manner, but helter-skelter in a tightly woven, dense

Table 2.5 The consequences of tropical deforestation

1 Reduced biological diversity
   (a) Species extinctions
   (b) Reduced capacity to breed improved crop varieties
   (c) Inability to make some plants economic crops
   (d) Threat to production of minor forest products

2 Changes in local and regional environments
   (a) More soil degradation
   (b) Changes in water flows from catchments
   (c) Changes in buffering of water flows by wetland forests
   (d) Increased sedimentation of rivers, reservoirs etc.
   (e) Possible changes in rainfall characteristics

3 Changes in global environments
   (a) Reduction in carbon stored in the terrestrial biote
   (b) Increase in carbon dioxide content of atmosphere
   (c) Changes in global temperature and rainfall patterns by greenhouse effects
   (d) Other changes in global climate due to changes in land surface processes

Source: from Grainger, 1992.

botanical fabric. It is, in Geertz's words (1963: 25) 'a miniaturized tropical forest'. Secondly, swidden agriculture normally involves a wide range of cultigens, thereby having a high diversity index like the rain forest itself. Thirdly, both swidden plots and the rain forest have high quantities of nutrients locked up in the biotic community (Douglas, 1969) compared to those in the soil. The primary concern of 'slash-and-burn' activities is not merely the clearing of the land, but rather the transfer of the rich store of nutrients locked up in the prolific vegetation of the rain forest to a botanical complex whose yield to people is a great deal larger. If the period of cultivation is not too long and the period of fallow is long enough, an equilibrated, non-deteriorating and reasonably productive farming regime can be sustained in spite of the rather impoverished soil base upon which it rests.

Unfortunately this system often breaks down, especially when population increase precludes the maintenance of an adequately long fallow period. When this happens, the rain forest cannot recuperate and is replaced by a more open vegetation assemblage ('derived savanna') which is often dominated by the notorious *Imperata* savanna grass which has turned so much of South-East Asia into a green desert. *Imperata cyclindrica* is a tall grass which springs up from rhizomes. Because of its rhizomes it is fire-resistant, but because it is a tall grass it helps to spread fire (Gourou, 1961).

One particular type of tropical forest ecosystem coming under increasing pressure from various human activities is the mangrove forest characteristic of intertidal zones. These ecosystems constitute a reservoir, refuge, feeding ground and nursery for many useful and unusual plants and animals (Mercer and Hamilton, 1984). In particular, because they export decomposable plant debris into adjacent coastal waters, they provide an important energy source and nutrient input to many tropical estuaries. In addition they can serve as buffers against the erosion caused by tropical storms – a crucial consideration in low-lying areas like Bangladesh. In spite of these advantages, mangrove forests are being degraded and destroyed on a large scale in many parts of the world, either through exploitation of their wood resources or because of their conversion to single-use systems such as agriculture, aquaculture, salt-evaporation ponds or housing developments. To give two examples: mangrove areas in the Philippines converted to fish ponds have increased from less than 90,000 hectares in the early 1950s to over 244,000 hectares in the early 1980s, while in Indonesia logging operations are claiming 200,000 hectares of mangrove each year.

On a global basis Richards (1991: 164) has calculated that since 1700 about 19 per cent of the world's forests and woodlands have been removed. Over the same period the world's cropland area has increased by over four and a half times, and between 1950 and 1980 it amounted to well over 100,000 square kilometres per year.

Deforestation is not an unstoppable or irreversible process. In the United States the forested area has increased substantially since the 1930s and 1940s. This 'rebirth of the forest' (Williams, 1988) has a variety of causes: new timber growth and planting were rendered possible because the old forest had been removed, forest fires have been suppressed and controlled, farmland has been abandoned and reverted to forest, and there has been a falling demand for lumber and lumber-derived products. It is, however, often very difficult to disentangle the relative importance of grazing impacts, fire and fluctuating climates in causing changes in the structure of forested landscapes. In some cases, as for example the Ponderosa pine forests of the American southwest, all three factors may have contributed to the way in which the park-like forest of the nineteenth century has become significantly denser and younger today (Savage, 1991).

Given the perceived and actual severity of tropical deforestation it is evident that strategies need to be developed to reduce the rate at which it is disappearing. Possible strategies include the following:

- *research, training and education* to give people a better understanding of how forests work and why they are important, and to change public opinion so that more people appreciate the uses and potential of forests;
- *land reform* to reduce the mounting pressures on landless peasants caused by inequalities in land ownership;

- *conservation of natural resources* by setting aside areas of rain forest as national parks or nature reserves;
- *restoration and reforestation* of damaged forests;
- *sustainable development*, namely development which, while protecting the habitat, allows a type and level of economic activity that can be sustained into the future with minimum damage to people or forest (e.g. selective logging rather than clear-felling; promotion of non-tree products; small-scale farming in plots within the forest);
- *control of the timber trade* (e.g. by imposing heavy taxes on imported tropical forest products and outlawing the sale of tropical hardwoods from non-sustainable sources);
- *'debt-for-nature' swaps* whereby debt-ridden tropical countries set a monetary value on their ecological capital assets (in this case forests) and literally trade them for their international financial debt;
- *involvement of local peoples* in managing the remaining rain forests;
- *careful control of international aid* and development funds to make sure that they do not inadvertently lead to forest destruction.

Having considered the importance of the three basic processes of fire, grazing and deforestation, we can now turn to a consideration of some of the major changes in vegetation types that have taken place over extensive areas as a result of such activities.

Secondary rain forest

When an area of rain forest that has been cleared for cultivation or timber exploitation is abandoned by humans, the forest begins to regenerate; but for an extended period of years the type of forest that occurs – secondary forest – is very different in character from the virgin forest it replaces. The features of such secondary forest, which is widespread in many tropical regions and accounts for as much as 40 per cent of the total forest area in the tropics, have been summarized by Richards (1952), Ellenberg (1979), Brown and Lugo (1990) and Corlett (1995).

First, secondary forest is lower and consists of trees of smaller average dimensions than those of primary forest; but since it is comparatively rare that an area of primary forest is clear-felled or completely destroyed by fire, occasional trees much larger than average are usually found scattered through secondary forest. Secondly, very young secondary forest is often remarkably regular and uniform in structure, though the abundance of small climbers and young saplings gives it a dense and tangled appearance which is unlike that of primary forest and makes it laborious to penetrate. Thirdly, secondary forest tends to be much poorer in species than primary, and is sometimes (though by no means always) dominated by a single species, or a small number of species. Fourthly, the dominant trees of secondary forest are light-demanding and intolerant of shade, most of the trees possess efficient dispersal mechanisms (having seeds or fruits well adapted for transport by wind or animals), and most of them can grow very quickly. The rate of net primary production of secondary forests exceeds that of the primary forests by a factor of two (Brown and Lugo, 1990). Some species are known to grow at rates of up to 12 metres in three years, but they tend to be short-lived and to mature and reproduce early. One consequence of their rapid growth is that their wood often has a soft texture and low density.

Secondary forests should not be dismissed as useless scrub. All but the youngest secondary forests are probably effective at preventing soil erosion, regulating water supply and maintaining water quality. They provide refuge for some flora and fauna and can provide a source of timber, albeit generally of inferior quality than that derived from primary forest.

The human role in the creation and maintenance of savanna

The savannas (plate 2.5) can be defined, following Hills (1965: 218–19), as:

a plant formation of tropical regions, comprising a virtually continuous ecologically dominant stratum of more or less xeromorphic plants, of which herbaceous plants, especially grasses and sedges, are frequently

Plate 2.5
A tropical savanna in the Kimberley District of north-west Australia

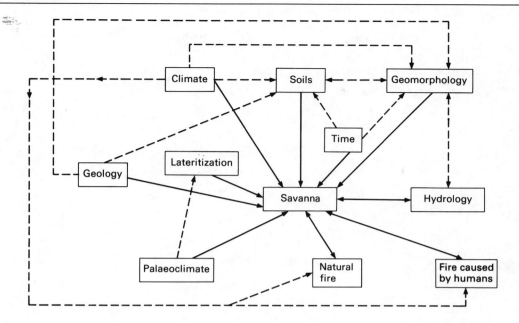

the principal, and occasionally the only, components, although woody plants often of the dimension of trees or palms generally occur and are present in varying densities.

Figure 2.8
The interrelated factors involved in the formation of savanna vegetation

They are extremely widespread in low latitudes (Harris, 1980), covering about 18 million square kilometres (an area about 2.6 million square kilometres greater than that of the tropical rain forest). Their origin has been the subject of great contention in the literature of biogeography, though most savanna research workers agree that no matter what savanna origins may be, the agent which seems to maintain them is intentional or inadvertent burning (Scott, 1977). As with most major vegetation types, a large number of interrelated factors are involved in causing savanna, and too many arguments about origins have neglected this fact (figure 2.8). Confusion has also arisen because of the failure to distinguish clearly between predisposing, causal, resulting and maintaining factors (Hills, 1965). It appears, for instance, that in the savanna regions around the periphery of the Amazon basin, the climate *predisposes* the vegetation towards the development of savanna rather than forest. The geomorphic evolution of the landscape may be a *causal* factor; increased laterite development a *resulting* factor; and fire a *maintaining* factor.

Originally, however, savanna was envisaged as a predominantly natural vegetation type of climatic origin (see e.g. Schimper, 1903), and the climatologist Köppen used the term 'savanna climate', implying that a specific type of climate is associated with all areas of savanna origin. According to supporters of this

Plate 2.6
The pindan bush of north-west Australia, composed of *Eucalyptus*, *Acacia* and grasses, is frequently burned. The pattern of the burning shows up clearly on Landsat imagery

theory, savanna is better adapted than other plant formations to withstand the annual cycle of alternating soil moisture: rain forests could not resist the extended period of extreme drought, while dry forests could not compete successfully with perennial grasses during the equally lengthy period of large water surplus (Sarmiento and Monasterio, 1975).

Other workers have championed the importance of edaphic (soil) conditions, including poor drainage, soils which have a low water-retention capacity in the dry season, soils with a shallow profile due to the development of a lateritic crust, and soils with a low nutrient supply (either because they are developed on a poor parent rock such as quartzite, or because the soil has undergone an extended period of leaching on an old land surface). Associated with soil characteristics, ages of land surfaces and degree of drainage is the geomorphology of an area (as stressed by e.g. Cole, 1963). This may also be an important factor in savanna development.

Some other researchers – for example, Eden (1974) – find that savannas are the product of former drier conditions (such as late Pleistocene aridity) and that, in spite of a moistening climate, they have been maintained by fire. He points to the fact that the patches of savanna in southern Venezuela occur in forest areas of similar humidity and soil infertility, suggesting that neither soil, nor drainage, nor climate was the cause. Moreover, the present 'islands' of savanna are characterized by species which are also present elsewhere in tropical American savannas and whose disjunct distribution conforms to the hypothesis of previous widespread continuity of that formation.

The importance of fire (plate 2.6) in maintaining and originating some savannas is suggested by the fact that many savanna trees are fire-resistant. Controlled experiments in Africa (Hopkins, 1965) demonstrate that some tree species, such as *Burkea africana* and *Lophira lanceolata*, withstand repeated burning better than others. There are also many observations of the frequency with which, for example, African herdsmen and agriculturists burn over much of tropical Africa and thereby maintain grassland. Certainly the climate of savanna areas is conductive to fire for, as Gillon (1983: 617) put it, 'Large scale grass fires are more likely to take place in areas having a climate moist enough to permit the production of a large amount of grass, but seasonally dry enough to allow the dried material to catch fire and burn easily.' On the other hand, Morgan and Moss (1965) express some doubts about the role of fire in western Nigeria, and point out that fire is not itself necessarily an independent variable:

The evidence suggests that some notions of the extent and destructiveness of savanna fires are rather exaggerated. In particular the idea of an annual burn which affects a large proportion of the area in each year could seem to be false. It is more likely that some patches, peculiarly susceptible to fire, as a result of edaphic or biotic influences upon the character of the community itself, are repeatedly burned, whereas others are hardly, if ever, affected . . . It is also important to note that there is no evidence anywhere along the forest fringe . . . to suggest that fire sweeps into the forest, effecting notable destruction of forest trees.

Some savannas are undoubtedly natural, for pollen analysis in South America shows that savanna vegetation was present before the arrival of human civilization. None the less even natural savannas, when subjected to human pressures, change their characteristics. For example, the inability of grass cover to maintain itself over long periods in the presence of heavy stock grazing may be documented from many of the warm countries of the world (Johannessen, 1963). Heavy grazing tends to remove the fuel (grass) from much of the surface. The frequency of fires is therefore significantly reduced, and tree and bush invasion take place. As Johannessen (p. 111) wrote: 'Without intense, almost annual fires, seedlings of trees and shrubs are able to invade the savannas where the grass sod has been opened by heavy grazing . . . the age and size of the trees on the savannas usually confirm the relative recency of the invasion.' In the case of the savannas of interior Honduras, he reports that they only had a scattering of trees when the Spaniards first encountered them, whereas now they have been invaded by an assortment of trees and tall shrubs.

Elsewhere in Latin America changes in the nature and density of population have also led to a change in the nature and distribution of savanna grassland. C. F. Bennett (1968: 101) noted that in Panama the decimation of the Indian population saw the re-establishment of trees:

Large areas which today are covered by dense forest were in farms, grassland or low second growth in the early sixteenth century when the Spaniards arrived. At that time horses were ridden with ease through areas which today are most easily penetrated by river, so dense has the tree growth become since the Indians died away.

Bush encroachment is a serious cause of rangeland deterioration. Overgrazing reduces the vigour of favourable perennial grasses, which tend to be replaced, as we have seen, by less reliable annuals and by woody vegetation. Annual grasses do not adequately hold or cover the soil, especially early in the growing season. Thus runoff increases and topsoil erosion occurs. Less water is then available in the topsoil to feed the grasses, so that the woody species, which depend on deeper water, become more competitive relative to the grasses which can only use water within their shallow root zones. The situation is exacerbated when there are few browsers in the herbivore population and, once established, woody vegetation competes effectively for light and nutrients. It is also extremely expensive to remove established dense scrub cover by mechanical means, though some success has been achieved by introducing animals that have a browsing or bulldozing effect, such as goats, giraffes and elephants. Experiments in Zimbabwe have shown that, if bush clearance is carried out, a threefold increase in the sustainable carrying capacity can take place in semi-arid areas (Child, 1985).

Another example of human interventions modifying savanna character is through their effects on savanna-dwelling mammals such as the elephant. They are what is known as a 'keystone species' because they exert a strong influence on many aspects of the environment in which they live (Waithaka, 1996). They diversify the ecosystems which they occupy and create a mosaic of habitats by browsing, trampling and knocking over bushes and trees. They also disperse seeds through their eating and defecating habits and maintain or create water holes by wallowing. All these roles are beneficial to other species. Conversely, where human interference prevents elephants from moving freely within their habitats and leads to their numbers exceeding the carrying capacity of the savanna, their effect can be environmentally catastrophic. Equally, if humans reduce elephant numbers in a particular piece of savanna, the savanna may become less diverse and less open, and its water holes may silt up. This will be to the detriment of other species.

In addition to its lowland savannas, Africa has a series of high-altitude grasslands called 'Afromontane grasslands'. They extend as a series of 'islands' from the mountains of Ethiopia to those of the Cape area of South Africa. Are they the result mainly of forest clearance by humans in the recent past? Or are they a long-standing and probably natural component of the pattern of vegetation (Meadows and Linder, 1993)? Are they caused by frost, seasonal aridity, excessively poor soils or an intensive fire regime? This is one of the great controversies of African vegetation studies.

Almost certainly a combination of factors has given rise to these grasslands. On the one hand current land management practices, including the use of fire, prevent forest from expanding. There has undoubtedly been extensive deforestation in recent centuries. On the other hand, pollen analysis from various sites in southern Africa suggests that grassland was present in the area as long ago as 12,000 BP. This would mean that much grassland is not derived from forest through very recent human activities.

The spread of desert vegetation on desert margins

One of the most contentious and important environmental issues of recent years has been the debate on the question of the alleged expansion of deserts (Middleton and Thomas, 1997).

The term 'desertification' was first used but not formally defined by Aubréville (1949), and for some years the term 'desertization' was also employed, for example by Rapp (1974: 3), who defined it as 'the spread of desert-like conditions in arid or semi-arid areas, due to man's influence or to climatic change'. An alternative expression, 'land aridization', has also been used by the Soviet pedologist, Kovda (1980: 15): 'The phrase "land aridization" means a complex of diverse processes and trends that reduce the effective moisture content over large areas and decrease the biological productivity of the soils and plants of an ecosystem.'

There has been some variability in how 'desertification' itself is defined. Some definitions stress the importance of human causes (e.g. Dregne, 1986: 6–7):

Desertification is the impoverishment of terrestrial ecosystems under the impact of man. It is the process of deterioration in these ecosystems that can be measured by reduced productivity of desirable plants, undesirable alterations in the biomass and the diversity of the micro and macro fauna and flora, accelerated soil deterioration, and increased hazards for human occupancy.

Others admit the possible importance of climatic controls but give them a relatively inferior role (e.g. Sabadell et al., 1982: 7):

The sustained decline and/or destruction of the biological productivity of arid and semi arid lands caused by man made stresses, sometimes in conjunction with natural extreme events. Such stresses, if continued or unchecked, over the long term may lead to ecological degradation and ultimately to desert-like conditions.

Yet others, more sensibly, are even-handed or open-minded with respect to natural causes (e.g. Warren and Maizels, 1976: 1):

A simple and graphic meaning of the word 'desertification' is the development of desert like landscapes in areas which were once green. Its practical meaning ... is a sustained decline in the yield of useful crops from a dry area accompanying certain kinds of environmental change, both natural and induced.

It is by no means clear how extensive desertification is or how fast it is proceeding. Indeed, the lack of agreement on the former makes it impossible to determine the latter. As Grainger (1990: 145) has remarked in a well-balanced review, 'Desertification will remain an ephemeral concept to many people until better estimates of its extent and rate of increase can be made on the basis of actual measurements.' He continues (p. 157): 'The subjective judgements of a few experts are insufficient evidence for such a major component of global environmental change ...

Monitoring desertification in the drylands is much more difficult than monitoring deforestation in the humid tropics, but it should not be beyond the ingenuity of scientists to devise appropriate instruments and procedures.'

The United Nations Environment Programme (UNEP) has played a central role in the promotion of desertification as a major environmental issue, as is made evident by the following remark by Tolba and El-Kholy (1992: 134):

Desertification is the main environmental problem of arid lands, which occupy more than 40 per cent of the total global land area. At present, desertification threatens about 3.6 billion hectares – 70% of potentially productive drylands, or nearly one-quarter of the total land area of the world. These figures exclude natural hyper-arid deserts. About one sixth of the world's population is affected.

However, in a recent book called *Desertification: exploding the myth*, Thomas and Middleton (1994) have discussed UNEP's views on the amount of land that is desertified. They state:

The bases for such data are at best inaccurate and at worst centred on nothing better than guesswork. The advancing desert concept may have been useful as a publicity tool but it is not one that represents the real nature of desertification processes.
                                        (Thomas and Middleton, 1994: 160)

Their views have been trenchantly questioned by Stiles (1995), but huge uncertainties do indeed exist.

There are relatively few reliable studies of the rate of supposed desert advance. Lamprey (1975) attempted to measure the shift of vegetation zones in the Sudan (see figure 2.9) and concluded that the Sahara had advanced by 90 to 100 kilometres between 1958 and 1975, an average rate of about 5.5 kilometres per year. However, on the basis of analysis of remotely sensed data and ground observation, Helldén (1985) found sparse evidence that this had in fact happened. One problem is that there may be very substantial fluctuations in biomass production from year to year. This has been revealed by meteorological satellite observations of green biomass production levels on the south side of the Sahara (Dregne and Tucker, 1988).

The spatial character of desertification is also the subject of some controversy (Helldén, 1985). Contrary to popular rumour, the spread of desert-like conditions is not an advance over a broad front in the way that a wave overwhelms a beach. Rather, it is like a 'rash' which tends to be localized around settlements. It has been likened to Dhobi's itch – a ticklish problem in difficult places. Fundamentally, as Mabbutt (1985: 2) has explained, 'the extension of desert-like conditions tends to be achieved through a process of accretion from without, rather than through

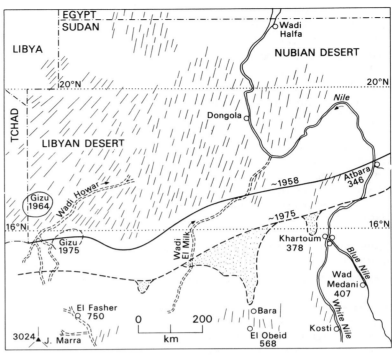

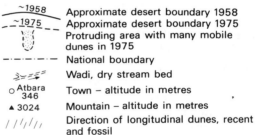

Figure 2.9
Desert encroachment in the
northern Sudan 1958–1975, as
represented by the position of
the boundary between sub-
desert scrub and grassland in
the desert (after Rapp et al.,
1976, figure 8.5.3)

expansionary forces acting from within the deserts.' This distinction is important in that it influences perceptions of appropriate
remedial or combative strategies, which are discussed in Goudie
(1990).

Figure 2.10 shows two models of the spatial pattern of desertification. One represents the myth of a desert spreading out-
ward over a wide front, while the other shows the more realistic
pattern of degradation around routes and settlements.

Woodcutting (plate 2.7) is an extremely serious cause of
vegetation decline around almost all towns and cities of the
Sahelian and Sudanian zones of Africa. Many people depend on
wood for domestic uses (cooking, heating, brick manufacture,
etc.), and the collection of wood for charcoal and firewood is an
especially serious problem in the vicinity of urban centres. This
is illustrated for Khartoum in Sudan in figure 2.11. Likewise,

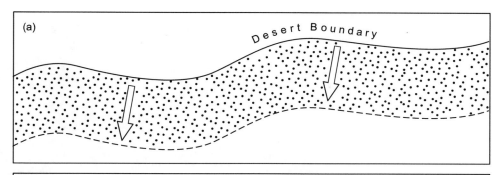

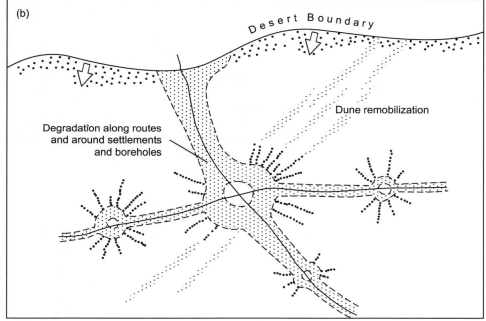

the installation of modern boreholes has enabled rapid multi-plication of livestock numbers and large-scale destruction of the vegetation in a radius of 15–30 kilometres around boreholes (figure 2.12). Given this localization of degradation, ameliora-tion schemes such as local tree planting may be partially effective, but ideas of planting green belts as a 'cordon sanitaire' along the desert edge (whatever that is) would not halt deterioration of the situation beyond this Maginot line (Warren and Maizels, 1977: 222). The deserts are not invading from without; the land is deteriorating from within.

There has been considerable debate as to whether the vegeta-tion change and environmental degradation associated with desertization are irreversible. In many cases, where ecological conditions are favourable because of the existence of such factors

Figure 2.10
Two models of the spatial pattern of desertification:
(a) the myth of desert migration over a wide front
(b) the more local development of degradation around routes, settlements and boreholes
(modified after Kadomura, 1994, figure 9)

Plate 2.7
In the Sahel zone of West Africa and on other desert margins human activities such as wood collection, over-grazing and cultivation in drought-prone areas have led to the process of desertization

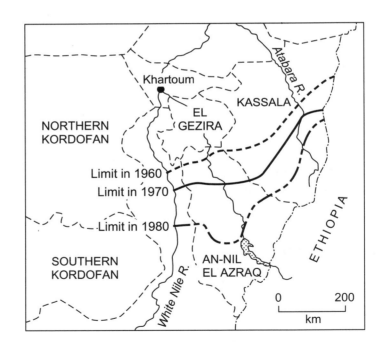

Figure 2.11
The expanding wood and charcoal exploitation zone south of Khartoum, Sudan (after Johnson and Lewis, 1995, figure 6.2)

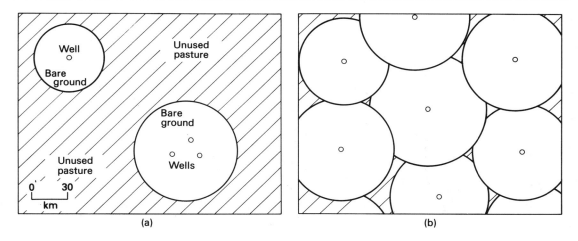

(a) (b)

as deep sandy soils or beneficial hydrological characteristics, vegetation recovers once excess pressures are eliminated. There is evidence of this in arid zones throughout the world where temporary or permanent enclosures have been set up (Le Houérou, 1977). The speed of recovery will depend on how advanced deterioration is, the size of the area which is degraded, the nature of the soils and moisture resources, and the character of local vegetation. It needs to be remembered in this context that much desert vegetation is adapted to drought and to harsh conditions, and that it often has inbuilt adaptations which enable a rapid response to improved circumstances.

None the less, elsewhere experiments and observations of natural conditions tend to reveal that in certain specific circumstances recovery is slow and so limited that it may be appropriate to talk of 'irreversible desertization'. Le Houérou (1977: 419), for example, has pointed to such a case in North Africa:

In southern Tunisia, tracks made by tanks and wheeled vehicles of Allied and Axis armies are still apparent on the ground and in the devastated and unregenerated vegetation 35 years after the conclusion of the fighting. The perennial species have not re-established themselves in spite of several series of years with long-term, above-average rainfall in the 1950s, the late 1960s, and the early 1970s, although in this area grazing pressure is very low due to the absence of permanent water.

The causes of desertification are also highly controversial. The question has been asked whether this process is the result of temporary drought periods of high magnitude, whether it is due to long-term climatic change towards aridity (either as alleged post-glacial progressive desiccation or as part of a 200-year cycle), whether it is caused by anthropogenic climatic change, or

Figure 2.12
Relation between spacing of wells and over-grazing:
(a) Original situation. End of dry season 1. Herd size limited by dry season pasture. Small population can live on pastoral economy
(b) After well-digging but no change in traditional herding. End of dry season 2. Larger total subsistence herd and more people can live there until a situation when grazing is finished in a drought year; then the system collapses
(after Rapp, 1974)

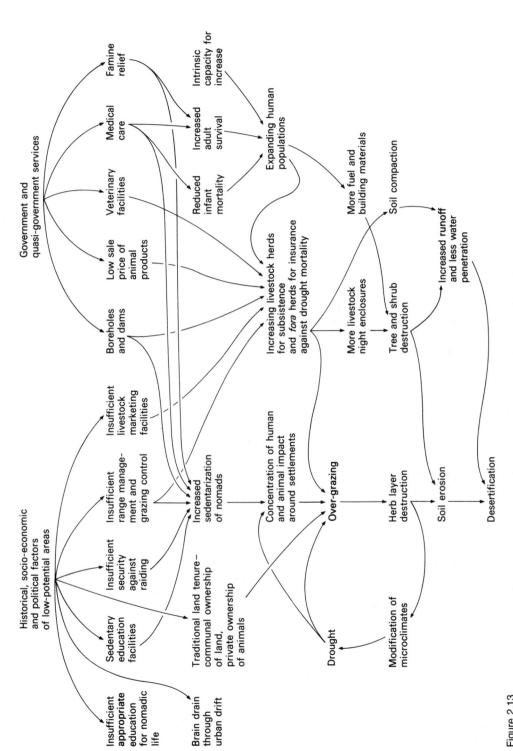

Figure 2.13
Some causal factors in desert encroachment in northern Kenya (after Lamprey, 1978, figure 2)

whether it is the result of human action degrading the biological environments in arid zones. There is little doubt that severe droughts do take place, and have taken place (Nicholson, 1978), and that their effects become worse as human and domestic animal populations increase. The devastating drought in the African Sahel from 1968 to 1984 caused greater ecological stress than the broadly comparable droughts of 1910–15 and 1944–8, largely because of the increasing anthropogenic pressures.

The venerable idea that climate is deteriorating through the mechanism of post-glacial progressive desiccation is now discredited (see Goudie, 1972a, for a critical analysis), though the idea that the Sahel zone is currently going through a 200-year cycle of drought has been proposed (Winstanley, 1973). However, numerous studies of available meteorological data (which in some cases date back as far as 130–150 years) do not allow any conclusions to be reached on the question of systematic long-term changes in rainfall, and the case for climatic deterioration – whether natural or aggravated by humans – is not proven. Indeed, in a judicious review, Rapp (1974: 29) wrote that after consideration of the evidence for the role of climatic change in desertization his conclusion was 'that the reported desertization northwards and southwards from the Sahara could not be explained by a general trend towards drier climate during this century'.

It is evident, therefore, that it is largely a combination of human activities (figure 2.13) with occasional series of dry years that leads to presently observed desertization. The process also seems to be fiercest not in desert interiors, but on the less arid marginal areas around them. It is in semi-arid areas – where biological productivity is much greater than in extremely arid zones, where precipitation is frequent and intense enough to cause rapid erosion of unprotected soils, and where humans are prone to mistake short-term economic gains under temporarily favourable climatic conditions for long-term stability – that the combination of circumstances particularly conductive to desert expansion can be found. It is in these marginal areas that dry farming and cattle rearing can be a success in good years, so that susceptible areas are ploughed and cattle numbers become greater than the vegetation can support in dry years. In this way, a depletion of vegetation occurs which sets in train such insidious processes as water erosion and deflation. The vegetation is removed by clearance for cultivation, by the cutting and uprooting of woody species for fuel, by overgrazing and by the burning of vegetation for pasture and charcoal.

These tendencies towards bad land-use practices result in part from the restrictions imposed on many nomadic societies through the imposition of national boundaries across their traditional migration routes, or through various schemes implemented for political and social reasons to encourage their establishment in settled communities. Some of their traditional grazing lands have been taken over by cash-crop farmers. In Niger, for example, there was a sixfold increase in the acreage of peanuts grown between 1934 and 1968. The traditional ability to migrate enabled pastoral nomads and their cattle to emulate the natural migration of such wild animals as wildebeest and kob, and thereby to make flexible use of available resources according to season and according to yearly variations in rainfall. They could also move away from regions that had become exhausted after a long period of use. As soon as migrations are stopped and settlements imposed, such options are closed, and severe degradation occurs (Sinclair and Fryxell, 1985).

The suggestion has sometimes been made not only that deserts are expanding because of human activity, but that the deserts themselves are created by human activity. There are authors who have suggested, for example, that the Thar Desert of India is a post-glacial and possibly post-medieval creation (see Allchin et al., 1977, for a critique of such views), while Ehrlich and Ehrlich (1970) have written: 'The vast Sahara desert itself is largely man-made, the result of over-grazing, faulty irrigation, deforestation, perhaps combined with a shift in the course of a jet stream.' Nothing could be further from the truth. The Sahara, while it has fluctuated greatly in extent, is many millions of years old, pre-dates human life, and is the product of the nature of the general atmospheric circulation, occupying an area of dry descending air.

The maquis of the Mediterranean lands

Around much of the Mediterranean basin there is a plant formation called maquis (plate 2.8). This consists of a stand of xerophilous non-deciduous bushes and shrubs which are evergreen and thick, and whose trunks are normally obscured by low-level branches. It includes such plants as holly oak (*Quercus ilex*), kermes oak (*Quercus coccifera*), tree heath (*Erica arborea*), broom heath (*Erica scoparia*) and strawberry trees (*Arbutus unedo*).

Some of the maquis may represent a stage in the evolution towards true forest in places where the climax has not yet been reached; but in large areas it represents the degeneration of forest. Considerable concern has been expressed about the speed with which degeneration to, and degeneration beyond, maquis is taking place as a result of human influences (Tomaselli, 1977), of which cutting, grazing and fire are probably the most important and long continued. Charcoal burners, goats and frequent outbreaks of fires among the resinous plants in the dry

Mediterranean summer have all taken their toll. Such aspects
of degradation in Mediterranean environments are discussed
in Conacher and Sala (1998).

On the other hand, in some areas, particularly marginal moun-
tainous and semi-arid portions of the Mediterranean basin, agricultural uses of the land have declined in recent decades, as local
people have sought easier and more remunerative employment.
In such areas scrubland, sometimes termed 'post-cultural shrub
formations', may start to invade areas of former cultivation (May,
1991), while maquis has developed to become true woodland.

There is considerable evidence that maquis vegetation is in
part adapted to, and in part a response to, fire. One effect of fire
is to reduce the frequency of standard trees and to favour species
which after burning send up a series of suckers from ground
level. Both *Quercus ilex* and *Quercus coccifera* seem to respond
in this way. Similarly, a number of species (for example, *Cistus
albidus*, *Erica arborea* and *Pinus halepensis*) seem to be distinctly
advantaged by fire, perhaps because it suppresses competition

Plate 2.8
The maquis of the
Mediterranean lands,
illustrated here in Corsica,
is a vegetation type in
which humans have played a
major role. It represents the
degeneration of the natural
forest cover

or perhaps because (as with the comparable chaparral of the south-west USA) a short burst of heat encourages germination (Wright and Wanstall, 1977).

The prairie problem

The mid-latitude grasslands of North America – the prairies – are another major vegetation type that can be used to examine the human impact, although, as in the case of savanna grasslands in the tropics, the human role is the subject of controversy (Whitney, 1994).

It was once fairly widely believed that the prairies were essentially a climatically related phenomenon. Workers like J. E. Weaver (1954) argued that under the prevailing conditions of soil and climate the invasion and establishment of trees were significantly hindered by the presence of a dense sod. High evapotranspiration levels combined with low precipitation were thought to give a competitive advantage to herbaceous plants with shallow, densely ramifying root systems, capable of completing their life cycles rapidly.

An alternative view was, however, put forward by Stewart (1956: 128): 'The fact that throughout the tall-grass prairie planted groves of many species have flourished and have reproduced seedlings during moist years and, furthermore, have survived the most severe and prolonged period of drought in the 1930s suggests that there is no climatic barrier to forests in the area.'

Other arguments along the same lines have been advanced. Wells (1965), for example, has pointed out that in the Great Plains a number of woodland species, notably the junipers, are remarkably drought-resistant, and that their present range extends into the Chihuahua Desert where they often grow in association with one of the most xerophytic shrubs of the American deserts, the creosote bush (*Larrea divaricata*). He remarks (p. 247): 'There is no range of climate in the vast grassland climate of the central plains of North America which can be described as too arid for all species of trees native to the region.' Moreover, confirming Stewart, he points out that numerous plantations and shelter belts have indicated that trees can survive for at least 50 years in a 'grassland' climate. One of his most persuasive arguments is that, in the distribution of vegetation types in the plains, a particularly striking vegetational feature is the widespread but local occurrence of woodlands along escarpments and other abrupt breaks in topography remote from fluvial irrigation. A probable explanation for this is the fact that fire effects are greatest on flat, level surfaces, where there are high wind speeds and no interruptions to the course of fire. It has also been noted that where burning has been restricted there has been extension of woodland into grassland. The reasons why fire tends to promote the establishment of grassland have been summarized by Cooper (1961: 150–1).

In open country fire favours grass over shrubs. Grasses are better adapted to withstand fire than are woody plants. The growing point of dormant grasses from which issues the following year's growth lies near or beneath the ground, protected from all but the severest heat. A grass fire removes only one year's growth, and usually much of this is dried and dead. The living tissue of shrubs, on the other hand, stands well above the ground, fully exposed to fire. When it is burned, the growth of several years is destroyed. Even though many shrubs sprout vigorously after burning, repeated loss of their top growth keeps them small. Perennial grasses, moreover, produce seeds in abundance one or two years after germination; most woody plants require several years to reach seed-bearing age. Fires that are frequent enough to inhibit seed production in woody plants usually restrict the shrubs to a relatively minor part of the grassland area.

Thus, as with savanna, anthropogenic fires may be a factor which maintains, and possibly forms, grasslands in the Great Plains. Though again, following the analogy with savanna, it is possible that some of the American prairies may have developed in a post-glacial dry phase, and that with a later increase in rainfall re-establishment of forest cover was impeded by humans through their use of fire and by grazing animals. The grazing animals concerned were not necessarily domesticated, however, for Larson (1940) has suggested that some of the short-grass plains were maintained by wild bison. These, he believed, stocked the plains to capacity so that the introduction of domestic livestock, such as cattle, after the destruction of the wild game was merely a substitution so far as the effect of grazing on plants is concerned. There is indeed pollen analytical evidence that shows the presence of prairie in the western Mid-West over 11,000 years ago, prior to the arrival of human settlers (Bernabo and Webb, 1977). Therefore some of it, at least, may be natural.

Comparable arguments have attended the origins of the great Pampa grassland of Argentina. The first Europeans who penetrated the landscape were much impressed by the treeless open country and it was always taken for granted, and indeed became dogma, that the grassland was a primary climax unit. However, this interpretation was successfully challenged, notably by Schmieder (1927a, b) who pointed out that planted trees thrived, that precipitation levels were quite adequate to maintain tree growth and that, in topographically favourable locations such as the steep gullies (*barrancas*) near Buenos Aires, there were numerous endemic representatives of the former forest cover (*monte*). Schmieder believed that Pampa grasslands were

produced by a pre-Spanish aboriginal hunting and pastoral population, the density of which had been underestimated, but whose
efficiency in the use of fire was proven.

In recent centuries mid-latitude grasslands have been especially
rapidly modified by human activities. As Whitney (1994: 257)
points out, they have been altered in ways that can be summarized in three phrases: 'ploughed out', 'grazed out' and 'worn out'.
Very few areas of natural North American prairies remain. The
destruction of 'these masterpieces of nature' took less than a
century. The preservation of remaining areas of North American
grasslands is a major challenge for the conservation movement
(Joern and Keeler, 1995). On a global basis it has been estimated
that cultivation has led to the area of grassland in the world
now having been reduced by about 20 per cent from its pre-
agricultural extent (Graetz, 1994).

Post-glacial vegetational change in Britain and Europe

The classic interpretation of the vegetational changes of post-
glacial (Holocene) times – that is, over the past 11,000 or so
years – has been in terms of climate. That changes in the vegetation of Britain and other parts of Western Europe took place
was identified by pollen analysts and palaeo-botanists, and these
changes were used to construct a model of climatic change: the
Blytt–Sernander model. There was thus something of a chicken-
and-egg situation: vegetational evidence was used to reconstruct
past climates, and past climates were used to explain vegetational
change. More recently, however, the importance of humanly
induced vegetation change in the Holocene has been described
thus by Behre (1986: vii):

Human impact has been the most important factor affecting vegetation change, at least in Europe, during the last 7,000 years. With the
onset of agriculture, at the so-called Neolithic revolution, the human
role changed from that of a passive component to an active element
which impinged directly on nature. This change had dramatic consequences for the natural environment and landscape development.
Arable and pastoral farming, the actual settlements themselves and the
consequent changes in the economy significantly altered the natural
vegetation and created the cultural landscape with its many different
and varying aspects.

Recent palaeo-ecological research has indicated the possibility
that certain features of the post-glacial pollen record in Britain
can be attributed to the action of Mesolithic and Neolithic
peoples. It is, for example, possible that the expansion of the
alder (*Alnus glutinosa*) was not so much a consequence of
supposed wetness in the Atlantic period as a result of Mesolithic

colonizers. Their removal of natural forest cover helped the spread of alder by reducing competition, as possibly did the burning of reed swamp. It may also have been assisted in its spread by the increased runoff of surface waters occasioned by deforestation and burning of catchment areas. Furthermore, the felling of alder itself promotes vegetative sprouting and cloning, which could result in its rapid spread in swamp forest areas (Moore, 1986).

In the Yorkshire Wolds of northern England pollen analysis suggests that Mesolithic peoples may have caused forest disturbance as early as 8900 BP. They may even have so suppressed forest growth that they permitted the relatively open landscapes of the early post-glacial to persist as grasslands even when climatic conditions favoured forest growth (Bush, 1988).

One vegetational change which has occasioned particular interest is the fall in *Ulmus* pollen – the so-called elm decline – which appears in all pollen diagrams from north-west Europe, though not from North America. A very considerable number of radiocarbon dates have shown that this decline was approximately synchronous over wide areas and took place at about 5000 BP (Pennington, 1974). It is now recognized that various hypotheses can be advanced to explain this major event in vegetation history: the original climatic interpretation; progressive soil deterioration; the spread of disease; or the role of people.

The climatic interpretation of the elm decline as being caused by cold, wet conditions has been criticized on various grounds (Rackham, 1980: 265):

A deterioration of climate is inadequate to explain so sudden, universal and specific a change. Had the climate become less favourable for elm, this would not have caused a general decline in elm and elm alone; it would have wiped out elm in areas where the climate had been marginal for it, but would not have affected elm at the middle of its climatic range unless the change was so great as to affect other species also. A climatic change universal in Europe ought to have some effect on North American elms.

Humans may have contributed to soil deterioration and affected elms thereby. Troels-Smith (1956), however, postulated that around 5,000 years ago a new technique of keeping stalled domestic animals was introduced by Neolithic peoples, and that these animals were fed by repeated gathering of heavy branches from those trees known to be nutritious – elms. This, it was held, reduced enormously the pollen production of the elms. In Denmark it was found that the first appearance of the pollen of a weed, Ribwort plaintain (*Plantago lanceolata*), always coincided with the fall in elm pollen levels, confirming the association with human settlements.

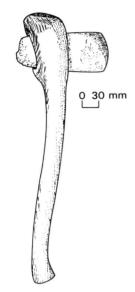

0 30 mm

Figure 2.14
A Neolithic chert axe-blade
from Denmark, of the type
which has been shown to be
effective at cutting forest in
experimental studies (after
Cole, 1970, figure 25)

Experiments have also shown that Neolithic peoples, equipped with polished stone axes, could cut down mature trees and clear by burning a fair-sized patch of established forest within about a week. Such clearings were used for cereal cultivation. In Denmark a genuine chert Neolithic axe was fitted into an ashwood shaft (figure 2.14). Three men managed to clear about 600 square metres of birch forest in four hours. Remarkably, more than 100 trees were felled with one axe-head, which had not been sharpened for about 4,000 years (Cole, 1970: 38).

Rackham, however, doubts whether humans alone could have achieved the sheer extent of change in such a short period (see also Peglar and Birks, 1993), and postulates that epidemics of elm disease may have played a role, aided by the fact that the cause of the disease, a fungus called *Ceratocystis*, is particularly attracted to pollarded elms (Rackham, 1980: 266).

Heathland is a vegetation type characteristic of temperate, oceanic conditions on acidic substrates, and is composed of ericoid low shrubs, which form a closed canopy at heights usually less than 2 metres. Trees and tall shrubs are absent or scattered.

Some heathlands are natural: for example, communities at altitudes above the forest limit on mountains and those on exposed coasts. There are also well-documented examples of heath communities which appear naturally in the course of plant succession, for example, where *Calluna vulgaris* (heather) colonizes *Ammophila arenaria* and *Carex arenaria* on coastal dunes.

Figure 2.15
The lowland heath region of
Western Europe (modified after
Gimingham and de Smidt,
1983, figure 2)

However, at low and medium altitudes on the western fringes of Europe between Portugal and Scandinavia (figure 2.15) extensive areas of heathland occur. The origin of these lands is strongly disputed (Gimingham and de Smidt, 1983). Some areas were once thought to have developed where there were appropriate edaphic conditions (for example, well-drained loess or very sandy, poor soils), but pollen analysis showed that most heathlands occupy areas which were formerly tree-covered. This evidence alone, however, did not settle the question whether the change from forest to heath might have been caused by Holocene climatic change. However, the presence of human artefacts and buried charcoal, and the fact that the replacement of forest by heath has occurred at many different points in time between the Neolithic and the late nineteenth century, suggest that human actions established, and then maintained, most of the heathland areas. In particular, fire is an important management tool for heather in locations such as upland Britain, since the value of *Calluna* as a source of food for grazing animals increases if it is periodically burned.

The area covered by heathland in Western Europe reached a peak around 1860, but since then there has been a very rapid decline. For example, by 1960 there had been a 60–70 per cent reduction in Sweden and Denmark (Gimingham, 1981). Reductions in Britain averaged 40 per cent between 1950 and 1984, and this was a continuation of a more long-term trend (figure 2.16). The reasons for this fall are many, and include unsatisfactory burning practices, peat removal, drainage, fertilization, replacement by improved grassland, conversion to forest, and sand and gravel abstraction. In England, the Dorset heathlands that were such a feature of Hardy's Wessex novels are now a fraction of their former extent (plate 2.9).

Thus far we have considered the human impact on general assemblages of vegetation over broad zones. However, in turning to questions such as the range of individual plant species, the human role is no less significant.

Introduction, invasion and explosion

People are important agents in the spread of plants and other organisms (Bates, 1956). Some plants are introduced deliberately by humans to new areas; these include crops, ornamental and miscellaneous landscape modifiers (trees for reafforestation, cover plants for erosion control, etc.). Indeed, some plants, such as bananas and breadfruit, have become completely dependent on people for reproduction and dispersal, and in some cases they have lost the capacity for producing viable seeds and depend on

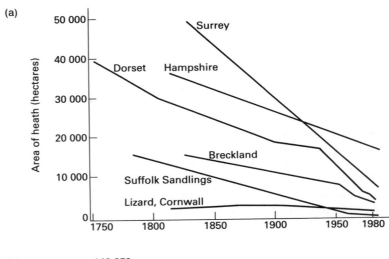

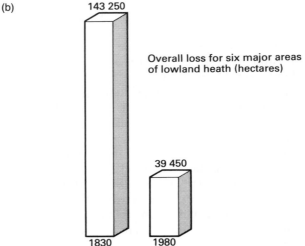

Figure 2.16
Losses of lowland heath in
southern England (from Nature
Conservancy Council, 1984)

human-controlled vegetation propagation. Most cultigens are
not able to survive without human attention, partly because of
this low capacity for self-propagation, but also because they
cannot usually compete with the better-adapted native vegetation.

However, some domesticated plants have, when left to their
own devices, shown that they are capable of at least ephemeral
colonization, and a small number have successfully naturalized
themselves in areas other than their supposed region of origin
(Gade, 1976). Examples of such plants include several umbelli-
ferous annual garden crops (fennel, parsnip and celery) which,
though native to Mediterranean Europe, have colonized waste
lands in California. The Irish potato, which is native to South
America, grows unaided in the mountains of Lesotho. The peach

Plate 2.9
Heaths, like these of Canford
in Dorset, southern England,
have been created by humans
but have become valued
habitats. Many of them,
however, are now under threat
from housing expansion,
reclamation, afforestation and
other pressures

(in New Zealand), the guava (in the Philippines), coffee (in Haiti)
and the coconut palm (on Indian Ocean island strands) are perennials that have established themselves as wild-growing populations, though the last-named is probably within the hearth region
of its likely domestication. In Paraguay, orange trees (originating
in South-East Asia and the East Indies) have demonstrated their
ability to survive in direct competition with natural vegetation.

Plants that have been introduced deliberately because they
have recognized virtues (Jarvis, 1979) can be usefully divided
into an economic group (for example, crops, timber trees, etc.),
and an ornamental or amenity one. In the British Isles, Jarvis
believes that the great bulk of deliberate introductions before
the sixteenth century had some sort of economic merit, but that
only a handful of the species introduced thereafter were brought
in because of their utility. Instead, plants were introduced increasingly out of curiosity or for decorative value.

A major role in such deliberate introductions was played by
European botanic gardens and those in the colonial territories
from the sixteenth century onwards. Many of the 'tropical'
gardens (such as those in Calcutta, Mauritius and Singapore)
were often more like staging posts or introduction centres than
botanic gardens in the modern sense (Heywood, 1989).

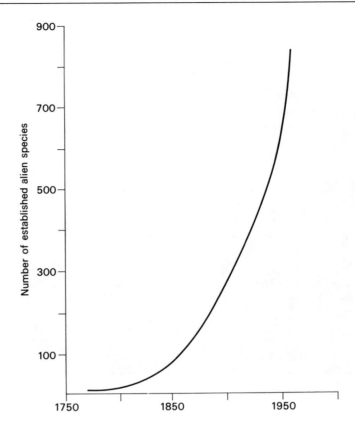

Figure 2.17
Estimation of the establishment
of alien plant species in
California, USA since 1750
(modified after Frenkel, 1970,
figure 3)

Many plants, however, have been dispersed accidentally as a
result of human activity: some by adhesion to moving objects,
such as individuals themselves or their vehicles; some among
crop seed; some among other plants (like fodder or packing
materials); some among minerals (such as ballast or road metal);
and some by the carriage of seeds for purposes other than planting (as with drug plants). As illustrated in figure 2.17, based on
California, the establishment of alien species proceeds rapidly.
In the Pampa of Argentina, Schmieder (1927b) estimates that
the invasion of the country by European plants has taken place
on such a large scale that at present only one-tenth of the plants
growing wild in the Pampa are native.

The accidental dispersal of such plants and organisms can
have serious ecological consequences (see Williamson, 1996,
for a recent analysis). In Britain, for instance, many elm trees
died in the 1970s because of the accidental introduction of the
Dutch elm disease fungus which arrived on imported timber at
certain ports, notably Avonmouth and the Thames Estuary ports
(Sarre, 1978). There are also other examples of the dramatic
impact of some introduced plant pathogens (Von Broembsen,

Table 2.6 Alien plant species on oceanic islands

Island	Number of native species	Number of alien species	% of alien species in flora
New Zealand	1,200	1,700	58.6
Campbell Island	128	81	39.0
South Georgia	26	54	67.5
Kerguelen	29	33	53.2
Tristan da Cunha	70	97	58.6
Falklands	160	89	35.7
Tierra del Fuego	430	128	23.0

Source: from data in Moore, 1983.

1989). The American chestnut *Castanea dentata* was, following the introduction of the chestnut blight fungus *Cryphonectria parasitica* in ornamental nursery material from Asia late in the 1890s, almost eliminated throughout its natural range in less than 50 years. In western Australia the great jarrah forests have been invaded and decimated by a root fungus, *Phytophthora cinnamomi*. This was probably introduced on diseased nursery material from eastern Australia, and the spread of the disease within the forests was facilitated by road building, logging and mining activities that involved movement of soil or gravel containing the fungus. More than three million hectares of forest have been affected.

Ocean islands have often been particularly vulnerable. The simplicity of their ecosystems inevitably leads to diminished stability, and introduced species often find that the relative lack of competition enables them to broaden their ecological range more easily than on the continents. Moreover, because the natural species inhabiting remote islands have been selected primarily for their dispersal capacity, they have not necessarily been dominant or even highly successful in their original continental setting. Therefore, introduced species may prove more vigorous and effective (Holdgate and Wace, 1961). There may also be a lack of indigenous species to adapt to conditions such as bare ground caused by humans. Thus introduced weeds may catch on.

Table 2.6 illustrates clearly the extent to which the flora of selected islands now contain alien species, with the percentage varying between about one-quarter and two-thirds of the total number of species present.

There are a number of major threats that invasive plants pose to natural ecosystems. These have been discussed by Cronk and Fuller (1995):

1 Replacement of diverse systems with single species stands of aliens, leading to a reduction in biodiversity, as for example

Plate 2.10
Feral animals (introduced
domestic stock that have gone
wild) have had a major impact
on large portions of Australia.
Feral buffalo, for example,
range widely in parts of
northern Australia

where Australian acacias have invaded the fynbos heathlands of South Africa.

2   Direct threats to native faunas by change of habitat.

3   Alteration of soil chemistry. For example, the African *Mesembryanthemum crystallinum* accumulates large quantities of salt. In this way it salinizes invaded areas in Australia and may prevent the native vegetation from establishing.

4   Alteration of geomorphological processes, especially rates of sedimentation and movement of mobile land forms (e.g. dunes and salt marshes).

5   Plant extinction by competition.

6   Alteration of fire regime. For example, in Florida in the southern United States the introduction of the Australasian *Melaleuca quinquenervia* has increased the frequency of fires because of its flammability, and has damaged the native vegetation which is less well adapted to fire.

7   Alteration of hydrological conditions (e.g. reduction in groundwater levels caused by some species having high rates of transpiration).

The introduction of new animals (plate 2.10) can have an adverse effect on plant species. A clear demonstration of this comes from the atoll of Laysan in the Hawaii group. Rabbits and hares were introduced in 1903 in the hope of establishing a meat cannery. The number of native species of plants at this time was 25; by 1923 it had fallen to four. In that year all the rabbits and hares were systematically exterminated to prevent the island turning into a desert, but recovery has been slower than destruction. By 1930 there were nine species, and by 1961 16 species, on the island (Stoddart, 1968).

Pigs are another animal that has been introduced extensively to the islands of the Pacific and long-established feral populations are known on many islands. Like rabbits they have caused considerable damage, not least because of their non-fastidious eating habits and their propensity for rooting into the soil. This is a theme that is reviewed by Nunn (1991).

It has often been proposed that the introduction of exotic terrestrial mammals has had a profound effect on the flora of New Zealand. Among the reasons that have been put forward for this belief are that the absence of native terrestrial mammalian herbivores permitted the evolution of a flora highly vulnerable to damage from browsing and grazing, and that the populations of wild animals (including deer and opossums) that were introduced in the nineteenth century grew explosively because of the lack of competitors and predators. It has, however, proved difficult to determine the magnitude of the effects which the introduced mammals had on the native forests (Veblen and Stewart, 1982).

There are many other examples of ecological explosions caused by humans creating new habitats. Some of the most striking are associated with the establishment of artificial lakes in place of rivers. Riverine species which cannot cope with the changed conditions tend to disappear, while others that can exploit the new sources of food, and reproduce themselves under the new conditions, multiply rapidly in the absence of competition (Lowe-McConnell, 1975). Vegetation on land flooded as the lake waters rise decomposes to provide a rich supply of nutrients which allow explosive outgrowth of organisms as the new lake fills. In particular, floating plants may form dense mats of vegetation, which in turn support large populations of invertebrate animals, may cause fish deaths by deoxygenating the water, and can create a serious nuisance for turbines, navigators and fishermen. On Lake Kariba in Central Africa there were dramatic growths in the communities of the South American water fern (*Salvinia molesta*), bladder-wort (*Utricularia*) and the African water lettuce (*Pistia stratiotes*); on the Nile behind the Jebel Aulia Dam there was a huge increase in the number of water hyacinths (*Eichhornia crassipes*); and in the Tennessee valley lakes there was a massive outbreak of the Eurasian water-millfoil (*Myriophyllum*).

Roads have been of major importance in the spread of plants. As Frenkel (1970) has pointed out in a valuable survey of this aspect of anthropogenic biogeography:

By providing a route for the bearers of plant propagules – man, animal and vehicle – and by furnishing, along their margins, a highly specialized habitat for plant establishment, roads may facilitate the entry of plants into a new area. In this manner roads supply a cohesive directional component, cutting across physical barriers, linking suitable habitat to suitable habitat.

Roadsides tend to possess a distinctive flora in comparison with the natural vegetation of an area. As Frenkel has again written:

Roadsides are characterized by numerous ecological modifications including: treading, soil compaction, confined drainage, increased run-off, removal of organic matter and sometimes additions of litter or waste material of frequently high nitrogen content (including urine and faeces), mowing or crushing of tall vegetation but occasionally the addition of wood chips or straw, substrate maintained in an ecologically open condition by blading, intensified frost action, rill and sheetwash erosion, snow deposition (together with accumulated dirt, gravel, salt and cinder associated with winter maintenance), soil and rock additions related to slumping and rock-falls, and altered microclimatic conditions associated with pavement and right-of-way structures. Furthermore, road rights-of-way may be used for driving stock in which case unselective, hurried but often close grazing may constitute an additional modification. Where highway landscaping or stability of cuts and fills is a concern, exotic or native plants may be planted and nurtured.

(1970: 1)

The speed with which plants can invade roadsides is impressive. A study by Helliwell (1974) demonstrated that the M1 motorway in England, less than 12 years after its construction, had on its cuttings and embankments not only the 30 species which had been deliberately sown or planted, but more than 350 species that had not been introduced there.

Railways have also played their role in plant dispersal. The classic example of this is provided by the Oxford ragwort, *Senecio squalidus*, a species native to Sicily and southern Italy. It spread from the Oxford Botanical Garden (where it had been established since at least 1690) and colonized the walls of Oxford. Much of its dispersal to the rest of Britain (Usher, 1973) was achieved by the Great Western Railway, in the vortices of whose trains and in whose cargoes of ballast and iron ore the plumed fruits were carried. The distribution of the plant was very much associated with railway lines, railway towns and waste ground.

Indeed, by clearing forest, cultivating, depositing rubbish and many other activities, humans have opened up a whole series of environments which are favourable to colonization by a particular group of plants. Such plants are generally thought of as weeds. In fact, it has often been said that the history of weeds is the history of human society (though the converse might equally be true), and that such plants follow people like flies follow a ripe banana or a gourd of unpasteurized beer (see Harlan, 1975b).

One weed which has been causing especially severe problems in upland Britain in the 1980s is bracken (*Pteridium aquilinum*), although the problems it poses through rapid encroachment are of wider geographical significance. Indeed, Taylor (1985: 53) maintains that it 'may justifiably be dubbed as the most successful international weed of the twentieth century', for 'it is found, and is mostly expanding, in all of the continents.' This tolerant, aggressive opportunist follows characteristically in the wake of evacuated settlement, deforestation or reduced grazing pressure, and estimated encroachment rates in the upland parts of the UK average 1 per cent per annum. This encroachment results from reduced use of bracken as a resource (e.g. for roofing) and from changes in grazing practices in marginal areas. As bracken is hostile to many other plants and animals, and generates toxins, including some carcinogens, this is a serious issue.

Air pollution and its effects on plants

Air pollutants exist in gaseous or particulate forms. The gaseous pollutants may be separated into primary and secondary forms. The primary pollutants, such as sulphur dioxide, most oxides of nitrogen, and carbon monoxide, are those directly emitted into the air from, for example, industrial sources. Secondary air pollutants, like ozone and products of photochemical reactions, are formed as a consequence of subsequent chemical processes in the atmosphere, involving the primary pollutants and other agents such as sunlight. Particulate pollutants consist of very small solid or liquid-suspended droplets (for example, dust, smoke, and aerosolic salts), and contain a wide range of insoluble components (for example, quartz) and soluble components (for example, various common cations, together with chloride, sulphate and nitrate).

Some of the air pollutants humans have released into the atmosphere have had detrimental impacts on plants (Yunus and Iqbal, 1996); sulphur dioxide, for example, is toxic to them. This was shown by Cohen and Rushton (in Barrass, 1974), who grew plants in containers of similar soil in different parts of Leeds, England, in 1913. They found a close relationship between the amount of sulphate in the air and in the plants, and in the yield obtained (figure 2.18). In the more polluted areas of the city the leaves were blackened by soot and there was a smaller leaf area. The whole theme of urban vegetation has been studied in the North American context by Schmid (1975).

Lichens are also sensitive to air pollution and have been found to be rare in central areas of cities such as Bonn, Helsinki,

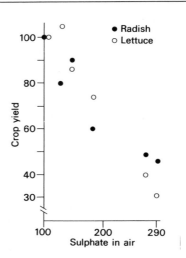

Figure 2.18
The effect of air quality on plant growth in Leeds, England, in 1913. Values for yield and for air sulphate in a low pollution area are taken as 100 and other values are scaled in proportion (data of Cohen and Rushton in Barrass, 1974, p. 187)

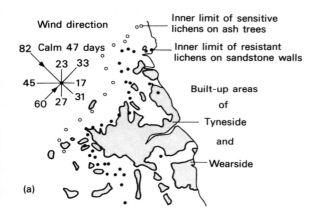

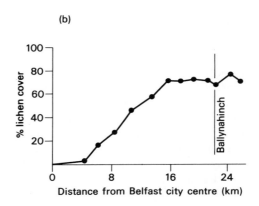

Figure 2.19
(a) Air pollution in north-east England and its impact upon growth area for lichens (after Gilbert in Barrass, 1974, figure 73)
(b) The increase of lichen cover on trees outside the city of Belfast, Northern Ireland (after Fenton in Mellanby, 1967, figure 3)

Stockholm, Paris and London. There appears to be a zonation of lichen types around big cities as shown for north-east England (figure 2.19a) and Belfast (figure 2.19b). Moreover, when a healthy lichen is transplanted from the country to a polluted atmosphere, the algae component gradually deteriorates and then the whole plant dies (see Gilbert, 1970). Overall it has been calculated (Rose, 1970) that more than one-third of England and Wales, extending in a belt from the London area to Birmingham, broadening out to include the industrial Midlands and most of Lancashire and West Yorkshire, and reaching up to Tyneside, has lost nearly all its epiphytic lichen flora, largely because of sulphur dioxide pollution.

Hawksworth (1990: 50), indeed, believes 'the evidence that sulphur dioxide is the major pollutant responsible for impoverishment of lichen communities over wide areas of Europe is now overwhelming,' but he also points out that sulphur dioxide is not necessarily always the cause of impoverishment. Other factors, such as fluoride or photochemical smog, acting independently or synergistically, can also be significant in particular locations.

Local concentrations of industrial fumes also kill vegetation (Freedman, 1995). In the case of the smelters of the Sudbury mining district of Canada, 2 million tonnes of noxious gases annually affect an area of 1,900 square kilometres and White pine now only exists on about 7–8 per cent of the productive land. Likewise in the lower Swansea valley, Wales, the fumes from a century of coal-burning resulted in almost complete destruction of the vegetation, with concomitant soil erosion. The area became a virtual desert. In Norway a number of the larger Norwegian aluminium smelters built immediately after the Second World War were sited in deep, narrow, steep-sided valleys at the heads of fjords. The relief has not proved conductive to the rapid dispersal of fumes, particularly of fluoride. In one valley a smelter with a production of 1,110,000 tonnes per year causes the death of pines (*Pinus sylveststris)* for over 13 kilometres in each direction up and down the valley. To about 6 kilometres from the source all pines are dead. Birch, however, seems to be able to withstand these conditions, and to grow vigorously right up to the factory fence (Gilbert, 1975).

Photochemical smog is also known to have adverse effects on plants both within cities and on their outskirts. In California, Ponderosa pines (*Pinus ponderosa*) in the San Bernadino mountains as much as 129 kilometres to the east of Los Angeles have been extensively damaged by smog. In summer months in Britain ozone concentrations produced by photochemical reactions have reached around 17 pphm (parts per hundred million), compared with a maximum of 4 pphm associated with clean air; while in Los Angeles ozone concentrations may reach 70 pphm (Marx, 1975). Fumigation experiments in the United States show that plant injury can occur at levels only marginally above the natural maximum and well within the summertime levels now known to be present in Britain and the United States. Ozone appears to reduce photosynthesis and to inhibit flowering and germination (Smith, 1974). It also seems to predispose conifers to bark-beetle infestation and to microbial pathogens.

Vegetation will also be adversely affected by excessive quantities of suspended particulate matter in the atmosphere. The particles, by covering leaves and plugging plant stomata, reduce both the absorption of carbon dioxide from the atmosphere and the intensity of sunlight reaching the interior of the leaf. Both tendencies may suppress the growth of some plants. This and other consequences of air pollution are well reviewed by Elsom (1992).

The adverse effects of pollution on plants are not restricted to air pollution: water and soil pollution can also be serious. Excessive amounts of heavy metals may prove toxic to them (Hughes et al., 1980) and, as a consequence, distinctive patterns of plant species may occur in areas contaminated with the waste from copper, lead, zinc and nickel mines (Cole and Smith, 1984). Heavy metals in soils may also be toxic to microbes, and especially to fungi which may in turn change the environment by reducing rates of leaf-litter decomposition (Smith, 1974). In many areas it has proved extremely difficult to undertake effective re-establishment of plant communities on mining spoil tips, though toxicity is only one of the problems. Kent (1982) has listed some of the other obstacles that have been encountered in reclaiming colliery spoil:

1 Erosion of spoil causes instability that hinders plant growth.
2 Black shale surfaces may create undesirable high surface temperatures.
3 Spontaneous combustion.
4 Concentrations of clay-rich sediments can produce areas of waterlogging or compaction.
5 Areas of coarse sediment with an open structure can create conditions of extreme soil drought.
6 Rock spoil may lack key nutrients (particularly phosphorus).
7 The weathering of pyrites may create toxicity, acidity and release other toxic elements (for example, aluminium).

Other types of industrial effluents may smother and poison some species. Salt marshes, mangrove swamps and other kinds of wetlands are particularly sensitive to oil spills, for they tend to be anaerobic environments in which the plants must ventilate their root systems through pores or openings that are prone to coating and clogging (Lugo et al., 1981). The situation is especially serious if the system is not subjected to flushing by, for example, frequent tidal inundation. There are many case studies of the consequences of oil spills. For example, at the Fawley Oil Refinery on Southampton Water in England, Dicks (1977) has shown how the salt marsh vegetation has been transformed by the pollution created by films of oil, and how, over extensive areas, *Spartina anglica* has been killed off.

Forest decline

Forest decline (plate 2.11), now often called *Waldsterben* or *Waldschäden* (the German words for 'forest death' and 'forest decline'), is an environmental issue that attained considerable

Plate 2.11
Forest decline is a serious
threat to various types of mid-
latitude forest, and these trees
in the Hartzgebirge Forest,
Germany, display some of the
major symptoms. However, the
causes of dieback are still a
matter of considerable debate

prominence in the 1980s. The common symptoms of this
phenomenon (modified from World Resources Institute, 1986,
table 12.1) are:

1 *Growth-decreasing symptoms*:
    discoloration and loss of needles and leaves;
    loss of feeder-root biomass (especially in conifers);
    decreased annual increment (width of growth rings);
    premature ageing of older needles in conifers;
    increased susceptibility to secondary root and foliar pathogens;
    death of herbaceous vegetation beneath affected trees;

Table 2.7 Results from forest damage surveys in Europe: percentage of trees with >25% defoliation (all species)

	Mean of 1993/94
Austria	8
Belarus	33
Belgium	16
Bulgaria	26
Croatia	24
Czech Republic	56
Denmark	35
Estonia	18
Finland	14
France	8
Germany	24
Greece	22
Hungary	21
Italy	19
Latvia	33
Lithuania	26
Luxembourg	29
Netherlands	22
Norway	26
Poland	52
Portugal	7
Romania	21
Slovak Republic	40
Slovenia	18
Spain	16
Switzerland	20
UK	15

Source: data in *Acid News* 5, 1995, p. 7.

prodigious production of lichens on affected trees;
death of affected trees.

2   *Abnormal growth symptoms*:
active shedding of needles and leaves while still green, with
   no indication of disease;
shedding of whole green shoots, especially in spruce;
altered branching habit;
altered morphology of leaves.

3   *Water-stress symptoms*:
altered water balance;
increased incidence of wet wood disease.

The decline is widespread in much of Europe and is particularly severe in Poland and the Czech Republic (see table 2.7). The process is now undermining the health of North America's high elevation eastern coniferous forests (World Resources Institute, 1986, chapter 12). In Germany it was the white fir, *Abies alba*, which was afflicted initially, but since then the symptoms have spread to at least ten other species in Europe, including Norway spruce (*Picea abies*), Scots pine (*Pinus sylvestris*), European larch (*Larix decidua*), and seven broad-leaved species.

Many hypotheses have been put forward to explain this dieback (Wellburn, 1988): poor forest management practices, ageing of stands, climatic change, severe climatic events (such as the severe drought in Britain during 1976), nutrient deficiency, viruses, fungal pathogens and pest infestation. However, particular attention is being paid to the role of pollution, either by gaseous pollutants (sulphur dioxide, nitrous oxide or ozone), acid deposition on leaves and needles, soil acidification and associated aluminium toxicity problems and excess leaching of nutrients (for example, magnesium), over-fertilization by deposited nitrogen, and accumulation of trace metals or synthetic organic compounds (e.g. pesticides, herbicides) as a result of atmospheric deposition.

The arguments for and against each of these possible factors have been expertly reviewed by Innes (1987), who believes that in all probability most cases of forest decline are the result of the cumulative effects of a number of stresses. He draws a distinction between predisposing, inciting and contributing stresses (p. 25):

Predisposing stresses are those that operate over long time scales, such as climatic change and changes in soil properties. They place the tree under permanent stress and may weaken its ability to resist other forms of stress. Inciting stresses are those such as drought, frost and short-term pollution episodes, that operate over short time scales. A fully healthy tree would probably have been able to cope with these, but the presence of predisposing stresses interferes with the tree's mechanisms of natural recovery. Contributing stresses appear in weakened plants and are frequently classed as secondary factors. They include attack by some insect pests and root fungi. It is probable that all three types of stress are involved in the decline of trees.

An alternative categorization of the stresses leading to forest decline has been proposed by Nihlgård (1997). He also has three classes of stress that have a temporal dimension: predisposing factors (related primarily to pollution and long-term climatic change); triggering factors (including droughts, frosts and inappropriate management or choice of trees); and mortal factors such as pests, pathogenic fungi and extreme weather events, which can lead to plant death (figure 2.20).

As with many environmental problems, interpretation of forest decline is bedevilled by a paucity of long-term data and detailed surveys. Given that forest condition oscillates from year to year in response to variability in climatic stress (e.g. drought, frost, wind throw) it is dangerous to infer long-term trends from short-term data (Innes and Boswell, 1990).

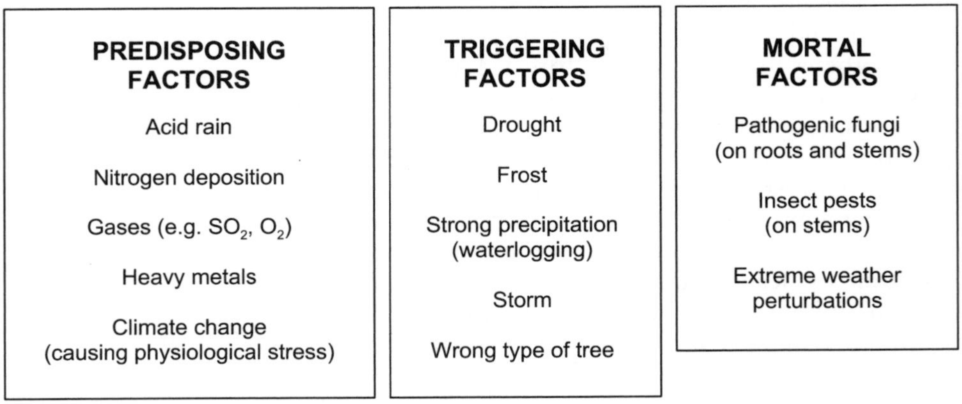

PREDISPOSING FACTORS	TRIGGERING FACTORS	MORTAL FACTORS
Acid rain	Drought	Pathogenic fungi (on roots and stems)
Nitrogen deposition	Frost	
Gases (e.g. SO_2, O_2)	Strong precipitation (waterlogging)	Insect pests (on stems)
Heavy metals	Storm	Extreme weather perturbations
Climate change (causing physiological stress)	Wrong type of tree	

Figure 2.20
Some of the different stress variables used to explain forest decline (modified from Nihlgård, 1997, figure 24-1)

There may also be differences in causation in different areas. Thus, while widespread forest death in eastern Europe may result from high concentrations of sulphur dioxide combined with extreme winter stress, this is a much less likely explanation in Britain, where sulphur dioxide concentrations have shown a marked decrease in recent years. Indeed, in Britain, Innes and Boswell (1990: 46) suggest that the direct effects of gaseous pollutants appear to be very limited.

It is also important to recognize that some stresses may be particularly significant for a particular tree species. Thus, in 1987 a survey of ash trees in Great Britain showed extensive dieback over large areas of the country. Almost one-fifth of all ash trees sampled showed evidence of this phenomenon. Hull and Gibbs (1991) indicated that there was an association between dieback and the way the land is managed around the tree, with particularly high levels of damage being evident in trees adjacent to arable land. Uncontrolled stubble burning, the effects of drifting herbicides, and the consequences of excessive nitrate fertilizer applications to adjacent fields were seen as possible mechanisms. However, the prime cause of dieback was seen to be root disturbance and soil compaction by large agricultural machinery. Ash has shallow roots and if these are damaged repeatedly the tree's uptake of water and nutrients might be seriously reduced, while broken root surfaces would be prone to infection by pathogenic fungi.

Innes (1992: 51) also suggests that there has been some modification in views about the seriousness of the problem since the mid-1980s:

The extent and magnitude of the forest decline is much less than initially believed. The use of crown density as an index of tree health has resulted in very inflated figures for forest 'damage' which cannot now

be justified . . . If early surveys are discounted on the basis of inconsist-
ent methodology . . . then there is very little evidence for a large-scale
decline of tree health in Europe.

. . . the term 'forest decline' is rather misleading in that there are
relatively few cases where entire forest ecosystems are declining. Forest
ecosystems are dynamic and may change through natural processes.

He also suggests that the decline of certain species has been
associated with climatic stress for as long as records have been
maintained.

Miscellaneous causes of plant decline

Some of the main causes of plant decline have already been
referred to: deforestation, grazing, fire and pollution. However,
there are many records of species being affected by other forms
of human interference. For example, casual flower-picking has
resulted in local elimination of previously common species,
and has been held responsible for decreases in species such as
the primrose (*Primula vulgaris*) on a national scale in England.
In addition, serious naturalists or plant collectors using their
botanical knowledge to seek out rare, local and unusual species
can cause the eradication of rare plants in an area.

More significantly, agricultural 'improvements' mean that
many types of habitat are disappearing or that the range of such
habitats is diminishing. Plants associated with distinctive habitats
suffer a comparable reduction in their range. This is certainly
the case for two British plants (figure 2.21a and b), the corncockle
(*Agrostemma githago*) and the pasque flower (*Pulsatilla vulgaris*).
The former was characteristic of unmodified cereal fields, while
the latter was characteristic of traditional chalk downland and
limestone pastures. These calcareous grasslands were semi-natural
swards formed by centuries of grazing by sheep and rabbits,
and covered large areas underlain by Cretaceous chalk and
Jurassic limestone until the Napoleonic Wars. Since then (figure
2.22) they have been subjected to increasing amounts of cultivation, reclamation and reseeding, and the loss of such grasslands
between 1934 and 1972 has been estimated at around 75 per
cent (Nature Conservancy Council, 1984). Other plants have
suffered a reduction in range because of drainage activities (see
figure 2.23).

The introduction of pests, either deliberately or accidentally,
can also lead to a decrease in the range and numbers of a particular species. Reference has already been made to the decline
in the fortunes of the elm in Britain because of the unintentional
establishment of Dutch Elm disease. There are, however, many
cases where 'pests' have been introduced deliberately to check

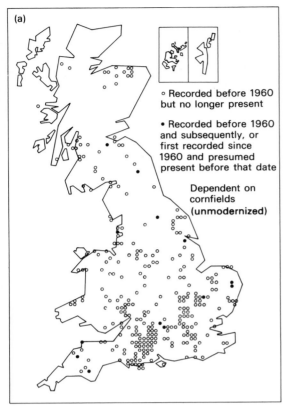

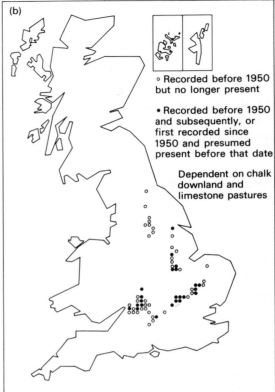

Figure 2.21
Reduction in the range of
species related to habitat loss:
(a) corncockle (*Agrostemma
    githago*)
(b) pasque flower (*Pulsatilla
    vulgaris*) (after Nature
    Conservancy Council,
    1977, figures 8 and 9)

Figure 2.22
The loss of lowland chalk
grasslands in Dorset, southern
England, as depicted by the
area remaining (in hectares)
(from Nature Conservancy
Council, 1984, figure 12.2.1.2)

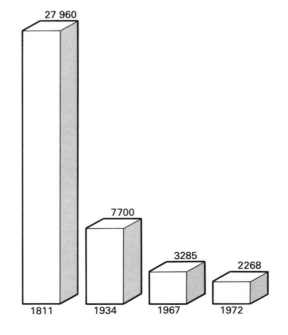

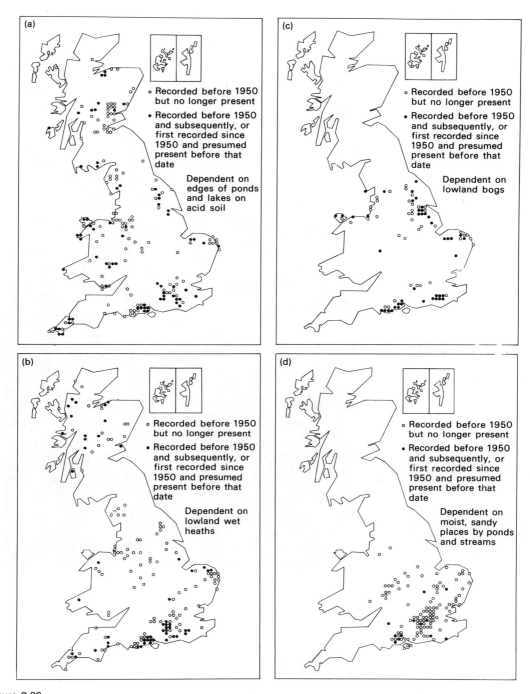

Figure 2.23
Reduction in the range of species related to habitat loss associated with drainage activities:
(a) pillwort (*Pilularia globulifera*)
(b) marsh clubmoss (*Lycopodiella inundata*)
(c) marsh gentian (*Gentiana pneumonanthe*)
(d) small fleabane (*Pulicaria vulgaris*) (after Nature Conservancy Council, 1977, figures 4–7)

Plate 2.12
The prickly pear (*Opuntia*) is a
plant that has been introduced
from the Americas to Africa
and Australia. It has often
spread explosively. Recently it
has been controlled by the
introduction of moths and
beetles, an example of
biological control

the explosive invasion of a particular plant. One of the most
spectacular examples of this involves the history of the prickly
pear (*Opuntia*), imported into Australia from the Americas
(plate 2.12). It was introduced some time before 1839 (Dodd,
1959), and spread dramatically. By 1900, it covered 4 million
hectares, and by 1925 more than 24 million. Of the latter figure
approximately one-half was occupied by dense growth (1,200–
2,000 tonnes per hectare) and other more useful plants were
excluded. To combat this menace one of *Opuntia*'s natural
enemies, a South American moth, *Cactoblastus*, was introduced
to remarkable effect. 'By the year 1940 not less than 95 per cent

of the former 50 million acres (20 million ha) of prickly pear in Queensland had been wiped out' (Dodd, 1959: 575).

A further good example comes from Australia. By 1952 the aquatic fern, *Salvinia molesta*, which originated in south-eastern Brazil, appeared in Queensland and spread explosively as a result of clonal growth, accompanied by fragmentation and dispersal. Significant pests and parasites appear to have been absent. Under optimal conditions *Salvinia* has a doubling time for biomass production of only 2.5 days. In June 1980 possible control agents from the *Salvinia*'s native range in Brazil – the black, long-snouted weevil (*Cyrobagous* sp.) – were released on to Lake Moon Darra (which carried an infestation of 50,000 tonnes fresh weight of *Salvinia*, covering an area of 400 hectares). By August 1981 there was estimated to be less than 1 tonne of the weed left on the lake.

Finally, the growth of leisure activities is placing greater pressure on increasingly fragile communities, notably in tundra and high-altitude areas. These areas tend to recover slowly from disturbance, and both the trampling of human feet and the actions of vehicles can be severe (see e.g. Bayfield, 1979).

The change in genetic and species diversity

The application of modern science, technology and industry to agriculture has led to some spectacular progress in recent decades through such developments as the use of fertilizers and the selective breeding of plants and animals. The latter has caused some concern, for in the process of evolution domest-icated plants have become strikingly different from their wild progenitors. Plant species that have been cultivated for a very long time and are widely distributed demonstrate this particu-larly clearly. Crop evolution through the millennia has been shaped by complex interactions reflecting the pressures of both artificial and natural selection. Alternate isolation of stocks followed by migration and seed exchanges brought distinctive stocks into new environments and permitted new hybridiza-tions and the recombination of characteristics. Great genetic diversity resulted.

There are fears, however, that since the Second World War the situation has begun to change (Harlan, 1975a). Modern plant-breeding programmes have been established in many parts of the developing world in the midst of genetically rich centres of diversity. Some of these programmes, associated with the so-called Green Revolution, have been successful, and new, uni-form high-yielding varieties have begun to replace the wide range of old, local strains that have evolved over the millennia. This may lead to a serious decline in the genetic resources which could potentially serve as reservoirs of variability. Ehrlich et al. (1977: 344) have warned:

Aside from nuclear war, there is probably no more serious environmental threat than the continued decay of the genetic variability of crops. Once the process has passed a certain point, humanity will have permanently lost the coevolutionary race with crop pests and diseases and will no longer be able to adapt crops to climatic change.

New, high-yielding crop varieties need continuous development if they are to avoid the effects of crop pests, and Ehrlich and Ehrlich (1982: 65) have summarized the situation thus:

The life of a new cultivated wheat variety in the American Northwest is about five years. The rusts (fungi) adapt to the strain, and a new resistant one must be developed. That development is done through artificial selection: the plant breeder carefully combines genetic types that show promise of giving resistance.

The impact of human activities on species diversity, while clearly negative on a global scale (as evidenced by extinction rates) is not so cut-and-dried on the local scale. Under certain conditions chronic stress caused by humans can lead to extremely high numbers of coexisting species within small areas. By contrast, site enrichment or fertilization can result in a decline of species density (Peet et al., 1983).

Certain low-productivity grasslands, which have been grazed for long periods, have high species densities in Japan, the UK and the Netherlands. The same applies to Mediterranean scrub vegetation in Israel, and to savanna in Sri Lanka and North Carolina (United States). Studies have confirmed that species densities may increase in areas subject to chronic mowing, burning, domestic grazing, rabbit grazing or trampling. It is likely that in such ecosystems humans encourage a high diversity of plant growth by acting as a 'keystone predator', a species which prevents competitive exclusion by a few dominant species.

Grassland enrichment experiments employing fertilizers have suggested that in many cases high growth rates result in the competitive exclusion of many plants. Thus the tremendous increase in fertilizer use in agriculture has had in retrospect the predictable result of a widespread decrease in the species diversity of grasslands.

One other area in which major developments may take place in the coming years, with implications for plant life, is the field of genetic engineering. This involves the manipulation of DNA, the basic chromosomal unit that exists in all cells and contains genetic information that is passed on to subsequent generations. Recombinant DNA technology (also known as *in vitro* genetic manipulation and gene cloning) enables the insertion of foreign DNA (containing the genetic information necessary to confer a particular target characteristic) into a vector. Fears have been expressed that this technology could produce pathogens that might interact detrimentally with naturally occurring species.

3 Human Influence on Animals

Introduction

The range of impacts that humans have had on animals, though large, can be grouped conveniently into five main categories: domestication, dispersal, extinction, expansion and contraction. As with plants, people have helped to disperse animals deliberately, though many have also been dispersed accidentally.

The number of animals that accompany people without their leave is enormous, especially if we include the clouds of micro-organisms that infest their land, food, clothes, shelter, domestic animals and their own bodies.

Humans have also domesticated many animals, to the extent that, as with many plants, those animals depend on humans for their survival and, in some cases, for their reproduction. The extinction of animals by human predators has been extensive over the past 20,000 years, and in spite of recent interest in conservation continues at a high rate. In addition, the presence of humans has led to contraction in the distribution and welfare of many animals (because of factors like pollution), though in other cases human alteration of the environment and modification of competition has favoured the expansion of some species, both numerically and spatially. Humans now have the highest biomass of any animal species, some 200 million tonnes (Myers, 1979: ix).

Domestication of animals

We have already referred briefly in chapter 1 to one of the great themes in the study of human influence on nature: domestication. This has been one of the most profound ways in which humans have affected animals, for during the ten or eleven millennia that have passed since this process was initiated the animals that human societies have selected as useful to them have undergone major changes. Relatively few animal species have been domesticated in comparison with plant species, but for those that have been the consequences are so substantial that the differences between breeds of animals of the same species often exceed those between different species under natural conditions. A cursory and superficial comparison of the tremendous range of shapes and sizes of modern dog breeds (as, for example, between a wolfhound and a chihuahua) is sufficient to establish the extent of alteration brought about by domestication, and the speed at which domestication has accelerated the process of evolution. In particular, humans have changed and enhanced the characteristics for which they originally chose to domesticate animals. For example, the wild ancestors of cattle gave no more than a few hundred millilitres of milk; today the best milk cow can yield up to 15,000 litres of milk during its lactation period. Likewise, sheep have changed enormously (Ryder, 1966). Wild sheep have short tails, while modern domestic sheep have long tails which may have arisen during human selection of a fat tail. Wild sheep also have an overall brown colour, whereas domestic sheep tend to be mainly white. Moreover, the woolly undercoats of wild sheep have developed at the expense of bristly outer coats. With the ancestors of domestic sheep, wool (which served as protection for the skin and as insulation) consisted mainly of thick rough hairs and a small amount of down; the total weight of wool grown per year probably never reached 1 kilogram. The wool of present-day fine-fleeced sheep consists of uniform, thin down fibres and the total yearly weight may now reach 20 kilograms. Wild sheep also undergo a complete spring moult, while domestic sheep rarely shed wool.

Indeed, one of the most important consequences or manifestations of the domestication of animals consists of a sharp change in the seasonal biology. Whereas wild ancestors of domesticated beasts are often characterized by relatively strict seasonal reproduction and moulting rhythms, most domesticated species can reproduce at almost any season of the year and tend not to moult to a seasonal pattern.

The most important centre of animal domestication, shown in figure 1.6, was south-west Asia (cattle, sheep, goats and pigs), but other centres were important (e.g. the chicken in south-east Asia, the turkey in the Andes and the horse in the Ukraine).

Dispersal and invasions of animals

Most textbooks on zoogeography devote much time to dividing the world into regions with distinctive animal life – the great 'faunal realms' of Wallace and subsequent workers. This pattern of wildlife distribution evolved slowly over geological time. The most striking dividing line between such realms is probably that between Australasia and Asia – Wallace's Line. Australasia developed its distinctive fauna – its relative absence of placental mammals, its well-developed marsupials, and the egg-laying monotremes (echidna and platypus) – because of its isolation from the Asian land mass.

Modern societies, by moving wildlife from place to place, consciously or otherwise, are breaking down these classic distinct faunal realms. Highly adaptable and dispersive forms are spreading, perhaps at the expense of more specialized organisms. Humans have introduced a new order of magnitude into the distances over which dispersal takes place, and through the transport by design or accident of seeds or other propagules, through the disturbance of native plant and animal communities and of their habitat, and by the creation of new habitats and niches, invasion and colonization by adventive species is facilitated.

Di Castri (1989) has identified three main stages in the process of biological invasions stimulated by human actions (table 3.1). In the first stage, covering several millennia up to about AD 1500, human historical events favoured invasions and migrations primarily within the Old World. The second stage commenced about AD 1500, with the discovery, exploration and colonization of new territories, and the initiation of 'the globalization of exchanges'. It shows the occurrence of flows of invaders from, to and within the Old World. The third stage, which covers only the last 100–150 years, sees an even more extensive 'multifocal globalization' and an increasing rate of exchanges. The Eurocentric focus has diminished.

Deliberate introductions of new animals to new areas has been carried out for many reasons (Roots, 1976): for food, for sport, for revenue, for sentiment, for control of other pests, and for aesthetic purposes. Such deliberate actions probably account, for instance, for the widespread distribution of trout (see figure 3.1). There have been many accidental introductions, especially since the development of ocean-going vessels. The rate is increasing, for whereas in the eighteenth century there were few ocean-going vessels of more than 300 tonnes, today there are thousands. Because of this, in the words of Elton (1958: 31), 'we are seeing one of the great historical convulsions in the world's fauna and flora.' Indeed, many animals are introduced with vegetable products, for 'just as trade followed the flag, so animals have followed the plants.'

One of the most striking examples of the accidental introduction of animals given by Elton is the arrival of some chafer beetles, *Popillia japonica* (the Japanese beetle), in New Jersey in a consignment of plants from Japan. From an initial population of about one dozen beetles in 1916, the centre of population

Table 3.1 Human-history driving forces in the Old World as related to biological invasions

Before AD 1500	After AD 1500	From last century in a worldwide perspective
Forest clearing	Exploration, discovery and early colonization by Europeans of other territories and continents	Improvement of transportation systems (roads, railways, internal navigation canals)
Primeval agriculture	Establishment of new market economies and crossroads places (e.g. Amsterdam, London) favouring the 'globalization' of trade exchanges	Large engineering works for irrigation and hydropower
Sheep and cattle raising	Large 'colonies' under the rule of Europeans, often entailing introduction of European-like agriculture and increasing *inter alia* intertropical exchanges	Opening of inter-oceanic canals (e.g. Suez, Panama, Volga-Don)
Migrations and nomadism	Revolution of food customs in wealthy Europe (e.g. increased use of tea, coffee, chocolate, rice, sugar, potatoes, maize, beef and lamb)	Aircraft transportation
Inshore coastal traffic	Increased demand in Europe for products such as cotton, tobacco, wool, etc.	World wars and displacement of human populations
Settlements of islands (e.g. Corsica)	Negro slavery: Indian and Chinese migrations	'Decolonization': international aid to newly independent countries following 'western' patterns
Intensification of agriculture by ploughing	Missionary establishments	Emergence of multinational companies
Offshore traffic and trade	Occupation by Russians of northern areas and part of central Asia, up to Siberia	Tropical deforestation and resettlement schemes
Coastal 'colonies' (e.g. Phoenician and Greek colonies)	Intentional introduction into the Old World of exotic species through activities of acclimatization societies, botanical gardens and zoos, and for agricultural, forestry, fishery or ornamental purposes	Afforestation of arid lands with exotic species
Building up of large empires (e.g. Persian, Roman, Arab, Mogul) with considerable expansion of communication and transportation systems	Large-scale emigration from the Old World due to persecution during religious conflicts, civil and independence wars, and to increased demography, unemployment and famine	Environment impacts decreasing ecosystem resilience
Long-ranging wars and military expansion		Increased urbanization and creation of ruderal habitats
Invasions of German and Asian people, mainly from east to west		International interdependence of markets
Long-distance shipping trade		Release of genetically engineered organisms
Establishment of 'market economies' (e.g. Venice) covering the 'known world' up to the Far East		

Source: Di Castri, 1989, table 1.2.

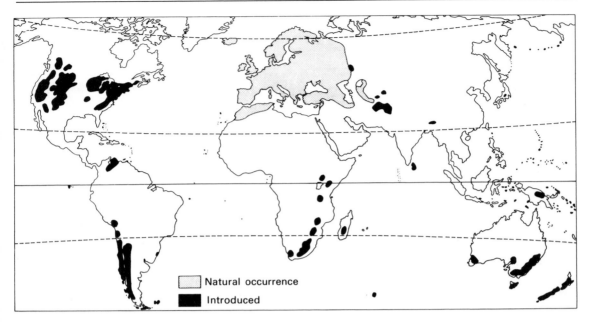

Figure 3.1
The original area of distribution
of the brown trout and areas
where it has been artificially
naturalized (after MacCrimmon
in Illies, 1974, figure 3.2)

grew rapidly outwards to cover many thousands of square kilometres in only a few decades (figure 3.2).

A more recent example of the spread of an introduced insect
in the Americas is provided by the Africanized honey bee
(Rinderer et al., 1993). A number of these were brought to
Brazil from South Africa in 1957 as an experiment and some
escaped. Since then (figure 3.3) they have moved northwards to
Central America and Texas, spreading at a rate of 300–500
kilometres per year, and competing with established populations
of European honey bees.

Some animals arrive accidentally with other beasts that are
imported deliberately. In northern Australia, for instance, water
buffalo were introduced (McKnight, 1971) and brought their
own bloodsucking fly, a species which bred in cattle dung and
transmitted an organism sometimes fatal to cattle. Australia's
native dung beetles, accustomed only to the small sheep-like
pellets of the grazing marsupials, could not tackle the large dung
pats of the buffalo. Thus untouched pats abounded and the flies
were able to breed undisturbed. Eventually African dung beetles
were introduced to compete with the flies (Roots, 1976).

A study conducted in California (Moyle, 1976) gives an
indication of the causes and significance of fish introductions
in the state, for no fewer than 50 of the 133 fish species are introduced, and the rate of introduction appears to be increasing
(figure 3.4). The prime reason for introductions, both authorized
and unauthorized, has been sport fishing, but introductions have

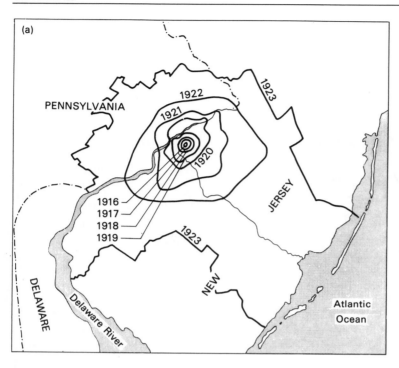

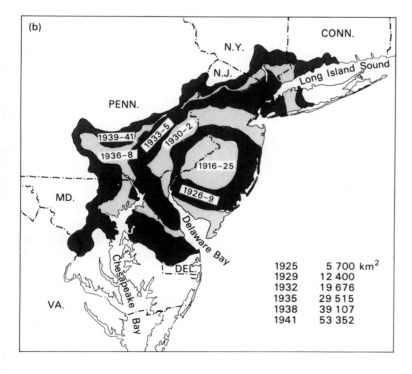

Figure 3.2
The spread of the Japanese
beetle, *Popillia japonica*, in the
eastern USA:
(a) from its point of
    introduction in New Jersey,
    1916–23 (after Elton, 1958,
    figure 14, and Smith and
    Hadley, 1926)
(b) from its point of
    introduction to elsewhere
    1916–41 (after Elton, 1958,
    figure 15, and US Bureau
    of Entomology, 1941)

Year	Area
1925	5 700 km²
1929	12 400
1932	19 676
1935	29 515
1938	39 107
1941	53 352

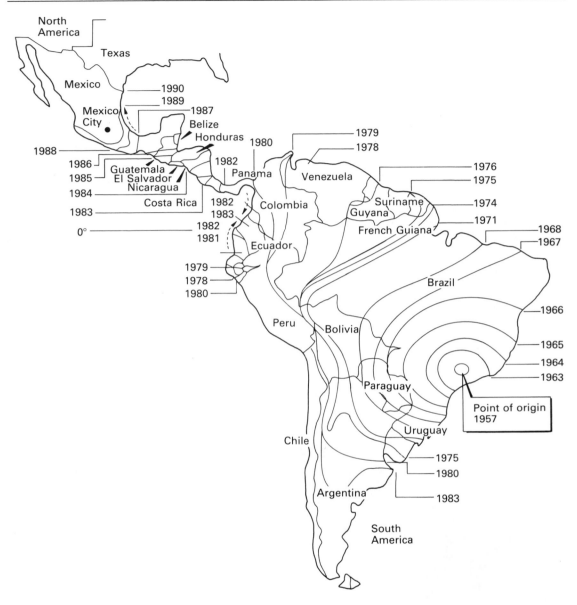

Figure 3.3
The spread of the Africanized honey bee in the Americas between 1957 (when it was introduced to Brazil) and 1990 (modified after Texas Agricultural Experiment Station in *Christian Science Monitor*, September 1991)

also occurred through the provision of forage for game fish, the escape of live bait, the release of pets, and the provision of fish for the biological control of insects.

While domesticated plants have, in most cases, been unable to survive without human help, the same is not so true of domesticated animals. There are a great many examples of cattle, horses (see, for example, McKnight, 1959), donkeys and goats which have effectively adapted to new environments and have

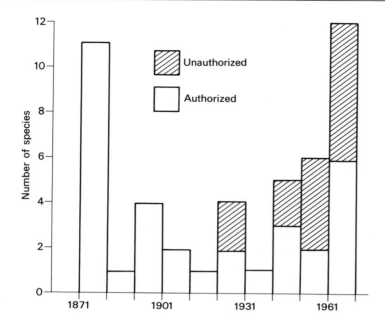

Figure 3.4
The introduction of fish
species (adapted from Moyle,
1976); 1871, when the graph
begins, was the period of the
railroad arrival, the US Fish
Commission and the California
Fish Commission

virtually become wild (feral). Frequently they have both ousted
native animals and, particularly in the case of goats on ocean
islands, caused desertization. Sometimes, however, introduced
animals have spread so thoroughly and rapidly, and have led to
such a change in the environment to which they were introduced, that they have sown the seeds of their own demise. Reindeer, for example, were brought from Lapland to Alaska in
1891–1902 to provide a new resource for the Eskimos, and the
herds increased and spread to over half a million animals. By
the 1950s, however, there was less than a twentieth of that
number left, since the reindeer had been allowed to eat the
lichen supplies that are essential to winter survival; as lichen
grows very slowly their food supply was drastically reduced
(Elton, 1958: 129).

The accidental dispersal of animals can be facilitated by means
other than transport on ships or introduction with plants. This
applies particularly to aquatic life, which can be spread through
human alteration of waterways by methods such as canaliza-
tion. Wooster (1969) gives the example of the way in which the
construction of the Suez Canal has enabled the exchange of
animals between the Red Sea and the eastern Mediterranean.
Initially, the high salinity of the Great Bitter Lake acted to prevent movement, but the infusion of progressively fresher waters
(figure 3.5) through the Suez Canal has meant that this barrier
has gradually become less effective. Some 39 Red Sea fish immigrants have now been identified in the Mediterranean; they are

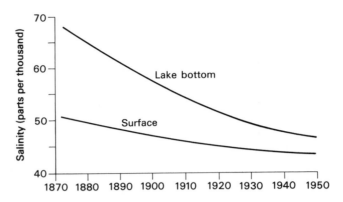

Figure 3.5
Decrease in the salinity of the
Great Bitter Lake, Egypt,
resulting from the intrusion of
fresher water by way of the
Suez Canal (after Wooster,
1969)

especially important in the Levant basin, where they comprise
about 12 per cent of the fish population. Menacing jellyfish
(*Rhopilema nomadica*) have invaded Levantine beaches (Spanier
and Galil, 1991). This type of movement has recently been
termed 'Lessepsian migration'. The construction of the Welland
Canal, linking the Atlantic with the Great Lakes, has permitted
similar movements with more disastrous consequences. Much
of the native fish fauna has been displaced by alewife (*Alosa
pseudoharengus*) through competition for food, and by the
sea-lamprey (*Petromyzon marinum*) as a predator, so that once
common Atlantic salmon (*Salmo salar*), lake trout (*Salvelinus
namaycush*) and lake herring (*Leucichtys artedi*) have been nearly
exterminated (Aron and Smith, 1971).

Can one make any generalizations about the circumstances that
enable successful invasion by exotic vertebrates? Brown (1989)
suggests that there may be 'five rules of biological invasions':

Rule 1
    'Isolated environments with a low diversity of native species
    tend to be differentially susceptible to invasion.'
Rule 2
    'Species that are successful invaders tend to be native to
    continents and to extensive, non-isolated habitats within
    continents.'
Rule 3
    'Successful invasion is enhanced by similarity in the physical
    environment between the source and target areas.'
Rule 4
    'Invading exotics tend to be more successful when native spe-
    cies do not occupy similar niches.'
Rule 5
    'Species that inhabit disturbed environments and those with a
    history of close association with humans tend to be successful
    in invading man-modified habitats.'

Table 3.2 Dependence of bird biomass and diversity on urbanization

	City (Helsinki)	Near rural houses	Uninhabited forest
Biomass (kg km^{-2})	213	30	22
No. of birds km^{-2}	1,089	371	297
Number of species	21	80	54
Diversity	1.13	3.40	3.19

Source: data from Nuorteva, 1971, modified after Jacobs, 1975, table 2.

Human influence on the expansion of animal populations

Although most attention tends to be directed towards the decline in animal numbers and distribution brought about by human agency, there are many circumstances where alterations of the environment and modification of competition have favoured the expansion of some species. Such expansion is not always welcome or expected, as Marsh (1864: 34) appreciated:

Insects increase whenever the birds which feed upon them disappear. Hence in the wanton destruction of the robin and other insectivorous birds, the *bipes implumis*, the featherless biped, man, is not only exchanging the vocal orchestra which greets the rising sun for the drowsy beetle's evening drone, and depriving his groves and his fields of their fairest ornament, but he is waging a treacherous warfare on his natural allies.

Human actions, however, are not invariably detrimental, and even great cities may have effects on animal life which can be considered desirable or tolerable. This has been shown in the studies of bird populations in several urban areas. For example, Nuorteva (in Jacobs, 1975) examined the bird fauna in the city of Helsinki (Finland), in agricultural areas near rural houses, and in uninhabited forests (see table 3.2). The city supported by far the highest biomass and the highest number of birds, but exhibited the lowest number of species and the lowest diversity. In the artificially created rural areas the number of species, and hence diversity, were much higher than in the uninhabited forest, and so was biomass. Altogether, human civilization appeared to have brought about a very significant increase of diversity in the whole area: there were 37 species in city and rural areas that were not found in the forest. Similarly, after a detailed study of suburban neighbourhoods in west-central California, Vale and Vale (1976) found that in suburban areas the number of bird species and the number of individuals increased with time. Moreover, when compared to the pre-suburban habitats adjacent to the suburbs, the residential areas were found to support a larger number of both species and individuals. Horticultural activities appear to provide more luxuriant and more diverse habitats than do pre-suburban environments.

Some beasts other than the examples of birds given here are also favoured by urban expansion (Schmid, 1974). Animals that can tolerate disturbance, are adaptable, utilize patches of open or woodland-edge habitat, creep about inside buildings, tap people's food supply surreptitiously, avoid recognizable competition with humans, or attract human appreciation and esteem, may increase in the urban milieu. For these sorts of reasons the north-east megalopolis of the United States hosts thriving populations of squirrels, rabbits, racoons, skunks and opossums, while some African cities are now frequently blessed with the scavenging attention of hyenas.

Jacobs (1975) provides another apposite example of how humans can increase species diversity inadvertently. The saline Lake Nakuru in Kenya before 1961 was not a particularly diverse ecosystem. There were essentially one or two species of algae, one copopod, one rotifer, corrixids, notonectids and some 500,000 flamingos belonging virtually to one species. In 1962, however, a fish (*Tilapia grahami*) was introduced in order to check mosquitoes. In the event it established itself as a major consumer of algae, and its numbers increased greatly. As a consequence, some 30 species of fish-eating birds (pelicans, anhingas, cormorants, herons, egrets, grebes, terns and fish-eagles) have colonized the area to make Lake Nakuru a much more diverse system. None the less, such fish introductions are not without their potential ecological costs. In Lake Atitlan in Central America the introduction of the game fish *Micropterus salmoides* (largemouth bass) and *Pomoxis nigromaculatus* (black crappie), both voracious eaters, led to the diminution of local fish and crab populations. Similarly, the introduction to the Gatun Lake in the Panama Canal Zone in 1967 of the cichlid fish, *Cichla ocellaris* (a native of the Amazon river), led to the elimination of six of the eight previously common fish species within five years, and the tertiary consumer populations such as the birds, formerly dependent on the small fishes for food, appeared less frequently (Zaret and Paine, 1973).

Human economic activities may lead to a rise in the number of examples of a particular habitat which can lead to an expansion in the distribution of certain species, though often humans have prompted a contraction in the range of a particular species because of the removal or modification of its preferred habitat. In Britain the range of the little ringed plover (*Charadrius dubius*), a species virtually dependent on anthropogenic habitats, principally wet gravel and sandpits, has greatly expanded as mineral extraction has been accelerated in recent decades (see figure 3.6).

Indeed, it needs to be stressed that the changes of habitat brought about by urbanization, industry (Davis, 1976) and mining need not be detrimental. Ratcliffe (1974: 368) has described the present situation:

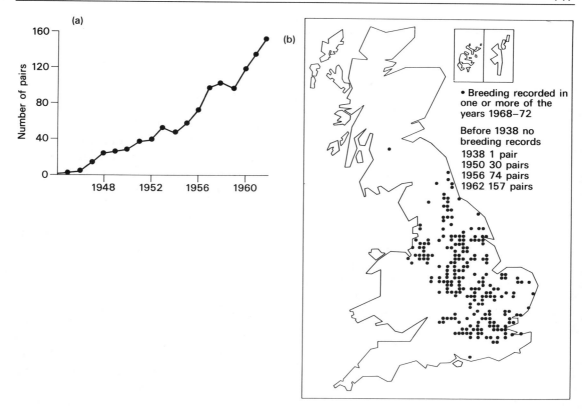

Figure 3.6
The changing range of the
little ringed plover (*Charadrius
dubius*):
(a) the increase in the number
    summering in Britain (after
    Murton, 1971, figure 6)
(b) the increase in the range,
    related to habitat change,
    especially as a result of
    the increasing number of
    gravel pits (after Nature
    Conservancy Council,
    1977, figure 11)

I think it is true to say that while mineral exploitation has caused some damage to nature conservation interests in Britain, this has been mostly on a local and minor scale, and on the whole the gains have outweighed the losses, especially in the creation of interesting new habitats.

He does, however, express some doubts about the future:

The anxiety about future mineral extraction stems from the scale and method of operation. Many mineral workings in the past were mainly subterranean, and surface disturbance was limited largely to waste tipping. Future workings are likely to use strip or opencast mining techniques increasingly, and the demand to rehabilitate new workings is often disadvantageous from the wildlife angle.

Very often human activities do not lead to species diversity, but to important increases in numbers of individuals, by creating new and favourable environments. Two especially serious examples of this from the human point of view are the explosions in the prevalence of both mosquitoes and bilharzia snails as a result of the extension of irrigation.

One of the most remarkable examples of the consequences of creating new environments is provided by the European rabbit

(*Oryctolagus cuniculus*). Introduced into Britain in early medieval times, and originally an inhabitant of the western Mediterranean lands, it was kept for food and fur in carefully tended warrens. Agricultural improvements, especially to grassland, together with the increasing decline in the numbers of predators such as hawks and foxes, brought about by game-guarding landlords (Sheail, 1971), enabled the rabbit to become one of the most numerous of mammals in the British countryside. By the early 1950s there were 60–100 million rabbits in Britain. Frequently, as many contemporary reports demonstrated, it grazed the land so close that in areas of light soil, like the Breckland of East Anglia, or in coastal dune areas, wind erosion became a serious problem. Similarly, the rabbit flourished in Australia, especially after the introduction of the merino sheep which created favourable pasture-lands. Erosion in susceptible lands like the Mallee was severe. Both in England and Australia an effective strategy developed to control the rabbit was the introduction of a South American virus, *Myxoma*.

Some familiar British birds have benefited from agricultural expansion. Formerly, when Britain was an extensively wooded country, the starling (*Sturnus vulgaris*) was a rare bird. The lapwing (*Vanellus vanellus*) is yet another component of the grassland fauna of Central Europe which has benefited from agriculture and the creation of open country with relatively sparse vegetation (Murton, 1971). There is also a large class of beasts which profit so much from the environmental conditions wrought by humans that they become very closely linked to them. These animals are often referred to as synanthropes. Pigeons and sparrows now form permanent and numerous populations in almost all the large cities of the world; human food supplies are the food supplies for many synanthropic rodents (rats and mice); a once shy forest bird, the blackbird (*Turdus merula*), has in the course of a few generations become a regular and bold inhabitant of many gardens; and the squirrel (*Sciurus vulgaris*) in many places now occurs more frequently in parks than in forests (Illies, 1974: 101). The English house sparrow (*Passer domesticus*), a familiar bird in towns and cities, currently occupies approximately one-quarter of the earth's surface, and over the past one hundred years it has doubled the area that it inhabits as settlers, immigrants and others have carried it from one continent to another (figure 3.7). It is found around settlements both in Amazonia and the Arctic Circle (Doughty, 1978). The marked increase in many species of gulls in temperate regions over recent decades is largely attributable to their growing utilization of food scraps on refuse tips. This is, however, something of a mixed blessing, for most of the gull species which feed on urban rubbish dumps breed in coastal areas, and there is now good evidence to show that their greatly increased numbers are threatening other

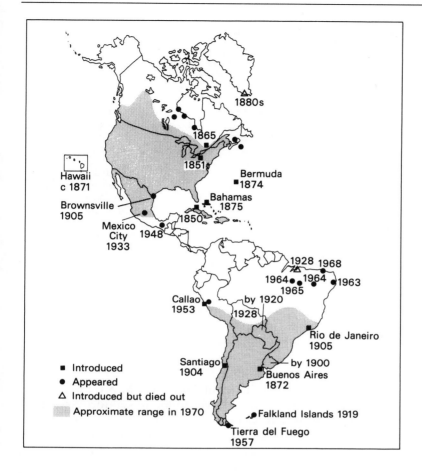

1880s

Hawaii
c 1871

1865

1851

Bermuda
1874

Brownsville
1905

Bahamas
1875

1850

Mexico
City   1948
1933

1928 1968

1964   1964   1963
1965

Callao
1953

by 1920

1928

Rio de Janeiro
1905

Santiago
1904

by 1900

Buenos Aires
1872

- ■ Introduced
- ● Appeared
- △ Introduced but died out
- ▨ Approximate range in 1970

Falkland Islands 1919

Tierra del Fuego
1957

Figure 3.7
The spread of the English
house sparrow in the New
World (after Doughty, 1978,
p. 14)

less common species, especially the terns (*Sterna* spp.). It may
seem incongruous that the niches for scavenging birds which
exploit rubbish tips should have been filled by sea-birds (Murton,
1971). But the gulls have proved ideal replacements for the
kites which humans had removed, as they have the same ability
to watch out for likely food sources from aloft and then to
hover and plunge when they spy a suitable victim.

Most of the examples given so far to illustrate how human
actions can lead to expansion in the numbers and distributions
of certain animal species have been used to make the point that
such expansion frequently occurs as an incidental consequence
of human activities. There are, of course, many ways in which
people have intentionally and effectively promoted the expansion
of particular species (table 3.3 presents data for some introduced
mammals in Britain). This may sometimes be done deliberately
to reduce the numbers of a species which has expanded as an
unwanted consequence of human actions. Perhaps the best-known
exemplification of this is biological control using introduced

Table 3.3 Introduced mammals in Britain

Species	Date of introduction of present stock	Reason for introduction
House mouse *Mus musculus*	Neolithic?	Accidental
Wild goat *Capra hircus*	Neolithic?	Food
Fallow deer *Dama dama*	Roman or earlier	Food, sport
Domestic cat *Felis catus*	Early Middle Ages	Ornament, pest control
Rabbit *Oryctolagus cuniculus*	Mid to late twelfth century	Food, sport
Black rat *Rattus rattus*	Thirteenth century?	Accidental
Brown rat *Rattus norvegicus*	Early eighteenth century (1728–9)	Accidental
Sika deer *Cervus nippon*	1860	Ornament
Grey squirrel *Sciurus carolinensis*	1876	Ornament
Indian muntjak *Muntiacus muntjak*	1890	Ornament
Chinese muntjak *Muntiacus reevesi*	1900	Ornament
Chinese water deer *Hydropetes inermis*	1900	Ornament
Edible dormouse *Glis glis*	1902	Ornament
Musk rat *Ondatra zibethica*	1929 (–1937)	Fur
Coypu *Myocastor coypus*	1929	Fur
Mink *Mustela vison*	1929	Fur
Bennett's wallaby *Macropus rufogriseus bennetti*	1939 or 1940	Ornament
Reindeer *Rangifer tarandus*	1952	Herding, ornament
Himalayan (Hodgson's) porcupine *Hystrix hodgsoni*	1969	Ornament
Crested porcupine *Hystrix cristata*	1972	Ornament
Mongolian gerbil *Meriones unguiculatus*	1973	Ornament

Source: from Jarvis, 1979, table 1, p. 188.

predators and parasites. Thus an Australian insect, the cotton-cushion scale, *Icerya purchasi*, was found in California in 1868. By the mid-1880s it was effectively destroying the citrus industry, but its ravages were quickly controlled by deliberately importing a parasitic fly and a predatory beetle (the Australian ladybird) from Australia. Because of the uncertain ecological effects of synthetic pesticides, such biological control has its attractions; but the importation of natural enemies is not without its own risks. For example, the introduction into Jamaica of mongooses to control rats led to the undesirable decimation of many native birds and small land mammals.

One further calculated method of increasing the numbers of wild animals, of conserving them, and of gaining an economic return from them, is game cropping. In some circumstances, because they are better adapted to, and utilize more components of, the environment, wild ungulates provide an alternative means of land use to domestic stock. The exploitation of the saiga antelope, *Saiga tatarica* (plate 3.1) in the CIS is the most successful story of this kind (Edington and Edington, 1977). Hunting, much overdone because of the imagined medical properties of its horns, had led to its near demise, but this was banned in the 1920s and a system of controlled cropping was instituted. Under this regime (which produces appreciable quantities of meat and leather) total numbers have risen from about 1,000 to

over 2 million, and in the process the herds have reoccupied most of their original range.

Causes of animal contractions and decline: pollution

The extreme effect of human interference with animals is extinction, but before that point is reached humans may cause major contractions in both animal numbers and animal distribution. This decline may be brought about partly by intentional killings for subsistence and commercial purposes, but much wildlife decline occurs indirectly (Doughty, 1974), for example, through pollution (Waldichuk, 1979; Holdgate, 1979) and habitat change or loss.

Particular concern has been expressed about the role of certain pesticides in creating undesirable and unexpected changes. The classic case of this concerns DDT and related substances. These were introduced on a worldwide basis after the Second World War and proved highly effective in the control of insects such as malarial mosquitoes. However, evidence accumulated that DDT was persistent, capable of wide dispersal, and reached high levels of concentration in certain animals at the top trophic levels (see figure 5.23, p. 244). The tendency for DDT to become more concentrated as one moves up the food chain is illustrated further by the data in table 3.4(c). DDT levels in river and estuary waters may be low, but the zooplankton and shrimps contain higher

Plate 3.1
The Saiga, populations of which were decimated by ruthless hunting, has now become more numerous in the CIS because of a system of controlled cropping

Table 3.4 Examples of biological magnification

(a) Enrichment factors for the trace element compositions of shellfish
compared with the marine environment

Element	Enrichment factor		
	Scallops	Oysters	Mussels
Ag	23,002	18,700	330
Cd	260,000	318,000	100,000
Cr	200,000	60,000	320,000
Cu	3,000	13,700	3,000
Fe	291,500	68,200	196,000
Mn	55,000	4,000	13,500
Mo	90	30	60
Ni	12,000	4,000	14,000
Pb	5,300	3,300	4,000
V	4,500	1,500	2,500
Zn	28,000	110,300	9,100

Source: Merlini, 1971, in King, 1975, table 8.8, p. 303.

(b) Mean methyl mercury concentrations in organisms from a contaminated
salt marsh

Organism	Parts per million
Sediments	<0.001
Spartina	<0.001–0.002
Echinoderms	0.01
Annelids	0.13
Bivalves	0.15–0.26
Gastropods	0.25
Crustaceans	0.28
Fish muscle	1.04
Fish liver	1.57
Mammal muscle	2.2
Bird muscle	3.0
Mammal liver	4.3
Bird liver	8.2

Source: Gardner et al., 1978, in Bryan, 1979.

(c) The concentration of DDT in the food chain

Source	Parts per million
River water	0.000003
Estuary water	0.00005
Zooplankton	0.04
Shrimps	0.16
Insects–*Diptera*	0.30
Minnows	0.50
Fundulus	1.24
Needlefish	2.00
Tern	2.80–5.17
Cormorant	26.40
Immature gull	75.50

Source: King, 1975, table 8.7, p. 301.

levels, the fish that feed on them higher levels still, while fish-eating birds have the highest levels of all.

DDT has a major effect on sea-birds. Their eggshells become thinned to the extent that reproduction fails in fish-eating birds. Similar correlations between eggshell thinning and DDT residue concentrations have been demonstrated for various raptorial land-birds. The bald eagle (*Haliaeetus leucocephalus*), peregrine falcon (figure 3.8, plate 3.2) (*Falco peregrinus*), and osprey (*Pandion haliaetus*) all showed decreases in eggshell thickness and population decline from 1947 to 1967, a decline which correlated with DDT usage and subsequent reproductive and metabolic effects upon the birds (Johnston, 1974).

Another example serves to make the same point about the 'magnification' effects associated with pesticides (Southwick, 1976: 45–6). At Clear Lake in California, periodic appearances of large numbers of gnats caused problems for tourists and residents. An insecticide, DDD, was applied at a low concentration (up to 0.05 ppm) and killed 99 per cent of the larvae. Following the application of DDD, Western grebes (*Alchmorphorus occidentalis*) were found dead, and tissue analysis showed concentrations of DDD in them of 1,600 ppm, representing a concentration of 32,000 times the application rate. DDD had accumulated in the insect-eating fish at levels of 40–2,500 ppm and the grebes feeding on the fish received lethal doses of the pesticide. There are also records elsewhere of trout reproduction ceasing when DDT levels build up (Chesters and Konrad, 1971).

However, appreciation of the undesirable side-effects of DDD and DDT brought about by ecological concentration has led to strict controls on their use. For example, DDT reached a peak in terms of utilization in the United States in 1959 (35×10^6kg) and by 1971 was down to 8.1×10^6kg. The monitoring of birds in Florida over the same period indicated a parallel decline in the concentration of DDT and its metabolites (DDD and DDE) in their fat deposits (Johnston, 1974).

Other substances are also capable of concentration. For example, heavy metals and methyl mercury may build up in marine organisms, and filter feeders like shellfish have a strong tendency to concentrate the metals from very dilute solutions, as is clear from table 3.4(a) and (b).

It is possible, however, that the importance of biological accumulation and magnification has been overstated in some textbooks. G. W. Bryan (1979) reviews the situation and notes that, although the absorption of pollutants from foods is often the most important route for bioaccumulation and transfer along food chains certainly occurs, this does not automatically mean that predators at high trophic levels will always contain the highest levels. He writes (p. 497):

Figure 3.8
Changing eggshell thickness
versus time, associated with
DDT use since the Second
World War (modified after
Hickey and Anderson, 1968)

Plate 3.2
Carnivores, including
peregrine falcons, are prone
to the effects of biological
magnification, whereby
pesticides may become
concentrated in their bodies.
One consequence of this has
been a reduction in eggshell
thickness, which in turn has
reduced the number of
successful hatchings

although, for a number of contaminants, concentrations in individual predators sometimes exceed those of their prey, when the situation overall is considered only the more persistent organochlorine pesticides, such as DDT and its metabolites and methyl mercury, show appreciable signs of being biologically magnified as a result of food-chain transfer.

Attempts in Israel to increase agricultural productivity by the chemical control of rodents and jackals have had certain unintended impacts on other beasts. Thallium-sulphate-coated wheat was used for rodent control but also afflicted birds of prey which lived on the rodents. All the birds found dead were discovered to contain large amounts of thallium. By 1955–6, breeding as well as wintering populations of those species which fed mainly on rodents had been almost entirely eliminated. The only bird of prey to succeed in keeping its number at about the same level was the short-toed eagle, *Circaetus gallicus*, which feeds entirely on reptiles. Attempts were also made to exterminate jackals (*Canis aureus*) by means of poisoned baits, mainly strychnine. The undesirable results of this included the poisoning (and serious decline) of many Griffon vultures (*Gyps fulvus*), hooded crows (*Corvus corone*), crested cuckoos (*Clamator glandarius*) and the mongoose (*Herpestes auropunctatus*). This in turn meant that there was an enormous increase in the number of hares (*Lepus europaeus*) and Palestine vipers (*Vipera xanthina palestinae*) – the latter thriving because of the reduction in the mongoose population (Mendelssohn, 1973).

Oil pollution is a serious problem for marine and coastal fauna and flora (plate 3.3), although some of it derives from natural seeps (Landes, 1973; Blumer, 1972). Sea-birds are especially vulnerable since oil clogs their feathers, while the ingestion of oil when birds attempt to preen themselves leads to enteritis and other complaints. Local bird populations may be seriously diminished though probably no extinction has resulted (Bourne, 1970). Fortunately, though oil is toxic it becomes less so with time, and the oil spilt from the *Torrey Canyon* in March 1967 was almost biologically inert when it was stranded on the Cornish beaches (Cowell, 1976). There are 'natural' oil-degrading organisms in nature. Much of the damage caused to marine life as a result of this particular disaster was caused not by the oil itself but by the use of 2.5 million gallons of detergents to disperse it (Smith, 1968).

It is possible that oil spills pose a particular risk in cold areas, for biodegradation processes achieved by microbes appear to be slow. This is probably because of a combination of low temperatures and limited availability of nitrogen, phosphorus and oxygen. The last of these is a constraint because, compared to temperate ecosystems, arctic tundra and coastal ecosystems are relatively stagnant – ice dampens re-aeration due to wave action

Plate 3.3
Oil spills are generally perceived as a major cause of pollution on coastlines. In January 1993 the tanker *Braer* ran aground on the Shetland Isles (Scotland) and large quantities of oil were liberated, causing some distress and mortality among sea birds

in marine ecosystems, while standing water in tundra soils limits inputs of oxygen to them. Detailed reviews are provided in Engelhardt (1985).

There is far less information available on the effects of oil spills in freshwater environments, though they undoubtedly occur. Freshwater bodies have certain characteristics which, compared with marine environments, tend to modify the effects of oil spills. The most important of these are the smaller and shallower dimensions of most rivers, streams and lakes, which means that dilution and spreading effects are not as vigorous and significant in reducing surface slicks as would be the case at sea. In addition, the confining dimensions of ponds and lakes are likely to cause organisms to be subjected to prolonged exposure to dissolved and dispersed hydrocarbons. Information on this problem is summarized in Vandermeulen and Hrudey (1987).

Other industrial pollutants have a clear impact in aquatic systems; only a few examples can be given here (from Southwick, 1976: 19–22). In Biscayne Bay, Florida, industrial and municipal wastes from Miami are thought to be responsible for a high prevalence of dermal tumours in several species of marine fish, while in Bellingham Bay, north of Seattle, sulphite wastes from pulp mills lead to the abnormal growth of oyster larvae. The lower Mississippi river, from which New Orleans obtains its drinking water, contains (even after water treatment) at least 36 chemical compounds of industrial origin known to be harmful to laboratory animals. Fish kills can sometimes eliminate massive populations. In San Diego harbour in 1962 an estimated 37.8 million fish were killed by pollution. Such kills are often produced by industrial accidents.

One particular type of aquatic ecosystem where pollution is an increasingly serious problem is the coral reef (see Kühlmann, 1988). Coral reefs (plate 3.4) are important because they are among the most diverse, productive and beautiful communities in the world. Although they cover an area of less than 0.2 per cent of the world's ocean beds, up to one-quarter of all marine species and one-fifth of known marine fish species live in coral reef ecosystems (World Resources Institute, 1998: 253). They also provide coastal protection, opportunities for recreation, and potential sources of substances like drugs. Accelerated sedimentation resulting from poor land management (Nowlis et al., 1997), together with dredging, is probably responsible for more damage to reef communities than all the other forms of human insult combined, for suspended sediments restrict the light penetration necessary for coral growth. Also, the soft, shifting sediments may not favour colonization by reef organisms. Sewage is the second worst form of stress to which coral reefs are exposed, for oxygen-consuming substances in sewage result in reduced levels of oxygen in the water of lagoons. The detrimental

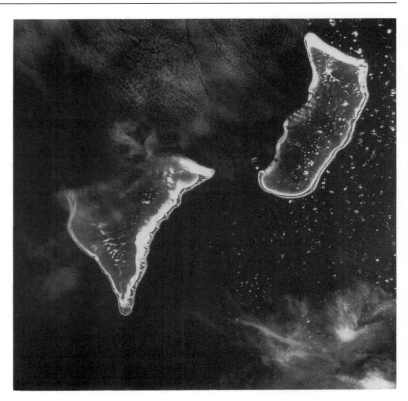

Plate 3.4
Coral reefs, like the Maldive
Islands shown on this Shuttle
photograph of part of the
Indian Ocean, are diverse,
productive and beautiful
habitats that are coming under
increasing human pressure

effects of oxygen starvation are compounded by the fact that
sewage may cause nutrient enrichment to stimulate algal growth,
which in turn can overwhelm coral. Poor land management can
also cause salinity levels to be reduced below the level of tolerance
of reef communities as a consequence of accelerated runoff of
freshwater from catchments draining into lagoons. One of the
reasons why all these stresses may be especially serious for reefs
is that corals live and grow for several decades or more, so that
it can take a long time for them to recover from damage.

As yet the effect of pollution from nuclear industries on plants
and animals seems to be limited (although its health effects on
humans are proven). Mellanby (1967: 63) has written: 'I do not
myself think that, except in the vicinity of nuclear test explosions, fall-out has caused measurable damage so far to animals
and plants . . . up to the present man-made radiation can seldom have had important ecological effects.'

Since the late nineteenth century it has been clear that industrial air pollution has had an adverse effect on domestic animals,
but there is less information about its influence on wildlife. None
the less, there is some evidence that it does affect wild animals
(Newman, 1979). Arsenic emissions from silver foundries are

known to have killed deer and wild rabbits in Germany; sulphur emissions from a pulp mill in Canada are known to have killed many song-birds; industrial fluorosis has been found in deer living in the United States and Canada; asbestosis has been found in free-living baboons and rodents in the vicinity of asbestos mines in South Africa; and oxidants from air pollution are recognized causes of blindness in bighorn sheep in the San Bernadino Mountains near Los Angeles in California.

In Britain considerable concern has arisen over the effects of lead poisoning on wildlife, particularly on wildfowl. Poisoning occurs in swans and mallards, for example, when they feed and seek grit, since they ingest spent shotgun pellets or discarded anglers' weights in the process. Such pellets are eroded in the bird's gizzard and the absorbed lead causes a variety of adverse physiological effects that can result in death: harm to the nervous system, muscular paralysis, anaemia, and liver and kidney damage (Mudge, 1983).

Soil erosion, by increasing stream turbidity (another type of pollution), may adversely affect fish habitats (Ritchie, 1972). The reduction of light penetration inhibits photosynthesis, which in turn leads to a decline in food and a decline in carrying capacity. Decomposition of the organic matter, which is frequently deposited with sediment, uses dissolved oxygen, thereby reducing the oxygen supply around the fish; sediment restricts the emergence of fry from eggs; and turbidity reduces the ability of fish to find food (though conversely it may also allow young fish to escape predators).

Turbidity has also been increased by mining waste (plate 3.5), by construction (Barton, 1977), and by dredging. In the case of some of the Cornish rivers in china-clay mining areas, river turbidities reach 5,000 mg per litre and trout are not present in streams so affected.

Indeed, the non-toxic turbidity tolerances of river fish are much less than that figure. Alabaster (1972), reviewing data from a wide range of sources, believes that in the absence of other pollution, fisheries are unlikely to be harmed at chemically inert suspended sediment concentrations of less than 25 mg per litre, that there should be good or moderate fisheries at 25–80 mg per litre, that good fisheries are unlikely at 80–400 mg per litre and that at best only poor fisheries would exist at more than 400 mg per litre.

The effect of mining on animals is referred to in the world's first mining textbook, *De Res Metallica* by Georgius Agricola (1556), which contains an excellent description of the destruction caused by mining in Germany (Down and Stocks, 1977: 7):

the strongest argument of the detractors is that the fields are devastated by mining operations . . . The woods and groves are cut down,

Plate 3.5
Gold mining in Brazil has caused serious deforestation in certain parts of the Amazonian rain forest, and this in turn has increased the sediment loads of Amazonian streams. Furthermore, the mining process itself delivers large quantities of sediment into stream channels

for there is need of an endless amount of wood for timber, machines and the smelting of metals. And when the woods and groves are felled, then are exterminated the beasts and birds, very many of which furnish a pleasant and agreeable food for man. Further, when the areas are washed, the water which has been used poisons the brooks and streams and either destroys the fish or drives them away.

Habitat change and animal decline

Habitat change is another major cause of animal decline. One of the most important habitat changes in Britain is the removal of many of the hedgerows that form the patchwork quilt so characteristic of large tracts of the country (Sturrock and Cathie, 1980) (plate 3.6). Their removal (figure 3.9) has taken place for a variety of reasons: to create larger fields, so that larger machinery can be employed and so that it spends less time turning; because as farms are amalgamated boundary hedges may become redundant; because as farms become more specialized (for example, just arable) hedges may not be needed to control stock or to provide areas for lambing and calving; and because drainage improvements may be more efficiently executed if hedges are absent. In 1962 there were almost 1 million kilometres of hedge

Plate 3.6
An aerial photograph of part of Suffolk in eastern England shows the traces of hedgerows that have been removed in the interests of modern farm technology. Hedgerows provide cover for many birds and animals

County (selected areas in each county only)

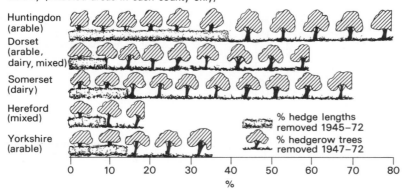

Huntingdon (arable)

Dorset (arable, dairy, mixed)

Somerset (dairy)

Hereford (mixed)

Yorkshire (arable)

% hedge lengths removed 1945–72
% hedgerow trees removed 1947–72

0   10   20   30   40   50   60   70   80
%

Figure 3.9
Recent changes in hedgerows in parts of England (from *New Society*, June 1979, p. 650)

in Britain, probably covering the order of 180,000 hectares, an area greater than that of the national nature reserves, but the figure before enclosure would not have been this high (see table 3.5). Bird numbers and diversity are likely to be severely reduced if the removal of hedgerows continues, for most British birds are essentially of woodland type and, since woods only cover about 5 per cent of the land surface of Britain, depend very much on hedgerows (Moore et al., 1967).

Table 3.5 Changes in length of hedgerows in three parishes in
Huntingdonshire, England over 600 years

Date	Length of hedge (km)	Date	Length of hedge (km)
Prior to 1364	32	1780	93
1364	40	1850	122
1500	45	1946	114
1550	52	1963	74
1680	74	1965	32

Source: from Moore et al., 1967.

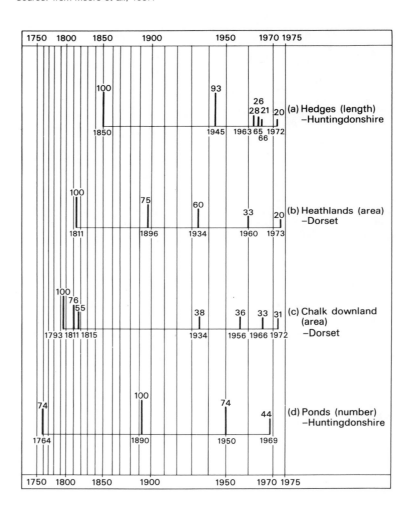

Figure 3.10
Examples of habitat loss in
England since the mid-
eighteenth century (after
Nature Conservancy Council,
1997, figure 3). All values are
expressed as percentages of
the largest value recorded in
each case

Other types of habitat have also been disappearing at a quick
rate with the agricultural changes of recent years (Nature
Conservancy Council, 1977) (see figure 3.10). For example, the
botanical diversity of old grasslands is often reduced by replacing
the lands with grass leys or by treating them with selective
herbicides and fertilizers; this can mean that the habitat does

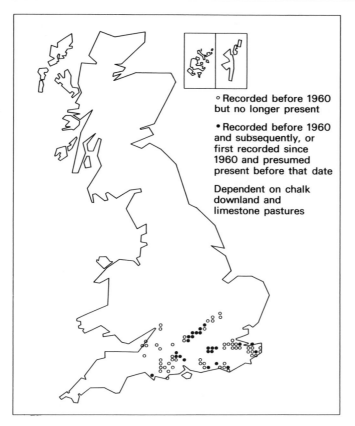

Figure 3.11
Reduction in the range of the
silver-spotted skipper butterfly
(*Hesperia comma*) as a result
of the reduction in the
availability of the chalk
downland and limestone
pastures upon which it
depends (after Nature
Conservancy Council, 1977,
figure 3.8)

not contain some of the basic requirements essential for many
species. One can illustrate this by reference to the larva of the
common blue butterfly (*Polyommatus icarus*) which feeds upon
bird's-foot trefoil (*Lotus corniculatus*). This plant disappears
when pasture is ploughed and converted into a ley, or when it is
treated with a selective herbicide. Once the plant has gone the
butterfly vanishes too because it is not adapted to feeding on the
plants grown in the ley of improved pasture. Figure 3.11 illustrates the way in which the distribution of the silver-spotted
skipper butterfly (*Hesperia comma*) has contracted with the
diminished availability of chalk downland and limestone pastures upon which it depends. Likewise, the large blue butterfly
(*Maculinea arion*) has decreased in Britain. Its larvae live solely
on the wild thyme (*Thymus druccei*), a plant which thrives on
close-cropped grassland. Since the decimation of the rabbit by
myxomatosis, conditions for the thyme have been less favourable
so that both thyme and the large blue butterfly have declined.
Another group of animals which has suffered from agriculture
is the species implicated in disease cycles involving farm stock.
Recently in Britain it has been thought necessary to eliminate
some badgers (*Meles meles*) from a number of farms in the west

Table 3.6 Approximate numbers of species occurring in equivalent habitats in unmodernized and modern farms

Group of animals	Unmodernized farms		Modern farms	
	Habitat	No. of species	Habitat	No. of species
	(a) Hedges* and semi-natural grass verges		(a) Wire fences with (i) sown grass (ii) semi-natural grass verges	
Mammals		20		(i) 5 (ii) 6
Birds		37		(i) 6 (ii) 9
Lepidoptera (butterflies only)		17		(i) 0 (ii) 8
	(b) Permanent pasture† (untreated)		(b) Grass leys	
Lepidoptera (butterflies only)		20		0
	(c) Permanent ponds and ditches		(c) Temporary ditches and piped water	
Mammals (aquatic)		2		0
Amphibia		5		2
Fish		9		0
Odonata (dragonflies)		11		0
Mollusca (gastropods only)		25		3

* Includes hedgerow trees.
† Includes grasslands on chalk.
Source: Nature Conservancy Council, 1977, table 2.

of England in an attempt to remove the threat of their transmitting tuberculosis to cattle.

Overall, as table 3.6 shows, the number of species occurring in England on unmodernized farms compared to modernized farms is relatively great.

The replacement of natural oak-dominated woodlands in Britain by conifer plantations, another major land-use change of recent decades, also has its implications for wildlife. It has been estimated that where this change has taken place the number of species of birds found has been approximately halved. This is illustrated by some data for Wales presented in table 3.7. Likewise, the replacement of upland sheep walks with conifer plantations has led to a sharp decline in raven numbers in southern Scotland and northern England. The raven (*Corvus corax*) obtains much of its carrion from open sheep country (Marquiss et al., 1978). Other birds that have suffered from planting moorland areas with forest plantations are miscellaneous types of wader, golden eagles, peregrine falcons and buzzards (Grove, 1983). A succinct summary of the present position of species changes brought about in Britain by these miscellaneous alterations to habitat is provided by the Nature Conservancy Council (1984), and its main findings are reproduced in figure 3.12.

Agricultural intensification on British farms also appears to have had a deleterious effect on species of bird, many of which have shown a downward trend in numbers over the last two decades (table 3.8). This is because of habitat changes – lack of

Table 3.7 Estimated numbers of breeding songbirds in two broad-leaved and two coniferous woods in Wales

| | Broad-leaved woods | | Coniferous woods | |
	(1) (275 trees/ha, relict)	(2) (375 trees/ha, relict)	(3) (450 trees/ha, planted)	(4) (275 trees/ha, planted)
Chaffinch	7	4	4	7
Wren	13	15	34	9
Robin	7	4	3	3
Dunnock	–	–	2	–
Goldcrest	2	4	9	31
Coal Tit	6	3	7	4
Blue Tit	5	2	–	–
Great Tit	6	2	–	–
Marsh Tit	–	–	–	–
Nuthatch	3	–	–	–
Tree Creeper	1	1	–	–
Blackbird	3	–	–	–
Song Thrush	–	–	–	–
Mistle Thrush	–	–	–	3
Wood Warbler	3	3	–	–
Willow Warbler	4	10	–	–
Blackcap	–	–	–	–
Chiffchaff	–	–	–	–
Pied Flycatcher	6	7	–	–
Redstart	2	2	–	–
TOTAL	68	57	59	57
No. of species	14	12	6	6
Index of diversity	2.47	2.21	1.31	1.37

Source: Edington and Edington, 1977, table 2.1, p. 8.

fallows, less mixed farming, new crops, modern farm management, biocide use, hedgerow removal, etc. (Fuller et al., 1991).

On a global basis the loss of wetland habitats (marshes, bogs, swamps, fens, mires, etc.) is a cause of considerable concern (Maltby, 1986; Williams, 1990). In all, wetlands cover about 6 per cent of the earth's surface (not far short of the total under tropical rain forest), and so they are far from being trivial, even though they tend to occur in relatively small patches. However, they also account for about one-quarter of the earth's total net primary productivity, have a very diverse fauna and flora, and provide crucial wintering, breeding and refuge areas for wildlife. According to some sources, the world may have lost half of its wetlands since 1900, and the United States alone has lost 54 per cent of its original wetland area, primarily because of agricultural developments. There are, however, other threats, including drainage, dredging, filling, peat removal, pollution and channelization.

In the former USSR, the ambitious ploughing-up of the so-called virgin lands of central Kazakhstan in the late 1950s and early 1960s led to the replacement of herbaceous steppe and

Losses of the heritage of nature

Habitat

Lowland herb-rich hay meadows: 95% now lacking significant wildlife interest and only 3% left undamaged by agricultural intensification.

Lowland grasslands of sheep walks. On chalk and Jurassic limestone: 80% loss, largely by conversion to arable or improved grassland (mainly since 1940), but some scrubbed over through lack of grazing.

Lowland heaths on acidic soils: 40% loss, largely by conversion to arable or improved grassland, afforestation and building; some scrubbed over through lack of grazing.

Limestone pavements in northern England: 45% damaged or destroyed, largely by removal of weathered surfaces for sale as rockery stone, and only 3% left completely undamaged.

Ancient lowland woods composed of native, broad-leaved trees: 30–50% loss, by conversion to conifer plantation or grubbing out to provide more farmland.

Lowland fens, valley and basin mires: 50% loss or significant damage through drainage operations, reclamation for agriculture and chemical enrichment of drainage water.

Lowland raised mires: 60% loss or significant damage through afforestation, peat-winning, reclamation for agriculture or repeated burning.

Upland grasslands, heaths and blanket bogs: 30% loss or significant damage through coniferous afforestation, hill land improvement and reclamation, burning and over-grazing.

Species

The large blue butterfly became extinct in 1979, but ten more species are vulnerable or even seriously endangered, and out of a total British list of 55 resident breeding species of butterfly another 13 have declined and contracted in range substantially since 1960.

Three or four of our 43 species of dragonflies have become extinct since 1953, six are vulnerable or endangered, and five have decreased substantially.

Four of our 12 reptiles and amphibians are endangered.

At least 36 breeding species of bird have shown appreciable long-term decline during the last 35 years as a result of habitat loss or deterioration, 30 in the lowlands and six in the uplands.

The otter has become rare or has disappeared in many parts of England and Wales.

Bats in general have decreased and several of our 15 species, notably the greater horseshoe and mouse-eared bats are at risk of extinction. Others such as Bechstein's, Leisler's and the barbastelle are rare, and even the pipistrelle is no longer common. Problems are food supply, destruction of breeding and hibernation roosts and pollution.

There have been local increases and extensions of range in some vertebrate populations (e.g. wild cat and pine marten), but the overall balance of change is on the debit side.

Figure 3.12
Summary of recent habitat and species changes in Britain (from Nature Conservancy Council, 1984, Summary of objectives and strategy, p. 7)

Table 3.8 Decline in farmland birds in Britain between 1970 and 1988 as shown by the Common Birds Census

Species	% decline
Corn bunting	69
Grey partridge	67
Tree sparrow	67
Lapwing	59
Bullfinch	58
Song thrush	54
Turtle dove	48
Linnet	36
Skylark	33
Spotted flycatcher	31
Blackbird	28

Source: from data in Fuller et al., 1991.

grass steppe by cultivated fields. As a result, the animal inhabitants of the steppe became increasingly restricted to smaller areas of suitable habitat. The population of the fur-bearing marmot or bobac (*Marmot bobac*) steadily declined from about three million in the middle 1950s to 318,000 in 1969. The animals are further threatened because the rise in the human population following the extension of agriculture has increased trapping activity to obtain pelts (Zimina et al., 1972).

Other causes of animal decline

One could list many other indirect causes of wildlife decline. Vehicle speed, noise and mobility upset remote and sensitive wildlife populations, and this problem has intensified with the rapid development in the use of off-road recreation vehicles. In California an estimated 1.2 million motorcycles and 500,000 dune buggies are regularly in use in the fragile desert ecosystems (Busack and Bury, 1974). The effects of roads are becoming ever more pervasive as a result of increasing traffic levels, vehicle speed and road width. A fine-meshed road network is a highly effective cause of habitat isolation, acting as a series of barriers to movement, particularly of small, cover-loving animals. As Oxley et al. (1974) have put it, 'a four-lane divided highway is as effective a barrier to the dispersal of small forest mammals as a body of fresh water twice as wide.' This barrier effect results from a variety of causes (Mader, 1984): roads interrupt microclimatic conditions; they are a broad band of emissions and disturbances; they are zones of instability due to cutting and spraying, etc.; they provide little cover against predators; and they subject animals to death and injury by moving wheels. Leisure activities in general may create problems for some species (Speight, 1973). A survey of the breeding status of the little tern (*Sterna albifrons*) in Britain gave a number of instances of breeding failure by the species, apparently caused by the presence of fishermen and bathers on nesting beaches. The presence of even a few people inhibited the birds from returning to their nests. In like manner, species building floating nests on inland waters, such as great crested grebes (*Podiceps cristatus*), are very prone to disturbance by water-skiers and power-boats.

New types of construction can cause problems. As Doughty (1974) has remarked, ' "wirescapes", tall buildings and towers can become to birds what dams, locks, canals and irrigation ditches are to the passage of fish.' The construction of canals can cause changes in aquatic communities.

Likewise, turbines associated with hydroelectric schemes can cause numerous fish deaths, either directly through ingesting them, or through gas-bubble disease. This resembles the 'bends' in divers

Table 3.9 Environmental effects of burning

(a) Change in number of species of breeding birds and small mammals after fire

	Foraging zone	Before fire	After fire	Gained (%)	Lost (%)
BIRDS	Grassland and shrub	48	62	38	8
	Tree trunk	25	26	20	16
	Tree	63	58	10	17
	TOTALS	136	146	21	14
SMALL MAMMALS	Grassland and shrub	42	45	17	10
	Forest	16	14	13	25
	TOTALS	58	59	16	14

(b) Change in density and trend of population of breeding birds and small mammals after fire

		Density (%)			Trend (%)		
	Foraging zone	Increase	Decrease	No change	Increase	Decrease	No change
BIRDS	Grassland and shrub	50	9	41	24	10	66
	Tree trunk	28	16	56	4	8	88
	Tree	24	19	57	6	6	88
	TOTALS	35	15	50	12	8	80
SMALL MAMMALS	Grassland and shrub	24	13	63	20	5	75
	Forest	23	42	35	0	11	89
	TOTALS	23	25	52	14	1	80

and is produced if a fish takes in water supersaturated with gases. The excess gas may come out of solution as bubbles which can lodge in various parts of the fish's body, causing injury or death. Supersaturation of water with gas can be produced in turbines or when a spillway plunges into a deep basin (Baxter, 1977).

It might also be thought that human use of fire could be directly detrimental to animals, though the evidence is not conclusive. Many forest animals appear to be able to adapt to fire, and fire also tends to maintain habitat diversity. Thus, after a general review of the available ecological literature, Bendell (1974) found that fires did not seem to produce as much change in birds or small mammals as one might have expected. Some of his data are presented in table 3.9. The amount of change is fairly small, though some increase in species diversities and animal densities is evident. Indeed, Vogl (1977) has pointed to the diversity of fauna associated with fire-affected ecosystems (p. 281): 'Birds and mammals usually do not panic or show fear in the presence of fire, and are even attracted to fires and smoking or burned landscapes . . . The greatest arrays of higher animal species, and the largest numbers per unit area, are associated with fire-dependent, fire-maintained, and fire-initiated ecosystems.'

Purposeful hunting, particularly when it involves modern firearms, means that the distribution of many animal species has

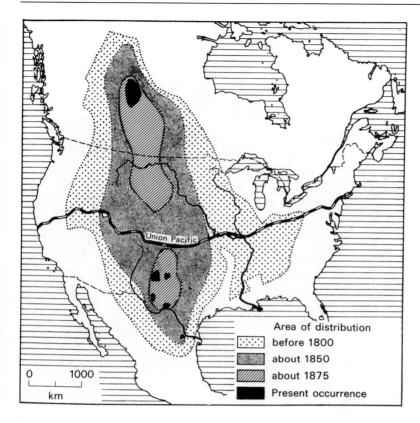

Figure 3.13
The former and present
distribution of the bison in
North America (after Ziswiler in
Illies, 1974, figure 3.1)

Area of distribution
before 1800
about 1850
about 1875
Present occurrence

Union Pacific

0   1000
km

become smaller very rapidly. This is clearly illustrated by a study
of the North American bison (figure 3.13), which on the eve of
European colonization still had a population of some 60 million
in spite of the presence of a small Indian population. By 1850
only a few dozen examples of bison still survived, though
conscious protective and conservation measures have saved it
from extinction.

Animals native to remote ocean islands have been especially
vulnerable since many have no flight instinct (plate 3.7). They
also provided a convenient source of provisions for seafarers.
Thus the last example of the dodo (*Raphus cucullatus*) (plate
3.8) was slaughtered on Mauritius in 1681, and the last Steller's
sea-cow (*Hydrodamalis gigas*) was killed on Bering Island in
1768 (Illies, 1974: 96). The land-birds, which are such a feature
of coral atoll ecosystems, rapidly become extinct when cats, dogs
and rats are introduced. The endemic flightless rail of Wake
Island, *Rallus wakensis*, became extinct during the prolonged
siege of the island in the Second World War, and the flightless rail
of Laysan, *Porzanula palmeri*, is also now extinct. Sea-birds are
more numerous on the atolls, but many colonies of terns, noddies
and boobies have been drastically reduced by vegetation clearance and by introduced rats and cats in recent times (Stoddart,

Plate 3.7
The Moa is an extinct land
bird from New Zealand. This
reconstructed version was
placed in the Dunedin Public
Gardens

1968). Other atoll birds, including the albatross, have been culled
because of the threat they pose to aircraft using military airfields
on the islands. Some birds, including the sulids, have declined
because they are large, vulnerable and edible (Feare, 1978).

The fish resources of the oceans are still exploited by hunting
and gathering techniques, and about 110 million tonnes of fish
are caught in the world each year, a more than fivefold increase
since 1950. There are countless examples of fish decline brought
about by reckless over-exploitation. Overall (see Ehrlich et al.,
1977: 358) it has been estimated, for instance, that whale stocks
have been reduced by at least half through uncontrolled hunting
during the past half-century (plate 3.9). Populations of the large
varieties, those that have been most heavily exploited, have been
cut to a tiny fraction of their former sizes, and the hunting,
especially by the Russians and the Japanese, still goes on. As
Fisher et al. (1969) have put it, 'It may not be long before the
blue whale joins the dinosaurs in the museum of oblivion.'
Chronic overfishing is a particular problem in shallow marginal
seas (e.g. the North Sea) and in the vicinity of coral reefs, where
the removal of herbivorous fish can enable algal blooms to
develop catastrophically (Hughes, 1994).

Plate 3.8
The dodo. Top: the skin and skull remains. Bottom: its reconstructed form. The last example disappeared over 300 years ago in Mauritius

Plate 3.9
Whale butchering, whaling station Hvalfijordur Bay, Iceland. If reckless over-exploitation continues some species will soon become extinct

The introduction of a new animal species can cause the decline of another, whether by predation, competition or hybridization. This last mechanism has been found to be a major problem with fish in California, where many species have been introduced by humans. Hybridization results when fish are transferred from one basin to a neighbouring basin, since closely related species of the type likely to hybridize usually exist in adjacent basins. One species can eliminate another closely related to it through genetic swamping (Moyle, 1976).

Animal extinctions in prehistoric and historic times

As the size of human populations has increased and technology has developed, humans have been responsible for the extinction of many species of animals. Indeed, it has often been maintained that there has been a close correlation between the curve of population growth since the mid-seventeenth century and the curve of the number of species that have become extinct (figure 3.14a, b).

However, although it is likely that European expansion overseas during that time has led to many extinctions, the dramatic increase in the rate of extinctions over the last few centuries that is shown in the figure may in part be a consequence of a dramatic increase in the documentation of natural phenomena.

A more recent attempt to look at the historical trends in extinction rates (figure 3.14c) for 20-year intervals from the year 1600 shows a rather different pattern, with a decline in rate for 1960–90. This may partly be because of a time-lag in the recording of the most recent extinction events, but it may also be the result of successful conservation efforts and the spread of the areal extent of protected areas. What is clear, however, is just what a large proportion of total extinctions have been recorded from islands.

Some workers, notably Martin (1967, 1974), believe that the human role in animal extinctions goes back to the Stone Age and to the Late Pleistocene. They believe that extinction closely follows the chronology of the spread of prehistoric civilization and the development of big-game hunting. They would also maintain that there are no known continents or islands in which accelerated extinction definitely pre-dates the arrival of human settlements. An alternative interpretation, however, is that the Late Pleistocene extinctions of big mammals were caused by the rapid and substantial changes of climate at the termination of the last glacial (Martin and Wright, 1967; Martin and Klein, 1984). Martin (1982) argues that the global pattern of extinctions of the large land mammals follows the footsteps of Palaeolithic settlements. He suggests that Africa and parts of southern Asia were affected first, with substantial losses at the end of the

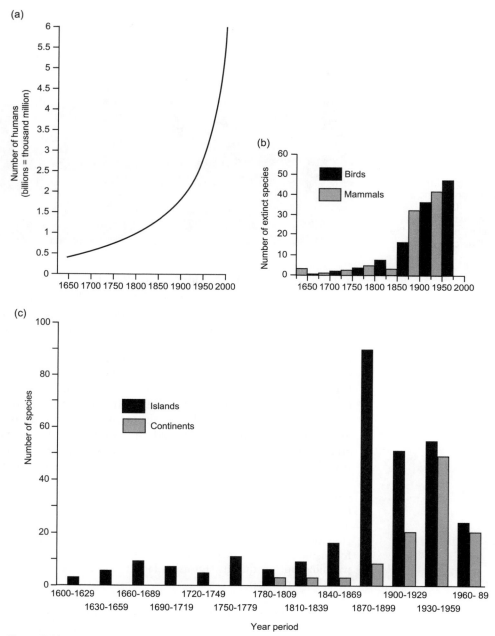

Figure 3.14
Time series of animal extinctions since the seventeenth century, in relation to human population increase (a). Figure (b) shows an early attempt to trace the number of extinctions in birds and mammals (after Ziswiler in National Academy of Sciences, 1972, figure 3.2), while figure (c) is a more recent attempt and displays a decreasing rate of animal extinctions in recent decades (after World Conservation Monitoring Centre, 1992, figure 16.5)

Plate 3.10
The inhabitants of Easter Island who carved and erected these famous monuments may, through deforestation, have sown the seeds of their own demise

Acheulean, around 200,000 years ago. Europe and northern Asia were affected between 20,000 and 10,000 years ago, while North and South America were stripped of large herbivores between 12,000 and 10,000 years ago. Extinctions continued into the Holocene on oceanic islands, and in the Galapagos Islands, for example, virtually all extinctions took place after the first human contact in AD 1535 (Steadman et al., 1991). Likewise, the complete deforestation of Easter Island in the Pacific (plate 3.10) between 1200 and 800 BP was an ecological disaster that led to the demise of much of the native flora and fauna and also precipitated the decline of the megalithic civilization that had erected the famous statues on the islands (Flenley et al., 1991). As Burney (1993: 536) has written:

For millennia after the late-Pleistocene extinctions, these seemingly fragile and ungainly ecosystems apparently thrived throughout the world. This is one of the soundest pieces of evidence against purely climatic explanations for the passing of Pleistocene faunas. In Europe, for instance, the last members of the elephant family survived climatic warming not on the vast Eurasian land mass, but on small islands in the Mediterranean.

The dates of the major episodes of Pleistocene extinction are crucial to the discussion (table 3.10). Martin (1967: 111–15) writes:

Radiocarbon dates, pollen profiles associated with extinct animal remains, and new stratigraphic and archaeological evidence show that, depending on the region involved, late-Pleistocene extinction occurred either after, during or somewhat before worldwide climatic cooling of the last maximum . . . of glaciation . . . Outside continental Africa and South East Asia, massive extinction is unknown before the earliest

Table 3.10 Dates of major episodes of generic extinction in the Late Pleistocene

Area	Date (years BP)
North America	11,000
South America	10,000
West Indies	Mid-post Glacial
Australia	13,000
New Zealand	900
Madagascar	800
Northern Eurasia	13,000–11,000
Africa and South-East Asia (?)	50,000–40,000

Source: data in Martin, 1967: 111.

known arrival of prehistoric man. In the case of Africa, massive extinction coincides with the final development of Acheulean hunting cultures which are widespread throughout the continent . . . The thought that prehistoric hunters ten to fifteen thousand years ago (and in Africa over forty thousand years ago) exterminated far more large animals than has modern man with modern weapons and advanced technology is certainly provocative and perhaps even deeply disturbing.

The nature of the human impact on animal extinctions can conveniently be classified into three types (Marshall, 1984): the 'blitzkrieg effect', which involves rapid deployment of human populations with big-game hunting technology so that there is very rapid demise of animal populations; the 'innovation effect', whereby long-established human population groups adopt new hunting technologies and erase fauna that have already been stressed by climatic changes; and the 'attrition effect', whereby extinction takes place relatively slowly after a long history of human activity because of loss of habitat and competition for resources.

The arguments that have been used in favour of an anthropogenic interpretation can be summarized as follows. First, massive extinction in North America seems to coincide in time with the arrival of humans in sufficient quantity and with sufficient technological skill in making suitable artefacts (the bifacial Clovis blades) to be able to kill large numbers of animals (Krantz, 1970). Second, in Europe the efficiency of Upper Palaeolithic hunters is attested by such sites as Solutré in France, where a late-Perigordian level is estimated to contain the remains of over 100,000 horses. Third, many beasts unfamiliar with people are remarkably tame and stupid in their presence, and it would have taken them a considerable time to learn to flee or seek concealment at the sight or scent of human life. Fourth, humans, in addition to hunting animals to death, may also have competed with them for particular food or water supplies. Fifth, Coe (1981, 1982) has suggested that the supposed preferential extinction of the larger mammals could also lend support to the role of human

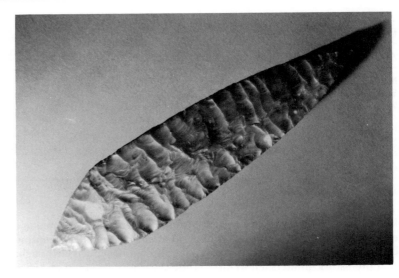

Plate 3.11
Prehistoric hunters may have
been very effective at killing
large numbers of mammals in
relatively short periods of time.
Upper Palaeolithic hunters
from Solutré in France used
superb 'laurel-leaf' points for
this purpose

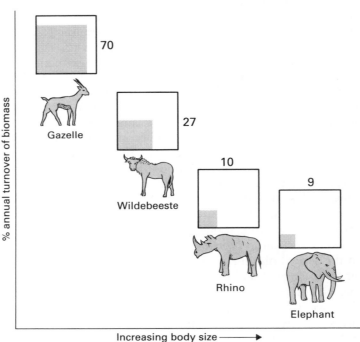

Figure 3.15
Decreasing rate of population
biomass turnover with
increasing body size in warm-
and cold-blooded vertebrates
(from Coe, 1982)

actions. He argues that while large body size has certain definite advantages, especially in terms of avoiding predation and being able to cover vast areas of savanna in search of food, large body size also means that these herbivores are required to feed almost continuously to sustain a large body mass. Furthermore, as the size and generation time of a mammal increases, the rate at which they turn over their biomass decreases (figure 3.15).

Hence a population of Thompson's gazelle will turn over up to 70 per cent of their biomass each year, a wildebeeste over 27 per cent, but a rhino only 10 per cent and the elephant just 9 per cent. The significance of this is that, since large mammals can only turn over a small percentage of their population biomass each year, the rate of slaughter that such a population can sustain in the face of even a primitive hunter is very low indeed.

Certain objections have been levelled against the climatic change model and these tend to support the anthropogenic model. It has been suggested, for instance, that changes in climatic zones are generally sufficiently gradual for beasts to be able to follow the shifting vegetation and climatic zones of their choice. Similar environments are available in North America today as were present, in different locations and in different proportions, during Late Pleistocene times. Second, it can be argued that the climatic changes associated with the multiple glaciations, interglacials, pluvials and interpluvials do not seem to have caused the same striking degree of elimination as those in the Late Pleistocene. A third difficulty with the climatic-cause theory is that animals like the mammoth occupied a broad range of habitats from Arctic to tropical latitudes, so that is unlikely all would perish as a result of a climatic change (Martin, 1982).

This is not to say, however, that the climatic hypothesis is without foundation or support. The migration of animals in response to rapid climatic change could be halted by geographical barriers such as high mountain ranges or seas. The relatively rich state of the African big mammalian fauna is due, according to this point of view, to the fact that the African biota is not, or was not, greatly restricted by an insuperable geographical barrier. Another way in which climatic change could cause extinction is through its influence on disease transmission. It has been suggested that during glacials animals would be split into discrete groups cut off by ice sheets but that, as the ice melted (before 11,000 BP in many areas), contacts between groups would once again be made, enabling diseases to which immunity might have been lost during isolation to spread rapidly. Large mammals, because of their low reproduction rates, would recover their numbers only slowly, and it was large mammals (according to Martin) that were the main sufferers in the Late Pleistocene extinctions.

The detailed dating of the European megafauna's demise lends further credence to the climatic model (Reed, 1970). The Eurasiatic boreal mammals, such as mammoth, woolly rhinoceros, musk ox and steppe bison, were associated with and adapted to the cold steppe which was the dominant environment in northern Europe during the glacial phases of the last glaciation. Each of them, especially the mammoth and the steppe bison, had been hunted by humans for tens of thousands of years, yet managed to survive through the last glacial. They seem to have disappeared, according to Reed, within the space of a few hundred years when warm conditions associated with the Allerød interstadial led to the restriction and near disappearance of their habitat.

Grayson (1977) has added to the doubts expressed about the anthropogenic overkill hypothesis. He suggests that the overkill theory, because it states that the end of the North American Pleistocene was marked by extraordinarily high rates of extinction of mammalian genera, requires terminal Pleistocene mammalian generic extinctions to have been relatively greater than generic extinctions within other classes of vertebrates at this time. When he examined the extinction of birds he found that an almost exactly comparable proportion became extinct at the end of the Pleistocene as one finds for the megafauna. Moreover, as the radiocarbon dates for early societies in countries like Australia are pushed back, it becomes increasingly clear that humans and several species of megafauna were living together for quite long periods, thereby undermining the notion of rapid overkill (Gillespie et al., 1978).

Further arguments can be marshalled against the view that humans as predators played a critical role in the Late Pleistocene extinctions in North America (Butzer, 1972: 509–10). There are, for example, relatively few Palaeo-Indian sites over an immense area, and the majority of these have a very limited cultural inventory. In addition there is no clear evidence that Palaeo-Indian subsistence was necessarily based, in the main part, on big-game hunting; if it was, only two genera were hunted intensively: mammoth and bison. A final point that militates against the argument that humans were primarily responsible for the waves of extinction is the survival of many big-game species well into the nineteenth century, despite a much larger and more efficient Indian population.

Thus the human role in the great Late Pleistocene extinctions is still a matter of debate. The problem is complicated because certain major cultural changes in human societies may have occurred in response to climatic change. The cultural changes may have assisted in the extinction process, and increasing numbers of technologically competent humans may have delivered the final *coup de grâce* to isolated remnants already doomed by rapid post-glacial environmental changes (Guilday, 1967). In this context it is worth noting that Haynes (1991) has suggested that at around 11,000 years ago conditions were dry in the interior of the United States and that Clovis hunters may as a result have found large game animals easier prey when concentrated around water holes and under stress.

Actual extinction may have occurred concurrently with a dwarfing in the size of animals, and this too is a subject of controversy. On the one hand, there are those who believe that the dwarfing could result from a reduction in food availability brought about by climatic deterioration. On the other, it can be maintained that this phenomenon derives from the fact that small animals, being more adept at hiding and being a less attractive target for a hunter, are more likely to survive human predators, so that a genetic selection towards reduced body size takes place (Marshall, 1984).

Although the debate about the cause of this extinction spasm has now persisted for a long time, there are still great uncertainties. This is particularly true with regard to the chronologies of extinction and of human colonization. This is particularly so in the context of North America, as noted by Grayson (1988: 118):

While most of us dealing with the extinctions issue have assumed that all of the extinctions occurred between 12,000 and 10,000 years ago, and were, by that assumption, drawn into thinking that the extinction event had occurred rapidly, there is surprisingly little hard evidence to support that assumption. Much effort needs to be placed into the construction of a detailed extinction chronology. Did the extinctions occur in a brief – say 500 year – period of time, with no major decreases in population numbers before then, or were they scattered across the millennia, with major population increases leading slowly to the final losses? Were mammoths so greatly diminished in numbers by 12,000 years ago that Clovis peoples could have made all the difference 500 years later?

What a difference to our proposed explanation if, as the current chronology perhaps suggests and surely does not deny, the extinctions were well under way prior to Clovis times.

The only way I can conceive of showing overkill to be incorrect is by showing that significant losses had already occurred by the time people had arrived south of the glacial ice. Even then, overkill would be falsified only for those taxa already extinct by the time of human occupation. Unfortunately, the statement that 35 genera of mammals became extinct in North America at about 11,000 years ago has about the same status as the statement that Clovis represents the first human occupation of the New World south of the glacial ice. We simply do not know if either is true, and there are reasons to suspect that both are false.

It is probable that the Late Pleistocene extinctions themselves had major ecological consequences. As Birks (1986: 49) has put it:

The ecological effects of rapid extinction of over 75% of the New World's large herbivores . . . must have been profound, for example on seed dispersal, browsing, grazing, trampling and tree regeneration . . . large grazers and browsers such as bison, mammoth and wooly rhinoceros may have been important in delaying or even inhibiting tree growth.

Modern-day extinctions

When considering modern-day extinctions the role of human agency is much less controversial (plate 3.12). None the less, some modern extinctions of species are natural. Extinction is a biological reality: it is part of the process of evolution. In any period, including the present, there are naturally doomed species, which are bound to disappear, either through overspecialization or through an incapacity to adapt themselves to climatic change and the competition of others, or because of natural cataclysms such as earthquakes, eruptions and floods. Fisher et al. (1969) believe that probably one-quarter of the species of birds and mammals that have become extinct since 1600 may have died out naturally. In spite of this, however, the rate of animal extinctions brought about by humans in the last 400 years is of a very high order when compared to the norm of geological time. In 1600 there were approximately 4,226 living species of mammals: since then 36 (0.85 per cent) have become extinct and at least 120 (2.84 per cent) are presently believed to be in some (or great) danger of extinction. A similar picture applies to birds. In 1600 there were about 8,684 living species; since then 94 (1.09 per cent) have become extinct and at least 187 (2.16 per cent) are presently, or have very lately been, in danger of extinction (Fisher et al., 1969).

Some beasts appear to be more prone to extinction than others, and a distinction is now often drawn between *r*-selected species and *k*-selected species. These are two ends of a spectrum. The former have a high rate of increase, short gestation periods, quick maturation, and the advantage that they have the ability either to react quickly to new environmental opportunities or to

Plate 3.12
In recent years many populations of African elephants have suffered extreme reductions in their numbers as a result of poaching for ivory. The situation is less grave in those countries of southern and central Africa where conservation policies have been rigorously enforced (e.g. South Africa, Zimbabwe and Botswana), but because of their large territorial requirements, long gestation periods and small litter sizes, elephants have many qualities that make them especially susceptible to extinction

Table 3.11 Characteristics of species affecting survival

Endangered	Safe
Large size	Small size
Predator	Grazer, scavenger, insectivore
Narrow habitat tolerance	Wide habitat tolerance
Valuable fur, oil, hide, etc.	Not a source of natural products
Restricted distribution	Broad distribution
Lives largely in international waters	
Migrates across international boundaries	Lives largely in one country
Reproduction in one or two vast aggregates	Reproduction by solitary pairs or in many small aggregates
Long gestation period	Short gestation period
Small litters	Big litters and quick maturation
Behavioural idiosyncracies that are non-adaptive today	Adaptive
Intolerance of the presence of humans	Tolerance of humans
Dangerous to humans, livestock, etc.	Perceived as harmless

make use of transient habitats (such as seasonal ponds). The life duration of individuals tends to be short, populations tend to be unstable, and the species may over-exploit their environment to their eventual detriment. Many pests come into this category. The other end of the spectrum, the k-selection species, are those which tend to be endangered or to become extinct. Their prime characteristics are that they are better adapted to physical changes in their environment (such as seasonal fluctuations in temperature and moisture) and live in a relatively stable environment (Miller and Botkin, 1974). These species tend to have much greater longevity, longer generation times and fewer offspring, but a higher probability of survival of young and adults. They have traded a high rate of increase and the ability to exploit transient environments for the ability to maintain more stable populations with low rates of increase, but correspondingly low rates of mortality and a closer adjustment to the long-term capacity of the environment to support their population.

Table 3.11, which is modified after Ehrenfeld (1972), attempts to bring together some of the characteristics of species which affect their survival in the human world. This is also a theme discussed in Jeffries (1997).

There are at least nine ecological or life history traits that have been proposed as factors which determine the sensitivity of an animal species to a reduction and fragmentation in habitat (World Conservation Monitoring Centre, 1992, p. 193):

Rarity
Several studies have found that the abundance of a species prior to habitat fragmentation is a significant predictor of extinction . . . This is only to be expected, since fewer individuals of a rare species than a common species are likely to occur in habitat fragments, and the

mechanisms of extinction mean that small populations are inherently more likely to become extinct than large.

Dispersal ability

If animals are capable of migrating between fragments or between 'mainland' areas and fragments, the effects of small population size may be partly or even greatly mitigated by the arrival of 'rescuers'. Species that are good dispersers may therefore be less prone to extinction in fragmented habitats than poor dispersers.

Degree of specialisation

Ecological specialists often exploit resources which are patchily distributed in space and time, and therefore tend to be rare. Specialists may also be vulnerable to successional changes in fragments and to the collapse of coevolved mutualisms or food webs.

Niche location

Species adapted to, or able to tolerate, conditions at the interface between different types of habitats may be less affected by fragmentation than others. For example, forest edge species may actually benefit from habitat fragmentation.

Population variability

Species with relatively stable populations are less vulnerable than species with pronounced population fluctuations, since they are less likely to decline below some critical threshold from which recovery becomes unlikely.

Trophic status

Animals occupying high trophic levels usually have small populations: e.g. insectivores are far fewer in number than their insect prey and, as noted above, rarer species are more vulnerable to extinction.

Adult survival rate

Species with naturally low adult survival rates may be more likely to become extinct . . .

Longevity

Long-lived animals are less vulnerable to extinction than short-lived.

Intrinsic rate of population increase

Populations which can expand rapidly are more likely to recover after population declines than those which cannot.

Possibly one of the most fundamental ways in which humans are causing extinction is by reducing the area of natural habitat available to a species (Turner, 1996). Even wildlife reserves tend to be small 'islands' in an inhospitable sea of artificially modified vegetation or urban sprawl. We know from many of the classic studies in true island biogeography that the number of species living at a particular location is related to area (see figure 3.16 for some examples); islands support fewer species than do similar areas of mainland, and small islands have fewer species than do large ones. Thus it may well follow that if humans destroy the greater part of a vast belt of natural forest, leaving just a small reserve, initially it will be 'supersaturated' with species, containing more than is appropriate to its area when at equilib-

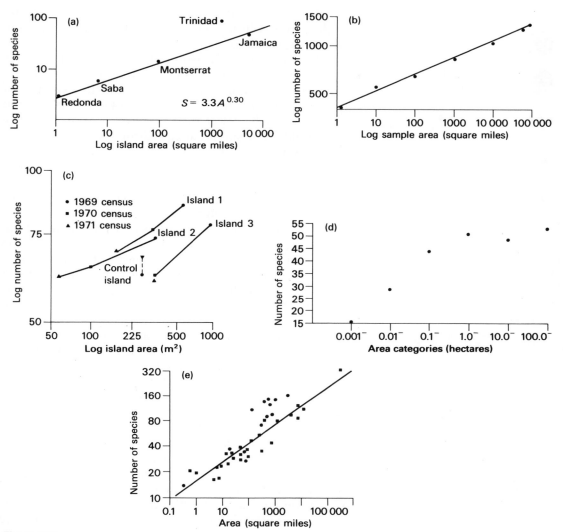

Figure 3.16
Some relationships between the size of 'islands' and numbers of species (after Gorman, 1979, figures 3.1, 3.2, 8.1 and 8.2):
(a) the number of amphibian and reptile species living on West Indian islands of various sizes. Trinidad, joined to South America 10,000 years ago, lies well above the species-area curve on other islands
(b) The area-species curve for the number of species of flowering plants found in a sample area of England
(c) The effect on the number of arthropod species of reducing the size of mangrove islands. Islands 1 and 2 were reduced after both the 1969 and the 1970 census. Island 3 was reduced only after the 1969 census. The control island was not reduced, the changes in species number being attributable to random fluctuation
(d) The number of species of bird living in British woods of various size categories
(e) The number of species of land birds living in the lowland rain forest on small islands in New Guinea, plotted as a function of island area. The squares represent islands which have not had a land connection to New Guinea and whose avifaunas are at equilibrium. The regression line is fitted through these points. The circles represent former land-bridge islands connected to New Guinea some 10,000 years ago. Note that the large ones have more species than one would expect for their size

Good   Not so good

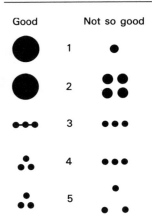

Figure 3.17
A set of general design
rules for nature reserves
based on theories of island
biogeography. The designs on
the left are preferable to those
on the right because they
should enjoy lower rates of
species extinction:
(1) A large reserve will hold
    larger populations with
    lower probabilities of
    going extinct than a
    small reserve
(2) A single reserve is
    preferable to a series of
    smaller reserves of equal
    total area, since these will
    support only small
    populations with relatively
    high probabilities of going
    extinct
(3–5) If reserves must be
    fragmented, then they
    should be connected by
    corridors of similar
    vegetation or be placed
    equidistant and as near to
    each other as possible.
    In this way immigration
    rates between the
    fragments will be
    increased, thereby
    maximizing the chances of
    extinct populations being
    replaced from elsewhere in
    the reserve complex (after
    Gorman, 1979, figure 8.4)

rium (Gorman, 1979). Since the population sizes of the species living in the forest will now be greatly reduced, the extinction rate will increase and the number of species will decline towards equilibrium. For this reason it may be a sound principle to make reserves as large as possible; a larger reserve will support more species at equilibrium by allowing the existence of larger populations with lower extinction rates. Several small reserves will plainly be better than no reserves at all, but they will tend to hold fewer species at equilibrium than will a single reserve of the same area (figure 3.17). If it is necessary to have several small reserves, they should be placed as close to each other as possible so that each may act as a source area for the others. In this way their equilibrial number of species will be raised due to increased immigration rates.

There are, however, situations when small reserves may have advantages over a single large reserve: they will be less prone to total decimation by some natural catastrophe such as fire; they may allow the preservation of a range of rare and scattered habitats; and they may allow the survival of a group of competitors, one of which would exclude the others from a single reserve.

Reduction in area leads to reduction in numbers, and this in turn can lead to genetic impoverishment through inbreeding (Frankel, 1984). The effect on reproductive performance appears to be particularly marked. Inbreeding degeneration is, however, not the only effect of small population size for, in the longer term, the depletion of genetic variance is more serious since it reduces the capacity for adaptive change. Space is therefore an important consideration, especially for those animals that require large expanses of territory. For example, the population density of the wolf is about one adult per 20 square kilometres, and it has been calculated that for a viable population to exist, one might need 600 individuals ranging over an area of 12,000 square kilometres. The significance of this is apparent when one realizes that most nature reserves are small: 93 per cent of the world's national parks and reserves have an area less than 5,000 square kilometres, and 78 per cent less than 1,000 square kilometres.

Equally, range loss, the shrinking of the geographical area in which a given species is found, often marks the start of a downward spiral towards extinction. Such a contraction in range may result from habitat loss or from such processes as hunting and capture. Particular concern has been expressed in this context about the pressures on primates, notably in South-East Asia. Of 44 species, 33 have lost at least half their natural range in the region. In two cases, those of the Javan leaf monkey and the Javan and grey gibbons, the loss of range is no less than 96 per cent. Recent figures produced by the International Union for the

Table 3.12 Wildlife habitat loss

(a) In Indomalayan nations

	Original wildlife habitat (km²)	Amount remaining (km²)	Habitat loss (%)
Bangladesh	1,142,777	68,567	94
Bhutan	34,500	22,770	34
Brunei	5,764	4,381	24
Burma	774,817	225,981	71
China	423,066	164,996	61
Hong Kong	1,066	32	97
India	3,017,009	615,095	80
Indonesia	1,446,433	746,861	49
Japan	320	138	57
Kampuchea	180,879	43,411	76
Laos	236,746	68,656	71
Malaysia	356,254	210,190	41
Nepal	117,075	53,855	54
Pakistan	165,900	39,816	76
Philippines	308,211	64,724	79
Sri Lanka	64,700	10,999	83
Taiwan	36,961	10,719	71
Thailand	507,267	130,039	74
Vietnam	332,116	66,423	80
TOTAL	8,169,860	2,487,683	68

(b) In Afrotropical nations, 1986

	Original wildlife habitat (km²)	Amount remaining (km²)	Habitat loss (%)
Angola	1,246,700	760,847	39
Benin	115,800	46,320	60
Botswana	585,400	257,576	56
Burkina Faso	273,800	54,760	80
Burundi	25,700	3,598	86
Cameroon	469,400	192,454	59
Central African Rep.	623,000	274,120	56
Chad	720,800	172,992	76
Congo	342,000	172,420	49
Côte d'Ivoire	318,000	66,780	79
Djibouti	21,800	11,118	49
Equatorial Guinea	26,000	12,740	51
Ethiopia	1,101,003	30,300	70
Gabon	267,000	173,550	35
Gambia	11,300	1,243	89
Ghana	230,000	46,000	80
Guinea	245,900	73,770	70
Guinea Bissau	36,100	7,942	78
Kenya	569,500	296,140	48
Lesotho	30,400	9,728	68
Liberia	111,400	14,482	87
Madagascar	595,211	148,803	75
Malawi	94,100	40,463	57
Mali	754,100	158,361	79
Mauritania	388,600	73,834	81

Table 3.12(b) *(cont'd)*

	Original wildlife habitat (km²)	Amount remaining (km²)	Habitat loss (%)
Mozambique	783,203	36,776	57
Namibia	823,200	444,528	46
Niger	566,000	127,880	77
Nigeria	919,800	229,950	75
Rwanda	25,100	3,263	87
Senegal	196,200	35,316	82
Sierra Leone	71,700	10,755	85
Somalia	637,700	376,243	41
South Africa	1,236,500	531,695	57
Sudan	1,703,000	510,900	70
Swaziland	17,400	7,656	56
Tanzania	886,200	505,134	43
Togo	56,000	19,040	66
Uganda	193,700	42,614	78
Zaire	2,335,900	1,051,155	55
Zambia	752,600	534,346	29
Zimbabwe	390,200	171,688	56
TOTAL	20,797,441	8,340,920	65

Source: IUCN/UNEP data in *World Resources 1988–9*, tables 6.4 and 6.5.

Conservation of Nature and the United Nations Environment Programme for wildlife habitat loss show the severity of the problem (table 3.12). In the Indomalayan countries 68 per cent of the original wildlife habitat has been lost, and the comparable figure for tropical Africa is 65 per cent. In these regions, only Brunei and Zambia have lost less than 30 per cent of their original habitat, while at the other end of the spectrum, Bangladesh, the most densely populated large country in the world, has suffered a loss of 94 per cent. Figure 3.18 shows the fragmentation and reduction in area of rain forest within two areas: Sumatra and Costa Rica.

Although habitat change and destruction are clearly major causes of extinction in the modern era, a remarkably important cause is the introduction of competitive species. When new species are deliberately or accidentally introduced to an area, they can cause the extermination of local fauna by preying on them or out-competing them for food and space. As we have seen elsewhere (see p. 139), island species have proved to be especially vulnerable. The World Conservation Monitoring Centre, in its analysis of the known causes of animal extinctions since AD 1600, believed that 39 per cent were caused by species introductions, 36 per cent by habitat destruction, 23 per cent by hunting, and 2 per cent for other reasons (*World Resources*, 1994–95, p. 149).

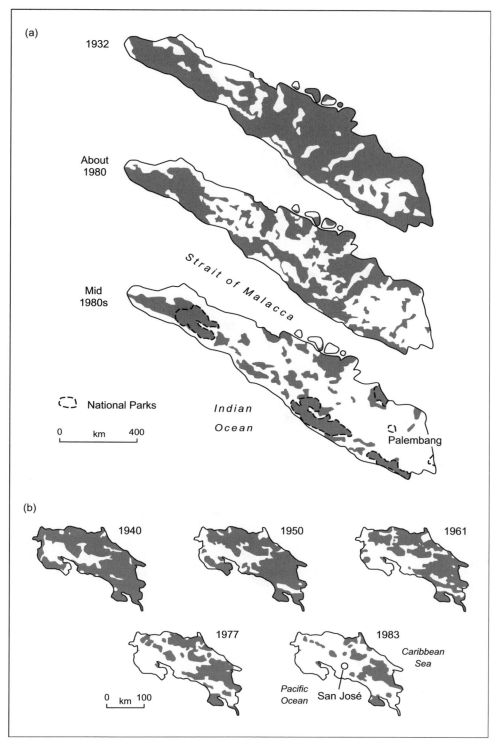

Figure 3.18
Progressive habitat fragmentation in the rain-forest environments of (a) Sumatra and
(b) Costa Rica (modified after Whitten et al., 1987 and Terborgh, 1992)

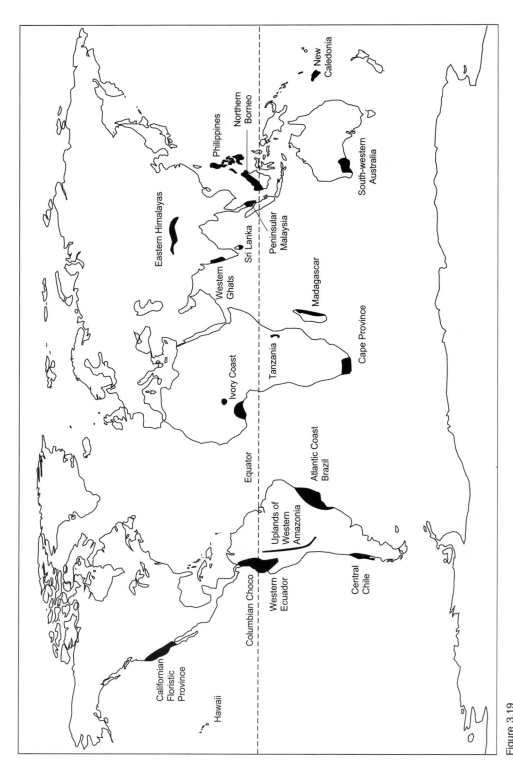

Figure 3.19
Forest and heathland hot spot areas. Hot spots are habitats with many species found nowhere else and in greatest danger of extinction from human activity (after Wilson, 1992, pp. 262–3)

The rate of future extinctions is the cause of very considerable concern. As the Global 2000 Report (Council on Environmental Quality and the Department of State, 1982) has suggested (pp. 327–9):

If present trends continue – as they certainly will in many areas – hundreds of thousands of species can be expected to be lost by the year 2000 . . . the extinctions projected for the coming decades will be largely human-generated and on a scale that renders natural extinction trivial by comparison. Efforts to meet human needs and rising expectations are likely to lead to the extinction of between one-fifth and one-seventh of all species (of plants and animals) over the next two decades.

The great bulk of these extinctions will take place because of the removal of areas of humid tropical moist forest, and the largest number of extinctions can be expected in the insect order.

There are certainly some particularly important environments in terms of their species diversity (Myers, 1990). Such biodiversity 'hot-spots' (figure 3.19) need to be made priorities for conservation. They include coral reefs, tropical forests (which support well over half the planet's species on only about 6 per cent of its land area), and some of the Mediterranean climate ecosystems (including the extraordinarily diverse Fynbos shrublands of the Cape region of South Africa). Some environments are crucial because of their high levels of species diversity or endemic species; others are crucial because their loss would have consequences elsewhere. This applies, for example, to wetlands which provide habitats for migratory birds and produce the nutrients for many fisheries.

According to Myers (1979: 31), 'during the last quarter of this century we shall witness an extinction spasm accounting for 1 million species.' This is a considerable proportion of the estimated number of species living in the world today, for which Myers gives a figure of between three and nine million. He has calculated that from AD 1600 to 1900 humans were accounting for the demise of one species every four years, that from 1900 onwards the rate increased to an average of around one per year, that at present the rate is about one per day, and that within a decade we could be losing one every hour. By the end of the century, our planet could lose anywhere between 20 and 50 per cent of its species (Lugo, 1988). The need to maintain *biodiversity* has become one of the crucial issues with which we must contend (Kramer et al., 1997).

4 The Human Impact on the Soil

Introduction

Humans live close to and depend on the soil. It is one of the thinnest and most vulnerable human resources and one upon which, both deliberately and inadvertently, humans have had a very major impact. Moreover, such an impact can occur with great rapidity in response to land-use change, new technologies or waves of colonization (see Russell and Isbell, 1986, for a review in the context of Australia).

Natural soil is the product of a whole range of factors and the classic expression of this is that of Jenny (1941):

$$s = f(cl, o, r, p, t \dots)$$

where s denotes any soil property, cl is the regional climate, o the biota, r the topography, p the parent material, t the time (or period of soil formation), and the dots represent additional, unspecified factors. In reality soils are the product of highly complex interactions of many interdependent variables, and the soils themselves are not merely a passive and dependent factor in the environment. None the less, following Jenny's subdivision of the classic factors of soil formation, one can see more clearly the effects humans have had on soil, whether detrimental or beneficial. These can be summarized as follows (adapted from the work of Bidwell and Hole, 1965):

1 *Parent material*
   Beneficial: adding mineral fertilizers; accumulating shells and bones; accumulating ash locally; removing excess amounts of substances such as salts.
   Detrimental: removing through harvest more plants and animal nutrients than are replaced; adding materials in amounts toxic to plants or animals; altering soil constituents in such a way as to depress plant growth.

2 *Topography*

Beneficial: checking erosion through surface roughening, land forming and structure building; raising land level by accumulation of material; land levelling.

Detrimental: causing subsidence by drainage of wetlands and by mining; accelerating erosion; excavating.

3 *Climate*

Beneficial: adding water by irrigation; rainmaking by seeding clouds; removing water by drainage; diverting winds, etc.

Detrimental: subjecting soil to excessive insolation, to extended frost action, to wind, etc.

4 *Organisms*

Beneficial: introducing and controlling populations of plants and animals; adding organic matter including 'night-soil'; loosening soil by ploughing to admit more oxygen; fallowing; removing pathogenic organisms, e.g. by controlled burning.

Detrimental: removing plants and animals; reducing organic content of soil through burning, ploughing, overgrazing, harvesting, etc.; adding or fostering pathogenic organisms; adding radioactive substances.

5 *Time*

Beneficial: rejuvenating the soil by adding fresh parent material or through exposure of local parent material by soil erosion; reclaiming land from under water.

Detrimental: degrading the soil by accelerated removal of nutrients from soil and vegetation cover; burying soil under solid fill or water.

Space constraints preclude following all these aspects of anthropogenic soil modification or, to use the terminology of Yaalon and Yaron (1966), of *metapedogenesis*. We will therefore concentrate on certain highly important changes which humans have brought about, especially chemical changes (such as salinization and lateritization), various structural changes (such as compaction), some hydrological changes (including the effects of drainage and the factors leading to peat-bog development), and, perhaps most important of all, soil erosion.

Salinity: natural sources

Many semi-arid and arid areas are naturally salty. By definition they are areas of substantial water deficit where evapotranspiration exceeds precipitation. Thus, whereas in humid areas there is sufficient water to percolate through the soil and to leach soluble materials from the soil and the rocks into the rivers and hence into the sea, in deserts this is not the case. Salts therefore tend to accumulate. This tendency is exacerbated by the fact

Table 4.1 Hydrological and chemical data for south-western Australia, illustrating the effects of deforestation on ground-water recharge and river-water chemistry

Catchment (type or name)	Chloride input in precipitation (kg ha⁻¹ yr⁻¹)	Chloride loss in stream flow (kg ha⁻¹ yr⁻¹)	Recharge before clearing (mm yr⁻¹)	Recharge with farming (mm yr⁻¹)
(a) Forested				
Julimar	53	78	–	–
Seldom seen	120	160	–	–
More seldom seen	120	140	–	–
Waterfall gully	110	180	–	–
North Dandalup	130	180	–	–
Davies	130	140	–	–
Yarragil	97	130	–	–
Harris	84	130	–	–
(b) Farmland				
Brockman	80	340	8	73
Wooroloo	78	420	4	61
Dale	24	460	0.8	24
Hotham	48	370	2	26
Williams	31	650	1	37
Collie East	50	740	2	60
Brunswick	110	350	70	500

Source: from data in Peck, 1975.

that many desert areas are characterized by closed drainage basins (endoreic drainage) which act as terminal evaporative sumps for rivers.

The amount of natural salinity varies according to numerous factors, one of which is the source of salts. Some of the salts are brought into the deserts by rivers, though it needs to be pointed out that most of these rivers do not have particularly high salt contents (generally less than 300 ppm). A second source of salts is the atmosphere – a source which in the past has often been accorded insufficient importance. Rainfall, coastal fogs and dust storms all transport significant quantities of soluble salts. This is shown by the data in table 4.1 for some forested catchments in south-western Australia, where up to 130 kilograms per hectare per year of chlorides are introduced in precipitation. Further soluble salts may be derived from the weathering and solution of bedrock. In the Middle East, for example, notably in Iran, there are extensive salt domes and evaporite beds within the bedrock which create locally high groundwater and surface-water salinity levels. In other areas, such as the Rift Valley of southern Ethiopia and northern Kenya, volcanic rocks may provide a large source of sodium carbonate (trona) to groundwater, while elsewhere the rocks in which groundwater occurs may contain salt because they are themselves ancient desert sediments. Even in the absence of such localized sources of highly saline groundwater it needs to be remembered that over a period of time most rocks will provide soluble products to groundwater, and in a closed hydrological system such salts will eventually accumulate to significant levels.

A further source of salinity may be marine transgressions. At times of higher sea levels, it has sometimes been proposed (see e.g. Godbole, 1972) that salts would have been laid down by the sea. Likewise in coastal areas, salts in groundwater aquifers may be contaminated by contact with sea water.

Human agency and increased salinity

It has already been stated that salinity is a normal characteristic of many desert areas. However, humans have increased the extent and degree of salinity in numerous different ways.

The extension of irrigation (see table 4.2), and the use of a wide range of different techniques for water abstraction and application, can lead to a build-up of salt levels in the soil through the mechanism of raising groundwater (plate 4.1) so that it is

Table 4.2 Estimates of the increasing area of irrigated land on a global basis

Year	Irrigated area (10^6 ha)
1900	44–8
1930	80
1950	94
1955	120
1960	168
1980	211
1990 (estimate)	240

Source: Goudie and Viles, 1997, table 1.3.

Plate 4.1
An irrigated field in southern Morocco. The application of large amounts of irrigation water causes groundwater levels to rise and high air temperatures lead to rapid evaporation of the water. This leads to the eventual build up of salts in the soil

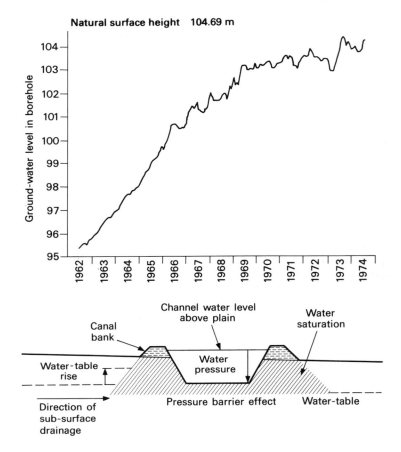

Figure 4.1
Piezometer record to show
groundwater level changes
in the Murray Valley, Victoria,
Australia, from 1962 to 1974
as a result of the extension
of irrigation (after Currey,
1977, figure 2.13)

Figure 4.2
The effect of an irrigation
channel in causing a local rise
in the water table, through the
development of a pressure
barrier which influences the
movement of sub-surface
drainage (after Currey, 1977,
figure 2.12)

near enough to the ground surface for capillary rise and subsequent evaporative concentration to take place. In the case of the
semi-arid northern plains of Victoria in Australia (figure 4.1),
for instance, the water table has been rising at around 1.5 metres
a year so that now in many areas it is almost within 3 metres of
the surface in clay soils (less for silty and sandy soils); capillary
forces bring moisture to the surface where evaporation takes
place (Currey, 1977). A survey of the problem in south-eastern
Australia is provided by Grieve (1987).

Second, many irrigation schemes require the addition of large
quantities of water over the soil surface. This is especially true
for rice cultivation. Such surface water is readily evaporated so
that salinity levels build up.

Third, the construction of large dams and barrages to control
water flow and to give a head of water creates large reservoirs
from which further evaporation can occur.

Fourth, notably in areas of soils with high permeability, water
seeps laterally and downwards from irrigation channels so that
further evaporation takes place (figure 4.2). Many distribution

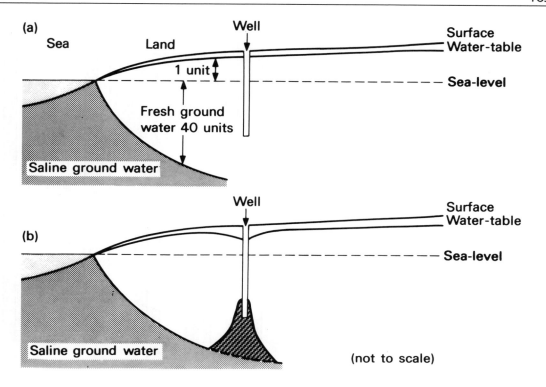

channels in a gravity scheme are located on the elevated areas of a flood plain or riverine plain to make maximum use of gravity. The elevated landforms selected are natural levees, river-bordering dunes and terraces, all of which are composed of silt and sand which may be particularly prone to seepage loss.

In coastal areas salinity problems are created by sea-water incursion brought about by overpumping. This can be explained as follows. Fresh water has a lower density than salt water, such that a column of sea water can support a column of fresh water approximately 2.5 per cent higher than itself (or a ratio of about 40:41). So where a body of fresh water has accumulated in a reservoir rock which is also open to penetration from the sea, it does not simply lie flat on top of the salt water but forms a lens, whose thickness is approximately 41 times the elevation of the piezometric surface above sea-level. This is called the Ghyben–Herzberg principle (see figure 4.3). The corollary of this rule is that if the hydrostatic pressure of the fresh water falls as a result of overpumping in a well, then the underlying salt water will rise by 40 units for every unit by which the fresh-water table is lowered.

This problem presents itself on the coast of Israel, in California (Banks and Richter, 1953), on the island of Bahrain, and in some of the small coastal dune aquifers of the United Arab

Figure 4.3
(a) The Ghyben-Herzberg relationship between fresh and saline groundwater
(b) The effect of excessive pumping from the well. The diagonal hatching represents the increasing incursion of saline water (after Goudie and Wilkinson, 1977, figure 63)

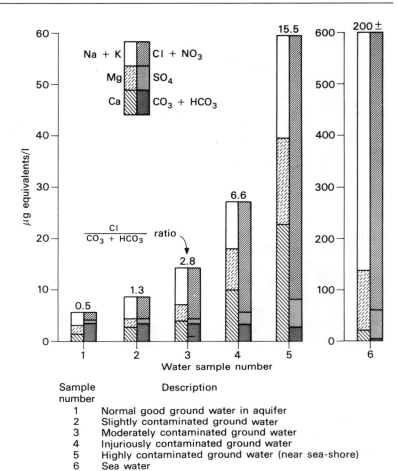

Figure 4.4
Evidence of salt-water intrusion
exemplified by chemical
analyses of a line of well-
waters from the Salinas valley,
California, extending from the
centre of the aquifer to the
coast. Chloride–bicarbonate
ratios are shown above each
quality diagram (after Simpson
in Todd, 1959, figure 12.14)

Sample number	Description
1	Normal good ground water in aquifer
2	Slightly contaminated ground water
3	Moderately contaminated ground water
4	Injuriously contaminated ground water
5	Highly contaminated ground water (near sea-shore)
6	Sea water

Emirates. A comparable situation arises in the case of the Nile delta where, as a result of the construction of the Aswan Dam, groundwater levels have dropped downstream, leading to the intrusion of coastal salt water which has salinated the soil above, thus rendering it less suitable for cultivation. This is one of the prices which have to be paid for the undoubted gains brought by rationalizing the flow of the Nile so that little of the water is wasted by draining out to sea.

The chemical changes in groundwater quality brought about by sea-water incursion may also be considerable. An example from the Salinas valley of California is shown in figure 4.4. The most notable feature, apart from the general overall increase in the total dissolved solids content approximately tenfold, is the change in the ratio of chloride to bicarbonate. As the coast is approached, as one might expect, the quantity of the chloride relative to bicarbonate rises from a ratio of 0.5 to around 200. Figure 4.5 shows the ways in which chlorine concentrations

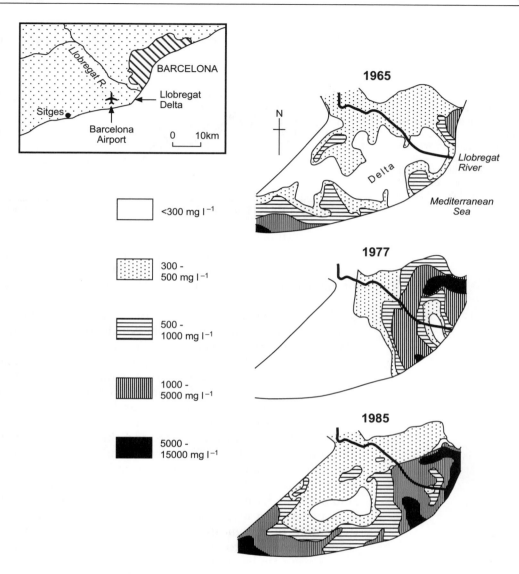

Figure 4.5
Changes in the chloride concentration of the Llobregat delta aquifer, Barcelona, Spain as a result of sea-water incursion caused by the over-pumping of groundwater (modified from Custodio et al., 1986)

have increased and spread in the Llobregat delta area of eastern Spain because of the incursion of sea water.

The problem of sea-water incursion as a result of overpumping of aquifers is not, however, restricted to arid and semi-arid areas. It was noted as early as the middle of the last century in London and Liverpool, England, and there are now many records of this phenomenon in Germany, the Netherlands, Japan and the eastern seaboard of the United States (Todd, 1959).

Increases in soil salinity are not confined to irrigated areas. In certain parts of the world salinization has resulted from vegetation clearance (Peck, 1978). The removal of native forest

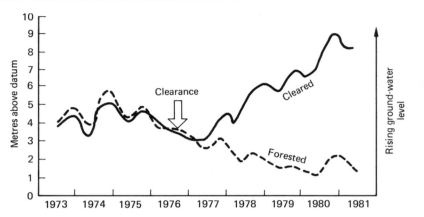

Figure 4.6
Comparison of hydrographs
recorded from the boreholes
in Wights (———) and Salmon
(------) catchments in Western
Australia. Both catchments
were forested until late in 1976
when Wights was cleared
(modified after Peck, 1983,
figure 1)

vegetation allows a greater penetration of rainfall into deeper soil layers, which causes groundwater levels to rise, creating seepage of sometimes saline water in low-lying areas. Through this mechanism an estimated 2×10^5 hectares of land in southern Australia, which at the start of European settlement supported good crops of pasture, is now suitable only for halophytic species. Similar problems exist also in North America, notably in Manitoba, Alberta, Montana and North Dakota.

The clearance of the native evergreen forest (predominantly *Eucalyptus* forest) in south-western Australia has led both to an increase in recharge rates of groundwater and to an increase in the salinity of the streams. The speed and extent of groundwater rise following such clearance of forest is shown in figure 4.6. Until late 1976, both the Wights and the Salmon catchments were forested, but then Wights was cleared. Whereas prior to late 1976 both catchments showed a similar pattern of groundwater fluctuation, after that date there was a marked divergence, amounting to 5.7 metres (Peck, 1983). Comparable studies have been undertaken in the State of Victoria (Williamson, 1983). Replanting of the eucalyptus cover has been found to cause groundwater levels to fall and groundwater salinity to decline. The recovery can be swift. In Western Australia Bari and Schofield (1992) found that after a cleared area was reforested the level of the water table fell by over 5 metres in just nine years. At the same time, groundwater salinity declined by 11 per cent.

The spread of salinity

Anthropogenic increases in salinity are not new. Jacobsen and Adams (1958) have shown that they were a problem in Mesopotamian agriculture after about 2400 BC. Individual fields, which in 2400 BC were registered as salt-free, can be seen in the records of ancient temple surveyors to have developed conditions of

Plate 4.2
The extension of irrigation in the Indus valley of Pakistan by means of large canals has caused widespread salination of the soils. Waterlogging is also prevalent. The white efflorescence of salt in the fields has been termed 'a satanic mockery of snow'

sporadic salinity by 2100 BC. Further evidence is provided by crop choice, for the onset of salinization strongly favours the adoption of crops which are most salt-tolerant. Counts of grain impressions in excavated pottery from sites in southern Iraq dated at about 3500 BC suggest that at that time the proportions of wheat and barley were nearly equal. A little more than 1,000 years later the less salt-tolerant wheat accounted for less than 20 per cent of the crop, while by about 2100 BC it accounted for less than 2 per cent of the crop. By 1700 BC the cultivation of wheat had been abandoned completely in the southern part of the alluvial plain. These changes in crop choice were accompanied by serious declines in yield which can also probably be attributed to salinity. At 2400 BC the yield was 2537 litres per hectare, by 2100 BC it was 1460, and by 1700 BC it was down to 897. It seems likely that this played an important part in the break-up of Sumerian civilization, though the evidence is not conclusive. Moreover, the Sumerians appear to have understood the problem and to have had coping strategies (Powell, 1985).

Another area where salinization has posed a severe problem is the Indus valley of Pakistan, where Sir John Marshall, the archaeologist, described the salt as 'a Satanic mockery of snow' (plate 4.2). Snelgrove (1967) reports that of the 25 million hectares of arable land in the country, 4.6 million are now mostly waterlogged or poorly drained, 1.9 million are predominantly severely saline, 4.5 million have saline patches, and some 0.4 million are being lost yearly through the spread of these conditions. The problem has largely followed the rise of groundwater levels by 0.15–0.60 metres a year brought about by the building of major irrigation systems in areas of somewhat sandy, silty and permeable soils during the last hundred or so years.

In summary, it has been estimated that the percentage of salt-affected and waterlogged soils amounts to 50 per cent of the irrigated area in Iraq, 23 per cent of all Pakistan, 50 per cent in the Euphrates valley of Syria, 30 per cent in Egypt and over 15 per cent in Iran (Worthington, 1977: 30).

Table 4.3 Salinization of irrigated cropland

Country	Percentage of irrigated lands affected by salinization
Algeria	10–15
Australia	15–20
China	15
Colombia	20
Cyprus	25
Egypt	30–40
Greece	7
India	27
Iran	<30
Iraq	50
Israel	13
Jordan	16
Pakistan	<40
Peru	12
Portugal	10–15
Senegal	10–15
Sri Lanka	13
Spain	10–15
Sudan	<20
Syria	30–35
USA	20–25

Source: FAO data as summarized in *World Resources* (1987, 1988): table 19.3.

Table 4.4 Global estimate of secondary salinization in the world's irrigated lands

Country	Cropped area (Mha)	Irrigated area (Mha)	Share of irrigated to cropped area (%)	Salt-affected land in irrigated area (%)	Share of salt-affected to irrigated land (%)
China	96.97	44.83	46.2	6.70	15.0
India	168.99	42.10	24.9	7.00	16.6
CIS	232.57	20.48	8.8	3.70	18.1
USA	189.91	18.10	9.5	4.16	23.0
Pakistan	20.76	16.08	77.5	4.22	26.2
Iran	14.83	5.74	38.7	1.72	30.0
Thailand	20.05	4.00	19.9	0.40	10.0
Egypt	2.69	2.69	100.0	0.88	33.0
Australia	47.11	1.83	3.9	0.16	8.7
Argentina	35.75	1.72	4.8	0.58	33.7
South Africa	13.17	1.13	8.6	0.10	8.9
Subtotal	842.80	158.70	18.8	29.62	20.0
World	1473.70	227.11	15.4	45.4	20.0

Source: Ghassemi et al., 1995, table 18. Reproduced by permission of CAB International and the University of New South Wales Press.

On a global basis, the calculations of Rozanov et al. (1991) make grim reading. They estimate (p. 120): 'From 1700 to 1984, the global area of irrigated land increased from 500,000 to 2,200,000 km², while at the same time some 500,000 km² were abandoned as a result of secondary salinization.'

During the 1950s the irrigated area was increasing at over 4 per cent annually, though the figure has now dropped to only about 1 per cent. The amount of salinized irrigated land varies from region to region (table 4.3), but in general ranges between 10 per cent and 50 per cent of the total. It needs to be noted, however, that there is a considerable range in these values according to the source of the data (compare table 4.4). This may in part reflect differences in the definition of the terms 'waterlogging' and 'salinization'.

Consequences of salinity

One consequence of the evaporative concentration of salts, and the pumping of saline waters back into rivers and irrigation canals from tubewells and other sources, is that river waters leading from irrigation areas show higher levels of dissolved salts. These, particularly when they contain nitrates, can make the water undesirable for human consumption.

A further problem is that, as irrigation water is concentrated by evapotranspiration, calcium and magnesium components tend to precipate as carbonates, leaving sodium ions dominant in the soil solution. The sodium ions tend to be absorbed by colloidal clay particles, deflocculating them and leaving the resultant structureless soil almost impermeable to water and unfavourable to root development.

The death of vegetation in areas of saline patches, due both to poor soil structure and to toxicity, creates bare ground which becomes a focal point for erosion by wind and water. Likewise, deflation from the desiccating surface of the Aral Sea causes large amounts of salt to be blown away in dust storms and to be deposited downwind. Some tens of millions of tonnes are being translocated by such means and their plumes are evident on satellite images (figure 4.7).

Probably the most serious impact of salinization is on plant growth. This takes place partly through its effect on soil structure, but more significantly through its effects on osmotic pressures and through direct toxicity.

When a water solution containing large quantities of dissolved salts comes into contact with a plant cell it causes a shrinkage of the protoplasmic lining. The phenomenon is due to the osmotic movement of the water, which passes from the cell towards the more concentrated soil solution. The cell collapses and the plant succumbs.

The toxicity effect varies with different plants and different salts. Sodium carbonate, by creating highly alkaline soil conditions, may damage plants by a direct caustic effect, while high nitrate may promote undesirable vegetative growth in grapes or

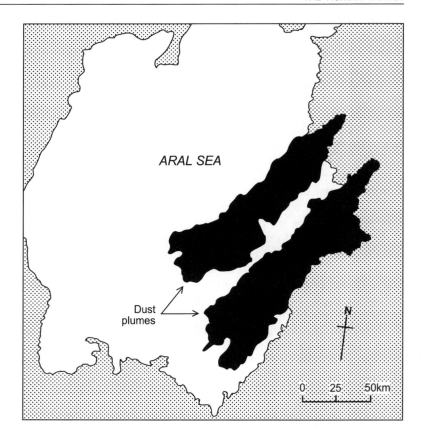

Figure 4.7
Dust plumes caused by the
deflation of salty sediments
from the drying floor of the
Aral Sea as revealed by a
satellite image (153/Meteor-
Priroda, 18 May 1975)
(modified after Mainguet,
1995, figure 4)

Table 4.5 Yield reduction for selected agricultural crops at different salinity levels, expressed by the electrical conductivity (mmho/cm) of saturated soil extracts

	% yield reduction		
Crop	10	25	50
Barley	11.9	15.8	17.5
Sugarbeet	10.0	13.0	16.0
Cotton	9.9	11.9	16.0
Wheat	7.1	10.0	14.0
Rice	5.1	5.9	8.0
Maize	5.1	5.9	7.0
Alfalfa	3.0	4.9	8.2

Source: from data in Carter, 1975.

sugarbeets at the expense of sugar content. Boron is injurious to many crop plants at solution concentrations of more than 1 or 2 ppm.

The expected yield reduction from salinity build-up is illustrated in table 4.5. The measure of soil salinity utilized is the electrical conductivity in mmho/cm of a saturated soil extract.

The tolerances of different plants to salinity vary greatly, but all suffer from increased salinity.

Reclamation of salt-affected lands

Because of the extent and seriousness of salinity, be the causes natural or anthropogenic, various reclamation techniques have been initiated. These can be divided into three main types: eradication, conversion and control.

Eradication predominantly involves the removal of salt either by improved drainage or by the addition of quantities of fresh water to leach the salt out of the soil. Both solutions involve considerable expense and pose severe technological problems in areas of low relief and limited fresh-water availability. Improved drainage can either be provided by open drains or by the use of tubewells (as at Mohenjo Daro, Pakistan) to reduce groundwater levels and associated salinity and waterlogging. A minor eradication measure, which may have some potential, is the biotic treatment of salinity through the harvesting of salt-accumulating plants such as *Suaeda fruticosa*.

Conversion involves the use of chemical methods to convert harmful salts into less harmful ones. For example, gypsum is frequently added to sodic soils to convert caustic alkali carbonates to soluble sodium sulphate and relatively harmless calcium carbonate:

$$Na_2CO_3 + CaSO_4 \leftrightarrow CaCO_3 + Na_2SO_4 \downarrow \text{leachable}$$

Some of the most effective ways of reducing the salinity hazard involve miscellaneous control measures, such as less wasteful and lavish application of water through the use of sprinklers rather than traditional irrigation methods; the lining of the canals to reduce seepage; the realignment of canals through less permeable soil; and the use of more salt-tolerant plants. As salinity is a particularly serious threat at the time of germination and for seedlings, various strategies can be adopted during this critical phase of plant growth: plots can be irrigated lightly each day after seeding to prevent salt build-up; major leaching can be carried out just before planting; and areas to be seeded can be bedded in such a way that salts accumulate at the ridge tops with the seed planted on the slope between the furrow bottom and the ridge top (Carter, 1975). A useful general review of methods for controlling soil salinity is given by Rhoades (1990).

Lateritization

In some parts of the tropics are extensive sheets of a material called laterite, an iron and/or aluminium-rich duricrust (see

Maignien, 1966 or Macfarlane, 1976). These iron-rich sheets result naturally, either because of a preferential removal of silica during the course of extensive weathering (leading to a *relative* accumulation of the sesquioxides of iron and aluminium), or because of an *absolute* accumulation of these compounds.

One of the properties of laterites is that they harden on exposure to air and through desiccation. Once hardened they are not favourable to plant growth. One particular way in which exposure may take place is by accelerated erosion, while forest removal may so cause a change in microclimate that desiccation of the laterite surface can take place. Indeed, one of the main problems with the removal of humid tropical rain forest is that soil hardening may occur. The phenomenon may occur in some, but by no means in all so-called tropical soils (Richter and Babbar, 1991). Should hardening occur, it tends to limit the extent of successful soil utilization and severely retards the re-establishment of forest. Although Vine (1968: 90) and Sanchez and Buol (1975) have rightly warned against exaggerating this difficulty in agricultural land use, there are records from many parts of the tropics of accelerated induration brought about by forest removal (Goudie, 1973). In the Cameroons, for example, around 2 metres of complete induration can take place in less than a century. In India, foresters have for a long time been worried by the role that plantations of teak (*Tectona grandis*) can play in lateritization. Teak is deciduous, demands light, likes to be well spaced (to avoid crown friction), dislikes competition from undergrowth and is shallow-rooted. These characteristics mean that teak plantations tend to expose the soil surface to erosive and desiccative forces more than does the native vegetation cover.

One of the main exponents of the role that human agency has played in lateritization in the tropical world has been Gourou (1961: 21–2). Although he may be guilty of exaggerating the extent and significance of laterite, Gourou gives many examples from low latitudes of falling agricultural productivity resulting from the onset of lateritization. It is worth quoting him at length:

On the whole, laterite is hostile to agriculture owing to its sterility and compactness. All tropical countries have not reached the same degree of lateritic 'suicide', but when the evolution has advanced a considerable way, man is placed in very strange conditions . . . Laterite is a pedological leprosy. Man's activities aggravate the dangers of laterite and increase the rate of the process of lateritization. To begin with, erosion when started by negligent removal of the forest simply wears away the friable and relatively fertile soil which would otherwise cover the laterite and support forest or crops . . . The forest checks the formation of the laterite in various ways. The trees supply plenty of organic matter and maintain a good proportion of humus in the soil. The action of capillary attraction is checked by the loosening of the soil; and the bases are retained through the absorbent capacity of humus. The forest slows down evaporation from the soil . . . it reduces percolation and consequently leaching. Lastly, the forest may improve the composition of the soil by fixing atmospheric dust.

Accelerated podzolization and acidification

There is an increasing amount of evidence that the introduction of agriculture, deforestation and pastoralism to parts of upland Western Europe promoted some major changes in soil character: notably an increase in the development of acidic and podzolized conditions, associated with the development of peat bogs. Climatic changes of the type envisaged by Blytt and Serander and later workers (see Goudie, 1972a: 94) may have played a role, as could progressive leaching of Devensian (last glacial) drifts during the passage of the Holocene. But the association in time and space of human activities with soil deterioration is becoming increasingly clear (Evans et al., 1975).

Replacing the natural forest vegetation with cultivation and pasture, human societies set in train various related processes, especially on base-poor materials. First, the destruction of deep-rooting trees curtailed the enrichment of the surface of the soil by bases brought up from the deeper layers. Second, the use of fire to effect forest clearance may have released nutrients in the form of readily soluble salts, some of which were inevitably lost in drainage, especially in soils poor in colloids (Dimbleby, 1974). Third, the taking of crops and animal products depleted the soil reserves to an extent probably greater than that arising from any of the manuring practices of prehistoric settlements. Fourth, as the soil degraded, the vegetation which invaded – especially bracken and heather – itself tended to produce a more acidic humus type of soil than the original mixed deciduous forest, and so continued the process.

Various workers now attribute much of the podzolization in upland Britain to such processes, and Dimbleby has concluded that 'although a few soils have been podzols since the Atlantic period, the majority are secondary, having arisen as a result of man's assault on the landscape, particularly in the Bronze Age' (see also Bridges, 1978).

The development of podzols, by impeding downward percolating waters, may have accelerated the formation of peats, which tend to develop where there is waterlogging through impeded drainage. Many peat bogs in highland Britain appear to coincide broadly in age with the first major land-clearance episodes (Moore, 1973; Merryfield and Moore, 1971). Another fact which would have contributed to their development is that when a forest canopy is removed (as by deforestation) the transpiration demand of the vegetation is reduced and less rainfall is intercepted, so that the supply of groundwater is increased, aggravating any waterlogging.

However, the role of natural processes must not be totally forgotten, and Ball's assessment would seem judicious (1975: 26):

It seems to be on balance that the highland trends in soil formation due to climate, geology and relief have been clearly running in the direction of leaching, acidity, podzolization, gleying and peat formation. For the British highlands generally, man has only intervened to hasten or slow the rate of these trends, rather than being in a position to alter the whole trend from one pedogenetic trend to another.

Moreover, it would be plainly misleading to stress only the deleterious effects of human actions on European soils. Traditional agricultural systems have often employed laborious techniques to augment soil fertility and to reduce such properties as undesirable acidity. In Britain, for example, the addition of chalk to light sandy land goes back at least to Roman times and the marl pits from which the chalk was dug are a striking feature of the Norfolk landscape, where Prince (1962) has identified at least 27,000 hollows. Similarly, in the Netherlands, Germany and Belgium there are soils which for centuries (certainly more than a thousand years) have been built up (often over 50 centimetres) and fertilized with a mixture of manure, sods, litter or sand. Such soils are called Plaggen soils (Pope, 1970). Plaggen soils also occur in Ireland, where the addition of sea-sand to peat was carried out in pre-Christian times. Likewise, before European settlement in New Zealand, the Maoris used thousands of tons of gravel and sand, carried in flax baskets, to improve soil structure (Cumberland, 1961).

Another type of soil which owes much to human influence is the category called 'paddy soil'. Long-continued irrigation, levelling and manuring of terraced land in China and elsewhere has changed the nature of the pre-existing soils in the area. Among the most important modifications that have been recognized (Gong, 1983) are an increase in organic matter, an increase in base saturation, and the translocation and reduction of iron and manganese.

Some soils are currently being acidified by air pollution and the deposition of acid precipitation (see also chapter 7). Many soils have a resistance to acidification because of their buffering capacity, which enables them to neutralize acidity. However, this resistance very much depends on soil type and situation, and soils which have a low buffering capacity because of their low calcium content (as, for example, on granite), and which are subjected to high levels of precipitation, may build up high levels of acidity.

The concept of critical loads has been developed. They are defined as exposures below which significant harmful effects on sensitive elements of the environment do not occur according to currently available knowledge. The critical load for sensitive forest soils on gneiss, granite or other slow-weathering rocks is often less than 3 kilograms of sulphur per hectare per year. In some of the more polluted parts of central Europe, the rates of sulphur deposition may be between 20 and 100 kilograms per hectare per year (Ågren and Elvingson, 1996).

The immediate impact of high levels of acid input to soils is to increase the exchange between hydrogen (H+) ions and the nutrient cations, such as potassium (K+), magnesium (M++) and calcium (Ca++). As a result of this exchange, such cations can be quickly leached from the soil, along with the sulphate from the acid input. This leaching leads to nutrient deficiency. Acidification also leads to a change in the rate at which dead organic matter is broken down by soil microbes. It can also render some ions, such as aluminium, more mobile and this has been implicated in the phenomenon of forest decline (see chapter 2).

Soil structure alteration

One of the most important features of a soil, in terms of both its suitability for plant growth and its inherent erodibility, is its structure. There are many ways in which humans can alter this, especially by compacting it with agricultural machinery, by the use of recreation vehicles and by changing its chemical character through irrigation. Soil compaction, which involves the compression of a mass of soil into a smaller volume, tends to increase the resistance of soil to penetration by roots and emerging seedlings, and limits oxygen and carbon dioxide exchange between the root zone and the atmosphere. Moreover, it reduces the rate of water infiltration into the soil, which may change the soil moisture status and accelerate surface runoff and soil erosion (Chancellor, 1977). For example, the effects of the passage of vehicles on some soil structural properties are shown in table 4.6. Excessive use of heavy agricultural machinery is perhaps the major cause of soil compaction, and most procedures in the cropping cycle, from tillage and seedbed preparation, through drilling, weeding and agrochemical application to harvesting, are now largely mechanized, particularly in the developed world. Most notable of all is the reduction that is caused in soil infiltration capacity, which may explain why vehicle movements can often lead to gully development. Whether one is dealing with primitive sledges (as in Swaziland), or with the latest recreational toys of leisured Californian adolescents, the effects may be comparable.

Table 4.6   Change to soil properties resulting from the passage of 100 motorcycles in New Zealand

Soil property	Total no. of sites	No. of sites with significant change*	Mean percentage at significant sites	No. of sites with significant increase	No. of sites with significant decrease	Main direction of change	Mean percentage change in direction
Infiltration capacity	16	16 (100%)	84.3	3	13	Decrease	78.1
Bearing capacity	21	10 (48%)	22.8	2	8	Decrease	18.6
Soil moisture	20	14 (70%)	15.5	5	9	Decrease	16.7
Dry bulk density	19	11 (58%)	13.6	9	2	Decrease	13.3

* Change is significant when greater than: 10% for infiltration capacity; 10% for bearing capacity; 5% for soil moisture; 5% for bulk density.
Source: Crozier et al., 1978, table 1.

Table 4.7   Rates of infiltration on grazed and ungrazed lands in America

	Rate of infiltration (mm h^{-1})	
Site	Ungrazed	Heavily grazed
Montana	2.5–66.0	5.1–15.2
Oklahoma	134.6–309.9	40.6–83.8
Colorado	40.6–83.8	20.3–30.5
Montana	109.2–185.4	20.3–96.5
Wyoming	30.5–38.1	17.8–30.5
Louisiana	45.7	17.8
Kansas	33.0	20.3
Arizona	40.6	30.5

Source: processed by author from data in Gifford and Hawkins, 1978.

Grazing is another activity that can damage soil structure through trampling and compaction. Heavily grazed lands tend to have considerably lower infiltration capacities than those found in ungrazed lands. This is indicated for some American examples in table 4.7. Trimble and Mendel (1995) have drawn particular attention to the capability that cows have to cause soil compaction. Given their large mass, their small hoof area, and the stress that may be imposed on the ground when they are scrambling up a slope, they are probably remarkably effective in compacting soils. The removal of vegetation cover and associated litter also changes infiltration capacity, since cover protects the soil from packing by raindrops and provides organic matter for binding soil particles together in open aggregates. Soil fauna that live on the organic matter assist this process by churning together the organic material and mineral particles. Dunne and Leopold have ranked the relative influence of different land-use types on infiltration (1978, table 6.2, after US Soil Conservation Service):

Highest infiltration	Woods, good
	Meadows
	Woods, fair
	Pasture, good
	Woods, poor
	Pasture, fair
	Small grains, good rotation
	Small grains, poor rotation
	Legumes after row crops
	Pasture, poor
	Row crops, good rotation
	(more than one quarter in hay or sod)
	Row crops, poor rotation (one quarter or
	less in hay or sod)
Lowest infiltration	Fallow

In general, experiments show that reafforestation improves soil structure, especially the pore volume of the soils (see e.g. Challinor, 1968). Ploughing is also known to produce a compacted layer at the base of the zone of ploughing (Baver et al., 1972). This layer has been termed the 'plough sole'. The normal action of the plough is to leave behind a loose surface layer and a dense subsoil where the soil aggregates have been pressed together by the sole of the plough. The compacting action can be especially injurious when the depth of ploughing is both constant and long-term, and when heavy machinery is used on wet ground (Greenland, 1977).

On the other hand, for many centuries farmers have achieved improvements in soil structure by deliberate practice, particularly with a view to developing the all-important crumb structure. In pre-Roman times people in Britain and France added lime to heavy clay soils, while the agricultural improvers of the eighteenth and nineteenth centuries improved the structure of sandy heath soils by adding clay, and of clay soils by adding calcium carbonate (marl). They also compacted such soils and added binding organic matter by breeding sheep and feeding them with turnips and other fodder plants (Russell, 1961).

In the modern era attempts have been made to reduce soil crusting by applying municipal and animal wastes to farm land, by adding chemicals such as phosphoric acid, and by adopting a cultivation system of the no-tillage type. The last practice is based on the idea that the use of herbicides has eliminated much of the need for tillage and cultivation in row crops; seeds are planted directly into the soil without ploughing, and weeds are controlled by the herbicides. With this method less bare soil is exposed and heavy farm machinery is less likely to create soil compaction problems (see Carlson, 1978, for some of these methods).

Soil structures may also be modified to increase water run-off, particularly in arid zones where the runoff obtained can augment the meagre water supply for crops, livestock, industrial and urban reservoirs, and groundwater recharge projects. In the Negev (in modern Israel) farming was practised in this way, especially in the Nabatean and the Romano-Byzantine periods (about 300 BC to AD 630), and attempts were made to induce runoff by clearing the surface gravel of the soil and heaping it into thousands of mounds. This exposed the finer silty soil beneath, facilitating soil crust formation by raindrop impact, decreasing infiltration capacity and reducing surface roughness, so that runoff increased (Evenari et al., 1971). Today a greater range of techniques are available for the same end: soils can be smoothed and compacted by heavy machinery, soil crusting can be promoted by dispersion of soil colloids with sodium salts, the permeability of the soil surface can be reduced by applying water-repellent materials, and soil pores can be filled with binders (Hillel, 1971).

Soil drainage and its impact

Soil drainage 'has been a gradual process and the environmental changes to which it has led have, by reason of that gradualness, often passed unnoticed' (Green, 1978: 171). To be sure, the most spectacular feats of drainage – arterial drainage – involving the construction of veritable rivers and large dike systems, as seen in the Netherlands and the Fenlands of eastern England, have received attention. However, more widespread than arterial drainage, and sometimes independent of it, is the drainage of individual fields, either by surface ditching or by underdrainage with tile pipes and the like. Green (1978) has attempted to map the areas of drained agricultural land. In Finland, Denmark, Great Britain, the Netherlands and Hungary, the majority of agricultural land is drained (figure 4.8).

In Britain underdrainage was promoted by government grants and reached a peak of about one million hectares per year in the 1970s in England and Wales. More recently government subsidies have been cut and the uncertain economic future of farming has led to a reduction in farm expenditure. Both tendencies have led to a reduction in field drainage, which may now be taking place at only 40,000 hectares per year (Robinson, 1990).

The drainage conditions of the soil have also frequently been altered by the development of ridge and furrow patterns created by ploughing (plate 4.3). Such patterns are a characteristic feature of many of the heavy soils of lowland England where large areas, especially of the Midland lowlands, are striped by long, narrow ridges of soil, lying more or less parallel to each other

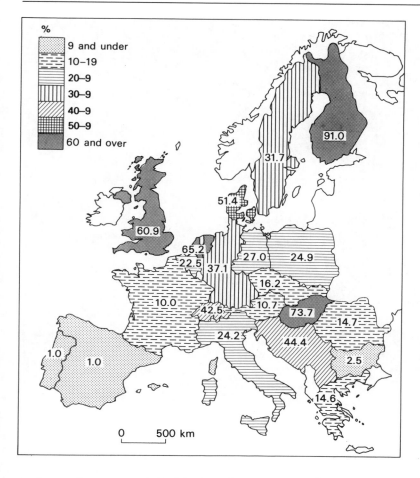

Figure 4.8
Percentage of drained
agricultural land in Europe.
There are no data for the
blank areas (after Green,
1978, figure 1)

and usually arranged in blocks of approximately rectangular
shape. They were formed by ploughing with heavy ploughs pulled
by teams of oxen; the precise mechanism has been described
with clarity by Coones and Patten (1986: 154):

The 'heavy' plough cut through the soil and turned the furrow-slice to
the right. Consequently, if the field were simply ploughed up and down
by working progressively across it from one side to the other, each
double run would merely produce a tiny ridge overlying an unploughed
ribbon of ground, the soil being thrown inwards from the two furrows
on either side. This would have been a pointless exercise, so instead
the field was ploughed as a number of separate units, each with its
own central axis. Ploughing started along the centre line and worked
gradually away from it, turning the soil inwards from both edges as
the plough went along one side and back down the other. Every sod
on each side was therefore turned the same way and laid against the
previous one. It has been suggested that once the plough team had
progressed to a point a certain distance from the original centre line,
the distance it was obliged to walk along the headlands to the other

Plate 4.3
Over large tracts of lowland England, the landscape is dominated by systems of ridge and furrow, as shown in this air photograph of Hollingdon, Buckinghamshire

side for the return run was such that it was easier to start a new circuit, based upon the axis of an adjacent unit. Repeated ploughing of exactly the same units year after year gradually led to the formation of adjacent ridges, separated from each other by furrows produced where the furrow-slices were turned away from each other.

Soil drainage has been one of the most successful ways in which communities have striven to increase agricultural productivity; it was practised by Etruscans, Greeks and Romans (Smith, 1976). Large areas of marshland and flood plain have been drained to human advantage. By leading water away, the water table is lowered and stabilized, providing greater depth for the root zone. Moreover, well-drained soils warm up earlier in the spring and thus permit earlier planting and germination of crops. Farming is easier if the soil is not too wet, since the damage to crops by winter freezing may be minimized, undesirable salts carried away, and the general physical condition of the soil improved. In addition, drained land may have certain inherent virtues: tending to be flat, it is less prone to erosion and more amenable to mechanical cultivation. It will also be less prone to drought risk than certain other types of land (Karnes, 1971). Paradoxically, by reducing the area of saturated ground, drainage can alleviate flood risk in some situations by limiting the extent of a drainage basin that generates saturation excess overland flow.

Conversely, soil drainage can have quite undesirable or unplanned effects. For example, some drainage systems, by raising drainage densities, can increase flood risk by reducing the distance over which unconcentrated overland flow (which is relatively slow) has to travel before reaching a channel (where flow is relatively fast). In central Wales, for instance, the establishment of drainage ditches in peaty areas to enable afforestation has tended to increase flood peaks in the rivers Wye and Severn (Howe et al., 1966).

Drainage can also cause long-term damage to soil quality. A fall in water level in organic soils can lead to the oxidation and eventual disappearance of peaty materials, which in the early stages of post-drainage use may be highly productive. This has occurred in the English Fenland and in the Everglades of Florida, where drainage of peat soils has led to a subsidence in the soil of 32 millimetres per year (Stephens, 1956).

The soil climate of neighbouring areas may also be modified when the water tables of drained land are lowered. This has been known to create problems for forestry in areas of marginal water availability.

Soil moisture content can also determine the degree to which soils are subjected to expansion and contraction effects, which in turn may affect engineering structures in areas with expansive soils (Holtz, 1983). Soils containing sodium montmorillonite type clays, when drained or planted with large trees, may dry out and cause foundation problems (Driscoll, 1983).

Some of the most contentious effects of drainage are those associated with the reduction of wetland wildlife habitats. In Britain this has become an important political issue, especially in the context of the Somerset Levels and the Halvergate Marshes.

Soil fertilization

The chemistry of soils has been changed deliberately by the introduction of chemical fertilizers. The employment of chemical fertilizers on a large scale is little more than 150 years old. In the early nineteenth century nitrates were first imported from Chile, and sulphate of ammonia was produced only after the 1820s as a by-product of coal gas manufacture. In 1843 the first fertilizer factory was established at Deptford Creek (London), but for a long time superphosphates were the only manufactured

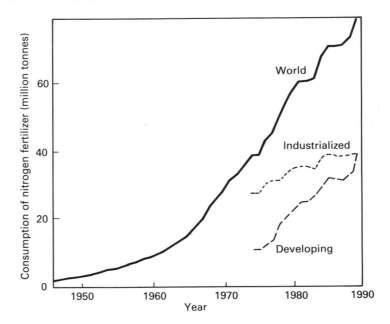

Figure 4.9
Global consumption of
nitrogen fertilizer over the last
four decades (after various
sources in Conway and Pretty,
1991, figure 4.1)

fertilizers in use. In the twentieth century synthetic fertilizers, particularly nitrates, were developed, notably by Scandinavian countries which used their vast resources of water power. Potassic fertilizers came into use much later than the phosphatic and nitrogenous; nineteenth-century farmers hardly knew them (Russell, 1961). The rise in nitrogen fertilizer consumption for the world as a whole over the last four decades is shown in figure 4.9.

The use of synthetic fertilizers has greatly increased agricultural productivity in many parts of the world, and remarkable increases in yields have been achieved. It is also true that in some circumstances proper fertilizer use can help minimize erosion by ensuring an ample supply of roots and plant residues, particularly on infertile or partially degraded soils (Bockman et al., 1990). On the other hand, the increasing use of synthetic fertilizers can create environmental problems such as water pollution, while their substitution for more traditional fertilizers may accelerate soil structure deterioration and soil erosion. One effect has been the increase in the water repellancy of some surface soil materials. This in turn can reduce soil infiltration rates and cause a consequential increase in erosion by overland flow (Conacher and Conacher, 1995: 39). However, fertilizers can promote soil acidity and may lead to deficiencies or toxic excesses of major nutrients and trace elements. They may also contain impurities, such as fluoride, lead, cadmium, zinc and uranium. Some of these heavy metals can inhibit water uptake and plant growth. They may also become concentrated in food crops, which can have important implications for human health.

Fires and soil

The importance and antiquity of fire as an agency through which
the environment is transformed requires that some attention be
given to the effects of fire on soil characteristics and on soil
erosion.

Fire has often been used intentionally to change soil properties,
and both the release of nutrients by fire and the value of ash
have long been recognized, notably by those involved in shifting
agriculture based on slash-and-burn techniques. Following cultivation, the loss of nutrients by leaching and erosion is very rapid
(Nye and Greenland, 1964), and this is why after only a few
years the shifting cultivators have to move on to new plots. Fire
rapidly alters the amount, form and distribution of plant nutrients in ecosystems, and, compared to normal biological decay
of plant remains, burning rapidly releases some nutrients into a
plant-available form. Indeed, the amounts of P, Mg, K and Ca
released by burning forest and scrub vegetation are high in relation to both the total and available quantities of these elements
in soils (Raison, 1979). In forests, burning often causes the pH
of the soil to rise by three units or more, creating alkaline conditions where formerly there was acidity. Burning also leads to
some direct nutrient loss by volatilization and convective transfer of ash, or by loss of ash to water erosion or wind deflation.
The removal of the forest causes soil temperatures to increase
because of the absence of shade, so that humus is often lost at a
faster rate than it is formed (Grigg, 1970).

Soil erosion: general considerations

Loss of soil humus, whether as a result of fire, drainage, deforestation or ploughing, is an especially serious manifestation of
human alteration of soil. As table 4.8 indicates, humus has many
beneficial effects on both the chemical and the physical properties of soil. Its removal by human activity can be a potent
contributory cause of soil erosion.

The scale of accelerated soil erosion that has been achieved by
human activities has been well summarized by Myers (1988: 6):

Since the development of agriculture some 12,000 years ago, soil erosion
is said by some to have ruined 4.3 million km² of agricultural lands, or
an area equivalent to rather more than one-third of today's crop-lands
. . . the amount of agricultural land now being lost through soil erosion,
in conjunction with other forms of degradation, can already be put at
a minimum of 200,000 km² per year.

In the late 1930s, towards the end of the so-called Dust Bowl
years, Sauer conducted a campaign against what he called the

Table 4.8 The beneficial properties of humus

Property	Explanation	Effect
(a) Chemical properties		
Mineralization	Decomposition of humus yields CO_2, NH_4^+, NO_3^-, PO_4^{3-}, and SO_4^{2-}	A source of nutrient elements for plant growth
Cation exchange	Humus has negatively charged surfaces which bind cations such as Ca^+ and K^+	Improved cation exchange capacity (CEC) of soil. From 20 to 70 per cent of the CEC of some soils (e.g. Mollisols) is attributable to humus
Buffer action	Humus exhibits buffering in slightly acid, neutral and alkaline ranges	Helps to maintain a uniform pH in the soil
Acts as matrix for biochemical action in soil	Binds other organic molecules electrostatically or by covalent bonds	Affects bioactivity persistence and biodegradability of pesticides
Chelation	Forms stable complexes with Cu^{2+}, Mn^{2+}, Zn^{2+}, and other polyvalent cations	May enhance the availability of micronutrients to higher plants
(b) Physical properties		
Water retention	Organic matter can hold up to 20 times its weight in water	Helps prevent drying and shrinking. May significantly improve the moisture-retaining properties of sandy soils
Combination with clay	Cements soil particles into structural units called aggregates	Improves aeration. Stabilizes structure. Increases permeability
Colour	The typical dark colour of many soils is caused by organic matter	May facilitate warming

Source: Swift and Sanenez, 1984, table 2.

'destructive exploitation in modern colonial expansion' (Sauer, 1938: 497), and placed soil erosion squarely into the context of geography:

We may well consider whether the theme of soil erosion should not be moved up to the first category of problems before the geographers of the world. It is very important for the future of mankind. It has critical significance for certain chapters of historical geography. The physical processes involved are poorly observed and generalized and their study will undoubtedly shake somewhat the rather lethargic present position of geomorphology . . . best of all the subject is suited to a 'hologeographic' approach in which the development of surface conditions of specific localization is examined as an interaction of identified physical and economic (i.e. Wirtschaft) processes.

That soil erosion is a major and serious aspect of the human role in environmental change is not to be doubted. There is a long history of weighty books and papers on the subject (see e.g. Marsh, 1864; Bennett, 1938; Jacks and Whyte, 1939; Morgan, 1995). Although many techniques have been developed to reduce the intensity of the problem (see Hudson, 1971) it appears to remain intractable. As L. J. Carter (1977: 409) has reported of the United States:

although nearly $15 billion has been spent on soil conservation since the mid-1930s, the erosion of croplands by wind and water ... remains one of the biggest, most pervasive environmental problems the nation faces. The problem's surprising persistence apparently can be attributed at least in part to the fact that, in the calculation of many farmers, the hope of maximizing short-term crop yields and profits has taken precedence over the longer term advantages of conserving the soil. For even where the loss of topsoil has begun to reduce the land's natural fertility and productivity, the effect is often masked by the positive response to heavy application of fertilizer and pesticides, which keep crop yields relatively high.

Although construction, urbanization, war, mining and other such activities are often significant in accelerating the problem, the prime causes of soil erosion are deforestation and agriculture. Pimentel (1976) estimated that in the United States soil erosion on agricultural land operates at a rate of about 30 tonnes per hectare per year, which is approximately eight times quicker than topsoil is formed. He calculated that water runoff delivers around 4 billion tonnes of soil each year to the rivers of the 48 contiguous states, and that three-quarters of this comes from agricultural land. He estimated that another billion tonnes of soil are eroded by the wind, a process which created the Dust Bowl of the 1930s. More recently, Pimentel et al. (1995) have argued that about 90 per cent of US cropland is losing soil above the sustainable rate, that about 54 per cent of US pasture land is overgrazed and subject to high rates of erosion, and that erosion costs about $44 billion each year. They argue that on a global basis soil erosion costs the world about $400 billion each year.

One serious consequence of accelerated erosion is the sedimentation that takes place in reservoirs, shortening their lives and reducing their capacity. Many small reservoirs, especially in semi-arid areas, appear to have an expected life of only 30 years of even less (see e.g. Rapp et al., 1972). Soil erosion also has serious implications for soil productivity. A reduction in soil thickness reduces available water capacity and the depth through which root development can occur. The water-holding properties of the soil may be lessened as a result of the preferential removal of organic material and fine sediment. Hardpans and duricrusts may become exposed at the surface, and provide a barrier to root penetration. Furthermore, splash erosion may cause soil compaction and crusting, both of which may be unfavourable to germination and seedling establishment. Erosion also removes nutrients preferentially from the soil. Some damage may be caused by associated excessive sedimentation, while wind erosion may lead to the direct sandblasting of crops. Finally, extreme erosion may lead to wholesale removal of both seeds and fertilizer. Stocking (1984) provides a useful review of these problems.

Plate 4.4
The removal of vegetation in Swaziland creates spectacular gully systems, which in southern Africa are called *dongas*. The smelting of local iron ores in the early nineteenth century required the use of a great deal of firewood which may have contributed to the formation of this example

Soil erosion associated with deforestation and agriculture

Forests protect the underlying soil from the direct effects of rainfall, generating what is generally an environment in which erosion rates tend to be low. The canopy plays an important role by shortening the fall of raindrops, decreasing their velocity and thus reducing kinetic energy. There are some examples of certain types (e.g. beech) in certain environments (e.g. maritime temperate) creating large raindrops, but in general most canopies reduce the erosion effects of rainfalls. Possibly more important than the canopy in reducing erosion rates in forest is the presence of humus in forest soils (Trimble, 1988), for this both absorbs the impact of raindrops and has an extremely high permeability. Thus forest soils have high infiltration capacities. Another reason that forest soils have an ability to transmit large quantities of water through their fabrics is that they have many macropores produced by roots and their rich soil fauna. Forest soils are also well aggregated, making them resistant to both wetting and water drop impact. This superior degree of aggradation is a result of the presence of considerable organic material, which is an important cementing agent in the formation of large water-stable aggregates. Furthermore, earthworms also help to produce large aggregates. Finally, deep-rooted trees help to stabilize steep slopes by increasing the total shear strength of the soils.

It is therefore to be expected that with the removal of forest, for agriculture or for other reasons, rates of soil loss will rise (plate 4.4) and mass movements will increase in magnitude and frequency. The rates of erosion that result will be particularly high if the ground is left bare; under crops the increase will be

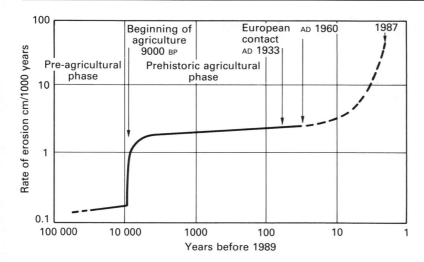

Figure 4.10
Rates of erosion in Papua New
Guinea in the Holocene
derived from rates of
sedimentation in Kuk Swamp
(after Hughes et al., 1991,
figure 5, with modifications)

less marked. Furthermore, the method of ploughing, the time of planting, the nature of the crop and the size of the fields will all have an influence on the severity of erosion.

It is seldom that we have reliable records of rates of erosion over a sufficiently long time-span to show just how much human activities have accelerated these effects. Recently, however, techniques have been developed which enable rates of erosion on slopes to be gauged over a lengthy time-span by means of dendrochronological techniques that date the time of root exposure for suitable species of tree. In Colorado in the United States, Carrara and Carroll (1979) found that rates over the last 100 years have been about 1.8 millimetres per year, whereas in the previous 300 years rates were between 0.2 and 0.5 millimetres per year, indicating an acceleration of about sixfold. This great jump has been attributed to the introduction of large numbers of cattle to the area about a century ago.

Another way of obtaining long-term rates of soil erosion is to look at rates of sedimentation on continental shelves and on lake floors. The former method was employed by Milliman et al. (1987) to evaluate sediment removal down the Yellow River in China during the Holocene. They found that, because of accelerated erosion, rates of sediment accumulation on the shelf over the last 2,300 years have been ten times higher than those for the rest of the Holocene (i.e. since around 10,000 BP).

Another good example of using long-term sedimentation rates to infer long-term rates of erosion is provided by Hughes et al.'s (1991) study of the Kuk Swamp in Papua New Guinea (figure 4.10). They identify low rates of erosion until 9000 BP, when, with the onset of the first phase of forest clearance, erosion rates increased from 0.15 centimetres per thousand years to about 1.2 centimetres per thousand years. Rates remained

Table 4.9 Runoff and erosion under various covers of vegetation in parts of Africa

Locality	Average annual rainfall (mm)	Slope (%)	Annual runoff (%)			Erosion ($t\ ha^{-1}\ yr^{-1}$)		
			A	B	C	A	B	C
Ouagadougou (Burkina Faso)	850	0.5	2.5	2–32	40–60	0.1	0.6–0.8	10–20
Sefa (Senegal)	1,300	1.2	1.0	21.2	39.5	0.2	7.3	21.3
Bouake (Ivory Coast)	1,200	4.0	0.3	0.1–26	15–30	0.1	1–26	18–30
Abidjan (Ivory Coast)	2,100	7.0	0.1	0.5–20	38	0.03	0.1–90	108–170
Mpwapwa* (Tanzania)	c.570	6.0	0.4	26.0	50.4	0	78	146

Note: A = forest or ungrazed thicket
   B = crop
   C = barren soil.
* From Rapp et al., 1972, figure 5, p. 259.
Source: after Charreau, table 5.5 p. 153 in Greenland and Lai, 1977.

relatively stable until the last few decades when, following European contact, the extension of anthropogenic grasslands, subsistence gardens and coffee plantations has produced a rate that is very markedly higher: 34 centimetres per thousand years.

A further good long-term study of the response rates of erosion to land use changes is provided by a study undertaken on the North Island of New Zealand by Page and Trustrum (1997). During the last 2,000 years of human settlement their catchment has undergone a change from indigenous forest to fern/scrub following Polynesian settlement (*c.*560 BP) and then a change to pasture following European settlement (AD 1878). Sedimentation rates under European pastoral land use are between five and six times the rates that occurred under fern/scrub and between eight and seventeen times the rate under indigenous forest. In a broadly comparable study, Sheffield et al. (1995) looked at rates of infilling of an estuary fed by a steepland catchment in another part of New Zealand. In pre-Polynesian times rates of sedimentation were 0.1 millimetres per year, during Polynesian times the rate climbed to 0.3 millimetres per year, and since European land clearance in the 1880s the rate has shot up to 11 millimetres per year.

In a more general sense there are plainly huge difficulties in estimating erosion rates in pre-human times, but in a recent analysis McLennan (1993) has estimated that the pre-human suspended sediment discharge from the continents was about 12.6×10^{15} grams per year, which is about 0.6 of the present figure.

Table 4.9, which is based on data from tropical Africa, shows the comparative rates of erosion for three main types of land use: trees, crops and barren soil. It is very evident from these data that under crops, but more especially when ground is left

Table 4.10 Debris-avalanche erosion in forest, clear-cut and roaded areas

Site	Period of records (years)	Area (%)	Area (km²)	No. of slides	Debris-avalanche erosion (m³ km⁻² yr⁻¹)	Rate of debris-avalanche erosion relative to forested areas
Stequaleho Creek, Olympic Peninsula						
Forest	84	79	19.3	25	71.8	× 1.0
Clear-cut	6	18	4.4	0	0	0
Road	6	3	0.7	83	11,825	× 165
TOTAL	–	–	24.4	108	–	–
Alder Creek, western Cascade Range, Oregon						
Forest	25	70.5	12.3	7	45.3	× 1.0
Clear-cut	15	26.0	4.5	18	117.1	× 2.6
Road	15	3.5	0.6	75	15,565	× 344
TOTAL	–	–	17.4	100	–	–
Selected drainages, Coast Mountains, south-west British Columbia						
Forest	32	88.9	246.1	29	11.2	× 1.0
Clear-cut	32	9.5	26.4	18	24.5	× 2.2
Road	32	1.5	4.2	11	282.5	× 25.2
TOTAL	–	–	276.7	58	–	–
H. J. Andrews Experimental Forest, western Cascade Range, Oregon						
Forest	25	77.5	49.8	31	35.9	× 1.0
Clear-cut	25	19.3	12.4	30	132.2	× 3.7
Road	25	3.2	2.0	69	1,772	× 49
TOTAL	–	–	64.2	130	–	–

Source: after Swanston and Swanson, 1976, table 4.

bare or under fallow, soil erosion rates are greatly magnified. At the same time, and causally related, the percentage of rainfall that becomes runoff is increased.

In some cases the erosion produced by forest removal will be in the form of widespread surface stripping. In other cases the erosion will occur as more spectacular forms of mass movement, such as mudflows, landslides and debris avalanches. Some detailed data on debris-avalanche production in North American catchments as a result of deforestation and forest road construction are presented in table 4.10. They illustrate the substantial effects created by clear-cutting and by the construction of logging roads. It is indeed probable that a large proportion of the erosion associated with forestry operations is caused by road construction, and care needs to be exercised to minimize these effects. The digging of drainage ditches in upland pastures and peat moors to permit tree planting in central Wales has also been found to cause accelerated erosion (Clarke and McCulloch, 1979), while the elevated sediment loads can cause reservoir pollution (Burt et al., 1983).

In general, the greater the deforested proportion of a river basin the higher the sediment yield per unit area will be. In the United States the rate of sediment yield appears to double for every 20 per cent loss in forest cover.

Table 4.11 Annual rates of soil loss (tonnes per hectare) under different land-use types in eastern England

Plot	Splash	Overland flow	Rill	Total
1 *Bare soil*				
Top slope	0.33	6.67	0.10	7.10
Mid-slope	0.82	16.48	0.39	17.69
Lower slope	0.62	14.34	0.06	15.02
2 *Bare soil*				
Top slope	0.60	1.11	–	1.71
Mid-slope	0.43	7.78	–	8.21
Lower slope	0.37	3.01	–	3.38
3 *Grass*				
Top slope	0.09	0.09	–	0.18
Mid-slope	0.09	0.57	–	0.68
Lower slope	0.12	0.05	–	0.17
4 *Woodland*				
Top slope	–	–	–	0.00
Mid-slope	–	0.012	–	0.012
Lower slope	–	0.008	–	0.008

Source: from Morgan, 1977.

Soil erosion resulting from deforestation and agricultural practice is often thought to be especially serious in tropical areas or semi-arid areas (see Moore, 1979, for a good case study). However, measurements by Morgan (1977) on sandy soils in the English East Midlands near Bedford indicate that rates of soil loss under bare soil on steep slopes can reach 17.69 tonnes per hectare per year, compared with 2.39 under grass and nothing under woodland (table 4.11), and subsequent studies have demonstrated that water-induced soil erosion is a substantial problem, in spite of the relatively low erosivity of British rainfall. Walling and Quine (1991: 123) have identified the following farming practices as contributing to this developing problem:

1 Ploughing up of steep slopes that were formerly under grass, in order to increase the area of arable cultivation.
2 Use of larger and heavier agricultural machinery which has a tendency to increase soil compaction.
3 Removal of hedgerows and the associated increase in field size. Larger fields cause an increase in slope length with a concomitant increase in erosion risk.
4 Declining levels of organic matter resulting from intensive cultivation and reliance on chemical fertilizers, which in turn lead to reduced aggregate stability.
5 Availability of more powerful machinery which permits cultivation in the direction of maximum slope rather than along the contour. Rills often develop along tractor and implement wheelings and along drill lines.

6   Use of powered harrows in seedbed preparation and the
    rolling of fields after drilling.
7   Widespread introduction of autumn-sown cereals to replace
    spring-sown cereals. Because of their longer growing season,
    winter cereals produce greater yields and are therefore more
    profitable. The change means that seedbeds are exposed with
    little vegetation cover throughout the period of winter rainfall.

Water is not the only active process creating accelerated
erosion in eastern England, though it is important (Evans and
Northcliffe, 1978). Ever since the 1920s dust storms have been
recorded in the Fenlands, the Brecklands, East Yorkshire (Radley
and Sims, 1967) and Lincolnshire (see e.g. Arber, 1946), and
they seem to be occurring with increasing frequency. The storms
result from changing agricultural practices, including the substitution of artificial fertilizers for farmyard manure, a reduction
in the process of 'claying', whereby clay was added to the peat
to stabilize it, the removal of hedgerows to facilitate the use of
bigger farm machinery and, perhaps most importantly, the
increased cultivation of sugar beet. This crop requires a fine
tilth and tends to leave the soil relatively bare in early summer
compared with other crops (Pollard and Miller, 1968).

Possibly the most famous case of soil erosion by deflation was
the Dust Bowl of the 1930s in the United States (see figure 4.11).
In part this was caused by a series of hot, dry years which
depleted the vegetation cover and made the soils dry enough to
be susceptible to wind erosion. The effects of this drought were
gravely exacerbated by years of overgrazing and unsatisfactory
farming techniques. However, perhaps the prime cause of the
event was the rapid expansion of wheat cultivation in the Great
Plains. The number of cultivated hectares doubled during the
First World War as tractors (for the first time) rolled out on to
the plains by the thousands. In Kansas alone wheat hectarage
increased from under 2 million hectares in 1910 to almost 5 million in 1919. After the war wheat cultivation continued apace,
helped by the development of the combine harvester and government assistance. The farmer, busy sowing wheat and reaping gold,
could foresee no end to his land of milk and honey; but the years
of favourable climate were not to last, and over large areas the
tough sod which exasperated the earlier homesteaders had given
way to friable soils of high erosion potential. Drought, acting
on damaged soils, created the 'black blizzards' (plate 4.5) which
have been so graphically described by Coffey (1978: 79–80):

There was something fantastic about a dust cloud that covered
1.35 m. square miles, stood three miles high and stretched from Canada to Texas, from Montana to Ohio – a cloud so colossal it obliterated
the sky . . . a four-day storm in May 1934 . . . transported some 300

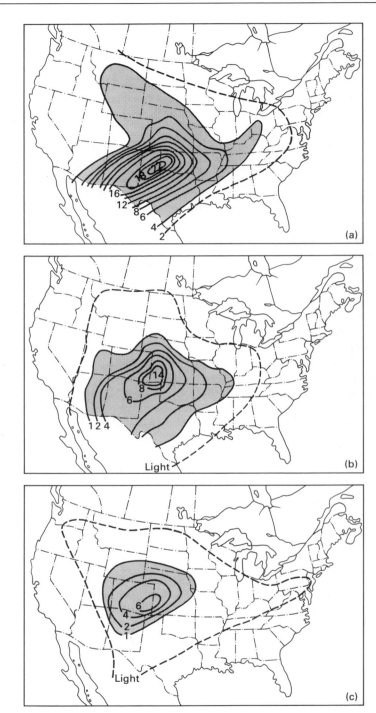

Figure 4.11
The concentration of dust
storms (number of days per
month) in the USA in 1939,
illustrating the extreme
localization over the High
Plains of Texas, Colorado,
Oklahoma and Kansas
(a) March
(b) April
(c) May
(after Goudie, 1983b)

Plate 4.5
The Dust Bowl created in 1930s America caused severe problems for the inhabitants of Cimarron County, Oklahoma

Plate 4.6
In the High Plains near Lubbock in Texas the effect of soil erosion and drifting was very evident in 1977. Note the vast fields and absence of windbreaks

million tonnes of dirt 1500 miles, darkened New York, Baltimore and Washington for five hours, and dropped dust not only on the President's desk in the White House, but also on the decks of ships some 300 miles out in the Atlantic . . . masses of dust began to billow into huge tumbling clouds ebony black at the base and muddy tan at the top, some so saturated with dust particles that ducks and geese caught in flight, suffocated; some turning the sky so black that chickens, thinking it night, would roost. Oklahoma counted 102 storms in the span of one year; North Dakota reported 300 in eight months.

Dust storms are still a serious problem in various parts of the United States; the Dust Bowl was not solely a feature of the 1930s (plate 4.6). Thus, for example, in the San Joaquin valley area of California in 1977 a dust storm caused extensive damage and erosion over an area of about 2,000 square kilometres. More than 25 million tonnes of soil were stripped from grazing land within a 24-hour period. While the combination of drought and a very high wind (as much as 300 kilometres per hour) provided the predisposing natural conditions for the stripping to occur, overgrazing and the general lack of windbreaks in the agricultural land played a more significant role. In addition, broad areas of land had recently been stripped of vegetation, levelled or ploughed up prior to planting. Other quantitatively less important factors included stripping of vegetation for urban expansion, extensive denudation of land in the vicinity of oilfields, and local denudation of land by vehicular recreation (Wilshire et al., 1981). One interesting observation made in the months after the dust storm was that in subsequent rain storms runoff occurred at an accelerated rate from those areas that had been stripped by the wind, exacerbating problems of flooding and initiating numerous gullies. Elsewhere in California dust yield has been considerably increased by mining operations in dry lake beds (Wilshire, 1980) and by disturbance of playas (Gill, 1996).

A comparable acceleration of dust storm activity has occurred in the CIS. After the 'virgin lands' programme of agricultural expansion in the 1950s, dust storm frequencies in the southern Omsk region increased on average by a factor of 2.5 and locally by factors of 5 to 6. Data on trends elsewhere are evaluated by Goudie (1983: 520) and Goudie and Middleton (1992).

Soil erosion produced by fire

Many fires are started by humans, either deliberately or accidentally, and because fires remove vegetation and expose the ground they tend to increase rates of soil erosion.

The burning of forests, for example, can, especially in the first years after the fire event, lead to high rates of soil loss (see table 4.12). Burnt forests often have rates of soil loss a whole order of magnitude higher than those of protected areas. Comparably large changes in soil erosion rates have been observed

Table 4.12 Soil erosion associated with *Calluna* (heather) burning on the North Yorkshire moors

Condition of Calluna or ground surface	Mean rate of litter accumulation (+) or erosion (−) (mm/year)	No. of observations
(a) *Calluna* 30–40 cm high. Complete canopy	+3.81	60
(b) *Calluna* 20–30 cm high. Complete canopy	+0.25	20
(c) *Calluna* 15–20 cm high. 40–100% cover	−0.74	20
(d) *Calluna* 5–15 cm high. 10–100% cover	−6.4	20
(e) Bare ground. Surface of burnt *Calluna*	−9.5	19
(f) Bare ground. Surface of peaty or mineral subsoil	−45.3	25

Source: data in Imeson, 1971.

to result from the burning of heather in the Yorkshire moors in northern England, and the effects of burning may be felt for the six years or more that may be required to regenerate the heather (*Calluna*). In the Australian Alps fire in experimental catchments has been found to lead to a greatly increased flow in the streams, together with a marked surge in the delivery of suspended load. Combining the two effects of increased flow rate and sediment yield, it was found that, after fire, the total sediment load was increased 1,000 times (Pereira, 1973). Likewise, watershed experiments in the chaparral scrub of Arizona, involving denudation by a destructive fire, indicated that whereas erosion losses before the fire were only 43 tonnes per square kilometre per year, after the fire they were between 50,000 and 150,000 tonnes per square kilometre per year. The causes of the marked erosion associated with chaparral burning are particularly interesting. There is normally a distinctive 'non-wettable' layer in the soils supporting chaparral. This layer, composed of soil particles coated by hydrophobic substances leached from the shrubs or their litter, is normally associated with the upper part of the soil profile (Mooney and Parsons, 1973), and builds up through time in the unburned chaparral. The high temperatures which accompany chaparral fires cause these hydrophobic substances to be distilled so that they condense on lower soil layers. This process results in a shallow layer of wettable soil overlying a non-wettable layer. Such a condition, especially on steep slopes, can result in severe surface erosion.

In chaparral terrain it is possible to envisage a fire-induced sediment cycle (Graf, 1988: 243). It starts with a fire that destroys the scrub and the root net, and changes surface soil properties in the way already discussed. After the fire, a precipitation event of low magnitude (with a return interval of around one or two years) is sufficient to induce extensive sheet and rill erosion, which removes enough soil to retard vegetation recovery. Eventually, a larger precipitation event occurs (with a return interval of around five to ten years) and, because of limited vegetation cover, produces severe debris slides. Slowly the vegetation cover re-establishes itself, and erosion rates diminish. However, in due course enough vegetation grows to create a fire hazard, and the whole process starts again.

Soil erosion associated with construction and urbanization

There are now a number of studies which illustrate clearly that urbanization can create significant changes in erosion rates.

The highest rates of erosion are produced in the construction phase, when there is a large amount of exposed ground and much disturbance produced by vehicle movements and excavations.

Wolman and Schick (1967) and Wolman (1967) have shown that the equivalent of many decades of natural or even agricultural erosion may take place during a single year in areas cleared for construction. In Maryland they found that sediment yields during construction reached 55,000 tonnes per square kilometre per year, while in the same area rates under forest were around 80–200 tonnes per square kilometre per year and those under farm 400 tonnes per square kilometre per year. New road cuttings in Georgia were found to have sediment yields up to 20,000–50,000 tonnes per square kilometre per year. Likewise, in Devon, England, Walling and Gregory (1970) found that suspended sediment concentrations in streams draining construction areas were two to ten times (occasionally up to 100 times) higher than those in undisturbed areas. In Virginia, Vice et al. (1969) noted equally high rates of erosion during construction and reported that they were ten times those from agricultural land, 200 times those from grassland and 2,000 times those from forest in the same area.

However, construction does not go on for ever, and once the disturbance ceases, roads are surfaced, and gardens and lawns are cultivated. The rates of erosion fall dramatically and may be of the same order as those under natural or pre-agricultural conditions (table 4.13). Moreover, even during the construction

Table 4.13 Rates of erosion associated with construction and urbanization

Location	Land use	Source	Rate $(t\ km^{-2}\ yr^{-1})$
1 Maryland, USA	Forest	Wolman (1967)	39
	Agriculture		116–309
	Construction		38,610
	Urban		19–39
2 Virginia, USA	Forest	Vice et al. (1969)	9
	Grassland		94
	Cultivation		1,876
	Construction		18,764
3 Detroit, USA	General non-urban	Thompson (1970)	642
	Construction		17,000
	Urban		741
4 Maryland, USA	Rural	Fox (1976)	22
	Construction		37
	Urban		337
5 Maryland, USA	Forest and grassland	Yorke and Herb (1978)	7–45
	Cultivated land		150–960
	Construction		1,600–22,400
	Urban		830
6 Wisconsin, USA	Agricultural	Daniel et al. (1979)	<1
	Construction		19.2
7 Tama New Town, Japan	Construction	Kadomura (1983)	c.40,000
8 Okinawa, Japan	Construction	Kadomura (1983)	25,000–125,000

phase several techniques can be used to reduce sediment removal, including the excavation of settling ponds, the seedings and mulching of bare surfaces, and the erection of rock dams and straw bales (Reed, 1980).

Attempts at soil conservation

Because of the adverse effects of accelerated erosion a whole array of techniques has now been widely adopted to conserve soil resources (Hudson, 1987). Some of the techniques, such as hill slope terracing, may be of some antiquity, and traditional techniques of a wide range of types have many virtues (see Critchley et al., 1994; Reij et al., 1996). The following are some of the main ways in which soil cover may be conserved:

1 *Revegetation*:
   (a)   deliberate planting;
   (b)   suppression of fire, grazing, etc., to allow regeneration.
2 *Measures to stop stream bank erosion.*
3 *Measures to stop gully enlargement*:
   (a)   planting of trailing plants, etc.;
   (b)   weirs, dams, gabions, etc.
4 *Crop management*:
   (a)   maintaining cover at critical times of year;
   (b)   rotation;
   (c)   cover crops.
5 *Slope runoff control*:
   (a)   terracing;
   (b)   deep tillage and application of humus;
   (c)   transverse hillside ditches to interupt runoff;
   (d)   contour ploughing;
   (e)   preservation of vegetation strips (to limit field width).
6 *Prevention of erosion from point sources like roads, feedlots*:
   (a)   intelligent geomorphic location;
   (b)   channelling of drainage water to non-susceptible areas;
   (c)   covering of banks, cuttings, etc., with vegetation.
7 *Suppression of wind erosion*:
   (a)   soil moisture preservation;
   (b)   increase in surface roughness through ploughing up clods or by planting windbreaks.

An alternative way of classifying soil conservation is provided by Morgan (1995). He identifies three main types of measure: agronomic, soil management and mechanical. The effects of these in relation to the main detachment and transport phases of erosion are shown in table 4.14.

There are some parts of the world where terraces (a mechanical measure) are one of the most prominent components of the

Table 4.14 Effect of various soil conservation practices on the detachment (D) and transport (T) phases of erosion

Practice	Rainsplash		Runoff		Wind	
	D	T	D	T	D	T
Agronomic measures						
Covering soil surface	*	*	*	*	*	*
Increasing surface roughness	–	–	*	*	*	*
Increasing surface depression storage	+	+	*	*	–	–
Increasing infiltration	–	–	+	*	–	–
Soil management						
Fertilizers, manures	+	+	+	*	+	*
Subsoiling, drainage	–	–	+	*	–	–
Mechanical measures						
Contouring, ridging	–	+	+	*	+	*
Terraces	–	+	+	*	–	–
Shelterbelts	–	–	–	–	*	*
Waterways	–	–	–	*	–	–

– no control; + moderate control; * strong control
D = detachment, T = transport.
Source: Morgan, 1995, table 7.1.

landscape (plate 4.7). This applies to many wine-growing areas, to some arid zone regions (such as Yemen and Peru), and to a wide selection of localities in the more humid tropics (Luzon, Java, Sumatra, Assam, Ceylon, Uganda, Cameroons, etc.). In areas subject to wind erosion other strategies may be necessary. Since soil blows only when it is dry, anything which conserves soil moisture is beneficial. Another approach to wind erosion is to slow down the wind by physical barriers, either in the form of an increased roughness of the soil surface brought about by careful ploughing, or by planted vegetative barriers, such as windbreaks and shelter-belts.

Some attempts at soil conservation have been particularly successful. For example, in Wisconsin, a study by Trimble and Lund (1982) showed that in the Coon Creek Basin erosion rates declined fourfold between the 1930s and the 1970s. One of the main reasons for this was the progressive adoption of contour-strip ploughing (figure 4.12). Indeed, a recent survey of soil erosion rates in the United States (Uri and Lewis, 1998) has proposed that as a result of conservation efforts, total soil erosion declined by 42 per cent between 1982 and 1997.

Attempts at soil conservation have not always been without their drawbacks. For example, the establishment of ground cover in dry areas to limit erosion may so reduce soil moisture because of accelerated evapotranspiration that the growth of the

Plate 4.7
The loess lands of China have been particularly susceptible to the development of large ravines. However, these conservation terraces, built by a production brigade, have effectively reduced the problem and allowed the land to become rehabilitated

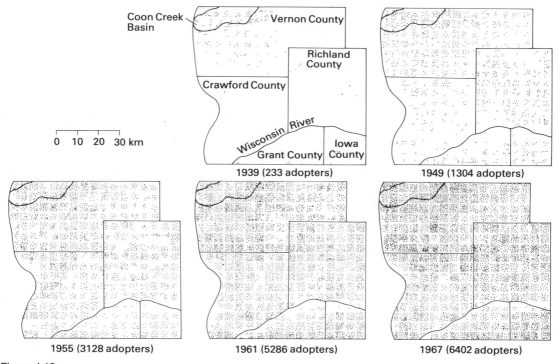

Figure 4.12
The spread of contour-strip soil conservation methods in Wisconsin, USA, between 1939 and 1967. One dot represents one adopter (after H. E. Johansen in Trimble and Lund, 1982, figure 22)

main crop is adversely affected. On a wider scale major affor-estation schemes can cause substantial runoff depletion in river catchments. Likewise, the provision of mulching is sometimes detrimental: in cool climates, reduced soil temperature shortens the growing season, while in wet areas, higher soil moisture may induce gleying and anaerobic conditions (Morgan, 1979: 60). Some terrace schemes have also had their shortcomings. They have been known to hold back so much water on hillsides that the soils have become saturated and landsliding has been induced. Similarly strip cropping, because it involves the farm-ing of small areas, is incompatible with highly mechanized agri-cultural systems, and insect infestation and weed control are additional problems which it has posed.

Soil conservation measures may not always be appropriate, for, paradoxically, soil erosion can be a useful phenomenon. Sanchez and Buol (1975), for instance, have pointed out that in recent volcanic areas soil erosion has enabled removal of the more weathered base-depleted material from the soil surface, exposing the more fertile, less weathered, base-rich material beneath. Likewise, in the Nochixtland area of southern Mexico, soil erosion has been utilized by local farmers to *produce* agri-cultural land. Severe gullies have cut in to steep valley-side slopes, and since the Spanish Conquest an average depth of 5 metres has been stripped from the entire surface area. The local Mixtec farmers, far from seeing this high rate of erosion as a hazard to be feared, have directed the flow of the eroded material to feed their fields with fertile soil and to extend their land. Over the past 1,000 years (see Whyte, 1977), the Mixtec cultivators have managed to use gully erosion to double the width of the main valley floors with flights of terraces. Judicious use of the phe-nomenon of gully erosion has enabled them to convert poor hilltop fields into the rich alluvial farmland below.

None the less it is undoubtedly true that manipulation of the soil is one of the most significant ways in which humans change the environment, and one in which they have had some of the most detrimental effects. Soil deterioration has led to many cases of what W. C. Lowdermilk once termed 'regional suicide', and the overall situation is at least as bleak as it was in the post-Dust Bowl years when Jacks commented: 'The organization of civilized societies is founded upon the measures taken to wrest control of the soil from wild Nature, and not until com-plete control has passed into human hands can a stable super-structure of what we call civilization be erected on the land' (Jacks and Whyte, 1939: 17).

The Human Impact on the Waters

5

Introduction

Because water is so important to human affairs, people have sought to control water resources in a whole variety of ways. Also, because water is such an important part of so many natural and human systems, its quantity and quality have undergone major changes as a consequence of human activities. We can quote Gleick (1993: 3):

As we approach the 21st century we must now acknowledge that many of our efforts to harness water have been inadequate or misdirected . . . Rivers, lakes, and groundwater aquifers are increasingly contaminated with biological and chemical wastes. Vast numbers of people lack clean drinking water and rudimentary sanitation services. Millions of people die every year from water-related diseases such as malaria, typhoid, and cholera. Massive water developments have destroyed many of the world's most productive wetlands and other aquatic habitats.

In recent decades human demand for fresh water has increased rapidly. Global water use has more than tripled since 1950, and now stands at 4,340 cubic kilometres per year – equivalent to eight times the annual flow of the Mississippi river. Annual irretrievable water losses have increased about sevenfold in the century (table 5.1a).

Deliberate modification of rivers

Although there are many ways in which humans influence water quantity and quality in rivers and streams – for example, by direct channel manipulation, modification of basin characteristics, urbanization and pollution – the first of these is of particularly great importance (Mrowka, 1974). Indeed, there are a great variety of methods of direct channel manipulation and many of them have a long history. Perhaps the most widespread of these

Table 5.1   Major changes in the hydrological environment

(a) *Irretrievable water losses (km³/yr)*

Users	1900	1940	1950	1960	1970	1980	1990	2000
Agriculture	409	679	859	1,180	1,400	1,730	2,050	2,500
Industry	3.5	9.7	14.5	24.9	38.0	61.9	88.5	117
Municipal supply	4.0	9.0	14	20.3	29.2	41.1	52.4	64.5
Reservoirs	0.3	3.7	6.5	23.0	66.0	120	170	220
TOTAL	417	701	894	1,250	1,540	1,950	2,360	2,900

(b) *Number of large dams (>15m high) constructed or under construction, 1950–86*

Continent	1950	1982	1986	Under construction 31.12.86
Africa	133	665	763	58
Asia	1,562	22,789	22,389	613
Australasia/Oceania	151	448	492	25
Europe	1,323	3,961	4,114	222
North and Central America	2,099	7,303	6,595	39
South America			884	69
TOTAL	5,268	35,166	36,327	1,036

Source: data provided by UNEP.

is the construction of dams and reservoirs (plate 5.1). The first recorded dam was constructed in Egypt some 5,000 years ago, but since that time the adoption of this technique has spread, variously to improve agriculture, to prevent floods, to generate power or to provide a reliable source of water.

The construction of large dams increased markedly, especially between 1945 and the early 1970s (Beaumont, 1978). Engineers have now built more than 36,000 dams around the world and, as table 5.1b shows, large dams (i.e. more than 15 metres high) are still being constructed at an appreciable rate, especially in Asia. In the late 1980s some 45 very large dams (more than 150 metres high) were under construction. Indeed, one of the most striking features of dams and reservoirs is that they have become increasingly large (Beckinsale, 1969). Thus in the 1930s the Hoover or Boulder Dam in the USA (221 metres high) was by far the tallest in the world and it impounded the biggest reservoir, Lake Mead, which contained 38 billion cubic metres of water. By the 1980s it was exceeded in height by at least 18 others, and some of these impounded reservoirs with four times the volume of Lake Mead.

Such large dams are capable of causing almost total regulation of the streams they impound but, in general, the degree to

Table 5.2 Peak flow reduction downstream from selected British reservoirs

Reservoir	% of catchment inundated	% of peak flow reduction
Avon, Dartmoor	1.38	16
Fernworth, Dartmoor	2.80	28
Meldon, Dartmoor	1.30	9
Vyrnwy, mid-Wales	6.13	69
Sutton Bingham, Somerset	1.90	35
Blagdon, Mendip	6.84	51
Stocks, Forest of Bowland	3.70	70
Daer, Southern Uplands	4.33	56
Camps, Southern Uplands	3.13	41
Catcleugh, Cheviots	2.72	71
Ladybower, Peak District	1.60	42
Chew Magna, Mendips	8.33	73

Source: after Petts and Lewin, 1979, table 1, p. 82.

Plate 5.1
The Kariba Dam on the Zambezi River between Zambia and Zimbabwe. Such large dams can provide protection against floods and water shortages, and generate a great deal of electricity. However, they can have a whole suite of environmental consequences

which peak flows are reduced depends on the size of the dam and the impounded lake in relation to catchment characteristics. In Britain, as table 5.2 shows, peak flow reduction downstream from selected reservoirs varies considerably, with some tendency for the greatest degree of reduction to occur in those catchments where the reservoirs cover the largest percentage of the area. When considering the magnitude of floods of different recurrence

Table 5.3   The ratios of post- to pre-dam discharges for flood magnitudes of
selected frequency

	Recurrence interval (years)			
Reservoir	1.5	2.3	5.0	10.0
Avon, R. Avon	0.90	0.89	0.93	1.02
Stocks, R. Hodder	0.83	0.86	0.84	0.95
Sutton Bingham, R. Yeo	0.52	0.61	0.69	0.79

Source: after Petts and Lewin, 1979, table 2, p. 84.

intervals before and after dam construction, it is clear that
dams have much less effect on rare events of high magnitude
(Petts and Lewin, 1979), and this is brought out in table 5.3.
None the less, most dams achieve their aim: to regulate river
discharge. They are also highly successful in fulfilling the needs
of surrounding communities: millions of people depend upon
them for survival, welfare and employment.

However, dams may have a whole series of environmental
consequences that may or may not have been anticipated (figure
5.1; and see Makkaveyev, 1972). Some of these are dealt with
in greater detail elsewhere, such as subsidence (p. 280), earthquake triggering (p. 323), the transmission and expansion in the
range of organisms, inhibition of fish migration (p. 89), the
build-up of soil salinity (p. 163), changes in groundwater levels
creating slope instability (p. 297) and waterlogging (p. 171).
Several of these processes may in turn affect the viability of the
scheme for which the dam was created.

A particularly important consequence of impounding a reservoir behind a dam is the reduction in the sediment load of the
river downstream. A clear demonstration of this effect has been
given for the South Saskatchewan River in Canada (table 5.4)
by Rasid (1979). During the pre-dam period, typified by 1962,
the total annual suspended loads at Saskatoon and Lemsford
Ferry were remarkably similar. As soon as the reservoir began
to fill late in 1963, however, some of the suspended sediment
began to be trapped, and the transitional period was marked by
a progressive reduction in the proportion of sediment which
reached Saskatoon. In the four years after the dam was fully
operational the mean annual sediment load at Saskatoon was
only 9 per cent that at Lemsford Ferry.

Even more dramatic are the data for the Colorado river in the
United States (figure 5.2). Prior to 1930 it carried around 125–
150 million tonnes of suspended sediment per year to its delta
at the head of the Gulf of California. Following a series of dams
the Colorado now discharges neither sediment or water to the
sea (Schwarz et al., 1991). There have also been marked changes

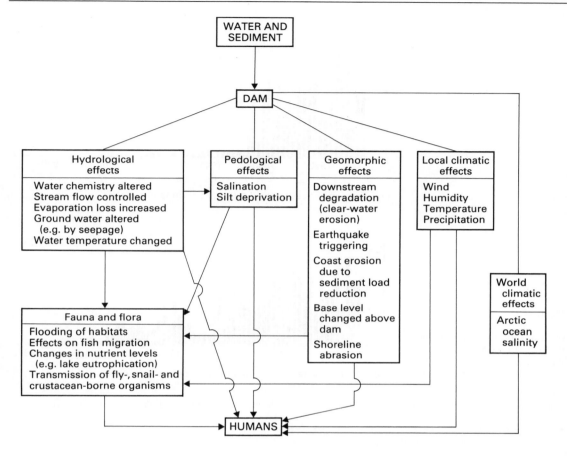

Table 5.4 Total yearly suspended load (thousand imperial tons) of the South Saskatchewan River at Lemsford Ferry and Saskatoon, 1962–70

Period of record		Lemsford Ferry	Saskatoon	Difference at Saskatoon (%)
Pre-dam				
1962		1,813	1,873	+3
Transitional				
1963		4,892	447	−8
1964		7,711	4,146	−46
1965		9,732	2,721	−72
1966		5,228	1,675	−68
	MEAN	6,891	3,255	−53
Post-dam				
1967		12,619	446	−96
1968		2,661	101	−96
1969		10,562	2,146	−80
1970		5,643	118	−98
	MEAN	7,871	703	−91

Source: Rasid, 1979, table 1.

Figure 5.1
Generalized representation of the possible effects of dam construction on human life and various components of the environment

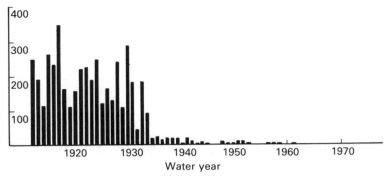

(a) Suspended-sediment discharge (millions of tons/yr)

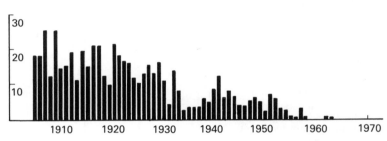

Figure 5.2
Historical (a) sediment and
(b) water discharge trends for
the Colorado river, USA (after
the United States Geological
Survey in Schwarz et al., 1991)

(b) Water discharge (millions of acre-feet/yr)

Table 5.5 Silt concentrations (in parts per million) in the Nile at Gaafra before and after the construction of the Aswan High Dam

BEFORE (averages for the period 1958–63)											
Jan.											*Dec.*
64	50	45	42	43	85	674	2702	2422	925	124	71
AFTER											
44	47	45	50	51	49	48	45	41	43	48	47
RATIO											
1.5	1.1	1.0	0.8	0.8	1.7	14.0	60.0	59.1	21.5	2.58	1.63

Source: Abu-Atta, 1978, p. 199.

in the amount of sediment passing along the Missouri and Mississippi rivers over the period 1938 to 1982. Downstream sediment loads have been reduced by about half over that period (figure 5.3a). Meade (1996) has attempted to compare the situation in the 1980s with that which existed before humans started to interfere with those rivers (*c.* AD 1700) (figure 5.3b).

Sediment retention is also well illustrated by the Nile (table 5.5), both before and after the construction of the great Aswan High Dam. Until its construction the late summer and autumn period of high flow was characterized by high silt concentrations, but since it has been finished the silt load is

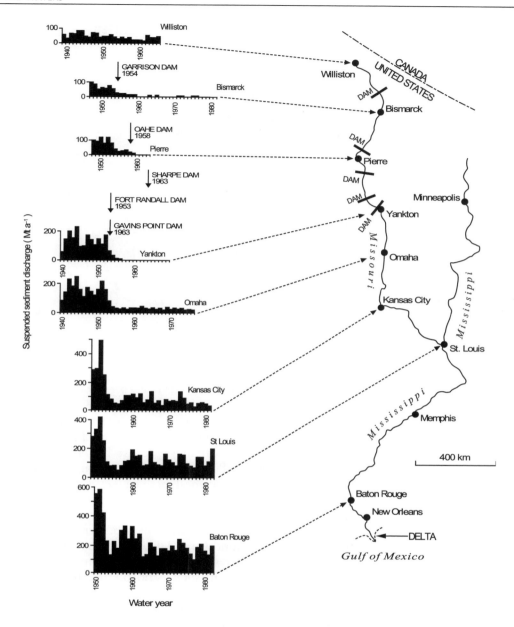

rendered lower throughout the year and the seasonal peak is removed. Petts (1985, table XVIII) indicates that the Nile now transports only 8 per cent of its natural load below the Aswan High Dam, although this figure seems to be exceptionally low. Other rivers for which data are available carry between 8 and 50 per cent of their natural suspended loads below dams.

Sediment removal in turn has various possible consequences, including a reduction in flood-deposited nutrients on fields, fewer

Figure 5.3

(a) Suspended sediment discharge on the Mississippi and Missouri rivers between 1939 and 1982 (after Meade and Parker, 1985, with modifications)

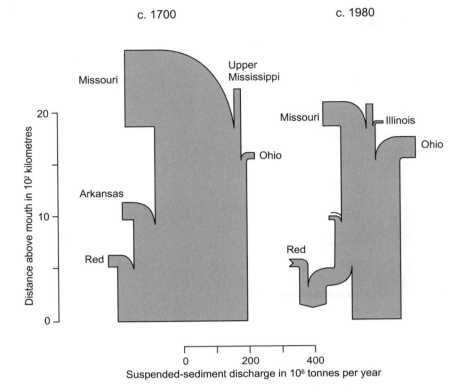

c. 1700   c. 1980

Figure 5.3
(b) Long-term average
discharges of suspended
sediment in the lower
Mississippi River c.1700
and c.1980

nutrients for fish in the south-east Mediterranean Sea, accelerated erosion of the Nile delta, and accelerated riverbed erosion
since less sediment is available to cause bed aggradation. The
last process is often called 'clear-water erosion' (see Beckinsale,
1972), and in the case of the Hoover Dam it affected the river
channel of the Colorado for 150 kilometres downstream by
causing incision. In turn, such channel incision may initiate a
rejuvenation of headward erosion in tributaries and may cause
the lowering of groundwater tables and the undermining of
bridge piers and other structures downstream of the dam. On
the other hand, in regions such as northern China, where modern dams trap silt, the incision of the river channel downstream
may alleviate the strain on levées and lessen the expense of levée
strengthening or heightening.

However, clear-water erosion does not always follow from
silt retention in reservoirs. There are examples of rivers where,
before impoundment, floods carried away the sediment brought
into the main stream by steep tributaries. Reduction of the peak
discharge after the completion of the dam leaves some rivers
unable to scour away the sediment that accumulates as large
fans of sand or gravel below each tributary mouth (Dunne and
Leopold, 1978). The bed of the main stream is raised, and if

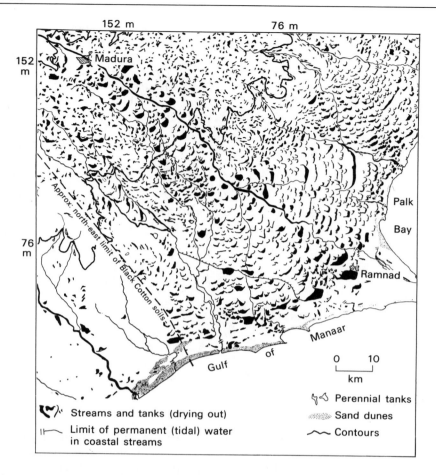

Figure 5.4
The Madurai–Ramanathapuram
tank country in south India
(after Spate and Learmonth,
1967, figure 25.12)

water intakes or other structures lie alongside the river they can be threatened again by flooding or channel shifting across the accumulating wedge of sediment. Rates of aggradation of a metre a year have been observed, and tens of kilometres of channel have been affected by sedimentation. One of the best-documented cases of aggradation concerns the Colorado river below Glen Canyon Dam. Since dam closure the extremes of river flow have been largely eliminated so that the ten years' recurrence interval flow has been reduced to less than one-third. The main channel flow is no longer capable of removing sediment provided by flash-flooding tributaries, and deposits up to 2.6 metres thick have accumulated within the upper Grand Canyon (Petts, 1985: 133).

Some landscapes in the world are dominated by dams, canals and reservoirs. Probably the most striking example of this (figure 5.4) is the 'tank' landscape of south-east India where myriads of little streams and areas of overland flow have been dammed by small earth structures to give what Spate (Spate and Learmonth, 1967: 778) has likened to 'a surface of vast overlapping fish-scales'.

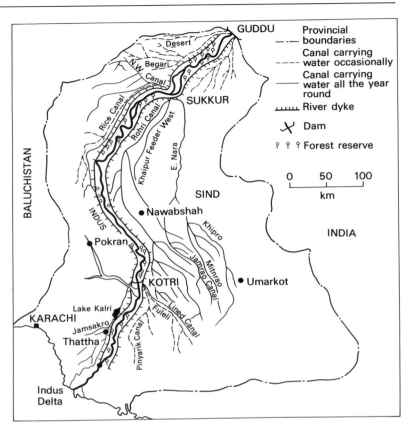

Figure 5.5
The irrigated areas in Sind
(Pakistan) along the Indus
valley (after Manshard, 1974,
figure 5.7)

In the northern part of the subcontinent, in Sind, the landscape changes wrought by hydrology are no less striking, with the mighty snow-fed Indus being controlled by large embankments (*bunds*) and interrupted by great barrages. Its waters are distributed over thousands of square kilometres by a canal network that has evolved over the past 4,000 years (figure 5.5).

Another landscape where equally far-reaching changes have been wrought is the Netherlands (plate 5.2). Coates (1976) has calculated that, before 1860, reclamation of that country from the sea, in the extension of drainage lines, involved the movement of 1,000 million cubic metres of material. The area is dominated by human constructions: canals, rivers, drains and lakes. As C. T. Smith (1978: 506) has put it: 'It can be said of the Netherlands with more truth than of any other country that it is a creation of man, for it owes its existence to the long continued struggle to drain and reclaim new land and to protect existing lands from the invasion of the sea.'

Another direct means of river manipulation is channelization. This involves the construction of embankments, dikes, levées and flood walls to confine flood waters; and improving the ability of channels to transmit floods by enlarging their capacity through straightening, widening, deepening or smoothing (table 5.6).

Plate 5.2
Polders have transformed the nature of the low-lying coastline of the Netherlands. Large areas have been reclaimed from the sea. The Afsluitdijk separates the Wazenezee and the Ijsselmeer

Table 5.6 Selected terminologies for the methods of river channelization in the USA and UK

American term	British equivalent	Method involved
Widening	Resectioning	Increase of channel capacity by manipulating width and/or depth variable
Deepening	Resectioning	
Straightening	Realigning	Increasing velocity of flow by steepening the gradient
Diking	Embanking	Raising of channel banks to confine floodwaters
Bank stabilization	Bank protection	Methods to control bank erosion e.g. gabions and concrete structures
Clearing and snagging	Pioneer tree clearance	Removal of obstructions from a watercourse, thereby decreasing the resistance and increasing the velocity of flow
	Control of aquatic plants	
	Dredging of sediments	
	Urban clearing	

Source: Brookes, 1985, table 1.

Some of the great rivers of the world are now lined by extensive embankment systems such as those that run for more than 1,000 kilometres alongside the Nile, 700 kilometres along the Hwang Ho, 1,400 kilometres by the Red River in Vietnam, and over 4,500 kilometres in the Mississippi valley (Ward, 1978). Like dams, embankments and related structures often fulfil their purpose; but they may also create some environmental problems and have some disadvantages. For example, they reduce natural storage for flood waters, both by preventing water from spilling on to much of the flood plain and by stopping bank storage in cases where impermeable flood walls are used. Likewise, the flow of water in tributaries may be constrained. In addition,

Figure 5.6
Comparison of the natural channel morphology and hydrology with that of a channelized stream, suggesting some possible ecological consequences (after Keller, 1976, figure 4)

NATURAL CHANNEL

Suitable water temperatures:
adequate shading; good cover for fish life; minimal variation in temperatures; abundant leaf material input

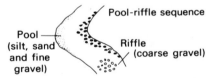

Pool-riffle sequence

Pool (silt, sand and fine gravel)

Riffle (coarse gravel)

Sorted gravels provide diversified habitats for many stream organisms

ARTIFICIAL CHANNEL

Increased water temperatures:
no shading; no cover for fish life; rapid daily and seasonal fluctuations in temperatures; reduced leaf material input

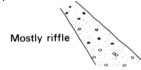

Mostly riffle

Unsorted gravels:
reduction in habitats; few organisms

POOL ENVIRONMENT

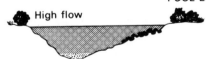

High flow

Diversity of water velocities:
high in pools, lower in riffles. Resting areas abundant beneath undercut banks or behind large rocks, etc.

Low flow

Sufficient water depth to support fish and other aquatic life during dry season

High flow

May have stream velocity higher than some aquatic life can withstand. Few or no resting places

Low flow

Insufficient depth of flow during dry seasons to support diversity of fish and aquatic life. Few if any pools (all riffle)

embankments may occasionally exacerbate the flood problem they were designed to reduce by preventing flood waters downstream of a breach from draining back into the channel once the peak has passed.

Channel improvement, designed to improve water flow, may also have unforeseen or undesirable effects. For example, the more rapid movement of water along improved channel sections can aggravate flood peaks further downstream and cause excessive erosion. The lowering of water tables in the 'improved' reach may cause overdrainage of adjacent agricultural land so that sluices must be constructed in the channel to maintain levels. On the other hand, lined channels may obstruct soil water movement (interflow) and shallow groundwater and so cause surface saturation. Brookes (1985) and Gregory (1985a) provide useful reviews on the impact of channelization.

Channelization may also have miscellaneous effects on fauna through the increased velocities of water flow, reductions in the extent of shelter in the channel bed, and reduced nutrient inputs due to the destruction of overhanging bank vegetation (see Keller, 1976). In the case of large swamps, like those of the Sudd in Sudan or the Okavango in Botswana, the channelization of rivers could completely transform the whole character of the swamp environment. Figure 5.6 illustrates some of the differences between natural and artificial channels.

Another type of channel modification is produced by the construction of bypass and diversion channels to carry excess flood water or to enable irrigation to take place. Such channels may be as old as irrigation itself. They may contribute to the salinity problems encountered in many irrigated areas.

Deliberate modification of a river regime can also be achieved by long-distance inter-basin water transfers (Shiklomanov, 1985), transfers necessitated by the unequal spatial distribution of water resources and by the increasing rates of water consumption. At present, the world water consumption for all human needs is 4,340 cubic kilometres a year; at the start of the twentieth century it was eight times less than this figure and in the early years of the new millennium it is expected to be 6,000 cubic kilometres a year. The total volume of water in the various transfer systems in operation and under construction on a global scale is about 300 cubic kilometres a year, with the largest countries in terms of volume of transfers being Canada, the CIS, the United States and India.

In future decades it is likely that many even greater schemes will be constructed (see figure 5.7); route lengths of some hundreds of kilometres will be common, and the water balances of many rivers and lakes will be transformed. This is already happening in the CIS (figure 5.8), where the operation of various anthropogenic activities of this type have caused runoff in the most intensely cultivated central and southern areas to decrease

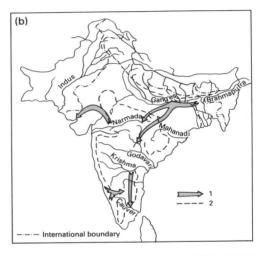

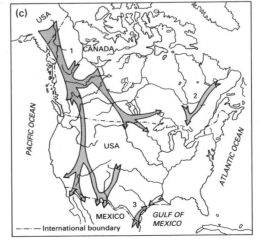

Figure 5.7
Some major schemes
proposed for large-scale
inter-basin water transfers:
(a) Projected water transfer
    systems in the CIS:
    1. From the Onega River
    and in future from Onega
    Bay. 2. From the Sukhona
    and Northern Dvina Rivers.
    3. From the Svir River and
    Lake Onega. 4. From the
    Pechora River. 5. From the
    Ob River. 6. From the
    Danube Delta
(b) Projected systems for
    water transfers in India.
    1. Scheme of the national
    water network. 2. Scheme
    of Grand Water Garland
(c) Some major projects for
    water transfers in North
    America: 1. NAWAPA.
    2. Grand Canal.
    3. Texas River Basins
    (after Shiklomanov, 1985,
    figures 12.6, 12.9 and
    12.11, in *Facets of
    Hydrology II*, ed. J. C.
    Rodda, by permission of
    John Wiley and Sons Ltd)

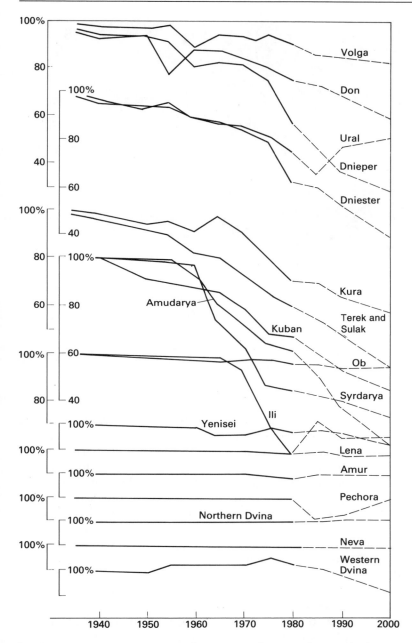

Figure 5.8
Changes in annual runoff in the CIS due to human activity during 1936–2000 (from Shiklomanov, 1985, figure 12.7, in *Facets of Hydrology II*, ed. J. C. Rodda, by permission of John Wiley and Sons Ltd)

by 30–50 per cent compared to normal natural runoff. At the same time, inflows into the Caspian and Aral Seas have declined sharply, so that their levels have fallen and their areas decreased (plate 5.3).

When one turns to the coastal portions of rivers, to estuaries, the possible effects of another human impact, dredging, can be as complex as the effects of dams and reservoirs upstream (La

Plate 5.3
The desiccation of the Aral
Sea, perhaps the greatest
environmental disaster of the
CIS, has in large measure
resulted from the transfer of
water from rivers flowing into it.
The fishing industry has
suffered greatly

Roe, 1977). Dredging and filling are certainly widespread and often desirable. Dredging may be performed to create and maintain canals, navigation channels, turning basins, harbours and marinas; to lay pipelines; and to obtain a source of material for fill or construction. Filling is the deposition of dredged materials to create new land. There are miscellaneous ecological effects of such actions. In the first place, filling directly disrupts habitats. Second, the generation of large quantities of suspended silt tends physically to smother bottom-dwelling plants and animals; tends to smother fish by clogging their gills; reduces photosynthesis through the effects of turbidity; and tends to lead to eutrophication by an increased nutrient release. Likewise, the destruction of marshes, mangroves and sea grasses by dredge and fill can result in the loss of these natural purifying systems. The removal of vegetation may also cause erosion. Moreover, as silt deposits stirred up by dredging accumulate elsewhere in the estuary they tend to create a 'false bottom'. Characterized by shifting, unstable sediments, the dredged bottom, fill deposits or spoil areas are only slowly – if at all – recolonized by fauna and flora. Furthermore, dredging tends to change the configuration of currents and the rate of fresh-water drainage, and may provide avenues for salt-water intrusion.

Urbanization and its effects on river flow

The process of urbanization has a considerable hydrological impact, in terms of controlling rates of erosion and the delivery of pollutants to rivers, and in terms of influencing the nature of runoff and other hydrological characteristics (Hollis, 1988). An attempt to generalize some of these impacts using a historical model of urbanization is summarized usefully by Savini and Kammerer (1961) and reproduced here in table 5.7.

Table 5.7 Stages of urban growth and their miscellaneous hydrological impacts

Stage	Impact
1 *Transition from pre-urban to early-urban stage:*	
(a) Removal of trees or vegetation	Decrease in transpiration and increase in storm flow
(b) Construction of scattered houses with limited sewerage and water facilities	
(c) Drilling of wells	Some lowering of water-table
(d) Construction of septic tanks, etc.	Some increase in soil moisture and perhaps some contamination
2 *Transition from early-urban to middle-urban stage:*	
(a) Bulldozing of land	Accelerated land erosion
(b) Mass construction of houses, etc.	Decreased infiltration
(c) Discontinued use and abandonment of some shallower wells	Rise in water-table
(d) Diversion of nearby streams for public supply	Decrease in runoff between points of diversion of disposal
(e) Untreated or inadequately treated sewerage into streams and wells	Pollution of streams and wells
3 *Transition from middle-urban to late-urban stage:*	
(a) Urbanization of area completed by addition of more buildings	Reduced infiltration and lowered water-table, higher flood peaks and lower low flows
(b) Larger quantitites of untreated waste into local streams	Increased pollution
(c) Abandonment of remaining shallow wells because of pollution	Rise in water-table
(d) Increase in population requiring establishment of new water supply and distribution systems	Increase in local stream flow if supply is from outside basin
(e) Channels of streams restricted at least in part to artificial channels and tunnels	Higher stage for a given flow (therefore increased flood damage) changes in channel geometry and sediment load
(f) Construction of sanitary drainage system and treatment plant for sewage	Removal of additional water from area
(g) Improvement of storm drainage system	
(h) Drilling of deeper, large-capacity industrial wells	Lowered water pressure, some subsidence, salt-water encroachment
(i) Increased use of water for air-conditioning	Overloading of sewers and other drainage facilities
(j) Drilling of recharge wells	Raising of water pressure surface
(k) Waste-water reclamation and utilization	Recharge to ground-water aquifers: more efficient use of water resources

Source: modified after Savini and Kammerer, 1961.

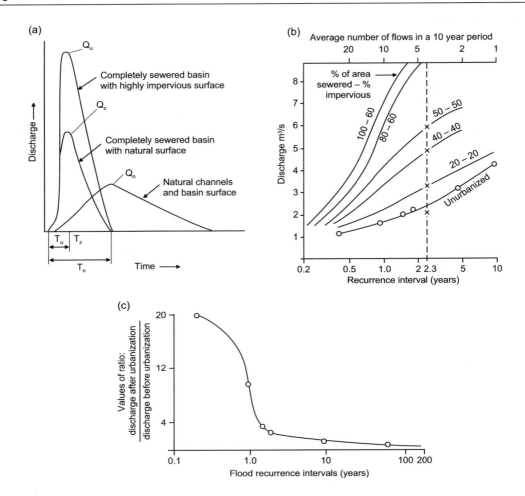

Figure 5.9
Some hydrological
consequences of urbanization:
(a) Effect of urban
    development on flood
    hydrographs. Peak
    discharges (Q) are higher
    and occur sooner after
    runoff starts (T) in basins
    that have been developed
    or sewered (after Fox,
    1979, figure 3)
(b) Flood frequency curves
    for a 1 square mile basin
    in various states of
    urbanization (after US
    Geological Survey in
    Viessman et al., 1977,
    figure 11.33)
(c) Effects on flood magnitude
    of paving 20 per cent of a
    basin (after Hollis, 1975)

One of the most important effects is the way in which urbanization affects flood runoff. Research both in the United States and in Britain has shown that, because urbanization produces extended impermeable surfaces of bitumen, tarmac, tiles and concrete, there is a tendency for flood runoff to increase in comparison with rural sites. City drainage densities may be greater than those in natural conditions (Graf, 1977) and the installation of sewers and storm drains accelerates runoff, as illustrated in figure 5.9a. The greater the area that is sewered, the greater is the discharge for a particular recurrence level (figure 5.9b). Peak discharges are higher and occur sooner after runoff starts in basins that have been affected by urbanization and the installation of sewers. Some runoff may be generated in urban areas because low vegetation densities mean that evapotranspiration is limited.

However, in many cases the effect of urbanization is greater on small floods; as the size of the flood and its recurrence interval increase, so the effect of urbanization diminishes (Martens, 1968;

Table 5.8 Measures for reducing and delaying urban storm runoff

Area	Reducing runoff	Delaying runoff
Large flat roof	Cistern storage Rooftop gardens Pool storage or fountain storage Sod roof cover	Ponding on roof by constricted drainpipes Increasing roof roughness (1) rippled roof (2) gravelled roof

Car parks	Porous pavement
	(1) gravel car parks
	(2) porous or punctured asphalt
	
	Concrete vaults and cisterns beneath car parks
	  in high value areas
	Vegetated ponding areas around car parks
	Gravel trenches
		Grassy strips on car parks
		Grassed waterways draining car parks
		Ponding and detention measures for
		  impervious areas
		(1) rippled pavement
		(2) depressions
		(3) basins

Residential	Cisterns for individual homes or groups of homes
	Gravel drives (porous)
	Contoured landscape
	Ground-water recharge
	(1) perforated pipe
	(2) gravel (sand)
	(3) trench
	(4) porous pipe
	(5) dry wells
	Vegetated depressions
		Reservoir or detention basin
		Planting a high-delaying grass (high
		  roughness)
		Grassy gutters or channels
		Increased length of travel of runoff by means
		  of gutters, diversions and so on

General	Gravel alleys Porous pavements	Gravel alleys

Source: after US Department of Agriculture, Soil Conservation Service, 1972, in Viessman et al., 1977: 569.

Hollis, 1975). A probable explanation for this is that, during a severe and prolonged storm event, a non-urbanized catchment may become so saturated and its channel network so extended that it begins to behave hydrologically as if it were an impervious catchment with a dense surface-water drain network. Under these conditions, a rural catchment produces floods of a type and size similar to those of its urban counterpart. Moreover, a further mechanism probably operates in the same direction, for in an urban catchment it seems probable that some throttling of flow occurs in surface-water drains during intense storms, tending to attenuate the very highest discharges. Thus, Hollis believes, while the size of small, frequent floods is increased many times by urbanization, large, rare floods (the ones likely to cause extreme damage) are not significantly affected by the construction of surburban areas within a catchment area (figure 5.9c). None the less, a whole series of techniques have been developed in an attempt to reduce and delay urban storm runoff (table 5.8), for Hollis's findings may not be universally applicable. For example, K. V. Wilson (1967) working in Jackson, Mississippi, found that the 50-year flood for an urbanized catchment was three times higher than that of a rural one.

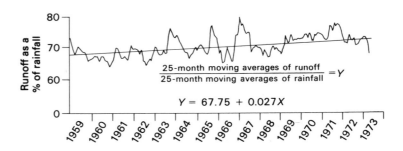

Figure 5.10
Trends in 25-month moving averages of runoff as a percentage of 25-month moving averages of rainfall for the River Tame, Lee Marston, England, 1957–74. The trend reflects increasing amounts of imported runoff (after Richards and Wood, 1977, figure 24.6)

Water transfers for industrial and municipal purposes are having an increasing impact on river regimes in urbanized societies (Richards and Wood, 1977). Under dry weather flow conditions, stream flow in urban catchments in the Midlands and south-east England is mainly imported water. This is illustrated by the River Tame which contains about 90 per cent effluent in its 95 per cent duration flow. This proportion is increasing as more water is imported (figure 5.10), reduced in quality and added to low natural base flows reduced by lack of recharge.

Deforestation and its effects on river flow

As we have already noted in chapter 1, one of the first major indications that humans could inadvertently adversely affect the environment was the observation that deforestation could create torrents and floods. The deforestation that gives rise to such flows can be produced both by felling and by fire (Scott, 1997).

The first experimental study, in which a planned land-use change was executed to enable observation of the effects of stream flow, began at Wagon Wheel Gap, Colorado, in 1910 (Pereira, 1973). Here stream flow from two similar watersheds of about 80 hectares each were compared for eight years. One valley was then clear-felled and the records were continued. After the clear-felling the annual water yield was 17 per cent above that predicted from the flows of the unchanged control valley. Peak flows also increased. Studies on two small basins in the Australian Alps (Wallace's Creek, 41 square kilometres; Yarrango Billy River, 224 square kilometres), which were burned over, showed that rain storms, which from previous records would have been expected to give rise to flows of 60–80 cubic metres per second, produced a peak of 370 cubic metres per second (a five or sixfold increase). Likewise, catchment experiments in Arizona have shown that when chaparral scrub is burned there is a tenfold increase in water yield.

Experiments with tropical catchments have shown typical maximum and mean annual stream-flow increases of 400–450 millimetres per year on clearance with increases in water yield of up to 6 millimetres per year for each percentage reduction in

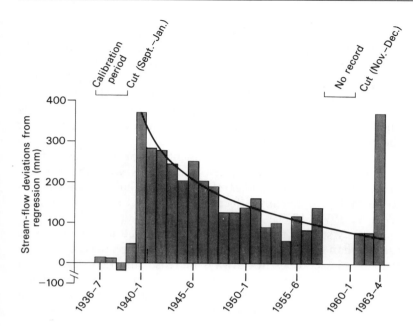

forest area above a 15 per cent change in cover characteristics (Anderson and Spencer, 1991: 49). Additional data for the effects of land-use change on annual runoff levels are presented for tropical areas in table 4.11.

As vegetation regenerates after a forest has been cut or burned, so stream flow tends to revert to normal, though the process may take some decades. This is illustrated in figure 5.11 which shows the dramatic effects produced on the Coweeta catchments in North Carolina by two spasms of clear-felling, together with the gradual return to normality in between.

The substitution of one forest type for another may also affect stream flow. This can again be exemplified from the Coweeta catchments, where two experimental catchments were converted from a mature deciduous hardwood forest cover to a cover of white pine (*Pinus strobus*). Fifteen years after the conversion, annual stream flow was found to be reduced by about 20 per cent (Swank and Douglass, 1974). The reason for this notable change is that the interception and subsequent evaporation of rainfall is greater for pine than it is for hardwoods during the dormant season.

Fears have also been expressed that the replacement of tall natural forests by eucalyptus will produce a decline in stream-flow. However, most current research does not support this contention, for transpiration rates from eucalyptus are similar to those from other tree species (except in situations with a shallow groundwater table), while their interception losses are if anything generally rather less than those from other tree species of similar height and planting density (Bruijnzeel, 1990).

Figure 5.11
The increase of water yield after clear-felling a forest: a unique confirmation from the Coweeta catchment in North Carolina, USA (after Pereira, 1973, figure 8)

The reasons why the removal of a forest cover and its replacement with pasture, crops or bare ground have such important effects on stream flow are many. A mature forest probably has a higher rainfall interception rate and a tendency to reduce rates of overland flow, and probably generates soils with a higher infiltration capacity and better general structure. All these factors will tend to produce both a reduction in overall runoff levels and less extreme flood peaks. However, with careful management the replacement of forest by other land-use types need not be detrimental in terms of either sediment loss or flood generation. In Kenya, for example, the tea plantations with shade trees, protective grass covers and carefully designed culverts were found to be 'a hydrologically effective substitute for natural forest' (Pereira, 1973: 127).

In many studies the runoff from clean-tilled land tends to be greater than that from areas under a dense crop cover, but tilling the soil surface does not always increase runoff (Gregory and Walling, 1973: 345). There are reports from the CIS suggesting that the reverse may be the case, and that autumn ploughing can decrease surface runoff, presumably because of its effect on surface detention and on soil structure.

Grazing practices also influence runoff, for heavy grazing can both compact the soil and cause vegetation removal (Trimble and Mendel, 1995). In general it tends to lead to an increase in runoff.

Changes in river-bank vegetation may have a particularly strong influence on river flow. In the south-west United States, for instance, many streams are lined by the salt cedar (*Tamaarix pentandra*). With roots either in the water table or freely supplied by the capillary fringe, these shrubs have full potential transpiration opportunity. The removal of such vegetation can cause large increases in stream flow. It is interesting to note that the salt cedar itself is an alien, native to Eurasia, which was introduced by humans. It has spread explosively in the south-west United States, increasing from about 4,000 hectares in 1920 to almost 400,000 hectares in the early 1960s (Harris, 1966). In the Upper Rio Grande valley in New Mexico, the salt cedar was introduced to try to combat anthropogenic soil erosion, but it spread so explosively that it came to consume approximately 45 per cent of the area's total available water (Hay, 1973).

Reforestation of abandoned farmlands reverses the effects of deforestation: increased interception and evapotranspiration can cause a decline in water yield. In parts of the eastern United States, farm abandonment and recolonization of the land by pines, spruce and cedar has been occurring throughout the twentieth century, and this has reduced stream flow by important amounts at a time when water supplies for some eastern cities were becoming critically short (Dunne and Leopold, 1978).

A process which is often associated with afforestation is peat drainage. The hydrological effects of this are the subject of controversy, since there are cases of both increased and decreased flood peaks after drainage. It has been suggested that differences in peat type alone might account for the different effects. Thus it is conceivable that the drainage of a *Sphagnum* catchment would lead to increased flooding since *Sphagnum* compacts with drainage, reducing its storage volume and its permeability. On the other hand, in the case of non-*Sphagnum* peat there would be relatively less change in structure, but there would be a reduction in moisture content and an increase in storage capacity, thereby tending to reduce flood flows. The nature of the peat is, however, but one feature to consider (Robinson, 1979). The intensity of the drainage works (depth, spacing, etc.) may also be important. In any case, there may be two (sometimes conflicting) processes operating as a result of peat drainage: the increased drainage network will facilitate rapid runoff, and the drier soil conditions will provide greater storage for rainfall. Which of these two tendencies is dominant will depend on local catchment conditions.

However, the impact of land drainage upon downstream flood incidence has long been a source of controversy. Much depends on the scale of study, the nature of land management and the character of the soil that has been drained. After a detailed review of experience in the UK, Robinson (1990) found that the drainage of heavy clay soils that are prone to prolonged surface saturation in their undrained state generally led to a reduction of large and medium flow peaks. He attributed this to the fact that their natural response is 'flashy' (with limited soil water storage available), whereas their drainage of permeable soils, which are less prone to such surface saturation, improves the speed of subsurface flow, thereby tending to increase peak flow levels.

It is not always easy to determine whether an increase in flood frequency or intensity is the result of land-use changes of the type we have been discussing, or whether some natural changes in rainfall have played a role. In central and southern Wales there is some clear evidence of changes in flood magnitude and frequency over recent decades, and this has sometimes been attributed to the increasing amount of afforestation that has been carried out since the First World War, and to the drainage of upland areas that this has necessitated. While in the Severn catchment this appears to be a partial explanation (Howe et al., 1967), in other basins the main cause of more frequent and intense floods appears to have been a marked increase in the magnitude and the frequency of heavy daily rainfalls. For example, in the case of the Tawe valley near Swansea, of 17 major floods since 1875, 14 occurred between 1929 and 1981 and only three during the 1875–1928 period. Of 22 notable widespread heavy rainfalls in the Tawe catchment since 1875, only two occurred during 1875–1928, but 20 from 1929 to 1981 (Walsh et al., 1982).

The human impact on lake levels

One of the results of human modification of river regimes is that lake levels have suffered some change, though it is not always possible to distinguish between the part played by humans and that played by natural climatic changes.

A lake basin for which there are particularly long records of change is the Valencia Basin in Venezuela (Böckh, 1973). It was the declining level of the waters in this lake which so struck the great German geographer von Humboldt in 1800. He recorded its level as being about 422 metres above sea level, and some previous observations on its level were made by Manzano in 1727, which established it as being at 426 metres. The 1968 level was about 405 metres, representing a fall of no less than 21 metres in about 240 years (figure 5.12). Humboldt believed that the cause of the declining level was the deforestation brought about by humans, and this has been supported by Böckh (1973), who points also to the abstraction of water for irrigation. This remarkable fall in level meant that the lake ceased to have an overflow into the River Orinoco. It has as a consequence become subject to a build-up in salinity, and is now eight times more saline than it was 250 years ago.

Even the world's largest lake, the Caspian, has been modified by human activities. The most important change was the fall of 3 metres in its level between 1929 and the late 1970s (see figure 5.13). This decline was undoubtedly partly the product of climatic change (Micklin, 1972), for winter precipitation in

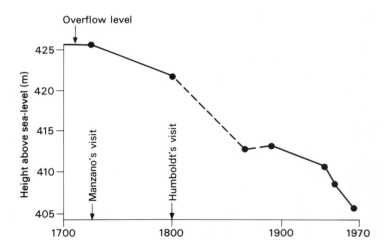

Figure 5.12
Variations in the level of Lake Valencia, Venezuela, to 1968 (after Bockh, 1973, figure 18.2)

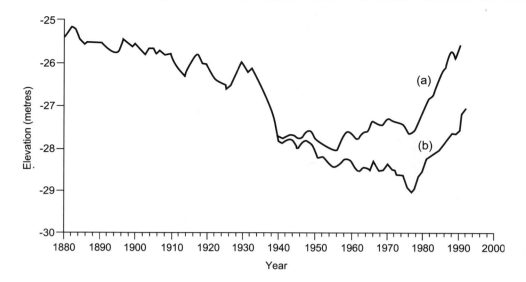

the northern Volga basin, the chief flow-generating area of the Caspian, was generally below normal for that period because of a reduction in the number of moist cyclones penetrating into the Volga basin from the Atlantic. None the less, human actions have contributed to this fall, particularly since the 1950s because of reservoir formation, irrigation, municipal and industrial with-drawals, and agricultural practices. In addition to the fall in level, salinity in the northern Caspian increased by 30 per cent since the early 1930s. A secondary effect of the changes in level has been a decline in fish numbers due to the disappearance of the shallows. These are biologically the most productive zones of the lake, providing a food base for the more valuable types of fish and also serving as spawning grounds for some species. There are plans to divert some water from northward-flowing rivers in Siberia towards the Volga to correct the decline in Caspian levels, but the possible climatic impacts of such action have caused some concern. In any event, an amelioration of climate since the late 1970s has caused some recovery in the level of the lake.

Perhaps the most severe change to a major inland sea is that taking place in the Aral Sea of the CIS (figure 5.14). Since 1960 the Aral Sea has lost more than 40 per cent of its area and about 60 per cent of its volume, and its level has fallen by more than 14 metres (Kotlyakov, 1991). This has lowered the artesian water table over a band 80–170 kilometres in width, has exposed 24,000 square kilometres of former lake bed to desiccation (plate 5.3) and has created salty surfaces from which salts are deflated to be transported in dust storms, to the detriment of soil quality. The mineral content of what remains has increased almost three-fold over the same period. It is probably the most dire ecological

Figure 5.13
Annual fluctuations in the level of the Caspian Sea, for the period 1880–1993. Curve (a) shows the changes in level which would have occurred but for anthropogenic influences, while curve (b) shows the actual observed levels (modified from World Meteorological Organization, 1995, figure 15.3)

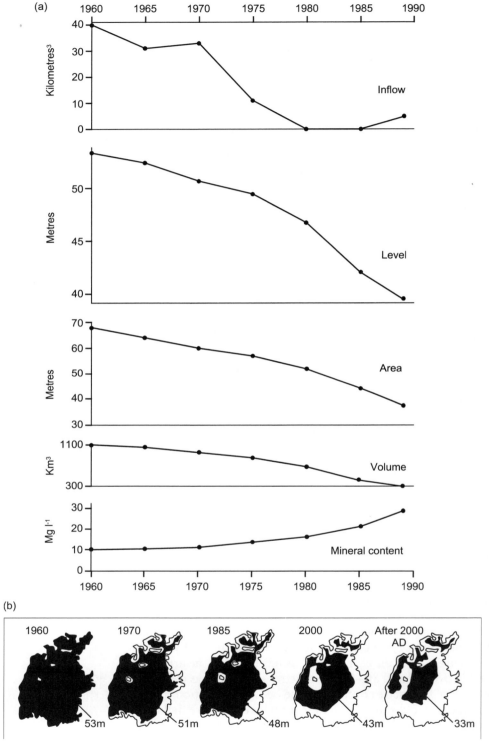

Figure 5.14
Changes in the Aral Sea: (a) 1960–1989 (from data in Kotlyakov, 1991) and (b) 1960 to after 2000 (after Hollis, 1978, p. 63)

tragedy to have afflicted the CIS, and as with the Caspian's decline much of the blame rests with excessive use of water which would otherwise replenish the sea.

Water abstraction from the River Jordan for irrigation purposes has caused a decline in the level of the Dead Sea. Under natural conditions fresh water from the Jordan constantly fed the less salty layer of the sea which occupied roughly the top 40 metres of the 320 metre deep body of water. Because the amount of water entering the lake via the Jordan was more or less equal to the quantity lost by evaporation, the lake maintained its stable, stratified state, with less salty water resting on the waters of high salinity. However, with the recent human-induced diminution in Jordan discharge, the sea's upper layer has receded because of the intense levels of evaporation. Its salinity has approached that of the older and deeper waters. It has now been established that as a consequence the layered structure has collapsed, creating a situation where there is increased precipitation of salts. Moreover, now that circulating waters carry oxygen to the bottom, the characteristic hydrogen-sulphide smell has largely disappeared (Maugh, 1979).

From time to time there have been proposals for major augmentation of lake volumes, either by means of river diversions or by allowing the ingress of sea water through tunnels or canals. Among such plans have been those to flood the salt lakes of the Kalahari by transferring water from the rivers of central Africa such as the Zambezi; the scheme to transfer Mediterranean water to the Dead Sea and to the Quattara Depression; and the Zaire–Chad scheme. Perhaps the most ambitious idea has been the so-called 'Atlantropa' project, whereby the Mediterranean would be empoldered, a dam built at the Straits of Gibraltar, and water fed into the new lake from the Zaire River (Cathcart, 1983).

Changes in groundwater conditions

In many parts of the world humans obtain water supplies by pumping from groundwater (see e.g. Drennan, 1979). This has two main effects: the reduction in the levels of water-tables and the replacement, in coastal areas, of fresh water by salt water. Environmental consequences of these two phenomena include ground subsidence and soil salinization. Increasing population levels and the adoption of new exploitation techniques (for example, the replacement of irrigation methods involving animal or human power by electric and diesel pumps) has increased these problems.

Some of the reductions in groundwater levels that have been caused by abstraction are considerable. Figure 5.15 shows the rapid increase in the number of wells tapping groundwater in

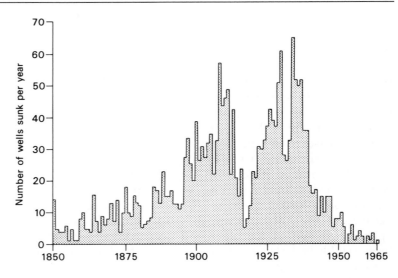

Figure 5.15
Construction of wells tapping
the confined aquifer below
London, 1850–1965

the London area from 1850 until after the Second World War, while figure 5.16 illustrates the widespread and substantial changes in groundwater conditions that resulted. The piezometric surface in the confined chalk aquifer has fallen by more than 60 metres over hundreds of square kilometres. Likewise, beneath Chicago, Illinois, pumping since the late nineteenth century has lowered the piezometric head by some 200 metres. The drawdown that has taken place in the Great Artesian Basin of Australia locally exceeds 80–100 metres (Lloyd, 1986).

The reductions in water levels that are taking place in Nebraska (plate 5.4) and the High Plains of Texas are some of the most serious, and threaten the long-term viability of irrigated agriculture in that area. Before irrigation development started in the 1930s, the High Plains groundwater system was in a state of dynamic equilibrium, with long-term recharge equal to long-term discharge. However, the groundwater is now being mined at a rapid rate to supply centre-pivot and other schemes. In a matter of only 50 years or less, the water level has declined by 30–50 metres in a large area to the north of Lubbock. The thickness of the aquifer has decreased more than 50 per cent in large parts of Castro, Crosby, Floyd, Hale, Lubbock, Parmer and Swisher counties, and already a decrease in the area irrigated by each well is taking place as a consequence of decreasing well yields (*The Cross Section*, Lubbock, 1982: 1).

Another example of excessive 'mining' of a finite resource is the exploitation of groundwater resources in the oil-rich kingdom of Saudi Arabia. Most of Saudi Arabia is desert, so climatic conditions are not favourable for rapid large-scale recharge of aquifers. Also, much of the groundwater that lies beneath the desert is a fossil resource, created during more humid conditions

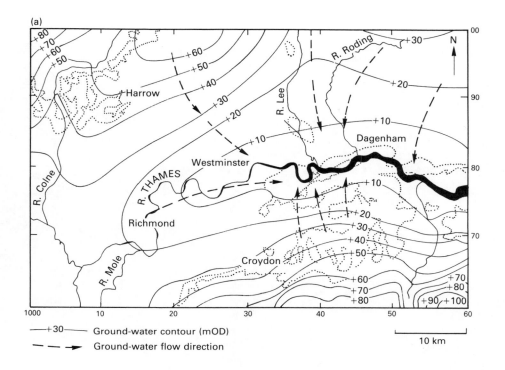

(a)

———+30——— Ground-water contour (mOD)

— — — ► Ground-water flow direction

10 km

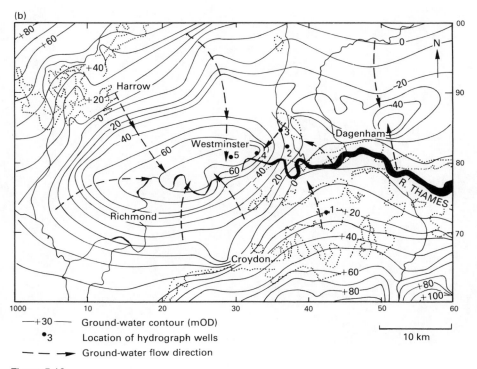

(b)

———+30——— Ground-water contour (mOD)

●3 Location of hydrograph wells

— — — ► Ground-water flow direction

10 km

Figure 5.16
Groundwater levels in the London area (a) prior to major development, (b) in 1965 and (c) in 1985 (from Wilkinson and Brassington, 1991, figures 4.5, 4.6, 4.7)

Plate 5.4
In Nebraska, USA, fields are irrigated by centre-pivot irrigation schemes which use groundwater. Groundwater levels have fallen rapidly in many areas because of the adoption of this type of irrigation technology

– pluvials – that existed in the Late Pleistocene, between 15,000 and 30,000 years ago. In spite of these inherently unfavourable circumstances, Saudi Arabia's demand for water is growing inexorably as its economy develops. In 1980 the annual demand was 2.4 billion cubic metres. By 1990 it had reached 12 billion cubic metres (a fivefold increase in just a decade), and it is expected to reach 20 billion cubic metres by 2010. Only a very small part of the demand can be met by runoff; over three-quarters of the supply is obtained from predominantly non-renewable groundwater resources. The drawdown on its aquifers is thus enormous. It has been calculated that by 2010 the deep aquifers will contain 42 per cent less water than in 1985. Much of the water is used ineffectively and inefficiently in the agricultural sector (Al-Ibrahim, 1991), to irrigate crops that could easily be grown in more humid regions and then imported.

However, there are situations where humans deliberately endeavour to increase the natural supply of groundwater by attempting artificial recharge of groundwater basins. Where the materials containing the aquifer are permeable (as in some alluvial fans, coastal sand dunes or glacial deposits) the technique of water-spreading is much used. In relatively flat areas river water may be diverted to spread evenly over the ground so that infiltration takes place. Alternative water-spreading methods may involve releasing water into basins which are formed by excavation, or the construction of dikes or small dams. On alluvial plains water can also be encouraged to percolate down to the water table by distributing it to a series of ditches or furrows. In some situations natural channel infiltration can be promoted by building small check dams down a stream course. In irrigated areas surplus water can be spread by irrigating with excess water during the dormant season. When artificial groundwater recharge is required in sediments with impermeable layers such water-spreading techniques are not effective, and the appropriate method may then be to pump water into deep pits or into wells. This last technique is used on the coastal plain of Israel, both to replenish the groundwater reservoirs when surplus irrigation water is available, and to attempt to diminish the problems associated with salt-water intrusion.

In some industrial areas, reductions in industrial activity have caused a recent reduction in groundwater abstraction, and as a consequence groundwater levels have begun to rise, a trend that is exacerbated by considerable leakage losses from ancient, deteriorating pipe and sewer systems. In cities like London, Liverpool and Birmingham an upward trend has already been identified (Brassington and Rushton, 1987; Price and Reed, 1989). In London, because of a 46 per cent reduction in groundwater abstraction, the water table in the Chalk and Tertiary beds has risen by as much as 20 metres. Such a rise has

Table 5.9 Possible effects of a rising ground-water level

Increase in spring and rivers flows
Re-emergence of 'dry springs'
Surface water flooding
Pollution of surface waters and spread of underground pollution
Flooding of basements
Increased leakage into tunnels
Reduction of slope and retaining wall stability
Reduction in bearing capacity of foundations and piles
Increased hydrostatic uplift and swelling pressures on foundations and
    structures
Swelling of clays
Chemical attack on foundations

Source: modified after Wilkinson and Brassington, 1991, table 4.3, p. 42.

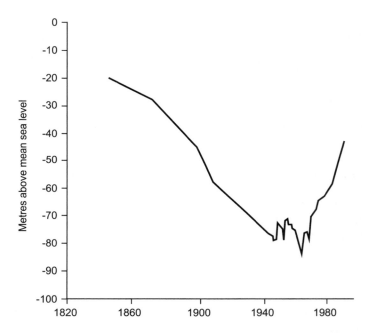

Figure 5.17
Changing groundwater levels
beneath Trafalgar Square,
Central London, showing the
sharp decline until the 1950s
and 1960s, and the substantial
rise since then (after UNEP
data)

numerous implications, which are listed in table 5.9. The trend
in groundwater levels beneath Trafalgar Square are shown in
figure 5.17.

There is, finally, a series of ways in which unintentional
changes in groundwater conditions can occur in urban areas. As
we have clearly seen, in cities surface runoff is increased by the
presence of impermeable surfaces. One consequence of this would
be that less water went to recharge groundwater. However,
there is an alternative point of view, namely that groundwater
recharge can be accelerated in urban areas because of leaking
water mains, sewers, septic tanks and soakaways (figure 5.18).
In cities in arid areas there is often no adequate provision for
storm runoff, and the (rare) increased runoff from impermeable
surfaces will infiltrate into the permeable surroundings. In some

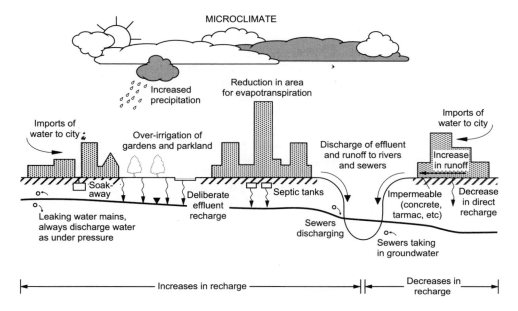

Figure 5.18
Urban effects on groundwater
recharge (after Lerner, 1990,
figure 2)

cities recharge may result from over-irrigation of parks and gardens. Indeed, where the climate is dry, or where large supplies of water are imported, or where pipes and drains are poorly maintained, groundwater recharge in urban areas is likely to exceed that in rural areas.

Water pollution

Water pollution is not new. In Britain during the reign of King George III a Member of Parliament reportedly wrote a letter to the Prime Minister complaining about the odour and appearance of the Thames. He wrote the letter, it is said, not with ink, but with water from the Thames itself (Strandberg, 1971).

Water pollution is frequently undesirable: it causes disease transmission through infection; it may poison humans and animals; it may create objectional odours and unsightliness; it may be the cause of the unsatisfactory quality even of treated water; and it may affect economic activities like shellfish culture.

The causes and forms of water pollution created by humans are many and can be classified into groups as follows (after Strandberg, 1971):

1   sewage and other oxygen-demanding wastes;
2   infectious agents;
3   organic chemicals;
4   other chemical and mineral substances;
5   sediments (turbidity);
6   radioactive substances;
7   heat (thermal pollution).

LAND AND CATCHMENT USE

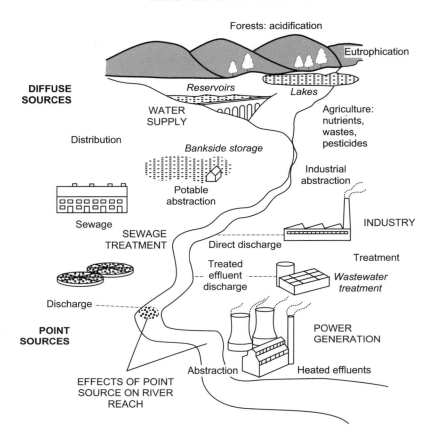

Figure 5.19
Diffuse and point sources of
pollution into river systems
(after Newson, 1992, figure 7.7)

Moreover, many human activities can contribute to changes in water quality, including agriculture, fire, urbanization, industry, mining, irrigation and many others. Of these, agriculture is probably the most important. Some pollutants have merely local effects, while others, such as acid rain or DDT, may have continental or even planetary implications.

It is possible to categorize water pollutants according to whether or not they are derived from 'point' or 'non-point' (also called 'diffuse') sources (figure 5.19). Municipal and industrial wastes tend to fall into the former category because they are emitted from one specific and identifiable place (e.g. a sewage pipe or industrial outfall). Pollutants from non-point sources include agricultural wastes, many of which enter rivers in a diffuse manner as chemicals percolate into groundwater or are washed off into fields, as well as some mining pollutants, uncollected sewage and some urban storm-water runoff.

It is not a simple matter to recognize trends in pollution. In general, long-term records are sparse, gauging stations are often moved, analytical methods change, and some trends in water

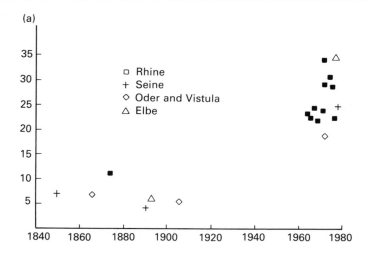

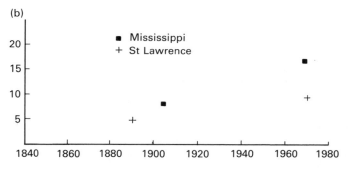

Figure 5.20
Trends in sulphur
concentrations (expressed
as mg/l) for (a) selected
European rivers and
(b) selected North American
rivers (after Husar and Husar,
1991, figures 24.15 and 24.16)

chemistry may be due to natural factors (including climatic
change) rather than to people. However, where long-term data
are available some of the changes in water chemistry are striking. In their study of the information collated on Lake Mich-
igan, the Mississippi river, the Ohio river and the Illinois river,
Ackerman et al. (1970) found that since the start of the twentieth century dissolved solids concentrations had risen markedly.
For chlorides the increase has varied from 100 per cent in
52 years to 300 per cent in 70 years, while for sulphates it
has increased from 80 per cent in 43 years to 155 per cent in
62 years. Figure 5.20 shows an historical data set, albeit imperfect, of changes in sulphur concentrations for some rivers in
Europe and North America. These data show an increase in the
European rivers by a factor of about 3, while that for the North
American rivers is about 1.5 (Husar and Husar, 1991). Some
major Chinese rivers have seen their nutrient concentrations
double in a decade (Zhang et al., 1995).

It is also plainly not a simple matter to try to estimate the
global figure for the extent of water pollution caused by humans.
For one thing we know too little about the natural long-term
levels of dissolved materials in the world's rivers. None the less,

Meybeck (1979) has calculated that about 500 million tonnes of dissolved salts reach the oceans each year as a result of human activity. These inputs have increased by more than 30 per cent the natural values for sodium, chloride and sulphate, and have created an overall global augmentation of river mineralization by about 12 per cent. Likewise, Peierls et al. (1991) have demonstrated that the quantity of nitrates in world rivers now appears to be closely correlated to human population density. Using published data for 42 major world rivers they found a highly significant correlation between annual nitrate concentration and human population density that explained 76 per cent of the variation in nitrate concentration for the 42 rivers. They maintain that 'human activity clearly dominates nitrate export from land.'

There are three main classes of chemical pollutant that deserve particular attention: nitrates and phosphates; metals; and synthetic and industrial organic pollutants.

Nitrates and phosphates, trends in the concentration of which are reviewed by Heathwaite et al. (1996), are an important cause of a process called eutrophication. Nitrates normally occur in drainage waters and are derived from soil nitrogen, from nitrogen-rich glacial deposits and from atmospheric deposition. Anthropogenic sources include synthetic fertilizers, sewage and animal wastes from feedlots. Land use changes (e.g. logging) can also increase nitrate inputs to streams. Phosphate levels are also rising in some parts of the world. Major sources include detergents, fertilizers and human wastes.

Metals, like nitrates and phosphates, occur naturally in soil and water. However, as the human use of metals has burgeoned, so has the amount of water pollution they cause. In addition, some metal ions reach river waters because they become more quickly mobilized as a result of acid rain. Aluminium is a notable example of this. From a human point of view, the metals of greatest concern are probably lead, mercury, arsenic and cadmium, all of which have adverse health effects. Other metals can be toxic to aquatic life, and these include copper, silver, selenium, zinc and chromium.

The anthropogenic sources of metal pollution include the industrial processing of ores and minerals, the use of metals, the leaching of metals from garbage and solid waste dumps, and animal and human excretions. Nriagu and Pacyna (1988) estimated the global anthropogenic inputs of trace metals into aquatic systems (including the oceans), and concluded that the sources producing the greatest quantities were, in descending order, the following (the metals produced by each source are listed in parentheses):

- domestic waste-water effluents (arsenic, chromium, copper, manganese, nickel);
- coal-burning power stations (arsenic, mercury, selenium);

- non-ferrous metal smelters (cadmium, nickel, lead, selenium);
- iron and steel plants (chromium, molybdenum, antimony, zinc);
- the dumping of sewage sludge (arsenic, manganese, lead).

However, in some parts of the world metal pollution may be derived from other sources. There is increasing evidence, for example, that in the western United States water derived from the drainage of irrigated lands may contain high concentrations of toxic or potentially toxic trace elements such as arsenic, boron, chromium, molybdenum and selenium. These can cause human health problems and poison fish and wildlife in desert wetlands (Lently, 1994).

Synthetic and industrial organic pollutants have been manu-factured and released in very large quantities since the 1960s. The dispersal of these substances into watercourses has resulted in widespread environmental contamination. There are many tens of thousands of synthetic organic compounds currently in use, and many are thought to be hazardous to life, even at quite low concentrations – concentrations possibly lower than those that can routinely be measured by commonly available analytical methods. Among these pollutants are synthetic organic pesticides, including chlorinated hydrocarbon insecticides (e.g. DDT). Some of these can reach harmful concentrations as a result of biological magnification in the food chain. Other important organic pol-lutants include PCBs, which have been used extensively in the electrical industry as di-electrics in large transformers and capacitors; PAHs, which result from the incomplete burning of fossil fuels; various organic solvents used in industrial and domestic processes; phthalates, which are plasticizers used, for example, in the production of polyvinyl chloride resins; and DBPs, which are a range of disinfection by-products. The long-term health effects of cumulative exposure to such substances are difficult to quantify. However, some work suggests that they may be implicated in the development of birth defects and certain types of cancer.

Chemical pollution by agriculture and other activities

Agriculture may be one, if not the most, important cause of pollution, either by the production of sediments or by the gen-eration of chemical wastes (Ministry of Agriculture, Fisheries and Food, 1976). With regard to the latter, it has been sug-gested that denitrification processes in the environment are incapable of keeping pace with the rate at which atmospheric nitrogen is being mobilized through industrial fixation processes and being introduced into the biosphere in the form of com-mercial fertilizers (Manners, 1978). Nitrogen, with phosphorus, tends to regulate the growth of aquatic plants and therefore the

Table 5.10   Characteristics of lakes experiencing 'cultural eutrophication'

Biological factors
(a) *Primary productivity*: usually much higher than in unpolluted water and is manifest as extensive algal blooms
(b) *Diversity of primary produces*: initially green algae increase, but blue-green algae rapidly become dominant and produce toxins. Similarly, macrophytes (e.g. reed maces) respond well initially but due to increased turbidity and anoxia (see below) they decline in diversity as eutrophication proceeds
(c) *Higher tropic level productivity*: overall decrease in response to factors given in this table
(d) *Higher tropic level diversity*: decreases due to factors given in this table. The species of macro- and micro-invertebrates which tolerate more extreme conditions increase in numbers. Fish are also adversely affected and populations are dominated by surface dwelling coarse fish such as pike and perch

Chemical factors
(a) *Oxygen content of bottom waters (hypolimnion)*: this is usually low due to algal blooms restricting oxygen exchange between the water and atmosphere. Oxygen-deficient conditions (anoxia) develop, especially at night when algae are not photosynthesizing. Thus seasonal and diurnal patterns of oxygen availability occur. The decay of algal blooms also produces anoxia
(b) *Salt content of water*: this can be very high and a further restriction on floral and faunal diversity

Physical factors
(a) *Mean depth of water body*: as infill occurs the depth decreases
(b) *Volume of hypolimnion*: varies
(c) *Turbidity*: this increases, as sediment input increases, and restricts the depth of light penetration which can become a limiting factor for photosynthesis. It is also increased if boating is a significant activity

Source: modified after Mannion, 1992, table 11.4.

eutrophication of inland waters. Excess nitrates can also cause health hazards to humans and animals, including methaemoglobinaemia in infants.

Eutrophication is the enrichment of waters by nutrients (Ryding and Rast, 1989). The process occurs naturally during, for example, the slow ageing of lakes, but it can be accelerated both by runoff from fertilized agricultural land and by the discharge of domestic sewage and industrial effluents (Lund, 1972). This process, often called 'cultural eutrophication', commonly leads to excessive growths of algae followed in some cases by a serious depletion of dissolved oxygen as the algae decay after death. Oxygen levels may become too low to support fish life, resulting in fish kills. Changes in diatom assemblages can also occur (see esp. Battarbee, 1977). The nature of these changes is summarized in table 5.10.

As agriculture has, in the developed world, become of an increasingly specialized and intensive nature, so the pollution impact has increased. The traditional mixed farm tended to be a more or less closed system which generated relatively few external impacts. This was because crop residues were fed to livestock or incorporated in the soil; and manure was returned to the land in amounts that could be absorbed and utilized. Many farms have now become more specialized, with the separation of crop and livestock activities; large numbers of stock may be kept on feedlots, silage may be produced in large silos and synthetic fertilizers may be applied to fields in large quantities (Conway and Pretty, 1991).

Increases in the application of nitrogenous fertilizers have almost amounted to a revolution (Green, 1976). The long-term trends in the UK are illustrated in figure 5.21a. The growth in fertilizer usage since the Second World War has been exponential, and in spite of the increasing costs of energy supplies and hydrocarbons in the 1970s world fertilizer production has continued to rise inexorably (table 5.11).

When one examines the evidence to see whether increasing nitrate fertilizer applications are indeed leading to widespread upward trends in nitrate concentrations in river and groundwaters, it is often conflicting (Manners, 1978; Rodda et al., 1976). This may be because soil and vegetation have a large storage capacity for nitrogen, while a great deal probably depends on local soil, climate and land management conditions. Further problems are presented by the inadequacy of long-term data. This complexity in the evidence is brought out strongly in Tomlinson's (1970) pioneer analysis of trends of nitrate concentrations in English rivers. In six rivers concentrations increased significantly with time, in eight the correlation coefficients were positive but not significant, in three cases the correlations were negative but not significant while in one case the correlation was negative and significant. However, the situation since Tomlinson has now become rather clearer (Royal Society Study Group, 1983) and, although there are considerable fluctuations from year to year, the trend in nitrate levels in English rivers is clearly apparent (figure 5.21b). They are 50 to 400 per cent higher than they were 20 years ago. The River Thames, which provides the major supply of London's water, has increased its mean annual nitrate concentration from around 11 milligrams per litre in 1928 to 35 milligrams per litre in the 1980s. Trends in nitrogen and phosphorus concentration in lowland regions of North America are not as great as those reported in lowland regions of Britain. This probably reflects the higher population density and greater intensity of land use in Britain (Heathwaite et al., 1996).

Also causing concern is the trend in nitrate concentrations in British groundwaters. Investigations have revealed a large quantity of nitrate in the unsaturated zone of the principal aquifers (mainly Chalk and Triassic sandstone), and this is slowly moving down towards the main groundwater body. The slow transit time means that in many water supply wells increased nitrate concentrations will not occur for 20 to 30 years, but they will then be above acceptable levels for human health (figure 5.21c).

However, even if an increase in nitrate levels is evident this may not necessarily be because of the application of fertilizers (Burt and Haycock, 1992). Some nitrate pollution may be derived from organic wastes. There has also been some anxiety in Britain over a possible decline in the amount of organic matter present in the soil, which could limit its ability to assimilate nitrogen. Moreover, the pattern of tillage may have affected the

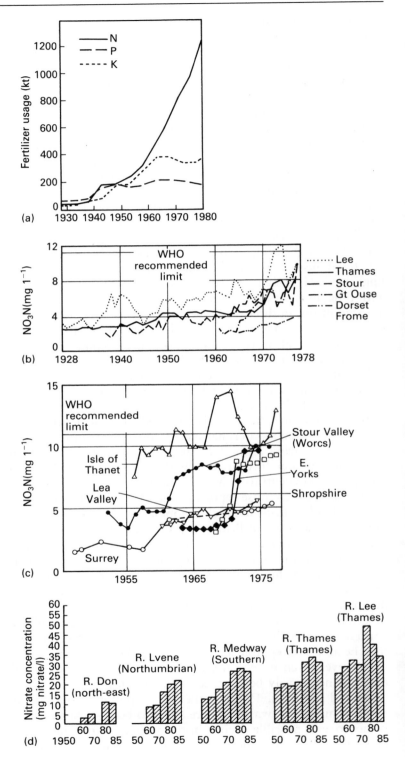

Figure 5.21
Nitrates in surface and
groundwaters in the UK:
(a) The trends in annual
    fertilizer usage in the UK
    during the period 1928–80
(b) Trends in mean annual
    nitrate concentration in five
    rivers for which long-term
    data are available
(c) Nitrate concentrations in
    selected public water
    supply abstraction
    boreholes in the Chalk and
    Triassic sandstone aquifers
    of the UK (Royal Society
    Study Group, 1983,
    figures 4, 18, and 31)
(d) Changing nitrate
    concentrations in five
    UK rivers. The averages
    are five-year means
    (from Department of the
    Environment statistics in
    Conway and Pretty, 1991,
    figure 4.8)

Table 5.11   World fertilizer production

	1970	1975	(Index 1970 = 100) 1980	1985	1988	Tonnes/km² 1988
Canada	100	184	301	403	371	2.6
USA	100	128	147	129	132	5.1
Japan	100	102	101	112	109	13.7
France	100	114	142	156	166	13.3
Germany	100	109	139	136	138	20.6
Italy	100	145	198	206	182	7.6
Netherlands	100	116	122	121	101	46.7
Spain	100	134	161	171	201	5.5
Sweden	100	115	111	111	103	7.6
UK	100	121	143	179	169	20.9
North America	100	130	152	138	140	4.6
OECD Pacific	100	98	99	112	114	2.1
OECD Europe	100	124	153	169	172	9.9
OECD	100	125	148	149	150	5.8
World	100	136	185	212	242	5.4

Source: OECD, 1991, figure 27.

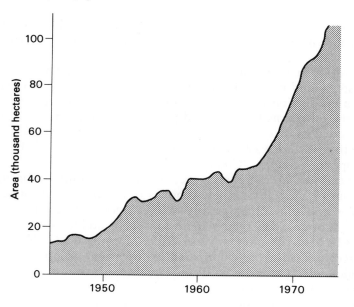

Figure 5.22
Annual total area of field
drainage by tile drains in
England and Wales (after
Green, 1976, figure 3)

liberation of nitrogen. The increased area and depth of modern
ploughing accelerates the decay of residues and may change the
pattern of water movement in the soil. Finally, tile drainage has
also expanded in area very greatly in recent decades in England
(figure 5.22). This has affected the movement of water through
the soil and hence the degree of leaching of nitrates and other
materials (Edwards, 1975).

Given the uncertainties surrounding the question of nitrate
pollution, and bearing in mind the major advances in agricultural productivity that they have permitted, attempts at

Figure 5.23
Biological concentration
occurs when relatively
indestructible substances
(DDT, for example) are
ingested by lesser organisms
at the base of the food
pyramid. An estimated 1,000
kg of plant plankton are
needed to produce 100 kg of
animal plankton. These in turn
are consumed by 10 kg of
fish, the amount needed by
one person to gain 1 kg. The
ultimate consumer (man or
woman) then takes in the DDT
taken in by 1,000 kg of the
lesser creatures at the base of
the food pyramid, when he or
she ingests enough fish to
gain 1 kg (after Strandberg,
1971, figure 2)

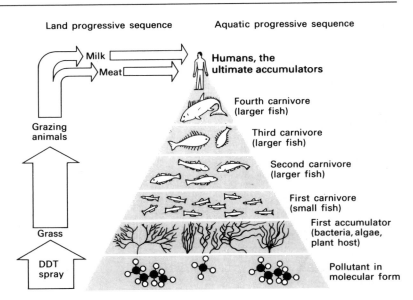

controlling their use have not been received with complete
favour. Indeed, as Viets (1971) has pointed out, if fertilizer
use were curtailed there would be less vegetation cover and
increased erosion and sediment delivery to rivers. This, he suggests, would necessitate an increase in land hectarage to maintain production levels which would also cause greater erosion;
at the same time the increased ploughing would lead to greater
nitrate loss from grasslands. Nevertheless some farmers are
inefficient and wasteful in their use of fertilizers and there is
scope for economy (Cooke, 1977).

Pesticides are another source of chemical pollution brought
about by agriculture. There is now a tremendous range of pesticides and they differ greatly in their mode of action, in the
length of time they remain in the biosphere, and in their toxicity.

Much of the most adverse criticism of pesticides has been
directed against the chlorinated hydrocarbon group of insecticides, which includes DDT and Dieldrin. These insecticides are
toxic not only to the target organism but to other insects too
(that is, they are non-specific). They are also highly persistent.
Appreciable quantities of the original application may survive
in the environment in unaltered form for years. This can have two
rather severe effects: global dispersal and the 'biological magnification' of these substances in food chains (Manners, 1978).
This last problem is well illustrated for DDT in figure 5.23.

In addition to the influence of nitrate fertilizers, the confining
of animals on feedlots results in tremendously concentrated
sources of nutrients and pollutants. Animal wastes in the United
States are estimated to be as much as 1.6 billion tonnes per
year, with 50 per cent of this amount originating from feedlots.

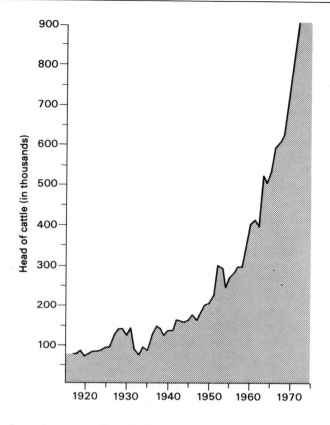

Figure 5.24
Cattle on feed lots in Colorado,
1917–1971 (after Bussing,
1972, figure 2)

It needs to be remembered that 1,000 head of beef cattle produce the equivalent organic load of 6,000 people (Sanders, 1972).
The number of cattle fed on feedlots in the United States has
increased rapidly (see figure 5.24), partly because per capita
consumption of meat has moved sharply upward, and partly
because the confinement of cattle on small lots, on which they
are fed a controlled selection of feeds, leads to the greatest possible weight gains with the least possible cost and time (Bussing,
1972). Cattle feedlot runoff is a high-strength organic waste,
high in oxygen-demanding material, and has caused many fish
kills in rivers in states such as Kansas.

Changes in surface-water chemistry may be produced by the
washing out of acids that pollute the air in precipitation. This is
particularly clear in the heavily industrialized area of north-western
Europe, where there is a marked zonation of acid rain with pH
values that often fall below 4 or 5. A broadly comparable situation exists in North America (Johnson, 1979). Very considerable
fears have been expressed about the possible effects that acid
deposition may have on aquatic ecosystems, and particularly on
fish populations. It is now generally accepted that increasing lake
and river acidification has caused fish kills and stock depletion.

Fishless lakes now occur in areas like the Adirondacks in the north-east United States. Species of fish vary in their tolerance to low pH, with rainbow trout being largely intolerant to values of pH below 6.0, while salmon, brown and brook trout are somewhat less sensitive. Values of 4.0–3.5 are lethal to salmonids, while values of 3.5 or less are lethal to most fish. Numerous lacustrine molluscs and crustaceans are not found even at weakly acid pH values of 5.8–6.0, and crayfish rapidly lose their ability to recalcify after moulting as the pH drops from 6.0 to 5.5 (Committee on the Atmosphere and the Biosphere, 1981).

Fish declines and deaths in areas afflicted by acid rain are not solely caused by water acidity per se. It has become evident that under conditions of high acidity certain metal ions, including aluminium, are readily mobilized, leading to their increasing concentration in fresh waters. Some of these metal ions are highly toxic and many mass mortalities of fish have been attributed to aluminium poisoning rather than high acidity alone. The aluminium adversely affects the operation of fish gills, causing mucus to collect in large quantities; this eventually inhibits the ability of the fish to take in necessary oxygen and salts. In addition, aluminium's presence in water can reduce the amount of available phosphates, an essential food for phytoplankton and other aquatic plants. This decreases the available food for fish higher in the food chain, leading to population decline (Park, 1987).

Mining can also create serious chemical pollution. In the coal strip-mining area of Kentucky, for example, Collier (1964) found that oxidation of the iron sulphide in the spoil banks and other parts of the mining area produced an excess of hydrogen ions. The ions then reacted with the mineral matter of the spoil bank, releasing soluble products at a rate 10 to 15 times greater than in the non-mined area. Such pollution was not entirely undesirable in its effects. Collier found that this release of minerals was good for tree growth, increasing the rate of development.

In Britain, coal seams and mudstones in the coal measures contain pyrite and marcasite (ferrous sulphide). In the course of mining the water table is lowered, air gains access to these minerals and they are oxidized. Sulphuric acid may be produced.

As a consequence, mine drainage waters may have pH values as low as 2 or 3. They have high sulphate and iron concentrations, they may contain toxic metals, and because of the reaction of the acid on clay and silicates in the rocks, they may also contain appreciable amounts of calcium, magnesium, aluminium and manganese. Following reactions with sediments or mixing with alkaline river waters, or when chemical or bacterial oxidation of ferrous compounds occurs, the iron may precipitate as ferric hydroxide. This may discolour the water and leave unsightly deposits (Rodda et al., 1976: 299). Some of the reactions involved in the development of acid mine drainage have been set down by Down and Stocks as follows (1977: 110–11):

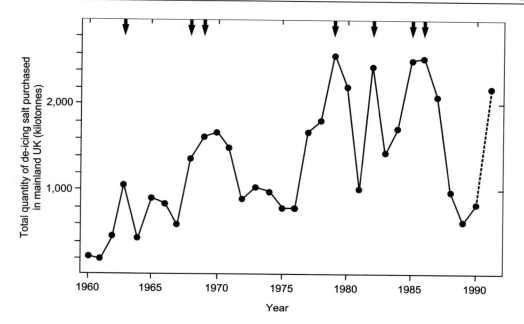

1 Oxidation of the sulphide (usually FeS_2) in a wet environment producing ferrous sulphate and sulphuric acid:

$$2FeS_2 + 2H_2O + 7O_2 = 2FeSO_4 + 2H_2SO_4$$

2 Ferrous sulphate, in the presence of sulphuric acid and oxygen, can oxidize to produce ferric sulphate, especially in the presence of the bacterium *Thiobacillus ferro-oxidans*:

$$4FeSO_4 + 2H_2SO_4 + O_2 = 2Fe_2(SO_4)_3 + 2H_2O$$

3 The ferric iron so produced combines with the hydroxyl $(OH)^-$ ions of water to form ferric hydroxide, which is insoluble in acid and precipitates:

$$Fe_2(SO_4)_3 + 6H_2O = 2Fe(OH)_3 + 3H_2SO_4$$

Figure 5.25
Estimates of the total quantity of de-icing salt purchased annually in mainland Britain during the period 1960–1991. Arrows highlight severe winters. Data provided by ICI in Dobson (1991, figure 1.1)

The nature of acid drainage from mines and the remedial measures that can be adopted to deal with it are discussed by Robb and Robinson (1995).

Rock salt (sodium chloride) has been used in increasing quantities since the Second World War to minimize the dangers to motorists and pedestrians from icy roads and pavements. With the rise in the number of vehicles there has been a tendency throughout Europe and North America for a corresponding increase in the use of salt for de-icing purposes (see e.g. Howard and Beck, 1993). Data for the UK between 1960 and 1991 are shown in figure 5.25 and demonstrate that while there is considerable inter-annual variability (related to weather severity), there has also been a general upward trend so that in the mid-1970s to end-1980s more than two million tonnes of de-icing

Table 5.12 Comparison of contaminant profiles for the urban surface runoff and raw domestic sewage, based on surveys throughout the USA

Constituent*	Urban surface runoff	Raw domestic sewage
Suspended solids	250–300	150–250
BOD[†]	10–250	300–350
Nutrients		
(a) Total nitrogen	0.5–5.0	25–85
(b) Total phosphorus	0.5–5.0	2–15
Coliform bacteria (MPN/100 ml)	10^4–10^6	10^6 or greater
Chlorides	20–200	15–75
Miscellaneous substances		
(a) Oil and grease	yes	yes
(b) Heavy metals	(10–100) times sewage conc.	traces
(c) Pesticides	yes	seldom
(d) Other toxins	potential exists	seldom

* All concentrations are expressed in mg l^{-1} unless stated ortherwise.
[†] Biochemical oxygen demand.
MPN – Most probable number.
Source: Burke, 1972, table 7.36.1.

salt were being purchased each year (Dobson, 1991). Data on de-icing salt application in North America are given by Scott and Wylie (1980). They indicate that the total use of de-icing salts in the United States increased at a nearly exponential rate between the 1940s and the 1970s, rising from about 200,000 tonnes in 1940 to approximately 9,000,000 tonnes in 1970. This represented a doubling time of about five years. Runoff from salted roads can cause ponds and lakes to become salinized, concrete to decay, and susceptible trees to die.

Similarly, storm-water runoff from urban areas may contain large amounts of contaminants, derived from litter, garbage, car-washings, horticultural treatments, vehicle drippings, industry, construction, animal droppings and the chemicals used for snow and ice clearance. Comparison of contaminant profiles for urban runoff and raw domestic sewage, based on surveys throughout the United States, indicates the importance of pollution from urban runoff sources (table 5.12). Further data for a variety of world cities are listed in table 5.13. In New York City some half a million dogs leave up to 20,000 tonnes of pollutant faeces and up to 3.8 million litres of urine in the city streets each year, all of which is flushed by gutters to storm-water sewers. Taking Britain as a whole, dogs deposit 1,000 tonnes of excrement and three million gallons of urine on the streets every day (Ponting, 1991).

None the less, it would be unfair to create the impression either that water pollution is a totally insoluble problem or that levels of pollution must inevitably rise. It is true that some components of the hydrological system – for example, lakes,

Table 5.13 Comparison of storm-water quality from various sources for selected parameters

Study location	BOD (mg/l)	Total solids (mg/l)	Suspended solids (mg/l)	Phosphate (mg/l)	Chloride (mg/l)	Faecal coliforms (MPN/100 ml)
Durham, North Carolina	2–232	194–8,860	27–7,340	0.15–2.5	3–390	7,000–86,000
Cincinnati, Ohio	1–173	–	5–1,200	0.02–7.3	5–705	500–76,000
Coshocton, Ohio	0.05–23	–	5–2,074	0.08	–	2–56,000
Detroit, Michigan	96–234	310–914	–	–	–	25,000–930,000
Seattle, Washington	10	–	–	4.3	–	16,000
Stockholm, Sweden	18–80	300–3,000	–	–	–	4,000–200,000
Pretoria, South Africa	30	–	–	–	–	240,000
Oxhey, Herts, UK	100	–	2,045	–	–	–
Moscow, Russia	18–285	–	100–3,500	–	–	–
Morristown, New Jersey	3–17	–	56–550	0.02–4.3	–	–
Ann Arbor, Michigan	28	–	2,080	0.8	–	100

MPN = Most probable number.
Source: Ellis, 1975.

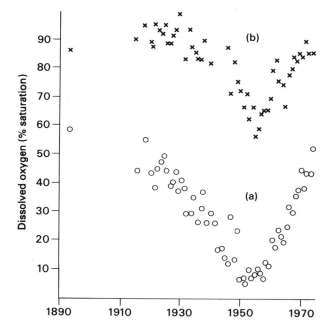

Figure 5.26
The average dissolved oxygen content of the River Thames at half-tide in the July–September quarter since 1890:
(a) 79 km below Teddington weir
(b) 95 km below Teddington Weir (after Gameson and Wheeler, 1977, figure 4)

because they may act as sumps – will prove relatively intractable to improvement, but in many rivers and estuaries striking developments have occurred in water quality in recent decades. This is clear in the case of the River Thames, which suffered a serious decline in its quality and in its fish as London's pollution and industries expanded. However, after about 1950, because of more stringent controls on effluent discharge, the downward trend in quality was reversed (see figure 5.26), and many fish, long absent from the Thames, are now returning (Gameson and Wheeler, 1977). Likewise, a long-term study of the levels of many heavy metals in sea water off the British Isles suggests

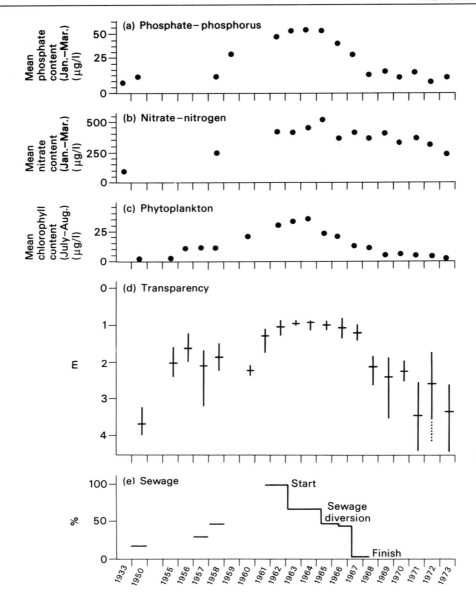

Figure 5.27
Changes in the state of Lake
Washington, USA, associated
with levels of untreated sewage
from 1933 to 1973. The relative
amount of treated sewage
entering the lake is shown as
a percentage of the maximum
rated capacity of the treatment
plants, 76 million litres per day
(after Edmonson, 1975, figure 3)

that levels are generally low and that concentrations do not
seem to be increasing (Preston, 1973).

A similar picture emerges if one considers the trend in the state
of Lake Washington near Seattle in the United States. Studies there
in 1933, 1950 and 1952 showed steady increases in the nutrient
content and in associated growth of phytoplankton (algae). These
resulted from the increasing levels of untreated sewage pumped
into the lake from the Seattle area. A sewage diversion project,
started in 1963 and completed in 1968 (see figure 5.27), resulted

in a significant decrease in the sewage input to the lake and was reflected in a rapid decline in phosphates, plankton and turbidity, and a less marked decline in nitrate levels. There also appears to have been some improvement in pollution levels in the Great Lakes since the mid-1970s (figure 5.28), as indicated by phosphorus loadings and levels of DDT and PCBs.

Deforestation and its effects on water quality

The removal of tree cover may not only affect river flow and stream-water temperature, it may also cause changes in stream-water chemistry. The reason for this is that in many forests large quantities of nutrients are cycled through the vegetation, and in some cases (notably in humid tropical rain forest) the trees are a great store of the nutrients. If the trees are destroyed the cycle of nutrients is broken and a major store is disrupted. The nutrients thus released may become available for crops, and shifting cultivators, for example, may utilize this fact and gain good crop yields in the first year after burning and felling forest. Some of the nutrients may, however, be leached out of the soils, and appear as dissolved load in streams. Large increases in dissolved load may have undesirable effects, including eutrophication (Hutchinson, 1973), soil salination and a deterioration in public water supply (Conacher, 1979).

A classic but, it must be added, extreme exemplification of this is the experiment that was carried out in the Hubbard Brook catchments in New Hampshire in the United States (Bormann et al., 1968). These workers monitored the effects of 'savage' treatment of the forest on the chemical budgets of the streams. The treatment was to fell the trees, leave them in place, kill lesser vegetation and prevent regrowth by the application of herbicide. The effects were dramatic: the total dissolved inorganic material exported from the basin was about 15 times larger after treatment, and, in particular, there was a very substantial increase in stream nitrate levels. Their concentration increased by an average of 50 times.

It has been argued, however, that the Hubbard Brook results are atypical (Sopper, 1975) because of the severity of the treatment. Most other forest clear-cutting experiments in the United States (table 5.14) indicate that nitrate/nitrogen additions to stream water are not so greatly increased. Under conventional clear-cutting the trees are harvested rather than left, all saleable material is removed, and encouragement is given to the establishment of a new stand of trees. Aubertin and Patric (1974), using such conventional clear-cutting methods rather than the drastic measures employed in Hubbard Brook, found that clear-cutting in their catchments in West Virginia had a negligible

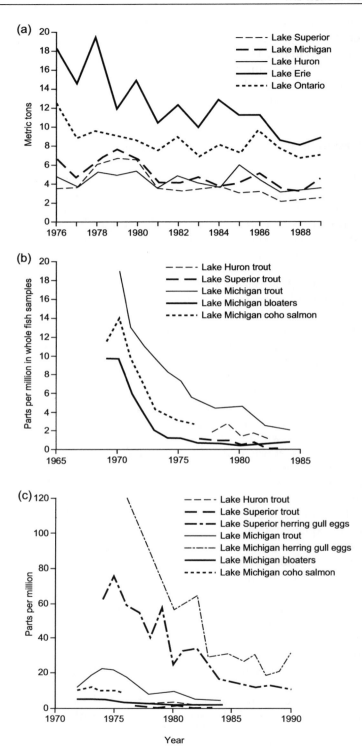

Figure 5.28
Changes in water pollution in
the Great Lakes of North
America:
(a) Estimated phosphorous
    loadings
(b) DDT levels
(c) PCB levels
(after Council on Environmental
Quality, 17th (1986) and 22nd
(1992) Annual Reports

Table 5.14 Nitrate–nitrogen losses from control and disturbed forest ecosystems

Site	Nature of disturbance	Nitrate–nitrogen loss (kg ha^{-1} y^{-1})	
		Control	Disturbed
Hubbard Brook	Clear-cutting without vegetation removal, herbicide inhibition of re-growth	2.0	97
Gale River (New Hampshire)	Commercial clear-cutting	2.0	38
Fernow (West Virginia)	Commercial clear-cutting	0.6	3.0
Coweeta (North Carolina)	Complex	0.05	7.3*
H. J. Andrews Forest (Oregon)	Clear-cutting with slash-burning	0.08	0.26
Alsea River (Oregon)	Clear-cutting with slash-burning	3.9	15.4
	MEAN	1.44	26.83

* This value represents the second year of recovery after a long-term disturbance. All other results for disturbed ecosystems reflect the first year after disturbance.
Source: modified after Vitousek et al., 1979.

effect on stream temperatures (they left a forest strip along the stream), pH and the concentration of most dissolved solids.

Even if clear-cutting is practised, there is some evidence of a rapid return to steady-state nutrient cycling because of quick regeneration. As Marks and Bormann (1972) have put it: 'Because terrestrial plant communities have always been subjected to various forms of natural disturbances, such as wind storms, fires, and insect outbreaks, it is only reasonable to consider recovery from disturbance as a normal part of community maintenance and repair.' Likewise Hewlett et al. (1984) found no evidence in the Piedmont region of the United States that clear-cutting would create such extreme nitrate loss that soil fertility would be impaired or water eutrophication caused. They point out that rapid growth of vegetation minimizes nutrient losses from the ecosystem by three main mechanisms. It channels water from runoff to evapotranspiration, thereby reducing erosion and nutrient loss; it reduces the rates of decomposition of organic matter through moderation of the microclimate so that the supply of soluble ions for loss in drainage water is reduced; and it causes the simultaneous incorporation of nutrients into the rapidly developing biomass so that they are not lost from the system.

In Britain, a comparison of water chemistry in forested and clear-cut areas in the Plynlimon area of mid-Wales has revealed certain trends (figure 5.29). The most notable of these is the increase in nitrate levels, which impacts on catchment acidification processes by causing a decrease in pH and releasing toxic metals such as aluminium. Also important is the increase in dissolved organic carbon, which causes discoloration of stream water (Institute of Hydrology, 1991).

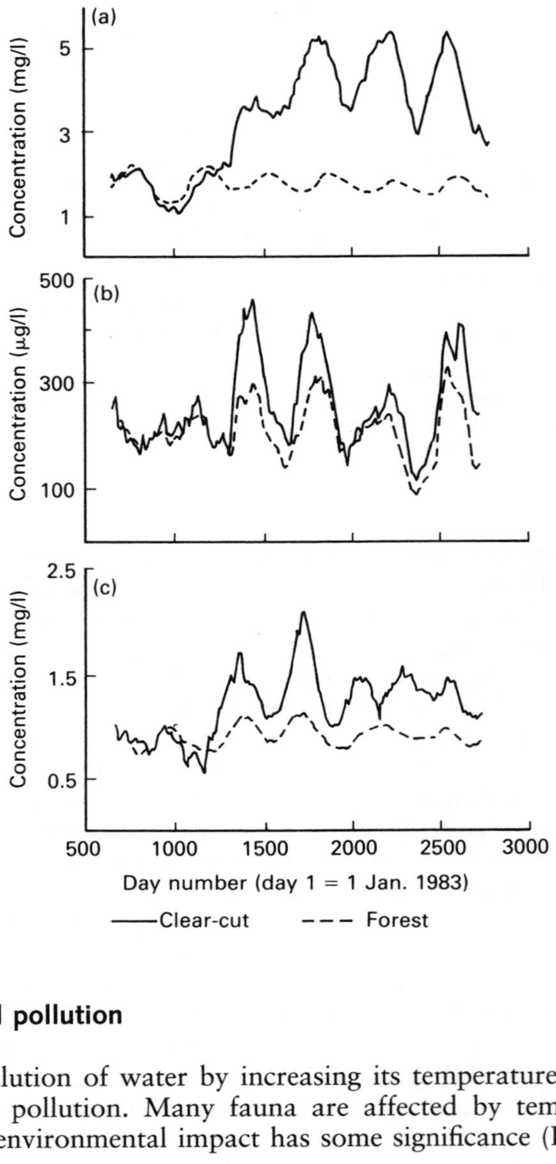

Figure 5.29
Changes in the chemical
composition of stream water
from clear-cut and forested
catchments in mid-Wales,
1983–1990:
(a) nitrate (NO_3)
(b) aluminium (Al)
(c) dissolved organic carbon
    (DOC)
(modified after Institute of
Hydrology Report, 1990/1,
figure 20)

Thermal pollution

The pollution of water by increasing its temperature is called thermal pollution. Many fauna are affected by temperature, so this environmental impact has some significance (Langford, 1990).

In industrial countries probably the main source of thermal pollution is condenser cooling water released from electricity generating stations. Water discharged from power stations has been heated some 6–9°C, but usually has a temperature of less than 30°C. The extent to which water affects river temperature depends very much on the state of flow. For example, below the Ironbridge power station in England, the Severn river undergoes a temperature increase of only 0.5°C during floods, compared with an 8°C increase at times of low flow (Rodda et al., 1976).

Table 5.15 Cooling water used per unit of power generation in the UK

Year	Temperature rise (°C)	Specific water circulation per unit of power generation (m³)
(a) *Coal and oil-fired*		
1950	7	0.21
1960	9	0.15
1970	10	0.12
1980 (estimated)	12–14	0.08–0.09
(b) *Nuclear*		
1962	8	0.33
1967	9	0.24
1980 (estimated)	12–14	0.09–0.14

Source: Royal Commission on Environmental Pollution, 1972, table 7, p. 21.

Over the past four decades or more (see Royal Commission on Environmental Pollution, 1972), both the increase in the capacity of individual electricity power-generating units and the improvements in their thermal efficiency have led to a diminution in the heat rejected in relation to the amount of cooling water per unit of production (table 5.15). Economic optimization of the generating plant has cut down the flow of cooling water per unit of electricity, though it has raised its temperature. This increased temperature rise of the cooling water is more than offset by the reduction achieved in the volume of water utilized. None the less, the expansion of generating capacity has meant that the total quantity of heat discharged has increased, even though it is much less than would be expected on a simple proportional basis.

Thermal pollution of streams may also follow from urbanization (Pluhowski, 1970). This results from various sources: changes in the temperature regime of streams brought about by reservoirs according to their size, their depth and the season; changes produced by the urban heat island effect; changes in the configuration of urban channels (for example, their width–depth ratio); changes in the degree of shading of the channel, either by covering it over or by removing natural vegetation cover; changes in the volume of storm runoff; and changes in the groundwater contribution. Pluhowski found that the basic effects of cities on river-water temperatures on Long Island, New York, was to raise temperatures in summer (by as much as 5–8°C) and to lower them in winter (by as much as 1.5–3.9°C).

Reservoir construction can also affect stream-water temperatures. Crisp (1977) found, for example, that as a result of the construction of a reservoir at Cow Green (Upper Teesdale, northern England) the temperature of the river downstream was modified. He noted a reduction in the amplitude of annual water

Table 5.16 Effect of forest-cutting on mean monthly maximum river-water temperature in the USA

Location	Temperature increases (°C)
Coweeta (North Carolina)	4
Fernow (West Virginia)	5.5
Penn State (Pennsylvania)	6
Hubbard Brook (New Hampshire)	6
H. J. Andrews (Oregon)	6.7
Coastal Range (Oregon)	8
MEAN	6.03

Source: from data in Sopper, 1975, table 2.

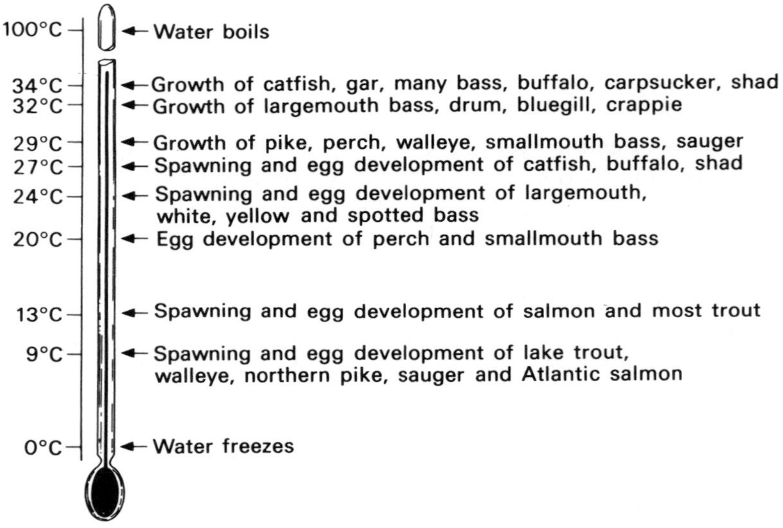

Figure 5.30
Maximum temperatures for the spawning and growth of fish. Heated waste water may be up to 5–10°C warmer than receiving waters and consequently the local fish populations cannot reproduce or grow properly (after Giddings, *Chemistry, Man and Environmental Change*, Harper and Row) from *Encounter with the Earth* by Leo F. Laporte, figure 13-2. Copyright © by permission of Harper & Row Publishers, Inc.)

100°C	← Water boils
34°C	← Growth of catfish, gar, many bass, buffalo, carpsucker, shad
32°C	← Growth of largemouth bass, drum, bluegill, crappie
29°C	← Growth of pike, perch, walleye, smallmouth bass, sauger
27°C	← Spawning and egg development of catfish, buffalo, shad
24°C	← Spawning and egg development of largemouth, white, yellow and spotted bass
20°C	← Egg development of perch and smallmouth bass
13°C	← Spawning and egg development of salmon and most trout
9°C	← Spawning and egg development of lake trout, walleye, northern pike, sauger and Atlantic salmon
0°C	← Water freezes

temperature fluctuations, and a delay in the spring rise of water temperature by 20–50 days and in the autumn fall by up to 20 days. In rural areas human activities can also cause significant modifications in river water temperature. Deforestation (table 5.16) is especially important (Lynch et al., 1984). Swift and Messer (1971), for example, examined stream temperature measurements during six forest-cutting experiments on small basins in the US southern Appalachians. They found that with complete cutting, because of shade removal, the maximum stream temperatures in summer went up from 19 to 23°C. They believed that such temperature increases were detrimental to temperature-sensitive fish like trout (see figure 5.30). On the other hand, the temporary shutdown of power plants may create severe cold-shock kills of fish in discharge-receiving waters in winter (Clark, 1977). Moreover, an increase in water temperature causes a decrease in the solubility of oxygen which is needed for the oxidation of biodegradable wastes. At the same time, the rate of oxidation is accelerated, imposing a faster oxygen demand on the smaller supply and thereby depleting the oxygen content of the water still further. Temperature also affects the lower organisms, such as plankton and crustaceans. In general, the more elevated the temperature, the less desirable the types of algae in water. In cooler waters diatoms are the predominant phytoplankton in water that is not heavily eutrophic; with the same nutrient levels, green algae begin to become dominant at higher temperatures and diatoms decline; at the highest water temperatures, blue-green algae thrive and often develop into heavy blooms. One further ecological consequence of thermal pollution is that the spawning and migration of many fish are triggered by temperature and this behaviour can be disrupted by thermal change.

Pollution with suspended sediments

Probably the most important effect that humans have had on water quality is the contribution to levels of suspended sediments in streams. This is a theme which is intimately tied up with that of soil erosion, on the one hand, and with channel manipulation by activities such as reservoir construction, on the other. The clearance of forest, the introduction of ploughing and grazing by domestic animals, the construction of buildings, and the introduction of spoil materials into rivers by mineral extraction industries have all led to very substantial increases in levels of stream turbidity. Frequently sediment levels are a whole order of magnitude higher than they would have been under natural conditions. However, the introduction of soil conservation measures, or a reduction in the intensity of land use, or the construction of reservoirs, can cause a relative (and sometimes absolute) reduction in sediment loads. All three of these factors have contributed to the observed reduction in turbidity levels monitored in the rivers of the Piedmont region of Georgia in the United States.

Marine pollution

It is not within the scope of this book to discuss human impacts on the oceans to any great extent. However, it is worth making a few points on this subject, which has been well reviewed by Clark (1997). At first sight, as Jickells et al. (1991: 313) point out, two contradictory thoughts may cross our minds on this issue:

The first is the observation of ocean explorers, such as Thor Heyerdahl, of lumps of tar, flotsam and jetsam, and other products of human society thousands of kilometers from inhabited land. An alternative, vaguer feeling is that given the vastness of the oceans (more than 1,000 billion billion litres of water!), how can man have significantly polluted them?

What is the answer to this conundrum? Jickells et al. (p. 330) draw a clear distinction between the open oceans and regional seas and in part come up with an answer:

The physical and chemical environment of the open oceans has not been greatly affected by events over the past 300 years, principally because of their large diluting capacity . . . Material that floats and is therefore not diluted, such as tar balls and litter, can be shown to have increased in amount and to have changed in character over the past 300 years.

In contrast to the open oceans, regional areas in close proximity to large concentrations of people show evidence of increasing concentrations of various substances that are almost certainly linked to human activities. Thus the partially enclosed North Sea and Baltic Sea show increases in phosphate concentrations as a result of discharges of sewage and agriculture. The same is true of the more enclosed Black Sea (Mee, 1992).

This can cause accelerated eutrophication – often called 'cultural eutrophication' – which can lead to excessive growths of algae. Coastal and estuary water are sometimes affected by algal foam and scum, often called 'red tides'. Some of these blooms are so toxic that consumers of seafood that has been exposed to them can be affected by diarrhoea, sometimes fatally.

The nature of red tides has recently been discussed by Anderson (1994), who points out that these blooms, produced by certain types of phytoplankton (tiny pigmented plants), can grow in such abundance that they change the colour of the seawater not only red but also brown or even green. They may be sufficiently toxic to kill marine animals such as fish and seals. Long-term studies at the local and regional level in many parts of the world suggest that these so-called red tides are increasing in extent and frequency as coastal pollution worsens and nutrient enrichment occurs more often.

Eutrophication also has adverse effects on coral reefs, as explained by Weber (1993: 49):

Initially, coral productivity increases with rising nutrient supplies. At the same time, however, corals are losing their key advantage over other organisms: their symbiotic self-sufficiency in nutrient poor seas. As eutrophication progresses, algae start to win out over corals for newly opened spaces on the reef because they grow more rapidly than corals when fertilized. The normally clear waters cloud as phytoplankton begin to multiply, reducing the intensity of the sunlight reaching the corals, further lowering their ability to compete. At a certain point, nutrients in the surrounding waters begin to overfertilize the corals' own zooxanthellae, which multiply to toxic levels inside the polyps. Eutrophication may also lead to black band and white band disease, two deadly coral disorders thought to be caused by algal infections. Through these stages of eutrophication, the health and diversity of reefs declines, potentially leading to death.

It is clear that pollution in the open ocean is, as yet, of limited biological significance. GESAMP (1990), an authoritative review of the state of the marine environment for the United Nations Environment Programme, reported (p. 1):

The open sea is still relatively clean. Low levels of lead, synthetic organic compounds and artificial radionuclides, though widely detectable, are biologically insignificant. Oil slicks and litter are common along sea lanes, but are, at present, of minor consequence to communities of organisms living in open-ocean waters.

On coastal waters it reported (p. 1):

The rate of introduction of nutrients, chiefly nitrates but sometimes also phosphates, is increasing, and areas of eutrophication are expanding, along with enhanced frequency and scale of unusual plankton blooms and excessive seaweed growth. The two major sources of nutrients to coastal waters are sewage disposal and agricultural run-off from fertilizer-treated fields and from intensive stock raising.

Attention is also drawn to the presence of synthetic organic compounds – chlorinated hydrocarbons, which build up in the fatty tissues of top predators such as seals which dwell in coastal waters. Levels of contamination are decreasing in northern temperate areas but rising in tropical and subtropical areas due to continued use of chlorinated pesticides there.

The world's oceans have been greatly contaminated by oil. The effects of this pollution on animals have been widely studied. The sources of pollution include tanker collisions with other ships, the explosion of individual tankers because of the build-up of gas levels, the wrecking of tankers on coasts through navigational or mechanical failure, seepage from offshore oil installations, and the flushing of tanker holds (figure 5.31a). There is no doubt that the great bulk of the oil and related materials polluting the oceans results from human action (Blumer, 1972), though not all the blame should be attributed to humans, for natural seepages are reasonably common (Landes, 1973). Paradoxically, human actions mean that many natural seepages have diminished as wells have drawn down the levels of hydrocarbons in the oil-bearing rocks. Likewise, the flow of asphalt that once poured from the Trinidad Pitch Lake into the Gulf of Paria has also ceased because mining of the asphalt has lowered the lake below its outlet.

Of the various serious causes of oil pollution, a more important mechanism than the much-publicized role of tanker accidents is the discharge of the ballast water taken into empty tankers to

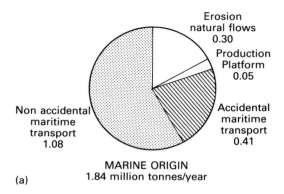

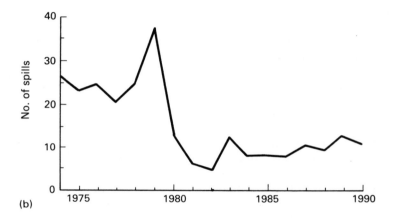

Figure 5.31
Causes and trends of oil
pollution in the oceans:
(a) the origin of oil pollution
(b) the number of accidental
    oil spills over 700 tonnes,
    1974–88
(from OECD Report, 1991,
figure 15, with modifications)

provide stability on the return voyage to the loading terminal
(Pitt, 1979). It is in this field, however, that technical developments have led to most amelioration. For example, the quantity
of oil in tanker ballast water has been drastically reduced by the
LOT (load on top) system, in which the ballast water is allowed
to settle so that the oil rises to the surface. The tank is then
drained until only the surface oil in the tank remains, and this
forms part of the new cargo. In the future large tankers will be
fitted with separate ballast tanks not used for oil cargoes, called
segregated or clean ballast systems. Given such advances it
appears likely that an increasing proportion of oil pollution will
be derived from accidents involving tankers of ever-increasing
size, such as the spillage of 300 million litres from the wreck
of the *Amoco Cadiz* in 1978. Even here, however, some progress has been made. Since 1980 there has been a reduction
in the number of major oil spills, partly as a consequence of
diminished long-distance oil transport to Western Europe
(figure 5.31b).

Human Agency in Geomorphology

6

Introduction

The human role in creating landforms and modifying the operation of geomorphological processes such as weathering, erosion and deposition is a theme of great importance though one that, particularly in the Western world, has not received the attention it deserves. Over the last three decades there have been some useful general surveys (see e.g. Brown, 1970; Jennings, 1966; Haigh, 1978; Nir, 1983) and, with the increasing concern for the relevance and application of geomorphological studies to human problems (see e.g. Cooke and Doornkamp, 1990; Hails, 1977), the situation is likely to change. Moreover, the development of process studies in post-Second World War geomorphology has led to a relative reduction in the importance of evolutionary and historical geomorphology within the discipline. Examination of processes, and in particular of their rate of operation, has served to highlight the part played by humans in modifying the physical landscape (Goudie, 1995).

The range of the human impact on both forms and process is considerable. For example, in table 6.1 there is a list of some anthropogenic landforms together with some of the causes of their creation. There are very few spheres of human activity which do not, even indirectly, create landforms. It is, however, useful to recognize that some features are produced by direct anthropogenic processes. These tend to be more obvious in their form and origin and are frequently created deliberately and knowingly. They include landforms produced by constructional activity (such as tipping), by excavation, by travel, by hydrological interference, and by farming (Spencer and Hale, 1961). Hillsides have been terraced in many parts of the world for many centuries, notably in the arid and semi-arid highlands of the New World (Donkin, 1979). Haigh (1978) has classified anthropogenic landforming processes as follows:

Table 6.1 Some anthropogenic landforms

Feature	Cause
Pits and ponds	Mining, marling
Broads	Peat extraction
Spoil heaps	Mining
Terracing, lynchets	Agriculture
Ridge and furrow	Agriculture
Cuttings and sunken lanes	Transport
Embankments	Transport, river and coast management
Dikes	River and coast management
Mounds	Defence, memorials
Craters	War, *qanat* construction
City mounds (*tells*)	Human occupation
Canals	Transport, irrigation
Reservoirs	Water management
Subsidence depressions	Mineral and water extraction
Moats	Defence
Banks along roads	Noise abatement

1   *Direct anthropogenic processes*
    1.1   Constructional:
        tipping – loose, compacted, molten, graded, moulded, ploughed, terraced
    1.2   Excavational:
        digging, cutting, mining, blasting of cohesive or non-cohesive materials
        cratered
        trampled, churned
    1.3   Hydrological interference:
        flooding, damming, canal construction, dredging, channel modification
        draining
        coastal protection
2   *Indirect anthropogenic processes*
    2.1   Acceleration of erosion and sedimentation:
        agricultural activity and clearance of vegetation
        engineering, especially road construction and urbanization
        incidental modifications of hydrological regime
    2.2   Subsidence – collapse, settling:
        mining
        hydraulic
        thermokarst
    2.3   Slope failure – landslide, flow, accelerated creep:
        loading
        undercutting
        shaking
        lubrication
    2.4   Earthquake generation:
        loading (reservoirs)
        lubrication (fault plane).

Landforms produced by indirect anthropogenic processes are often less easy to recognize, not least because they tend to involve, not the operation of a new process or processes, but the acceleration of natural processes. They are the result of environmental changes brought about inadvertently by human technology. None the less, it is probably this indirect and inadvertent modification of process and form which is the most crucial aspect of anthropogeomorphology. By removing natural vegetation cover – through the agency of cutting, burning and grazing (see Trimble and Mendel, 1995) – humans have accelerated erosion and sedimentation. Sometimes the results will be obvious, for example when major gully systems rapidly develop; other results may have less immediate effect on landform but are, nevertheless, of great importance. By other indirect means humans may create subsidence features and problems, trigger off mass movements like landslides, and even influence the operation of phenomena like earthquakes.

Finally there are situations where, through a lack of understanding of the operation of processes and the links between different processes and phenomena, humans may deliberately and directly alter landforms and processes and thereby set in train a series of events which were not anticipated or desired. There are, for example, many records of attempts to reduce coast erosion by important and expensive engineering solutions, which, far from solving erosion problems, only exacerbated them.

Landforms produced by excavation

Of the landforms produced by direct anthropogenic processes those resulting from excavation are widespread, and may have some antiquity. For example, Neolithic peoples in the Breckland of East Anglia in England used antler picks and other means to dig a remarkable cluster of deep pits in the chalk. The purpose of this was to obtain good-quality non-frost-shattered flint to make stone tools. The cratered area, called Grimes Graves, is well displayed on aerial photographs, and is illustrated in figure 6.1. In many parts of Britain chalk has also been excavated to provide marl for improving acidic, light, sandy soils, and Prince (1962, 1964) has made a meticulous study of the 27,000 pits and ponds in Norfolk that have resulted mainly from this activity – an activity particularly prevalent in the eighteenth century. It is often difficult in individual cases to decide whether the depressions are the results of human intervention, for solutional and periglacial depressions are often evident in the same area, but pits caused by human action do tend to have some distinctive features: irregular shape, a track leading into them, proximity to roads, etc. (see the debate between Prince and Sperling et al., 1979).

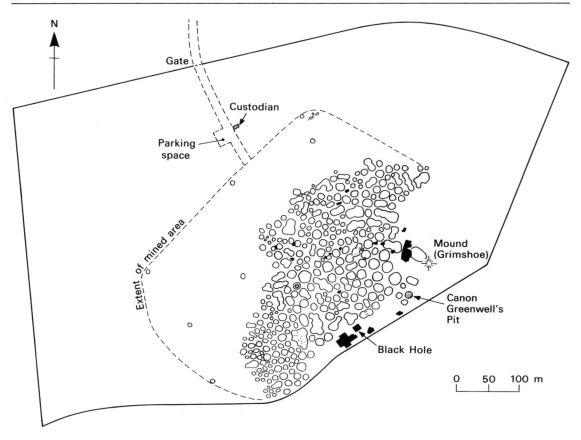

Figure 6.1

Flint mines at Grimes Graves, Norfolk, England, dug during Neolithic times (around 5,000 years ago) (after Clark, 1963)

Difficulties of identifying the true origin of excavational features were also encountered in explaining the Broads, a group of 25 fresh-water lakes in the county of Norfolk in England (plate 6.1). They are of sufficient area, and depth, for early workers to have precluded a human origin. Later it was proposed instead that they were natural features caused by uneven alluviation and siltation of river valleys which were flooded by the rapidly rising sea level of the Holocene (Flandrian) transgression. It was postulated by Jennings (1952) that the Broads were initiated as a series of discontinuous natural lakes, formed beyond the limits of a thick estuarine clay wedge laid down in Romano-British times by a transgression of the sea over earlier valley peats. The waters of the Broads were thought to have been ponded back in natural peaty hollows between the flanges of the clay and the marginal valley slopes, or in tributary valleys whose mouths were blocked by the clay.

It is now clear, however, that the Broads are the result of human work (Lambert et al., 1970). Some of them have rectilinear and steep boundaries, most of them are not mentioned in

Plate 6.1
The Norfolk Broads of eastern
England were originally
thought to be natural lakes.
However, recent researches
by historical geographers have
shown that for the most part
they result from the activities
of medieval peat diggers

early topographic books, and archival records indicate that peat-cutting (turbary) was widely practised in the area. On these and other grounds it is believed that peat-diggers, before AD 1300, excavated 25.5 million cubic metres of peat and so created the depressions in which the lakes have formed. The flooding may have been aided by sea-level change. Comparably extensive peat excavation was also carried on in the Netherlands, notably in the fifteenth century.

Other excavational features result from war, especially craters caused by bomb or shell impact. Regrettably, human power to create such forms is increasing. It has been calculated (Westing and Pfeiffer, 1972) that between 1965 and 1971, 26 million craters, covering an area of 171,000 hectares, were produced by bombing in Indo-China. This represents a total displacement of no less than 2.6 billion cubic metres of earth, a figure much greater than calculated as being involved in the peaceable creation of the Netherlands.

Some excavation is undertaken on a large scale for purely aesthetic reasons, when nature offends the eye (Prince, 1959), while in many countries where land is scarce, whole hills are levelled and extensive areas stripped to provide fill for harbour reclamation. One of the most spectacular examples of this kind of activity was the deliberate removal of steep-sided hills in the centre of Brazil's Rio de Janeiro, for housing development. In

Bahrain, an island in the Arabian-Persian Gulf, a substantial part of the desert surface has been stripped to provide fill for land reclamation in the commercial and industrial area in the north of the island.

An excavational activity of a rather specialized kind is the removal of limestone pavements (plate 6.2) – areas of exposed limestone which were stripped by glaciers and then moulded into bizarre shapes by solutional activity – for ornamental rock gardens. These pavements, which consist of arid, bare rock surfaces (clints) bounded by deep, humid fissures (grikes), have both an aesthetic and a biological significance (Ward, 1979). They occur in northern England, notably in the Pennines and in the Morecambe Bay area, and provide a habitat for various rare plants (including *Actaea spicata*, *Ribes spicatum and Dynopteris villarii*). A recent survey indicates that the pavements, which cover only about 2,150 hectares in Britain, have been severely damaged. On only 3 per cent of 537 pavement units recorded could no damage be detected, while the proportion considered to be 95 per cent or more intact was only 13 per cent.

The most important cause of excavation is still mineral extraction (plate 6.3), producing open-pit mines, strip mines, quarries for structural materials, borrow pits along roads, and similar features (Doerr and Guernsely, 1956). Of these, 'without question, the environmental devastation produced by strip mining exceeds in quantity and intensity all of the other varied forms of man-made land destruction' (Strahler and Strahler, 1973: 284). This form of mining (plate 6.4) is a particular environmental problem in the states of Pennsylvania, Ohio, West Virginia, Kentucky and Illinois. If, because of increasing pressure on available energy resources, the production of coal is expanded then

this particular geomorphological impact may also increase. The same applies to the probable exploitation of oil shales, a potential source of oil that at present is relatively untapped. This can be exploited by open-pit mining, by traditional room and pillar mining, and by underground in situ pyrolysis. Vast reserves exist but the amount of excavation required will probably be about three times the amount of oil produced (on a volume basis), suggesting that the extent of both the excavation and the subsequent dumping of overburden and waste will be considerable (Routson et al., 1979). Some of the waste, that produced by the retorting of the shale to release the oil, contains soluble salts and potentially harmful trace elements, which limit the speed of ground reclamation (Petersen, 1981).

An attempt to provide a general picture of the importance of excavation in the creation of the landscape of Britain was given by Sherlock (1922). He estimated (see table 6.2) that up until the time in which he wrote, human society had excavated around 31 billion cubic metres of material in the pursuit of its economic activities. That figure must now be a gross underestimate, partly because Sherlock himself was not in a position to appreciate the anthropogenic role in creating features like the Norfolk Broads, and partly because, since his time, the rate of excavation has greatly accelerated. Sherlock's 1922 study covers a period when

Plate 6.3
The Rössing uranium mine near Swakopmund in Namibia, southern Africa. The excavation of such mines involves the movement of prodigious amounts of material

Plate 6.4
Strip-mining for coal in
Missouri, USA, is a particularly
unsightly example of the
production of landforms by
human agency

earth-moving equipment was still ill-developed. None the less,
on the basis of his calculations, he was able to state that 'at the
present time, in a densely peopled country like England, Man is
many times more powerful, as an agent of denudation, than all
the atmospheric denuding forces combined' (p. 333). The most

Table 6.2 Total excavation of material in Great Britain until 1922

Activity	Approximate volume (m³)
Mines	15,147,000,000
Quarries and pits	11,920,000,000
Railways	2,331,000,000
Manchester Ship Canal	41,154,000
Other canals	153,800,000
Road cuttings	480,000,000
Docks and harbours	77,000,000
Foundations of buildings and street excavation	385,000,000
TOTAL	30,534,954,000

Source: Sherlock, 1922, p. 86.

notable change since Sherlock wrote has taken place in the production of aggregates for concrete which are, in descending order of importance, sand and gravel, crushed limestone, artificial or manufactured aggregates, and crushed sandstone. Demand for these materials in the UK grew from 20 million tonnes per annum in 1900, to 50 million tonnes in 1948 and 276 million tonnes in 1973, an increase in per capita consumption from 0.6 tonnes per year to about 5 tonnes per year (Jones, 1983).

For the world as a whole, the annual movement of soil and rock resulting from mineral extraction may be as high as 3,000 billion tonnes (Holdgate et al., 1982: 186). By comparison it has been estimated that the amount of sediment carried into the ocean by the world's rivers each year amounts to 24 billion tonnes per year (Judson, 1968).

Landforms produced by construction and dumping

The process of constructing mounds and embankments and the creation of dry land where none previously existed is long-standing. In Britain the mound at Silbury Hill (plate 6.5) dates back to prehistoric times, and the pyramids of central America, Egypt and the Far East are even more spectacular early feats of landform creation. Likewise in the Americas, Native Indians, prior to the arrival of Europeans, created large numbers of mounds of different shapes and sizes for temples, burials, settlement and effigies (Denevan, 1992: 377). In the same way, hydrological management has involved, over many centuries, the construction of massive banks and walls – the ultimate result being the landscape of the present-day Netherlands. Transport developments have also required the creation of large constructional landmarks, but probably the most important features are those resulting from the dumping of waste materials, especially those derived from mining (see figure 6.2). It has been calculated

Plate 6.5
In prehistoric times some striking constructional landforms were created. Top: Silbury Hill, Wiltshire, England. Bottom: a tell in the Middle East produced by the accumulation of the debris of years of human occupation

that there are at least 2,000 million tonnes of shale lying in pit heaps in the coalfields of Britain (Richardson, 1976). In the Middle East and other areas of long-continued human urban settlement the accumulated debris of life has gradually raised the level of the land surface, and occupation mounds (tells) are a fertile source of information to the archaeologist. Today, with the technical ability to build that humans have, even estuaries may be converted from ecologically productive environments into suburban sprawl (figure 6.3) by the processes of dredging

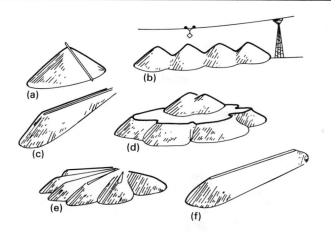

Figure 6.2
Some shapes produced by
shale tipping:
(a) conical resulting from
    MacClane tipping
(b) multiple cones tipped
    from aerial ropeways
(c) high fan-ridge by tramway
    tipping over slopes
(d) high plateau mounds
    topped with cones
(e) low multiple fan ridges
    by tramway tipping
(f) lower ridge by tramway
    tipping
(after Haigh, 1978, figure 2.1)

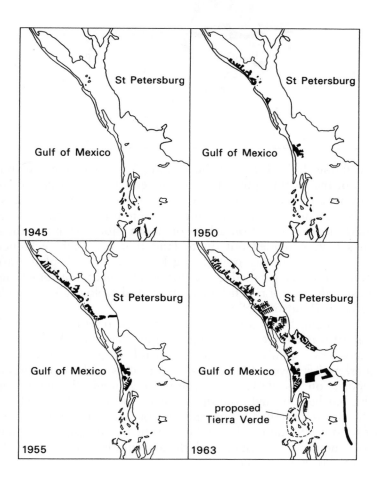

Figure 6.3
Boca Ciega Bay on the Gulf
of Mexico near St Petersburg
is rapidly being converted
by dredging and filling into
a typical suburban sprawl
(after Lauf, in Wagner, 1974,
figure 7.4)

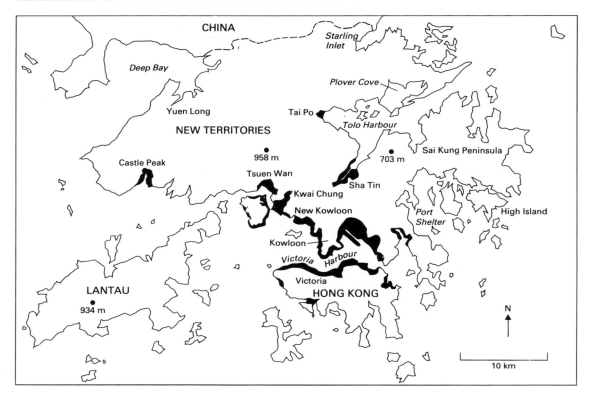

Figure 6.4
Map of Hong Kong showing
main urban reclamation areas
shaded black (reprinted from
Hudson, 1979, figure 1, in
Reclamation Review, 2, 3–16,
permission of Pergamon Press
Ltd © Pergamon Press Ltd)

and filling. Indeed, one of the striking features of the distribution of the world's population is the tendency for large human concentrations to occur near vast expanses of water. Of those cities with more than four million inhabitants, more than three-quarters are situated on a lake or ocean shore. Many of these cities have extended out on to land that has been reclaimed from the sea (for example, Hong Kong: figure 6.4, plate 6.6), thereby providing valuable sites for development, but sometimes causing the loss of rich fishing grounds and ecologically valuable wetlands (Hudson, 1979).

The ocean floors are also being affected because of the vast bulk of waste material that humankind is creating. Waste-solid disposal by coastal cities is now sufficiently large to modify shorelines, and it covers adjacent ocean bottoms with characteristic deposits on a scale large enough to be geologically significant. This has been brought out dramatically by Gross (1972: 3174), who has undertaken a quantitative comparison of the amount of solid wastes dumped into the Atlantic by humans in the New York metropolitan region with the amount of sediment brought into the ocean by rivers:

The discharge of waste solids exceeds the suspended sediment load of any single river along the U.S. Atlantic coast. Indeed, the discharge of

wastes from the New York metropolitan region is comparable to the estimated suspended-sediment yield (6.1 megatons per year) of all rivers along the Atlantic coast between Maine and Cape Hatteras, North Carolina.

Not only are the rates of sedimentation high, but the anthropogenic sediments tend to contain abnormally high contents of such substances as carbon and heavy metals (Goldberg et al., 1978).

Many of the features created by excavation in one generation are filled in by another, since phenomena like water-filled hollows produced by mineral extraction are often both wasteful of land and suitable locations for the receipt of waste. The same applies to natural hollows such as karstic or ground-ice depressions. Watson (1976), for example, has mapped the distribution of hollows, which were largely created by marl diggers in the lowlands of south-west Lancashire and north-west Cheshire, as they were represented on mid-nineteenth-century topographic maps (figure 6.5a). When this distribution is compared with the present-day distribution for the same area (figure 6.5b), it is evident that a very substantial proportion of the holes have

Plate 6.6
Hong Kong is one of the most densely populated territories on earth, and its landscapes have been transformed by coastal reclamation and by carving out ledges on steep slopes. Debris flows and landslips may result

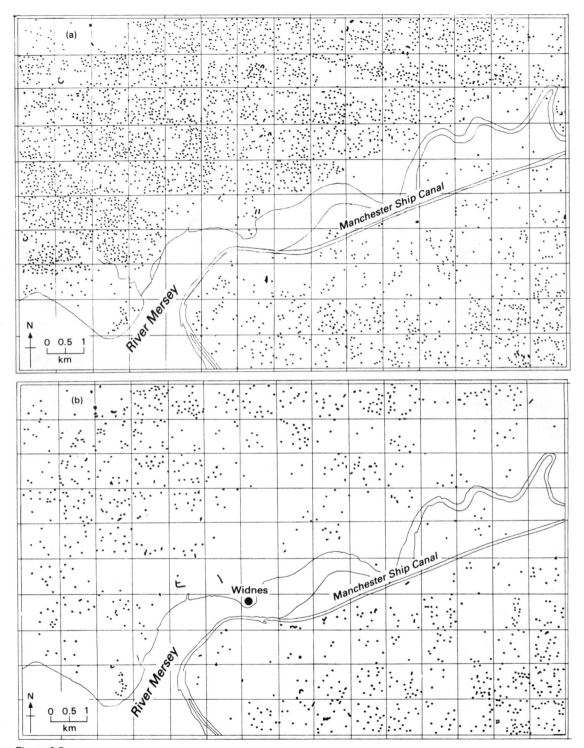

Figure 6.5
The distribution of pits and ponds in a portion of north-western England (a) in the mid-nineteenth century (b) in the mid-twentieth century (after Watson, 1976)

been infilled and obliterated by humans, with only 2,114 out of 5,380 remaining. Hole densities have fallen from 121 to 47 per square kilometre.

Accelerated sedimentation

An inevitable consequence of the accelerated erosion produced by human activities has been accelerated sedimentation. This has been heightened by the deliberate addition of sediments to stream channels as a result of the need to dispose of mining and other wastes.

In a classic study, G. K. Gilbert (1917) demonstrated that hydraulic mining in the Sierra Nevada mountains of California led to the addition of vast quantities of sediments into the river valleys draining the range (plate 6.7). This in itself raised their bed levels, changed their channel configurations and caused the flooding of lands that had previously been immune. Of even greater significance was the fact that the rivers transported increased quantities of debris into the estuarine bays of the San Francisco system, and caused extensive shoaling which in turn diminished the tidal prism of the bay. Gilbert calculated the volume of shoaling produced by hydraulic mining since the discovery of gold to be 846 million cubic metres. Comparably serious sedimentation of bays and estuaries has also been caused by human activity on the eastern coast of America. As Gottschalk (1945: 219) wrote:

Both historical and geological evidence indicates that the preagricultural rate of silting of eastern tidal estuaries was low. The history of sedimentation of ports in the Chesapeake Bay area is an epic of the effects of uncontrolled erosion since the beginning of the wholesale land clearing and cultivation more than three centuries ago.

He has calculated that at the head of the Chesapeake Bay, 65 million cubic metres of sediment were deposited between 1846 and 1938. The average depth of water over an area of 83 square kilometres was reduced by 0.76 metres. New land comprising 318 hectares was added to the state of Maryland and, as Gottschalk remarked, 'the Susquehanna River is repeating the history of the Tigris and Euphrates.' Much of the material entrained by erosive processes on upper slopes as a result of agriculture in Maryland, however, was not evacuated as far as the coast. Costa (1975) has suggested, on the basis of the study of sedimentation, that only about one-third of the eroded material left the river valleys. The remainder accumulated on floodplains as alluvium and colluvium at rates of up to 1.6 centimetres per year. Similarly, Happ (1944), working in Wisconsin, carried out an

Plate 6.7
Gold-washing in a long sluice
at Yale, British Columbia, in
the 1890s. Such activities
caused rivers to become
choked with debris

intensive augering survey of floodplain soil and established that,
since the development of agriculture, floodplain aggradation
had proceeded at a rate of approximately 0.85 centimetres per
year. He noted that channel and floodplain aggradation had
caused the flooding of low alluvial terraces to be more frequent,
more extensive and deeper. The rate of sedimentation has since
declined (Trimble, 1976), because of less intensive land use and
the institution of effective erosion control measures on farmland
(see also Magilligan, 1985).

Table 6.3 Accelerated sedimentation in Britain in prehistorical and historical times

Location	Source	Evidence and date
Howgill Fells	Harvey et al. (1981)	Debris cone production following tenth century AD introduction of sheep farming
Upper Thames basin	Robinson and Lambrek (1984)	River alluviation in Late Bronze Age and early Iron Age
Lake District	Pennington (1981)	Accelerated lake sedimentation at 5000 BP as a result of Neolithic agriculture
Mid-Wales	Macklin and Lewin (1986)	Floodplain sedimentation (Capel Bangor unit) on Breidol as a result of early Iron Age sedentary agriculture
Brecon Beacons	Jones et al. (1985)	Lake sedimentation increase after 5000 BP at Llangorse due to forest clearance
Weald	Burrin (1985)	Valley alluviation from Neolithic onwards until early Iron Age
Bowland Fells	Harvey and Renwick (1987)	Valley terraces at 5000–2000 BP (Bronze or Iron Age settlement) and after 1000 BP (Viking settlement)
Southern England	Bell (1982)	Fills in dry valleys: Bronze Age and Iron Age
Callaly Moor (Northumberland)	Macklin et al. (1991)	Valley fill sediments of late Neolithic to Bronze Age

Such valley sedimentation is by no means restricted to the newly settled terrains of North America. There is increasing evidence to suggest that silty valley fills in Germany, France and Britain, many of them dating back to the Bronze Age and the Iron Age, are the result of accelerated slope erosion produced by the activities of early farmers (Bell, 1982). Indeed, in recent years, various studies have been undertaken with a view to assessing the importance of changes in sedimentation rate caused by humans at different times in the Holocene in Britain. Among the formative events that have been identified are initial land clearance by Mesolithic and Neolithic peoples; agricultural intensification and sedentarization in the late Bronze Age; the widespread adoption of the iron plough in the early Iron Age; settlement by the Vikings; and the introduction of sheep farming (table 6.3).

A core from Llangorse Lake in the Brecon Beacons of Wales (Jones et al., 1985) provides excellent long-term data on changing sedimentation rates:

Period (years BP)	Sedimentation rate (100cm/year)
9000–7500	3.5
7500–5000	1.0
5000–2800	13.2
2800–AD 1840	14.1
c.AD 1840–present	59.0

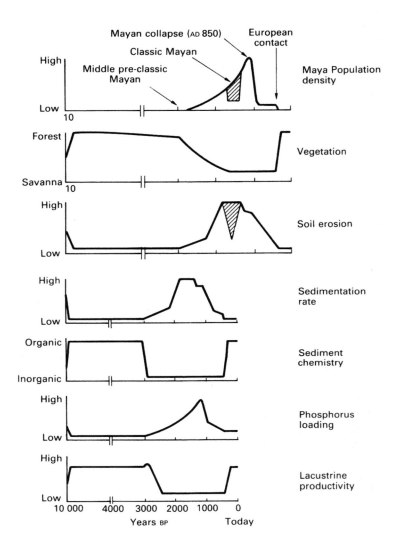

Figure 6.6
The human and environmental
history of the Peten Lakes,
Guatemala. The hatched areas
indicate a phase of local
population decline (modified
after Binford et al., 1987)

The thirteenfold increase in rates after 5000 BP seems to have
occurred rapidly and is attributed to initial forest clearance. The
second dramatic increase of more than fourfold took place in
the last 150 years and is a result of agricultural intensification.

The work of Binford et al. (1987) on the lakes of the Peten
region of northern Guatemala (Central America), an area of
tropical lowland dry forest, is also instructive with respect to
early agricultural colonization. Combining studies of archaeology and lake sediment stratigraphy, they were able to reconstruct the diverse environmental consequences of the growth
of Mayan civilization (figure 6.6). This civilization showed a
dramatic growth after 3000 years BP, but collapsed in the ninth

Table 6.4 Data on rates of erosion and sedimentation in lakes in the last two centuries

Dates	Activity	Rates of erosion in catchment (R. Cober) as determined from lake sedimentation rates ($t\ km^{-2}\ y^{-1}$)
(a) Loe Pool (Cornwall) (from O'Sullivan et al., 1982)		
1860–1920	Mining and agriculture	174
1930–6	Intensive mining and agriculture	421
1937–8	Intensive mining and agriculture	361
1938–81	Agriculture	12
(b) Seeswood Pool (Warwickshire) (from Foster et al., 1986)		
1765–1853		7.0
1854–80		12.2
1881–1902		8.1
1903–19		9.6
1920–5		21.6
1926–33		16.1
1934–47		12.7
1948–64		12.0
1965–72		13.9
1973–7		18.3
1978–82		36.2

century AD. The hypotheses put forward to explain this collapse include warfare, disease, earthquakes and soil degradation. The population has remained relatively low ever since, and after the first European contact (AD 1525) the region was virtually depopulated. The period of Mayan success saw a marked reduction in vegetation cover, an increase in lake sedimentation rates and in catchment soil erosion, an increased supply of inorganic silts and clays to the lakes, a pulse of phosphorus derived from human wastes, and a decrease in lacustrine productivity caused by high levels of turbidity.

In the last two centuries rates of sedimentation in lake basins have changed in different ways in different basins according to the differing nature of economic activities in catchments. Some data from various sources are listed for comparison in table 6.4. In the case of the Loe Pool in Cornwall (south-west England) rates of sedimentation were high while the mining industry was active, but fell dramatically when mining was curtailed. In the case of Seeswood Pool in Warwickshire, a dominantly agricultural catchment area in central England, the highest rates have occurred since 1978 in response to various land management changes, such as larger fields, continuous cropping and increased dairy herd size. In other catchments, pre-afforestation ploughing may have caused sufficient disturbance to bring about accelerated sedimentation. For example, Battarbee et al. (1985b) looked at sediment cores in the Galloway area of south-west Scotland and found that in Loch Grannoch the introduction of ploughing in the catchment caused an increase in sedimentation from 0.2 to 2.2 centimetres per year.

Sedimentation has severe economic implications because of the role it plays in reducing the effective lifetime of reservoirs. In the tropics, capacity depletion through sedimentation is commonly around 2 per cent per annum (Myers, 1988: 14), so that the expected useful life of the Paute hydroelectric project in Ecuador (cost US$600 million) is 32 years, that of the Mangla Dam in Pakistan (cost US$600 million) is 57 years, and that of the Tarbela Dam, also in Pakistan, is just 40 years.

Ground subsidence

Not all ground subsidence is caused by humans. For example, limestone solutional processes can, in the absence of humankind, create a situation where a cavern collapses to produce a surface depression, and permafrost will sometimes melt to produce a thermokarst depression without human intervention. None the less, ground subsidence can be caused or accelerated by humans in a variety of ways: by the transfer of subterranean fluids (such as oil, gas and water); by the removal of solids through underground mining or by dissolving solids and removing them in solution (for example, sulphur and salt); by the disruption of permafrost; and by the compaction or reduction of sediments because of drainage and irrigation (Johnson, 1991 and Barends et al., 1995).

Some of the most dangerous and dramatic collapses have occurred in limestone areas because of the dewatering of limestone caused by mining activities. In the far west Rand of South Africa (plate 6.8), gold mining has required the abstraction of water to such a degree that the local water table has been lowered by more than 300 metres. The fall of the water table caused miscellaneous clays and other materials filling the roofs of large caves to dry out and shrink so that they collapsed into the underlying void. One collapse created a depression 30 metres deep and 55 metres across, killing 29 people. In Alabama in the southern United States, water-level decline consequent upon pumping has had equally serious consequences in a limestone terrain; and Newton (1976) has estimated that since 1900 about 4,000 induced sink-holes or related features have been formed, while fewer than 50 natural sink-holes have been reported over the same interval. Sink-holes that may result from such human activity are also found in Georgia, Florida, Tennessee, Pennsylvania and Missouri.

In some limestone areas, however, a reverse process can operate. The application of water to overburden above the limestone may render it more plastic so that the likelihood of collapse
is increased. This has occurred beneath reservoirs, such as the
May Reservoir in central Turkey. Williams (1993) provides a
survey of the diverse effects of human activities on limestone
terrains.

The process can be accelerated by the direct solution of susceptible rocks. For example, collapses have occurred in gypsum
bedrock because of solution brought about by the construction
of a reservoir. In 1893 the MacMillan Dam was built on the
Pecos River in New Mexico, but within 12 years the whole river
flowed through caves which had developed since construction.
Both the San Fernando and Rattlesnake Dams in California
suffer severe leakage for similar reasons.

Plate 6.8
A gaping chasm in the
Transvaal (Gauteng), South
Africa, became the world's
biggest grave in August 1964.
The sink-hole opened because
groundwater conditions had
been altered by mining. It
swallowed up a family of five,
their servants and homes

Plate 6.9
In recent years the city of Venice has suffered from increasingly severe winter floods. The pumping of groundwater for industrial purposes has caused subsidence to occur

Subsidence produced by oil abstraction is an increasing problem in some parts of the world. The classic area is Los Angeles, where 9.3 metres of subsidence occurred as a result of exploitation of the Wilmington oilfield between 1928 and 1971. The Inglewood oilfield displayed 2.9 metres of subsidence between 1917 and 1963. Some coastal flooding problems occurred at Long Beach because of this process. Similar subsidence has been recorded from the Lake Maracaibo field in Venezuela (Prokopovich, 1972) and from some Russian fields (Nikonov, 1977).

A more widespread problem is posed by groundwater abstraction for industrial, domestic and agricultural purposes (plate 6.9).

Table 6.5 Ground subsidence

Location		Amount (m)		Rate (mm/y)
(a) Ground subsidence produced by oil and gas abstraction				
Azerbaydzhan, USSR	2.5	(1912–62)		50
Atravopol, USSR	1.5	(1956–62)		125
Wilmington, USA	9.3	(1928–71)		216
Inglewood, USA	2.9	(1917–63)		63
Maracaibo, Venezuela	5.03	(1929–90)		84
(b) Ground subsidence produced by ground-water abstraction				
London, England	0.06–0.08	(1865–1931)		0.91–1.21
Savannah, Georgia (USA)	0.1	(1918–55)		2.7
Mexico City	7.5	–		250–300
Houston, Galveston, Texas	1.52	(1943–64)		60–76
Central Valley, California	8.53	–		
Tokyo, Japan	4	(1892–1972)		500
Osaka, Japan	>2.8	(1935–72)		76
Niigata, Japan	>1.5	–		–
Pecos, Texas	0.2	(1935–66)		6.5
South-central Arizona	2.9	(1934–77)		96
Bangkok, Thailand	0.5	–		100
Shanghai, China	2.62	(1921–65)		60

Source: data in Cooke and Doornkamp, 1974; Prokopovich, 1972; Rosepiler and Reilinger, 1977; Holzer, 1979; Nutalaya and Rau, 1981; Johnson, 1991.

Table 6.5 presents some data for such subsidences from various parts of the world. The ratios of subsidence to water-level decline are strongly dependent on the nature of the sediment composing the aquifer, so that ratios range from 1:7 for Mexico City to 1:80 for the Pecos in Texas, and to less than 1:400 for London, England (Rosepiler and Reilinger, 1977). The extent of subsidence that has taken place in the United States as a result of groundwater abstraction has recently been reassessed on the basis of geodetic evidence (Chi and Reilinger, 1984). To the major areas previously identified, namely:

south-central Arizona
Savannah (Georgia)
Pecos and Houston–Galveston (Texas)
Denver (Colorado)
San Joaquin Valley, Santa Clara Valley, Saugus Basin, Los Angeles Basin
Bunker Hill–San Timoteo (California)
Milford (Utah)
Raft River Valley (Idaho)
Baton Rouge and New Orleans (Louisiana)
Las Vegas (Nevada)

they have added:

Ventura, Ontario and San Pedro–Santa Monica (California)
Monroe and Alexandria (Louisiana)
Jackson (Mississippi).

In Japan subsidence has now emerged as a major problem
(Nakano and Matsuda, 1976). In 1960 only 35.2 square kilometres of the Tokyo lowland was below sea level, but continuing subsidence meant that by 1974 this had increased to 67.6
square kilometres, exposing a total of 1.5 million people to
major flood hazard.

The subsidence caused by mining (which led to court cases in
England as early as the fifteenth century as a consequence of
associated damage to property) is perhaps the most familiar. Its
importance varies according to such factors as the thickness of
seam removed, its depth, the width of working, the degree of
filling with solid waste after extraction, the geological structure,
and the method of working adopted (Wallwork, 1974). In general terms, the vertical displacement by subsidence is less than
the thickness of the seam being worked, and decreases with an
increase in the depth of mining. This is because the overlying
strata collapse, fragment and fracture, so that the mass of rock
fills a greater space than it did when naturally compacted. Consequently, the surface expression of deep-seated subsidence may
be equal to little more than one-third of the thickness of the
material removed. Subsidence associated with coal mining may
disrupt surface drainage and the resultant depressions then become
permanently flooded. In England, the valleys of the Aire and Calder
in the Castleford–Knottingley area, the Anker Valley in east
Warwickshire, and the Wigan–Leigh area of Lancashire provide
examples of subsidence lakes formed on stream floodplains.

Coal-mining regions are not the only areas where subsidence
problems are serious. In Cheshire, north-west England, salt is
extracted from two major seams, each about 30 metres thick.
Moreover, these seams occur at no great depth, the uppermost
being at about 70 metres below the surface. A further factor to
be considered is that the rock salt is highly soluble in water, so
the flooding of mines may cause additional collapse. These three
conditions – thick seams, shallow depth and high solubility –
have produced optimum conditions for subsidence and many
subsidence lakes called 'flashes' have developed (Wallwork, 1956).
Some of these are illustrated in figure 6.7. The main railway line
between Crewe and Manchester at Elton subsided no less than
4.9 metres between 1892 and 1956 (Wallwork, 1960). Likewise,
salt extraction at Windsor, Canada, is known to have produced
a water-filled depression 150 metres in diameter and about 8
metres deep (Martinez, 1971).

Some subsidence is created by a process called hydrocompaction, which is explained thus. Moisture-deficient, unconsolidated,

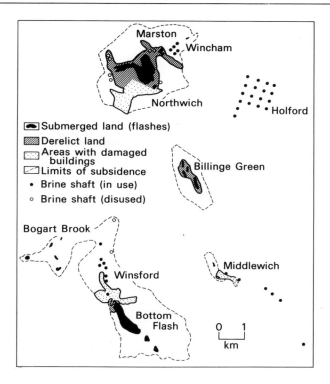

Figure 6.7
Subsidence in the salt area
of mid-Cheshire, England, in
1954 (after Wallwork, 1956,
figure 3)

low-density sediments tend to have sufficient dry strength to support considerable effective stresses without compacting. However, when such sediments, which may include alluvial fans or loess, are thoroughly wetted for the first time (for example, by percolating irrigation water) the inter-granular strength of the deposits is diminished, rapid compaction takes place, and ground surface subsidence follows. Unequal subsidence can create problems for irrigation schemes.

Land drainage can promote subsidence of a different type, notably in areas of organic soils. The lowering of the water table makes peat susceptible to oxidation and deflation so that its volume decreases. One of the longest records of this process, and one of the clearest demonstrations of its efficacy, has been provided by the measurements at Holme Fen Post in the English Fenlands. Approximately 3.8 metres of subsidence occurred between 1848 and 1957 (Fillenham, 1963), with the fastest rate occurring soon after drainage had been initiated (figure 6.8). The present rate averages about 1.4 centimetres per year (Richardson and Smith, 1977). At its maximum natural extent the peat of the English Fenland covered around 1,750 square kilometres. Now only about one-quarter (430 square kilometres) remains.

A further type of subsidence, sometimes associated with earthquake activity, results from the effects on the earth's crust of

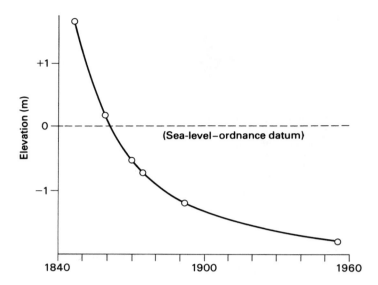

Figure 6.8
The subsidence of the English Fenlands peat in Holme Fen Post from 1842 to 1960 following drainage (from data in Fillenham, 1963)

Table 6.6 Hydro-isostatic subsidence caused by the creation of reservoirs

Reservoir	Maximum downwarping (cm)	Period
Lake Mead, USA	20.1	1950–63
Kariba, Central Africa	12.7	1959–68
Koyna, India	8–14	1962–8
Bratsk, Siberia	5.6	1961–6
Krasnoyarsk, Siberia	3	1967–71
Plyava, Baltic	0.5	1965–70

Source: from data in Nikonov, 1977.

large masses of water impounded behind reservoirs. Seismic effects can be generated in areas with susceptible fault systems and this may account for earthquakes recorded at Koyna (India) and elsewhere. The process whereby a mass of water causes coastal depression is called hydro-isostasy, and the degree of subsidence that has been monitored for various large reservoirs is listed in table 6.6.

In permafrost areas (plate 6.10) ground subsidence is associated with thermokarst development, thermokarst being irregular, hummocky terrain produced by the melting of ground ice, permafrost. The development of thermokarst is due primarily to the disruption of the thermal equilibrium of the permafrost and an increase in the depth of the active layer. This is illustrated in figure 6.9. Following French (1976: 106), consider an undisturbed tundra soil with an active layer of 45 centimetres. Assume also that the soil beneath 45 centimetres is supersaturated permafrost and yields on a volume basis upon thawing 50 per cent excess water and 50 per cent saturated soil. If the top

Plate 6.10
In Alaska, and other areas which are underlain by permafrost, ground subsidence caused by thermokarstic processes presents many engineering problems. Buildings sometimes have to be abandoned

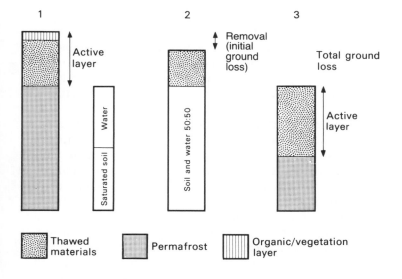

Figure 6.9
Diagram illustrating how the disturbance of high ice content terrain can lead to permanent ground subsidence. 1–3 indicate stages before, immediately after and subsequent to disturbance (after Mackay in French, 1976, figure 6.1)

15 centimetres were removed, the equilibrium thickness of the active layer, under the bare ground conditions, might increase to 60 centimetres. As only 30 centimetres of the original active layer remains, 60 centimetres of the permafrost must thaw before the active layer can thicken to 60 centimetres, since 30 centimetres of supernatant water will be released. Thus, the surface subsides 30 centimetres because of thermal melting associated with the degrading permafrost, to give an overall depression of 45 centimetres.

Thus the key process involved in thermokarst subsidence is the state of the active layer and its thermal relationships. When, for example, surface vegetation is cleared for agricultural or constructional purposes the depth of thaw will tend to increase. The movement of tracked vehicles has been particularly harmful to surface vegetation and deep channels may soon result from permafrost degradation. Similar effects may be produced by the siting of heated buildings on permafrost, and by the laying of oil, sewer and water pipes in or on the active layer (Ferrians et al., 1969; Lawson, 1986).

Thus subsidence is a diverse but significant aspect of the part humans play as geomorphological agents. The damage caused on a worldwide basis can be measured in billions of dollars each year (Coates, 1983): among the effects are broken dams, cracked buildings, offset roads and railways, fractured well casings, deformed canals and ditches, bridges that need relevelling, saline encroachment and increased flood damage.

Arroyo trenching, gullies and peat haggs

In the south-western United States many broad valleys and plains became deeply incised with valley-bottom gullies (arroyos) over a short period between 1865 and 1915, with the 1880s being especially important (Cooke and Reeves, 1976). This cutting had a rapid and detrimental effect on the flat, fertile and easily irrigated valley floors, which are the most desirable sites for settlement and economic activity in a harsh environment.

Many students of this phenomenon have believed that thoughtless human actions caused the entrenchment, and the apparent coincidence of white settlement and arroyo development tended to give credence to this viewpoint. The range of actions that could have been culpable is large: timber-felling, overgrazing, cutting grass for hay in valley bottoms, compaction along well-travelled routes, channelling of runoff from trails and railways, disruption of valley-bottom sods by animals' feet, and the invasion of grasslands by miscellaneous types of scrub.

On the other hand, study of the long-term history of the valley fills shows that there have been repeated phases of aggradation and incision and that some of these took place before the influence of humans could have been a significant factor. This has prompted debate as to whether the arrival of white communities was in fact responsible for this particularly severe phase of environmental degradation. Huntington (1914), for example, argued that valley filling would be a consequence of a climatic shift to more arid conditions. These, he believed, would cause a reduction in vegetation which in turn would promote rapid removal of soil from devegetated mountain slopes during storms, and overload streams with sediment. With a return to humid conditions vegetation would be re-established, sediment yields would be reduced and entrenchment of valley fills would take place. Bryan (1928) put forward a contradictory climatic explanation. He argued that a slight move towards drier conditions, by depleting vegetation cover and reducing soil infiltration capacity, would produce significant increases in storm runoff which would erode valleys. Another climatic interpretation was advanced by Leopold (1951), involving a change in rainfall intensity rather than quantity. He indicated that a reduced frequency of low-intensity rains would weaken the vegetation cover, while an increased frequency of heavy rains at the same time would increase the incidence of erosion. Support for this contention comes from the work of Balling and Wells (1990) in New Mexico. They attributed early twentieth-century arroyo trenching to a run of years with intense and erosive rainfall characteristics that succeeded a phase of drought conditions in which the productive ability of the vegetation had declined. It is also possible, as Schumm et al. (1984) have pointed out, that arroyo incision could result from neither climatic change nor human influence. It could be the result of some natural geomorphological threshold being crossed.

It is therefore clear that the possible mechanisms that can lead to alternations of cut-and-fill of valley sediments are extremely complex, and that any attribution of all arroyos in all areas to human activities may be a serious oversimplification of the problem (figure 6.10). In addition, it is possible that natural environmental changes, such as changes in rainfall characteristics, have operated at the same time and in the same direction as human actions.

In the Mediterranean lands there have also been controversies surrounding the age and causes of alternating phases of aggradation and erosion in valley bottoms. Vita-Finzi (1969) suggested that at some stage during historical times many of the streams in the Mediterranean area, which had hitherto been engaged primarily in downcutting, began to build up their beds. Renewed downcutting, still seemingly in operation today, has since incised the channels into the alluvial fill. He proposed that the reversal of the downcutting trend in the Middle Ages was both ubiquitous and confined in time, and that some universal and time-specific agency was required to explain it. He believed that devegetation by humans was not a medieval innovation and that some other mechanism was required. A solution he gave to account for the phenomenon was precipitation change during the climatic fluctuation known as the Little Ice Age (AD 1550–1850). This was not an interpretation which found favour with Butzer (1974). He reported that his investigations showed plenty of post-classical and pre-1500 alluviation (which could not therefore be ascribed to the Little Ice Age), and he doubted whether

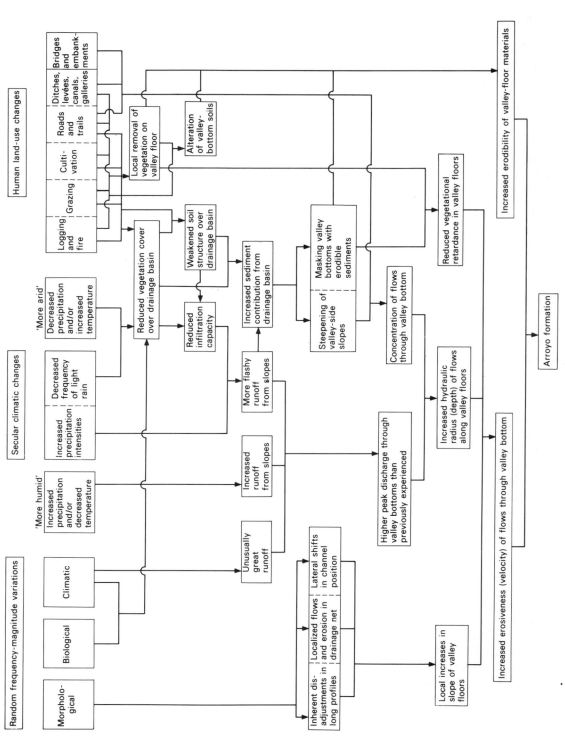

Figure 6.10
A model for the formation of arroyos (gullies) in the south-western USA (after Cooke and Reeves, 1976, figure 1.2)

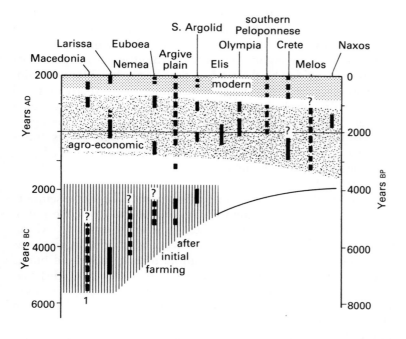

Figure 6.11
Chronology of Holocene
alluviation in Greece and the
Aegean. Broken bars are
dated uncertainly or represent
intermittent deposition (from
various sources in Van Andel
et al., 1990, figure 10)

Vita-Finzi's dating was precise enough to warrant a 1550–1850 date. Instead, he suggested that humans were responsible for multiple phases of accelerated erosion from slopes, and accelerated sedimentation in valley bottoms, from as early as the middle of the first millennium BC.

Butzer's interpretation has found favour with van Andel et al. (1990), who have detected an intermittent and complex record of cut and fill episodes during the late Holocene in various parts of Greece (figure 6.11). They believe that this evidence is compatible with a model of the control of timing and intensity of landscape destabilization by local economic and political conditions. This is a view shared in the context of the Algarve in Portugal by Chester and James (1991).

A further location with spectacular gullies, locally called *lavaka*, is Madagascar. Here too there have been debates about cultural versus natural causation. Proponents of cultural causes have argued that since humans arrived on the island in the last two thousand years, there has been excessive cattle grazing, removal of forest for charcoal and for slash-and-burn cultivation, devastating winter (dry season) burning of grasslands, and erosion along tracks and trails. However, the situation is more complex than that and the *lavaka* are polygenetic. Tectonism and natural climatic aridification may be at least as important as human agency, and given the climatic and soil types of the island many *lavaka* are a natural part of the landscape's evolution. Some of them also clearly predate primary (i.e. uncut) rain forest. The many factors, natural and cultural, involved in *lavaka* development are well reviewed by Wells and Andriamihaja (1993).

Another example of drainage incision that demonstrates the problem of disentangling the human from the natural causes of erosion is provided by the eroding peat bogs of highland Britain. Over many areas, including the Pennines of northern England and the Brecon Beacons of Wales, blanket peats are being severely eroded to produce pool and hummock topography, areas of bare peat and incised gullies (haggs). Many rivers draining such areas are discoloured by the presence of eroded peat particles, and sediment yields of organic material are appreciable (Labadz et al., 1991).

Some of the observed peat erosion may be an essentially natural process, for the high water content and low cohesion of undrained peat masses make them inherently unstable. Moreover, the instability must normally become more pronounced as peat continues to accumulate, leading to bog slides and bursts round margins of expanded peat blankets. Conway (1954) suggested that an inevitable end-point of peat build-up on high-altitude, flat or convex surfaces is that a considerable depth of unconsolidated and poorly humidified peat overlies denser and well-humidified peat, so adding to the instability. Once a bog burst or slide occurs, this leads to the formation of drainage gullies which extend back into the peat mass, slumping-off marginal peat downslope, and leading to the drawing off of water from the pools of the hummock and hollow topography of the watershed.

Tallis (1985) believes that there have been two main phases of erosion in the Pennines. The first, initiated 1,000–1,200 years ago, may have been caused by natural instability of the type outlined above. However, there has been a second stage of erosion, initiated 200–300 years ago, in which miscellaneous human activities appear to have been important. Radley (1962) suggested that among the pressures that had caused erosion were heavy sheep grazing, regular burning, peat cutting, the digging of boundary ditches, the incision of pack horse tracks, and military manoeuvres during the First World War. Other causes may include footpath erosion and severe air pollution (Tallis, 1965). In South Wales, there is some evidence that the blanket peats have degenerated as a result of contamination by particulate pollution (soot, etc.) during the industrial revolution (Chambers et al., 1979). On the other hand, in Scotland lake core studies indicate that severe peat erosion was initiated between AD 1500 and 1700, prior to air pollution associated with industrial growth, and Stevenson et al. (1990) suggest that this erosion initiation may have been caused either by the adverse climatic conditions of the Little Ice Age or by an increasing intensity of burning as land use pressures increased.

Plate 6.11
The ancient city of
Mohenjo-Daro in Pakistan
was excavated in the 1920s.
Irrigation has been introduced
into the area, causing
groundwater levels to be
raised. This has brought salt
into the bricks of the ancient
city, producing severe
disintegration

Accelerated weathering and the tufa decline

Although fewer data are available and the effects are generally
less immediately obvious, there is some evidence that human
activities have produced changes in the nature and rate of weathering (Winkler, 1970). The primary cause of this is probably
air pollution. It is clear that, as a result of increased emissions
of sulphur dioxide through the burning of fossil fuels, there
are higher levels of sulphuric acid in rain over many industrial
areas. This in itself may react with stones and cause their decay.
Chemical reactions involving sulphur dioxide can also generate
salts such as calcium sulphate and magnesium sulphate, which
may be effective in causing the physical breakdown of rock
through the mechanism of salt weathering.

Similarly, atmospheric carbon dioxide levels have been rising
steadily because of the burning of fossil fuels and deforestation.
Carbon dioxide may combine with water, especially at lower
temperatures, to produce weak carbonic acid which can dis-
solve limestone, marbles and dolomites. Weathering can also
be accelerated by changes in groundwater levels resulting from
irrigation. This can be illustrated by considering the Indus plain
in Pakistan (Goudie, 1977), where irrigation has caused the
water table to be raised by about 6 metres since 1922. This has
produced increased evaporation and salinization. The salts that
are precipitated by evaporation above the capillary fringe include
sodium sulphate, a very effective cause of stone decay. Indeed,
buildings such as the great archaeological site of Mohenjo-Daro
are decaying at a catastrophic rate (plate 6.11).

In other cases accelerated weathering has been achieved by
moving stone from one environment to another. Cleopatra's
Needle, an Egyptian obelisk in New York City, is an example

Table 6.7 Some possible mechanisms to account for the alleged Holocene tufa decline

Climatic/natural	Anthropogenic
Discharge reduction following rainfall decline leading to less turbulence	Discharge reduction due to overpumping, diversions, etc.
Degassing leads to less deposition	Increased flood scour and runoff variability of channels due to deforestation, urbanization, ditching, etc.
Increased rainfall causing more flood scour	Channel shifting due to deforestation of floodplains leads to tufa erosion
Decreasing temperature leads to less evaporation and more CO_2 solubility	Reduced CO_2 flux in system after deforestation
Progressive Holocene peat development and soil podzol development through time leads to more acidic surface waters	Introduction of domestic stock causes breakdown of fragile tufa structures
	Deforestation = less fallen trees to act as foci for tufa barrages
	Increased stream turbidity following deforestation reduces algal productivity

of rapid weathering of stone in an inhospitable environment. Originally erected on the Nile opposite Cairo about 1500 BC, it was toppled in about 500 BC by Persian invaders, and lay partially buried in Nile sediments until, in 1880, it was moved to New York. It immediately began to suffer from scaling and the inscriptions were largely obliterated within ten years because of the penetration of moisture which enabled frost-wedging and hydration of salts to occur.

During the 1970s and 1980s an increasing body of isotopic dates became available for deposits of tufa (secondary freshwater deposits of limestone, also known as travertine). Some of these dates suggest that over large parts of Europe, from Britain to the Mediterranean basin and from Spain to Poland, rates of tufa formation were high in the early and mid-Holocene, but declined markedly thereafter (Weisrock, 1986: 165–7). Vaudour (1986) maintains that since around 3000 BP 'the impact of man on the environment has liberated their disappearance', but he gives no clear indication of either the basis of this point of view or the precise mechanism(s) that might be involved. If the late Holocene reduction in tufa deposition is a reality then it is necessary to consider a whole range of possible mechanisms, both natural and anthropogenic (Nicod, 1986: 71–80; table 6.7). As yet the case for an anthropogenic role is not proven (Goudie et al., 1993).

Accelerated mass movements

There are many examples of mass movements being triggered by human actions (Selby, 1979). For instance, landslides can be

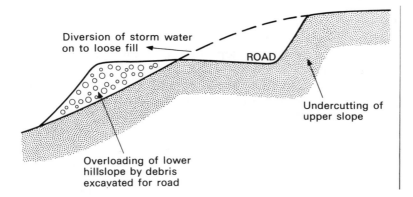

Diversion of storm water
on to loose fill

ROAD

Undercutting of
upper slope

Overloading of lower
hillslope by debris
excavated for road

Figure 6.12
Slope instability produced by
road construction

Table 6.8   Examples of methods of controlling mass movements

Type of movement	Method of control
Falls	Flattening the slope
	Benching the slope
	Drainage
	Reinforcement of rock walls by grouting with cement, anchor bolts
	Covering of wall with steel mesh
Slides and flows	Grading or benching to flatten the slope
	Drainage of surface water with ditches
	Sealing surface cracks to prevent infiltration
	Sub-surface drainage
	Rock or earth buttresses at foot
	Retaining walls at foot
	Pilings through the potential slide mass

Source: after R. F. Baker and H. E. Marshall, in Dunne and Leopold, 1978, table 15.16.

created either by undercutting or by overloading (figure 6.12). When a road is constructed, material derived from undercutting the upper hillside may be cast on to the lower hillslope as a relatively loose fill to widen the road bed. Storm water is then often diverted from the road on to the loose fill.

Because of the hazards presented by both natural and acceler- ated mass movements, humans have developed a whole series of techniques to attempt to control them. Such methods, many of which are widely used by engineers, are listed in table 6.8. These techniques are increasingly necessary, as human capacity to change a hillside and to make it more prone to slope failure has been transformed by engineering development. Excavations are going deeper, buildings and other structures are larger, and many sites which are at best marginally suitable for engineering projects are now being used because of increasing pressure on land. This applies especially to some of the expanding urban areas in the humid parts of low latitudes – Hong Kong, Kuala Lumpur, Rio de Janeiro and many others. It is very seldom that human agency deliberately accelerates mass movements; most are accidentally caused, the exception possibly being the deliberate triggering of a threatening snow avalanche (Perla, 1978).

The forces producing slope instability and landsliding can usefully be divided into disturbing forces and resisting properties (Cooke and Doornkamp, 1990: 113–14). The factors leading to an increase in shear stress (disturbing forces) can be listed as follows (modified after Cooke and Doornkamp, 1990: 113):

1   *Removal of lateral or underlying support:*
    undercutting by water (for example, river, waves), or glacier ice;
    weathering of weaker strata at the toe of the slope;
    washing out of granular material by seepage erosion;
    human cuts and excavations, drainage of lakes or reservoirs.
2   *Increased disturbing forces:*
    natural accumulations of water, snow, talus;
    pressure caused by human activity (for example, stockpiles of ore, tip-heaps, rubbish dumps, or buildings).
3   *Transitory earth stresses:*
    earthquakes;
    continual passing of heavy traffic.
4   *Increased internal pressure:*
    build-up of pore-water pressures (for example, in joints and cracks, especially in the tension crack zone at the rear or the slide).

Some of the factors are natural, while others (italicized) are affected by humans. Factors leading to a decrease in the shearing resistance of materials making up a slope can also be summarized (also modified after Cooke and Doornkamp, 1990: 113):

1   *Materials:*
    beds which decrease in shear strength if water content increases (clays, shale, mica, schist, talc, serpentine) (for example, *when local water table is artificially increased in height by reservoir construction*), or as a result of stress release (vertical and/or horizontal) following slope formation;
    low internal cohesion (for example, consolidated clays, sands, porous organic matter);
    in bedrock: faults, bedding planes, joints, foliation in schists, cleavage, brecciated zones, and pre-existing shears.
2   *Weathering changes:*
    weathering reduces effective cohesion, and to a lesser extent the angle of shearing resistance;
    absorption of water leading to changes in the fabric of clays (for example, loss of bonds between particles or the formation of fissures).

3   *Pore-water pressure increase:*
high groundwater table as a result of increased precipitation
or *as a result of human interference* (for example, *dam
construction*) (see 1 above).

Once again the italics show that there are a variety of ways in
which humans can play a role.

Some mass movements are created by humans piling up waste
soil and rock into unstable accumulations that fail spontaneously.
At Aberfan, in South Wales, a major disaster occurred when a
coal-waste tip 180 metres high began to move as an earth flow
(plate 6.12). The tip had been constructed not only as a steep
slope but also upon a spring line. This made an unstable configuration which eventually destroyed a school and claimed over
150 lives. Human-induced mass movements are also a problem
in Los Angeles, California. Some result from the denudation of
slope vegetation cover by fires, many of which are set by humans,
while others are triggered by infiltration of water from cesspools
and from irrigation water applied to lawns and gardens. In Hong
Kong, where a large proportion of the population is forced to
occupy steep slopes developed on deeply weathered granites and
other rocks, mass movements are a severe problem, and So (1971)
has shown that many of the landslides and washouts (70 per
cent of those in the great storm of June 1966, for example) were
associated with road sections and slopes artificially modified
through construction and cultivation.

One of the most serious mass movements partly caused by
human activity was that which caused the Vaiont Dam disaster
in Italy in 1963, in which 2,600 people were killed (Kiersch,
1965). Heavy antecedent rainfall conditions and the presence of
young, highly folded sedimentary rocks provided the necessary
conditions for a slip to take place, but it was the construction of
the Vaiont Dam itself which changed the local groundwater
conditions sufficiently to affect the stability of a rock mass on
the margins of the reservoir: 240 million cubic metres of ground
slipped, causing a rise in water level which overtopped the dam
and caused flooding and loss of life downstream. Comparable
slope instability resulted when the Franklin D. Roosevelt lake
was impounded by the Columbia river in the United States
(Coates, 1977), but the effects were, happily, less serious.

It is evident from what has been said about the predisposing
causes of the slope failure *triggered* by the Vaiont Dam that
human agency was able to have such an impact only because
the natural conditions were broadly favourable. Exactly the same
lesson can be learned from the accelerated landsliding in southern Italy. Nossin (1972) has demonstrated how, in Calabria, road
construction has triggered off (and been hindered by) landsliding,
but he has also stressed that the area is fundamentally susceptible

Plate 6.12
The village of Aberfan in
South Wales, where in 1966
a massive debris flow caused
severe loss of life when it
destroyed a school and
houses as it ran down from
a steep coal-waste tip

to such mass movement activity because of geological conditions. It is an area where recent rapid uplift has caused downcutting by rivers and the undercutting of slopes by erosion. It is also an area of incoherent metamorphic rocks, with frequent faulting. Further, water is often trapped by Tertiary clay layers, providing further stimulus to movement.

Although the examples of accelerated mass movements that have been given here are essentially associated with the effects of modern construction projects, more long-established activities, including deforestation and agriculture, are also highly important. For example, Innes (1983) has demonstrated, on the basis of lichenometric dating of debris-flow deposits in the Scottish Highlands, that most of the flows have developed in the last 250 years, and he suggests that intensive burning and grazing may be responsible.

However, as with so many environmental changes of that nature in the past, there are considerable difficulties in being certain about causation. This has been well expressed by Ballantyne (1991: 84):

Although there is growing evidence for Late Holocene erosion in upland Britain, the causes of this remain elusive. A few studies have presented evidence linking erosion to vegetation degradation and destruction due to human influence, but the validity of climatic deterioration as a cause of erosion remains unsubstantiated. This uncertainty stems from a tendency to link erosion with particular causes only through assumed coincidence in timing, a procedure fraught with difficulty because of imprecision in the dating of both putative causes and erosional effects. Indeed, in many reported instances, it is impossible to refute the possibility that the timing of erosional events or episodes may be linked to high magnitude storms of random occurrence, and bears little relation to either of the casual hypotheses outlined above.

Deliberate modification of channels

For purposes of both navigation and flood control humans have deliberately straightened many river channels. Indeed, the elimination of meanders contributes to flood control in two ways. First, it eliminates some overbank floods on the outside of curves, against which the swiftest current is thrown and where the water surface rises highest. Second, and more importantly, the resultant shortened course increases both the gradient and the flow velocity, and the flood waters erode and deepen the channel, thereby increasing its flood capacity.

It was for this reason that a programme of channel cutoffs was initiated along the Mississippi in the early 1930s. By 1940 it had lowered flood stages by as much as 4 metres at Arkansas City, Arkansas. By 1950 the length of the river between Memphis, Tennessee and Baton Rouge, Louisiana (600 kilometres down the valley) had been reduced by no less than 270 kilometres as a result of 16 cutouts.

The largest current scheme for channel modification is the Jonglei Canal in the Sudan (figure 6.13). This scheme, halted by political unrest, has involved the construction of a channel 360 km

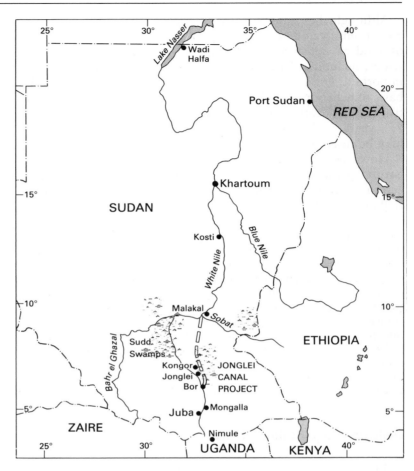

Figure 6.13
The diversion of the River Nile waters along the Jonglei Canal in Sudan. One of the major effects of this canal will be to reduce the degree of seasonal inundation of the Sudd and other floodplains. Currently, however, construction is delayed

long that will bypass the Sudd swamps where more than half of the White Nile's flow is lost by evapotranspiration every year. This, if completed, will provide an extra 4.75 billion m³ of water each year for Sudan and Egypt, but its impact on the life of the local pastoral people and on the ecology of the floodplains is the cause of considerable misgivings (Charnock, 1983).

Some landscapes have become dominated by artificial channels, normally once again because of the need for flood alleviation and drainage. This is especially evident in an area like the English Fenlands where straight constructed channels contrast with the sinuous courses of original rivers such as the Great Ouse.

Non-deliberate river-channel changes

There are thus many examples of the intentional modification of river-channel geometry by humans – by the construction of embankments, by channelization, and by other such processes.

Table 6.9 Accidental channel changes

Phenomenon	Cause
Channel incision	Clear-water erosion below dams caused by sediment removal
Channel aggradation	Reduction in peak flows below dams Addition of sediment to streams by mining, agriculture, etc.
Channel enlargement	Increase in discharge level produced by urbanization
Channel diminution	Discharge decrease following water abstraction or flood control
Channel diminution	Trapping and stabilizing of sediment by artificially introduced plants

Table 6.10 Causes of riverbed degradation

Type	Primary cause	Contributory cause
Downstream progressing	Decrease of bed-material discharge	Dam construction Excavation of bed material Diversion of bed material Change in land use Storage of bed material
	Increased water discharge	Diversion of flow Rare floods
	Decrease in bed-material size Other	River emerging from lake Thawing of permafrost
Upstream progressing	Lower base level	Drop in lake level Drop in level of main river Excavation of bed material
	Decrease in river length	Cutoffs Channelization Stream capture Horizontal shift of base level
	Removal of control point	Natural erosion Removal of dam

Source: after Galay, 1983.

However, major changes in the configuration of channels can be achieved accidentally (table 6.9), because of human-induced changes in either stream discharge or sediment load: both parameters affect channel capacity (Park, 1977). The causes of observed cases of riverbed degradation are varied and complex and result from a variety of natural and human changes (table 6.10). A useful distinction can be drawn between degradation that proceeds downstream, and that which proceeds upstream, but in both cases the complexity of causes is evident.

Deliberate channel straightening causes various types of sequential channel adjustment both within and downstream

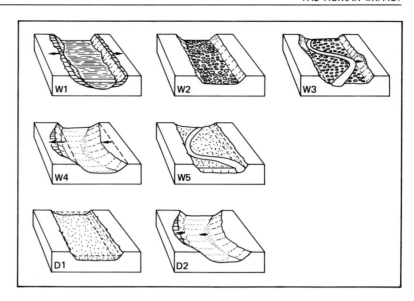

Figure 6.14
Principal types of adjustment
in straightened river channels
(after Brookes, 1987, figure 4).
For an explanation of the
different types see text

from straightened reaches, and the types of adjustment vary according to such influences as stream gradient and sediment characteristics. Brookes (1987) has recognized five types of change *within* the straightened reaches (types W1 to W5) and two types of change downstream (types D1 and D2). They are illustrated in figure 6.14.

Type W1	is degradation of the channel bed, which results from the fact that straightening increases the slope by providing a shorter channel path. This in turn increases its sediment transport capability.
Type W2	is the development of an armoured layer on the channel bed by the more efficient removal of fine materials as a result of the increased sediment transport capability referred to above.
Type W3	is the development of a sinuous thalweg in streams which are not only straightened but also widened beyond the width of the natural channel.
Type W4	is the recovery of sinuosity as a result of bank erosion in channels with high slope gradients.
Type W5	is the development of a sinuous course by deposition in streams with a high sediment load and a relatively low valley gradient.
Types D1 and D2	result from deposition downstream as the stream tries to even out its gradient, the

deposition occurring as a general raising of
the bed level, or as a series of accentuated
point bar deposits.

It is now widely recognized that the urbanization of a river
basin results in an increase in the peak flood flows in a river. It
is also recognized that the morphology of stream channels is
related to their discharge characteristics, and especially to the
discharge at which bank full flow occurs. As a result of urbanization, the frequency of discharges which fill the channel will
increase, with the effect that the beds and banks of channels in
erodible materials will be eroded so as to enlarge the channel
(Trimble, 1997b). This in turn will lead to bank caving, possible
undermining of structures, and increases in turbidity (Hollis and
Luckett, 1976).

Comparable changes in channel morphology result from discharge diminution produced by flood-control works and diversions for irrigation. This can be shown for the North Platte and
the South Platte in America, where both peak discharge and
mean annual discharge have declined to 10–30 per cent of their
pre-dam values. The North Platte, 762–1,219 metres wide in
1890 near the Wyoming–Nebraska border, has narrowed to
about 60 metres at present; the South Platte River was about
792 metres wide 89 kilometres above its junction with the North
Platte in 1897, but had narrowed to about 60 metres by 1959
(Schumm, 1977: 161). The tendency of both rivers has been to
form one narrow, well-defined channel in place of the previously
wide, braided channels, and, in addition, the new channel is
generally somewhat more sinuous than the old (figure 6.15).

Similarly, the building of dams can lead to channel aggradation upstream from the reservoir and channel deepening downstream because of the changes brought about in sediment loads
(figure 6.16). Some data on observed rates of degradation below
dams are presented in table 6.11. They show that the average rate
of degradation has been of the order of a few metres over a few
decades following closure of the dams. However, over time the
rate of degradation seems to become less or to cease altogether,
and Leopold et al. (1964: 455) suggest that this can be brought
about in several ways. First, because degradation results in a
flattening of the channel slope in the vicinity of the dam – the
slope may become so flat that the necessary force to transport
the available materials is no longer provided by the flow. Second,
the reduction of flood peaks by the dam reduces the competence
of the transporting stream to carry some of the material on its
bed. Thus if the bed contains a mixture of particle size the river
may be able to transport the finer sizes but not the larger, and
the gradual winnowing of the fine particles will leave an armour
of coarser material that prevents further degradation.

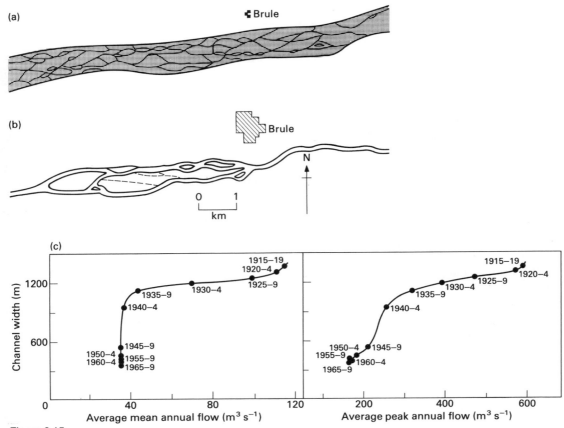

Figure 6.15
The configuration of the channel of the South Platte River at Brule, Nebraska, USA: (a) in 1897 and (b) in 1959. Such changes in channel form result from discharge diminution (c) caused by flood-control works and diversions for irrigation (after Schumm, 1977, figure 5.32 and Williams, 1978)

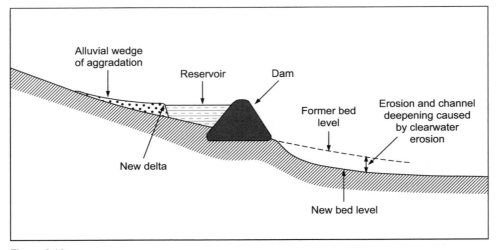

Figure 6.16
Diagrammatic long profile of a river showing the upstream aggradation and the downstream erosion caused by dam and reservoir construction

Table 6.11   Riverbed degradation below dams

River	Dam	Amount (m)	Length (km)	Time (years)
South Canadian (USA)	Conchos	3.1	30	10
Middle Loup (USA)	Milburn	2.3	8	11
Colorado (USA)	Hoover	7.1	111	14
Colorado (USA)	Davis	6.1	52	30
Red (USA)	Denison	2.0	2.8	3
Cheyenne (USA)	Angostura	1.5	8	16
Saalach (Austria)	Reichenhall	3.1	9	21
South Saskatchewan (Canada)	Diefenbaker	2.4	8	12
Yellow (China)	Samenxia	4.0	68	4

Source: from data in Galay, 1983.

Table 6.12   Channel capacity reduction below reservoirs

River	Dam	% channel capacity loss
Republican, USA	Harlan County	66
Arkansas, USA	John Martin	50
Rio Grande, USA	Elephant Buttre	50
Tone, UK	Clatworthy	54
Meavy, UK	Burrator	73
Nidd, UK	Angram	60
Burn, UK	Burn	34
Derwent, UK	Ladybower	40

Source: modified after Petts, 1979, table 1.

The overall effect of the creation of a reservoir by the construction of a dam is to lead to a reduction in downstream channel capacity (see Petts, 1979 for a review). This seems to amount to between about 30 and 70 per cent (see table 6.12).

Equally far-reaching changes in channel form are produced by land-use changes and the introduction of soil conservation measures. Figure 6.17 is an idealized representation of how the river basins of Georgia in the United States have been modified through human agency between 1700 (the time of European settlement) and the present. Clearing of the land for cultivation (figure 6.17b) caused massive slope erosion which resulted in the transfer of large quantities of sediment into channels and flood plains. The phase of intense erosive land use persisted and was particularly strong during the nineteenth century and the first decades of the twentieth century, but thereafter (figure 6.17c) conservation measures, reservoir construction and a reduction in the intensity of agricultural land use led to further channel changes (Trimble, 1974). Streams ceased to carry such a heavy sediment load, they became much less turbid, and incision took

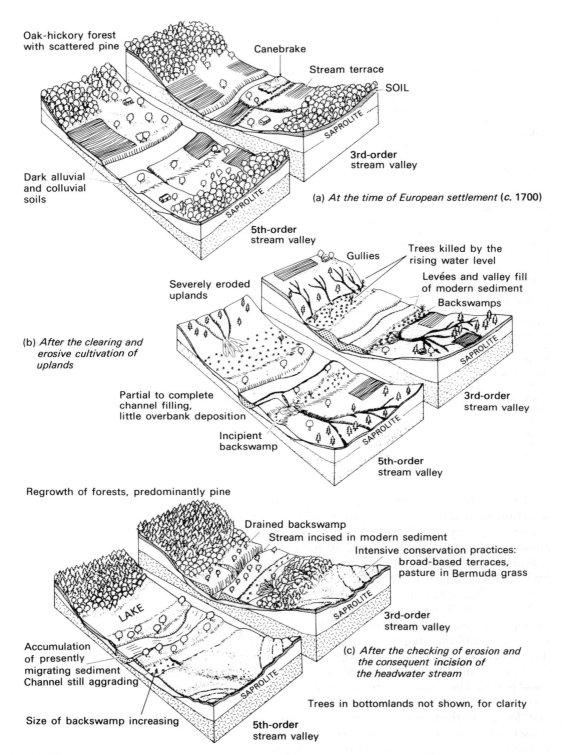

Oak-hickory forest
with scattered pine

Canebrake

Stream terrace

SOIL

SAPROLITE

3rd-order
stream valley

Dark alluvial
and colluvial
soils

SAPROLITE

(a) *At the time of European settlement* (c. 1700)

5th-order
stream valley

Gullies

Trees killed by the
rising water level

Levées and valley fill
of modern sediment

Backswamps

Severely eroded
uplands

SAPROLITE

(b) *After the clearing and
erosive cultivation of
uplands*

Partial to complete
channel filling,
little overbank deposition

3rd-order
stream valley

Incipient
backswamp

SAPROLITE

5th-order
stream valley

Regrowth of forests, predominantly pine

Drained backswamp

Stream incised in modern sediment

Intensive conservation practices:
broad-based terraces,
pasture in Bermuda grass

LAKE

SAPROLITE

3rd-order
stream valley

Accumulation
of presently
migrating sediment
Channel still aggrading

(c) *After the checking of erosion and
the consequent incision of
the headwater stream*

Trees in bottomlands not shown, for clarity

Size of backswamp increasing

SAPROLITE

5th-order
stream valley

Figure 6.17
Changes in the evolution of fluvial landscapes in the Piedmont of Georgia, USA, in response to land-use
change between 1700 and 1970 (after Trimble, 1974, p. 117, in S. W. Trimble, *Man-induced soil erosion on
the southern Piedmont*, Soil Conservation Society of America. © Soil Conservation Society of America)

place into the flood-plain sediments. By means of this active stream-bed erosion, streams incised themselves into the modern alluvium, lowering their beds by as much as 3–4 metres.

In the Platte catchment of south-west Winsconsin a broadly comparable picture of channel change has been documented by Knox (1977). There, as in the Upper Mississippi valley (Knox, 1987), it is possible to identify stages of channel modification associated with various stages of land use, culminating in decreased overbank sedimentation as a result of better land management in the last half century.

Other significant changes produced in channels include those prompted by accelerated sedimentation associated with changes in the vegetation communities growing along channels. The introduction of salt cedar in the southern United States has caused significant flood-plain aggradation. In the case of the Brazos river in Texas, for example, the plants encourage sedimentation by their damming and ponding effect. They clogged channels by invading sand banks and sand bars, and so increased the area subject to flooding. Between 1941 and 1979 the channel width declined from 157 to 67 metres, and the amount of aggradation was as much as 5.5 metres (Blackburn et al., 1983).

There is, however, a major question about the ways in which different vegetation types affect channel form. Are tree-lined banks more stable than those flowing through grassland? On the one hand tree roots stabilize banks and their removal might be expected to cause channel widening and shallowing (Brooks and Brierly, 1997). On the other hand, forests produce log-jams which can cause aggradation or concentrate flow on to channel banks, thereby leading to their erosion. These issues are discussed in Trimble (1997b) and Montgomery (1997).

Another organic factor which can modify channel form is the activity of grazing animals. These can break the banks down directly by trampling and can reduce bank resistance by removing protective vegetation and loosening soil (Trimble and Mendel, 1995).

Finally, the addition of sediments to stream channels by mining activity can cause channel aggradation. Mine wastes can clog channel systems (Gilbert, 1917; Lewin et al., 1983). Equally, the mining of aggregates from river beds themselves can lead to channel deepening (Bravard and Petts, 1996: 246–7).

Reactivation and stabilization of sand dunes

To George Perkins Marsh the reactivation and stabilization of sand dunes, especially coastal dunes, was a theme of great importance in his analysis of human transformation of nature. He devoted 54 pages to it:

The preliminary steps, whereby wastes of loose, drifting, barren sands are transformed into wooded knolls and plains, and finally through the accumulation of vegetable mould, into arable ground, constitute a conquest over nature which precedes agriculture – a geographical revolution – and therefore, an account of the means by which the change has been effected belongs properly to the history of man's influence on the great features of physical geography.

(1965: 393)

He was fascinated by 'the warfare man wages with the sand hills' and asked (1965: 410) 'in what degree the naked condition of most dunes is to be ascribed to the improvidence and indiscretion of man'.

His analysis showed quite clearly that most of the coastal dunes of Europe and North America had been rendered mobile, and hence a threat to agriculture and settlement, through human action, especially because of grazing and clearing. In Britain the cropping of dune warrens by rabbits was a severe problem, and a most significant event in their long history was the myxomatosis outbreak of the 1950s, which severely reduced the rabbit population and led to dramatic changes in stability and vegetative cover.

Appreciation of the problem of dune reactivation on mid-latitude shorelines, and attempts to overcome it, go back a long way (Kittredge, 1948). For example, the menace of shifting sand following denudation is recognized in a decree of 1539 in Denmark which imposed a fine on those who destroyed certain species of sand plants on the coast of Jutland. The fixation of coastal sand dunes by planting vegetation was initiated in Japan in the seventeenth century, while attempts at the reafforestation of the spectacular Landes dunes in south-west France began as early as 1717, and came to fruition in the nineteenth century through the plans of the great Bremontier: 81,000 hectares of moving sand had been fixed in the Landes by 1865. In Britain possibly the most impressive example of sand control is provided by the reafforestation of the Culbin Sands in north-east Scotland with conifer plantations (Edlin, 1976).

Human-induced dune instability is not, however, a problem restricted to mid-latitude coasts. In inland areas of Europe, clearing, fire and grazing have affected some of the late Pleistocene dune fields that were created on the arid steppe margins of the great ice sheets, and in eastern England the dunes of the Breckland presented problems on many occasions. There are records of carriages being halted by sand-blocked roads and of one village, Downham, being overwhelmed altogether.

However, it is possibly on the margins of the great subtropical and tropical deserts that some of the strongest fears are being expressed about sand-dune reactivation. This is one of the facets of the process of desertification and desertization. The increasing

Plate 6.13
An attempt to control dune
movement that threatens part
of the town of Walvis Bay in
Namibia. In this case, fences
have been erected to halt
dune movement

population levels of both humans and their domestic animals,
brought about by improvement in health and by the provision
of boreholes, has led to an excessive pressure on the limited
vegetation resources. As ground cover has been reduced, so dune
instability has increased. The problem is not so much that dunes
in the desert cores are relentlessly marching on to moister areas,
but that the fossil dunes, laid down during the more arid phase
peaking around 18,000 years ago, have been reactivated *in situ*.

A wide range of methods (plate 6.13) is available to attempt
to control drifting sand and moving dunes, as follows:

1 *Drifting sand:*
  enhancement of deposition of sand by the creation of large
    ditches, vegetation belts and barriers and fences;
  enhancement of transport of sand by aerodynamic stream-
    lining of the surface, change of surface materials or panel-
    ling to direct flow;
  reduction of sand supply, by surface treatment, improved
    vegetation cover or erection of fences;
  deflection of moving sand, by fences, barriers, tree belts, etc.
2 *Moving dunes:*
  removal by mechanical excavation;
  destruction by reshaping, trenching through dune axis or
    surface stabilization or barchan arms;
  immobilization, by trimming, surface treatment and fences.

In practice most solutions to the problem of dune instability
and sand blowing have involved the establishment of a vegetation
cover. This is not always easy. Species used to control sand dunes
must be able to endure undermining of their roots, burning,

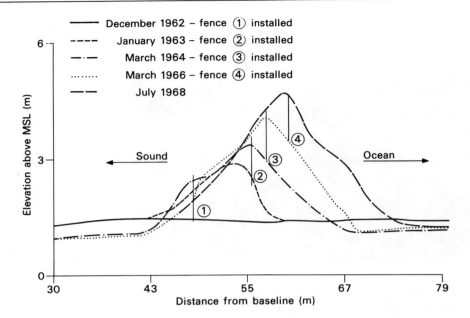

Figure 6.18
Sand accumulation using
the method of multiple fences
in North Carolina, USA.
This raised the dune height
approximately 4 metres over a
period of 6 years (after Savage
and Woodhouse in Goldsmith,
1978, figure 36)

abrasion and often severe deficiencies of soil moisture. Thus the
species selected need to have the ability to recover after partial
growth in the seedling stages, to promote rapid litter development, and to add nitrogen to the soil through root nodules.
During the early stages of growth they may need to be protected
by fences, sand traps and surface mulches. Growth can also be
stimulated by the addition of synthetic fertilizers.

In hearts of deserts sand dunes are naturally mobile because
of the sparse vegetation cover. Even here, however, humans
sometimes attempt to stabilize sand surfaces to protect settlements, pipelines, industrial plant and agricultural land. The use
of relatively porous barriers to prevent or divert sand movement
has proved comparatively successful, and palm fronds or chicken
wire have made adequate stabilizers. Elsewhere surfaces have
been strengthened by the application of high-gravity oil or by
salt-saturated water (which promotes the development of wind-
resistant surface crusts).

In temperate areas coastal dunes have been effectively stabilized by the use of various trees and other plants (Ranwell and
Boar, 1986). In Japan *Pinus thumbergii* has been successful,
while in the great Culbin Sands plantations of Scotland *P. nigra*
and *P. laricio* have been used initially, followed by *P. sylvestris*.
Of the smaller shrubs, *Hippophae* has proved highly efficient,
sometimes too efficient, at spreading. Its clearance from areas
where it is not welcome is difficult precisely because of some of
the properties that make it such an efficient sand stabilizer:
vigorous suckering growth and the rapid regrowth of cut stems

(Boorman, 1977). Different types of grass have also been employed, especially in the early stages of stabilization. These include two grasses which are moderately tolerant of salt: *Agropyron junceiforme*, now called *Elymus farctus* (sand twitch) and *Leymus arenarius* (lime grass). Another grass which is much used, not least because of its rapid and favourable response to burial by drifting sand, is *Ammophila arenaria* (marram).

Further stabilization of coastal dunes has been achieved by setting up sand fences. These generally consist of slats about 1.0–1.5 metres high, and have a porosity of 25–50 per cent. They have proved to be effective in building incipient dunes in most coastal areas. By installing new fences regularly, large dunes can be created with some rapidity (see figure 6.18). Alternative methods, such as using junk cars on the beaches at Galveston, Texas, have been attempted with little success.

Accelerated coastal erosion

Because of the high concentration of settlements, industries, transport facilities and recreational developments on coastlines, the pressures placed on coastal landforms are often acute (Nordstrom, 1994), and the consequences of excessive erosion serious. While most areas are subject to some degree of natural erosion and accretion, the balance can be upset by human activity in a whole range of different ways (table 6.13). However, humans seldom attempt to accelerate coastal erosion deliberately. More usually, it is an unexpected or unwelcome result of various economic projects. Frequently coastal erosion has been accelerated as a result of human efforts to reduce it.

One of the best forms of coastal protection is a good beach. If material is removed from a beach, accelerated cliff retreat may take place. Removal of beach materials may be necessary to secure valuable minerals, including heavy minerals, or to provide aggregates for construction. The classic example of the latter was the mining of 660,000 tonnes of shingle from the beach at Hallsands in Devon, England, in 1887 to provide material for the construction of dockyards at Plymouth. The shingle proved to be undergoing little or no natural replenishment and in consequence the shore level was reduced by about 4 metres. The loss of the protective shingle soon resulted in cliff erosion to the extent of 6 metres between 1907 and 1957. The village of Hallsands was cruelly attacked by waves and is now in ruins (plate 6.14).

Another common cause of beach and cliff erosion at one point is coast protection at another (plate 6.15), and a range of 'hard engineering' structures is available (figure 6.19). As already stated, a broad beach serves to protect the cliffs behind, and

Table 6.13 Mechanisms of human-induced erosion in coastal zones

Human-induced erosion zones	Effects
1 Beach mining for placer deposits (heavy minerals) such as zircon, rutile, ilmenite and monazite	1 Loss of sand from frontal dunes and beach ridges
2 Construction of groynes, breakwaters, jetties and other structures	2 Downdrift erosion
3 Construction of offshore breakwaters	3 Reduction in littoral drift
4 Construction of retaining walls to maintain river entrances	4 Interruption of littoral drift resulting in downdrift erosion
5 Construction of sea-walls, revetments, etc.	5 Wave reflection and accelerated sediment movement
6 Deforestation	6 Removal of sand by wind
7 Fires	7 Migrating dunes and sand drift after destruction of vegetation
8 Grazing of sheep and cattle	8 Initiation of blow-outs and transgressive dunes: sand drift
9 Off-road recreational vehicles (dune buggies, trail bikes, etc.)	9 Triggering mechanism for sand drift attendant upon removal of vegetative cover
10 Reclamation schemes	10 Changes in coastal configuration and interruption of natural processes, often causing new patterns in sediment transport
11 Increased recreational needs	11 Accelerated deterioration, and destruction, of vegetation on dunal areas, promoting erosion by wind and wave action

Source: Hails, 1977, table 9.11, p. 348.

beach formation is often encouraged by the construction of groynes. However, such structures sometimes merely displace the erosion (possibly in an even more marked form) further along the coast. This is illustrated in figure 6.20.

Piers or breakwaters can have similar effects to groynes. This has occurred at various places along the British coast: erosion at Seaford resulted from the Newhaven breakwater, while erosion at Lowestoft resulted from the pier at Gorleston. Figure 6.21 illustrates the changes in location of beach erosion achieved by the building of jetties or breakwaters at various points. At Madras in south-east India, for example, a 1,000 metre breakwater was constructed in 1875 to create a sheltered harbour on a notoriously inhospitable coast, dominated by sand transport from north to south. On the south side of the breakwater over 1 million square metres of new land formed by 1912, but erosion occurred for 5 kilometres north of the breakwater. At Ceara in Brazil, also in 1875, a detached breakwater was erected over a length of 430 metres more or less parallel to the shore. It was believed that by using a detached structure littoral drift would be able to move along the coast uninterrupted by the presence of a conventional structure built across the surf zone. This, however, proved to be a fallacy (Komar, 1976), since the removal of the wave action which provided the energy for transporting the

HALLSANDS IN 1894

After R. Hansford Worth

HALLSANDS IN 1904

The two pictures are from the same view-point, and the state of the tide is the same in both. By 1904 the beach had fallen about 10 feet

Plate 6.14
The village of Hallsands in south Devon, England, displays the effects of sediment removal and dredging. The top photo shows the original state of the fishing village and its beach, while the lower one shows the ravages consequent upon interference

littoral sands resulted in their deposition within the protected area.

Plate 6.16 shows the evolution of the coast at West Bay in Dorset, southern England, following the construction of a jetty. As time goes on the beach in the foreground builds outwards, while the cliff behind the jetty retreats and needs to be protected with a sea-wall. Likewise, the construction of some sea-walls (plate 6.17), erected to reduced coastal erosion and flooding, has had the opposite effect to the one intended (see figure 6.22). Given the extent to which artificial structures have spread along the coastlines of the world, this is a serious matter (Walker, 1988).

Problems of this type are exacerbated because there is now abundant evidence to suggest that much of the reservoir of sand

Plate 6.15
Coastal defences in
Weymouth, southern England.
The piecemeal emplacement
of expensive sea walls and
cliff protection structures is
often only of short-term
effectiveness and can cause
accelerated erosion downdrift

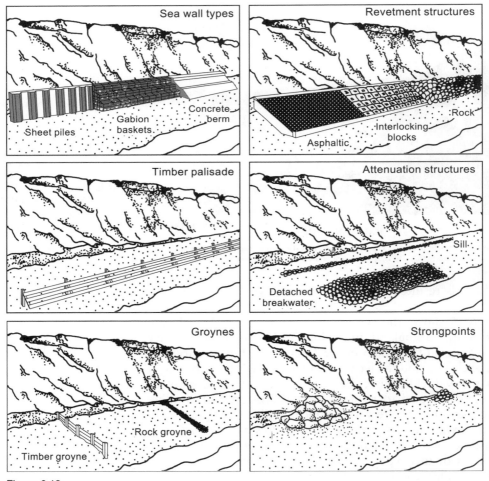

Figure 6.19
A selection of 'hard engineering' structures designed to afford coastal protection (modified from
A. H. Brampton, 'Cliff Conservation and Protection: Methods and Practices to Resolve Conflicts', in
J. Hooke, ed., *Coastal Defence and Earth Science Conservation* (Geological Society Publishing
House, 1998), figures 3.1, 3.2, 3.4, 3.5, 3.6 and 3.7)

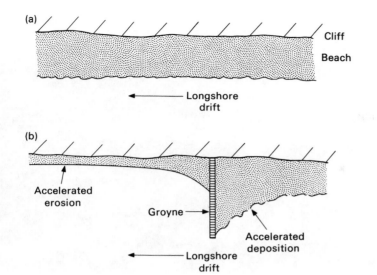

Figure 6.20
Diagrammatic illustration of the effects of groyne construction on sedimentation on a beach

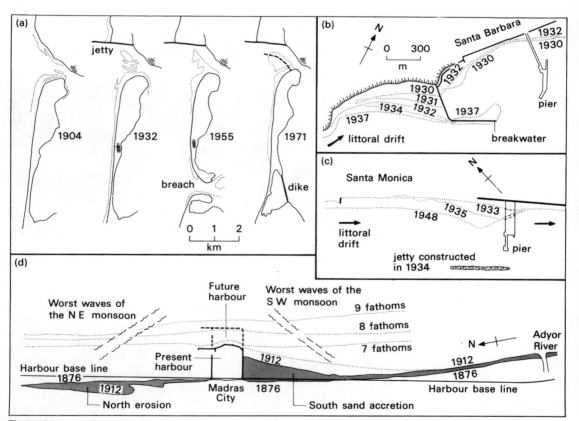

Figure 6.21
Examples of the effects of shoreline installations on beach and shoreline morphology:
(a) Erosion of Bayocean Spit, Tillamook Bay, Oregon, after construction of a north jetty in 1914–17. The heavy dashed line shows the position of the new south jetty under construction.
(b) The deposition-erosion pattern around the Santa Barbara breakwater in California
(c) Sand deposition in the protected lee of Santa Monica breakwater in California
(d) Madras Harbour, India, showing accretion on updrift side of the harbour and erosion on the downdrift side
(after Komar, *Beach Processes and Sedimentation*, p. 334, © 1976. Reprinted by permission of Prentice-Hall Inc.)

Plate 6.16
A jetty was built at West Bay
to facilitate entry to the
harbour. Top: in 1860 it had
had little effect on the
coastline. Centre: by 1900
sediment accumulation had
taken place in the foreground
but there was less sediment
in front of the cliff behind the
town. Bottom: by 1976 the
process had gone even
further and the cliff had to be
protected by a sea-wall. Even
this has since been severely
damaged by winter storms

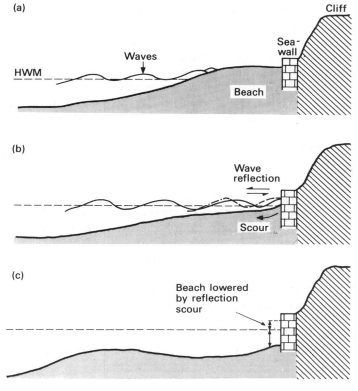

(a)

Cliff

Sea-wall

Waves

HWM

Beach

(b)

Wave reflection

Scour

(c)

Beach lowered by reflection scour

Plate 6.17
Near Brighton in southern England attempts have been made to reduce the rate of coast recession because of the threat to cliff-top housing. Defence mechanisms include a solid sea-wall and a large series of groynes. The reduction of erosion at this point may cause beach depletion elsewhere on the coast

Figure 6.22
Sea-walls and erosion:
(a) a broad, high beach prevents storm waves breaking against a sea-wall and will persist, or erode only slowly; but where the waves are reflected by the wall (b) scour is accelerated, and the beach is quickly removed and lowered (c) (modified after Bird, 1979, figure 6.3)

and shingle that creates beaches is in some respects a relict feature. Much of it was deposited on the continental shelf during the maximum of the last glaciation (around 18,000 BP), when sea level was about 120–40 metres below its present level. It was transported shoreward and incorporated in present-day beaches during the phase of rapidly rising post-glacial sea levels that characterized the Flandrian transgression until about 6000 BP. Since that time, with the exception of minor oscillations of the order of a few metres, world sea levels have been stable and much less material is, as a consequence, being added to beaches and shingle complexes. Therefore, according to Hails (1977: 322): 'in many areas, there is virtually no offshore supply to be moved onshore, except for small quantities resulting from seasonal changes.' It is because of these problems that many erosion prevention schemes now involve beach replenishment (by the artificial addition of appropriate sediments to build up the beach) or employ miscellaneous sand bypassing techniques (including pumping and dredging) whereby sediments are transferred from the accumulation side of an artificial barrier to the erosional side (King, 1974). Such methods of beach replenishment are reviewed by Bird (1996).

In some areas, however, sediment-laden rivers bring material into the coastal zone which becomes incorporated into beaches through the mechanism of longshore drift. Thus any change in the sediment load of such rivers may result in a change in the sediment budget of neighbouring beaches. When accelerated soil erosion occurs in a river basin the increased sediment load may cause coastal accretion and siltation. But where the sediment load is reduced through action such as the construction of large reservoirs, behind which sediments accumulate, coastal erosion may result. This is believed to be one of the less desirable consequences of the construction of the Aswan Dam on the Nile: parts of its delta have shown recently accelerated recession.

The Nile sediments, on reaching the sea, used to move eastward with the general anticlockwise direction of water movements in that part of the eastern Mediterranean, generating sand bars and dunes which contributed to delta accretion. About a century ago an inverse process was initiated and the delta began to retreat. For example, the Rosetta mouth of the Nile lost about 1.6 kilometres of its length from 1898 to 1954. The imbalance between sedimentation and erosion appears to have started with the delta barrages (1861) and then been continued by later works, including the Sennar Dam (1925), Gebel Aulia Dam (1937), Khasm el Girba Dam (1966), Roseires Dam (1966) and the High Dam itself. In addition, large amounts of sediment are retained in an extremely dense network of irrigation and drain channels that has been developed in the Nile delta itself (Stanley, 1996). Much of the Egyptian coast is now 'undernourished'

Table 6.14 Suspended loads of Texas rivers discharging into the Gulf of Mexico

| River | Suspended load (million tonnes) | | |
	1931–40	1961–70	%*
Brazos	350	120	30
San Bernard	1	1	100
Colorado	100	11	10
Rio Grande	180	6	3
TOTAL	631	138	20

* 1961–70 loads as % 1931–40 loads.
Source: modified from Hails, 1977, table 9.1 (after data from Stout et al. and Curtis et al.).

with sediment and, as a result of this overall erosion of the shoreline, the sand bars bordering Lake Manzala and Lake Burullus on the seaward side are eroded and likely to collapse. If this were to happen, the lakes would be converted into marine bays, so that saline water would come into direct contact with low-lying cultivated land and fresh-water aquifers.

Likewise in Texas, where over the last century four times as much coastal land has been lost as has been gained, one of the main reasons for this change is believed to be the reduction in the suspended loads of some of the rivers discharging into the Gulf of Mexico (table 6.14). The four rivers listed carried, in 1961–70, on average only about one-fifth of what they carried in 1931–40. Comparably marked falls in sediment loadings occurred elsewhere in the eastern United States (figure 6.23). Likewise, in France the once mighty Rhône only carries about 5 per cent of the load it did in the nineteenth century; and in Asia, the Indus discharges less than 20 per cent of the load it did before construction of large barrages over the last half-century (Milliman, 1990).

Construction of great levées on the Mississippi river since 1717 has also affected the Gulf of Mexico coast. The channelization of the river has increased its velocity, reduced overbank deposition of silt on to swamps, marshes and estuaries, and changed the salinity conditions of marshland plants (Cronin, 1967). As a result, the coastal marshes and islands have suffered from increased erosion or a reduced rate of development. This has been vividly described by Biglane and Lafleur (1967: 691):

Like a bullet through a rifle barrel, waters of the mighty Mississippi are thrust toward the Gulf between the confines of the flood control levées. Before the day of these man-made structures, these waters poured out over tremendous reaches of the coast . . . Freshwater marshes (salinities averaging 4–6%) were formed by deposited silts and vegetative covers of wire grass . . . As man erected his flood protection devices, these marshes ceased to form as extensively as before.

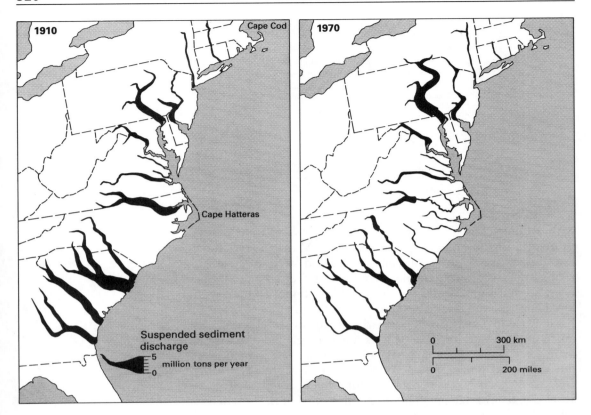

Figure 6.23
The decline in suspended sediment discharge to the eastern seaboard of the USA between 1910 and 1970 as a result of soil conservation measures, dam construction and land-use changes (after Meade and Trimble, 1974)

The changes between 1956 and 1978 are shown in figure 6.24. However, as with so many examples of environmental change, it is unlikely that just one factor, in this case channelization, is the sole cause of the observed trend. In their study of erosion loss in the Mississippi delta and neighbouring parts of the Louisiana coast, Walker et al. (1987) suggest that this loss is the result of a variety of complex interactions among a number of physical, chemical, biological and cultural processes. These processes include, in addition to channelization, worldwide sea-level changes, subsidence resulting from sediment loading by the delta of the underlying crust, changes in the sites of deltaic sedimentation as the delta evolves, catastrophic storm surges and subsidence resulting from subsurface fluid withdrawal.

In some areas anthropogenic vegetation modification creates increased erosion potential. This has been illustrated for the hurricane-afflicted coast of Belize, Central America (Stoddart, 1971). He showed that natural, dense vegetation thickets on low, sand islands (cays) acted as a baffle against waves and served as a massive sediment trap for coral blocks, shingle and sand transported during extreme storms. However, on many islands the natural vegetation had been replaced by coconut plantations.

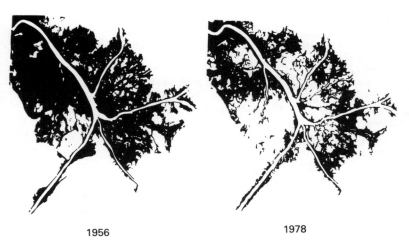

1956   1978

Figure 6.24
Comparison of the outlines
of the Mississippi Delta in
1956 and 1978 gives a clear
indication of the transformation
from marsh to open water.
Artificial controls upriver have
decreased the amount of
sediment carried by the river;
artificial levées along much
of the lower course have kept
flood-borne sediment from
replenishing the wetlands; and
in the active delta itself rock
barriers installed across breaks
similarly confine the river. The
Gulf of Mexico is intruding as
the marshland sinks or is washed
away (*National Geographic
Magazine*, August 1983)

These had an open structure easily penetrated by sea water, tended to have little or no ground vegetation (thus exposing the cay surface to stripping and channelling), and had a dense but shallow root net easily undermined by marginal sapping. Thus Stoddart found (p. 191) that 'where the natural vegetation had been replaced by coconuts before the storm (Hurricane Hattie), erosion and beach retreat led to net vertical decreases in height of 3–7 ft; whereas where natural vegetation remained, banking of storm sediment against the vegetation hedge led to a net vertical increase in height of 1–5 ft.'

Other examples of markedly accelerated coastal erosion and flooding result from anthropogenic degradation of dune ridges. Frontal dunes are a natural defence against erosion, and coastal changes may be long-lasting once they are breached. Many of those areas in eastern England which most effectively resisted the great storm and surge of 1952 were those where humans had not intervened to weaken the coastal dune belt.

Not all dune stabilization and creation schemes have proved desirable (Dolan et al., 1973). In North Carolina (see figure 6.25) the natural barrier-island system along the coastline met the challenge of periodic extreme storms, such as hurricanes, by placing no permanent obstruction in the path of the powerful waves. Under these natural conditions, most of the initial stress of such storms is sustained by relatively broad beaches (figure 6.25a). Since there is no resistance from impenetrable landforms, water can flow between the dunes (which do not form a continuous line) and across the islands, with the result that wave energy is rapidly exhausted. However, between 1936 and 1940, 1,000 kilometres of sand fencing was erected to create an artificial barrier dune along part of the outer banks, and 2.5 million trees and various grasses (especially *Ammophila breviligulata*) were

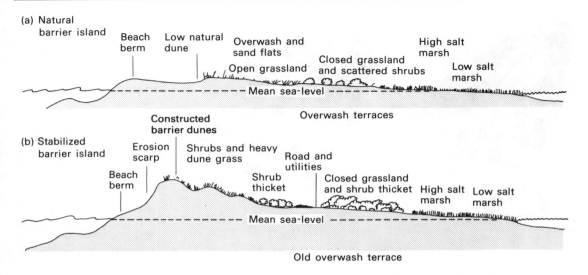

Figure 6.25
Cross-sections of two barrier islands in North Carolina, USA. The upper diagram (a) is typical of the natural systems and the lower (b) illustrates the stabilized systems (after Dolan et al., 1973, figure 4)

planted to create large artificial dunes. The altered barrier islands (see figure 6.25b) not only have the artificial barrier-dune system; they also have beaches that are often only 30 metres wide, compared with 140 metres for the unaltered islands. This beach-narrowing process, combined with the presence of a permanent dune structure, has created a situation in which high wave energy is concentrated in an increasingly restricted run-up area, resulting in a steeper beach profile, increased turbulence and greater erosion. Another problem associated with artificial dune stabilization is the flooding that occurs when north-east storms pile the water of the lagoon, Pamlico Sound, up against the barrier island. In the past, these surge waters simply flowed out between the low, discontinuous dunes and over the beach to the sea; but with the altered dune chain the water cannot drain off readily and vast areas of land are at times submerged.

General treatments of coastal problems and their management are provided by French (1997) and by Viles and Spencer (1995).

Changing rates of salt marsh accretion

In Britain in recent decades, the nature of some salt marshes, and the rate at which they accrete, has been transformed by a major vegetational change, namely the introduction of a salt marsh plant, *Spartina alterniflora*. This cord-grass appears to have been introduced to Southampton Water by accident from the east coast of North America, possibly in shipping ballast. The crossing of this species with the native *Spartina maritima* produced an invasive cord-grass of which there were two forms: *Spartina towsendii* and *Spartina anglica*, the latter of which is now the main species. It appeared first at Hythe on Southampton Water in 1870 and then spread rapidly to other salt marshes in Britain: partly because of natural spread and partly because of deliberate planting. For example, it reached Poole Harbour by c.1890 and the South Devon estuaries in the 1940s and early 1950s. By 1971 it had reached Walney on the Cumbria coast. Indeed, *Spartina* now reaches as far north as the island of Harris in the west of Scotland, and to the Cromarty Firth in the east (Doody, 1984).

The plant has often been effective at excluding other species and also at trapping sediment. Rates of accretion can therefore become very high. Ranwell (1964) gives rates of 8–10 centimetres per year for Bridgwater Bay (Bristol Channel) and 2 centimetres per year for Keysworth Marsh (Poole Harbour). There is evidence that this has caused progressive silting of estuaries such as the Dee (Marker, 1967) and the Severn (Page, 1982).

However, for reasons that are not fully understood, *Spartina* marshes have frequently suffered dieback, which has sometimes led to marsh recession. In the case of Poole Harbour and Beaulieu estuary, this may date back to the 1920s, but elsewhere on the south coast it has generally been rapid and extensive since about 1960 (Tubbs, 1984). Among the hypotheses that have been put forward to explain dieback are the role of rising sea level, pathogenic fungi, increased wave attack and the onset of waterlogging and anaerobic conditions on mature marsh. Some support for the latter view comes from the fact that in areas where the introduction has been more recent, for example in Wales and north-west England, the area of *Spartina* still appears to be increasing (Deadman, 1984).

The human impact on seismicity and volcanoes

The seismic and tectonic forces which mould the relief of the earth and cause such hazards to human civilization are one of the fields in which efforts to control natural events have had least success and where least has been attempted. None the less, the fact that humans have been able, inadvertently, to trigger off small earthquakes by nuclear blasts, as in Nevada (Pakiser et al., 1969), by injecting water into deep wells, as in Colorado, by mining, by building reservoirs and by fluid extraction, suggests that in due course it may be possible to 'defuse' earthquakes by relieving tectonic strains gradually in a series of non-destructive, low-intensity earthquakes. One problem, however, is that there is no assurance that an earthquake purposefully triggered by human action will be a small one, or that it will be restricted to a small area. The legal implications are immense.

The demonstration that increasing water pressures could initiate small-scale faulting and seismic activity was unintentionally

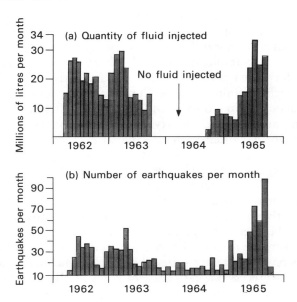

Figure 6.26
Correlation between quantity
of waste water pumped into
a deep well and the number
of earthquakes near Denver,
Colorado (after Birkeland and
Larson, 1978, p. 573)

demonstrated near Denver (Evans, 1966), where nerve-gas waste was being disposed of at great depth in a well in the hope of avoiding contamination of useful groundwater supplies. The waste was pumped in at high pressures and triggered off a series of earthquakes (see figure 6.26), the timing of which corresponded very closely to the timing of waste disposal in the well. It is also now thought that the pumping of fluids into the Inglewood oilfield, Los Angeles, to raise the hydrostatic pressure and increase oil recovery, may have been responsible for triggering the 1963 earthquake which fractured a wall of the Baldwin Hills reservoir. It appears that increased fluid pressure reduces the frictional force across the contact surface of a fault and allows slippage to occur, thereby causing an earthquake.

In general earthquake triggering has been related to fluid injection, but for reasons that are still obscure there may be some cases where increased seismicity has resulted from fluid abstraction (Segall, 1989).

The significance of these 'accidents' has now been verified experimentally at an oilfield at Rangley in Colorado, where variations in seismicity have been produced by deliberately controlled variations in the fluid pressure in a zone that is seismically active (Raleigh et al., 1976).

Perhaps the most important anthropogenically induced seismicity results from the creation of large reservoirs. Reservoirs impose stresses of significant magnitude on crustal rocks at depths rarely equalled by any other human construction. With the ever-increasing number and size of reservoirs the threat rises. There are at least six cases (Koyna, Kremasta, Hsinfengkiang,

Table 6.15 List of dams for which associated seismic phenomena are reported

Dam	Year of completion	Year of largest earthquake since dam completion
Marathon, Greece	1929/30	1938
Oued Fodda, Algeria	1932	–
Boulder, USA	1936	1939
Pieve di Cadore, Italy	1949	–
Clark Hill, USA	1952	1974
Kariba, Zimbabwe/Zambia	1959	1963
Grandval, France	1959	1963
Hsingfengkiang, China	1959	1962
Kurobe, Japan	1960	1961
Camarillas, Spain	1960	1961
Cannelles, Spain	1960	1962
Vaiont, Italy	1961	1963
Monteynard, France	1962	1963
Koyna, India	1962	1967
Benmore, New Zealand	1965	1966
Kremasta, Greece	1965	1966
Piastra, Italy	1965	1966
Contra, Switzerland	1965	1965
Najina Bašta, Yugoslavia	1966	1967
Grančarevo, Yugoslavia	1967	–
Oroville, USA	1968	1975
Kastraki, Greece	1968	–
Nurek. Tajik Republic	1969	1972
Vouglans. France	1970	1971
Karnafusa, Japan	1970	–
Talbingo, Australia	1971	1972
Schlegeis, Austria	1971	–
H. Verwoerd, South Africa	1972	–
Jocasses, USA	1972	1975
Keban. Turkey	1973	1974
Manic 3, Canada	1975	1975

Source: from data in Judd, 1974 and Milne, 1976, processed by author.

Kariba, Hoover and Marathon) where earthquakes of a magnitude greater than 5, accompanied by a long series of foreshocks and aftershocks, have been related to reservoir impounding. However, as table 6.15 shows, there are many more locations where the filling of reservoirs behind dams has led to appreciable, though less dramatic, levels of seismic activity. Detailed monitoring has shown that earthquake clusters occur in the vicinity of some dams after their reservoirs have been filled, whereas before construction activity was less clustered and less frequent. Similarly, there is evidence from Vaiont (Italy), Lake Mead (United States), Kariba (Central Africa), Koyna (India) and Kremasta (Greece) that there is a linear correlation between the storage level in the reservoir and the logarithm of the frequency of shocks. This is illustrated for Vaiont (figure 6.27a), Koyna (figure 6.27b) and Nurek (figure 6.27c).

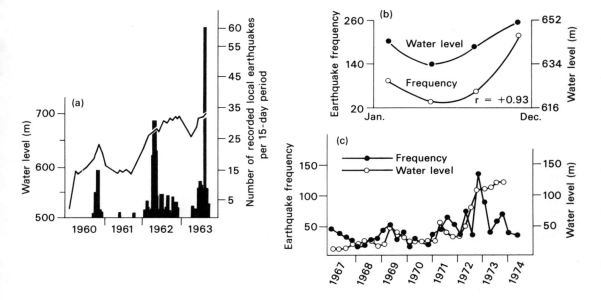

Figure 6.27
Relationships between
reservoir levels and
earthquake frequencies for:
(a) Vaiont Dam, Italy
(b) Koyna, India (these
curves show the 3-monthly
average of water level
and the total number of
earthquakes for the same
months from 1964 to 1968)
(c) The Nurek Dam, Tajikistan
(after Judd, 1974 and
Tajikistan Academy of
Sciences, 1975)

It is also apparent from Nurek that as the great reservoir has filled, so the depth of the more shallow-seated earthquakes appears to have increased.

One reason why dams induce earthquakes involves the hydro-isostatic pressure exerted by the mass of the water impounded in the reservoir, together with changing water pressures across the contact surfaces of faults. Given that the deepest reservoirs provide surface loads of only 20 bars or so, direct activation by the mass of the impounded water seems an unlikely cause (Bell and Nur, 1978) and the role of changing pore pressure assumes greater importance. Paradoxically, there are some possible examples of reduced seismic activity induced by reservoirs (Milne, 1976). One possible explanation of this is the increased incidence of stable sliding (fault creep) brought about by higher pore-water pressure in the vicinity of the reservoir.

However, the ability to prove an absolutely concrete cause-and-effect relationship between reservoir activity and earthquakes is severely limited by our inability to measure stress below depths of several kilometres, and some examples of induced seismicity may have been built on the false assumption that because an earthquake occurs in proximity to a reservoir it has to be induced by that reservoir (Meade, 1991).

Miscellaneous other human activities appear to affect seismic levels. In Johannesburg, South Africa, for example, gold mining and associated blasting activity have produced tens of thousands of small tremors, and there is a notable reduction in the number that occur on Sundays, a day of rest. In Staffordshire, England, coal mining has caused increased seismic activity. There are also cases where seismicity and faulting can be attributed to fluid extraction, for example, in the oilfields of Texas and California and the gasfields of the Po Valley in Italy.

When looking at the human impact on volcanic activity human impotence becomes apparent, though some success has been achieved in the control of lava flows. Thus in 1937 and 1947 the US Army attempted to divert lava from the city of Hilo, Hawaii, by bombing threatening flows, while elsewhere, where lava rises in the crater, breaching of the crater wall to direct lava towards uninhabited ground may be possible. In 1973 an attempt was made to halt the advance of lava with cold water during the Icelandic eruption of Krikjufell. Using up to 4 million litres of pumped waste per hour, the lava was cooled sufficiently to decrease its velocity at the flow front so that the chilled front acted as a dam to divert the still fluid lava behind (Williams and Moore, 1973).

7 The Human Impact on Climate and the Atmosphere

World climates

The climate of the world is now known to have fluctuated frequently and extensively in the three million or so years during which humans have inhabited the earth (Goudie, 1992). The bulk of these changes have nothing to do with human intervention. Climate has changed, and is currently changing, because of a wide range of different natural factors which operate over a variety of time-scales (figure 7.1).

No completely acceptable explanation of climatic change has ever been presented, and no one process acting alone can explain all scales of climate change. The complexity of possibly causative factors involved is daunting.

The complexity becomes evident if we follow the pathway of radiation derived from the ultimate force of climate – the sun. First of all, for reasons such as the varying tidal pull being exerted on the sun by the planets, the quality and quantity of solar radiation may change. It has been recognized that the sun's radiation changes both in quantity (through association with such familiar phenomena as sunspots, which are dark regions of lower surface temperature on the sun's surface) and in quality (through changes in the ultra-violet range of the solar spectrum). Cycles of solar activity have been established for the short term by many workers, with 11- to 22-year cycles particularly being noted. Sunspot cycles of 80–90 years have also been postulated. Observations of sunspots in historical times have also given a measure of solar activity, and one very striking feature is the near absence of sunspots between AD 1640 and 1710, a period sometimes called the Maunder Minimum. It is perhaps significant that this minimum occurred during some of the more extreme years of the Little Ice Age. Some evidence of possible longer-term solar effects comes from studies of the oscillation in the concentration of atmospheric ^{14}C (radiocarbon),

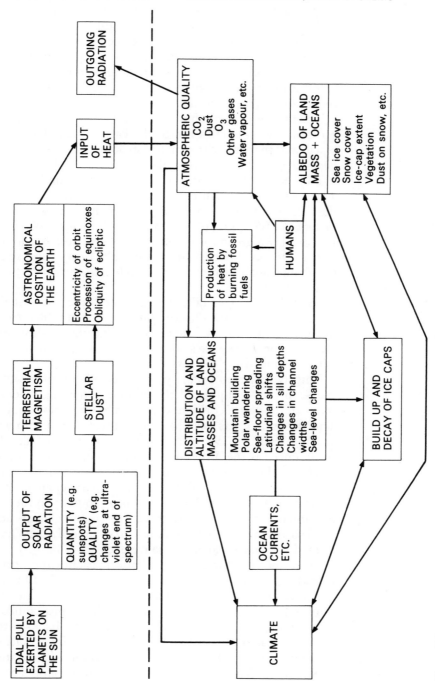

Figure 7.1
A schematic representation of some of the possible influences causing climatic change (after Goudie, 1992, figure 1)

which may in turn depend upon variations in the emission of solar radiation.

The receipt of such varying radiation at the earth's surface might itself vary because of the presence of fine interstellar matter (nebulae) through which the earth might from time to time pass, or which might interpose itself between the sun and our planet. This would tend to reduce the receipt of solar radiation. Likewise, the passage of the solar system through a dust lane bordering a spiral arm of the Milky Way galaxy might cause a temporary reduction in receipt of the sun's radiation output.

The receipt of incoming radiation will also be affected by the position and configuration of the earth. Such changes do take place, and there are three main astronomical factors which have been identified as of probable importance, all three occurring in a cyclic manner. First, the earth's orbit around the sun is not a perfect circle but an ellipse. If the orbit were a perfect circle then the summer and winter parts of the year would be equal in length. With greater eccentricity the length of the seasons will display a greater difference. Over a period of about 96,000 years, the eccentricity of the earth's orbit can 'stretch' by departing much further from a circle and then revert to almost true circularity.

Secondly, changes take place in the 'precession of the equinoxes', which means that the time of year at which the earth is nearest the sun varies. The reason is that the earth wobbles like a child's top and swivels round its axis. This cycle has a periodicity of about 21,000 years.

Thirdly, changes occur, with a periodicity of about 40,000 years, in the 'obliquity of the elliptic' – the angle between the plane of the earth's orbit and the plane of its rotational equator. This movement has been likened to the roll of a ship with a tilt varying from 21°39′ to 24°36′. The greater the tilt, the more pronounced is the difference between winter and summer.

These three cycles comprise what is often called the Milankovitch or Orbital Theory of climatic change. They have a temptingly close similarity in their periodicity to the durations of climatic change associated with the many glacials and interglacials of the last 1.6 million years. Indeed, they have been termed the 'pace-maker of the ice ages'.

Once the incoming solar radiation reaches the atmosphere, its passage to the surface of the earth is controlled by the gases, moisture and particulate matter that are present. Essential importance has been attached to the role of dust clouds emitted from volcanoes. These could increase the backscattering of incoming radiation and thus promote cooling. Volcanic dust veils produced by, for example, the eruption of Krakatoa in the 1880s and of Mount Pinatubo in 1991 caused global cooling for a matter of a few years. However, changing levels of volcanic activity are not the only way in which changes in atmospheric transparency might occur. For example, dust can be emplaced into the atmosphere by the wind erosion of fine-grained sediment and soil, and we know from the extensive deposits of wind-laid silts (loess) of glacial age that during the glacial maxima the atmosphere was probably very dusty, contributing to global cooling.

Carbon dioxide, methane, nitrous oxide, sulphur dioxide and water vapour can also modify the receipt of solar radiation. Particular attention has focused in recent years on the role of carbon dioxide (CO_2) in the atmosphere. This gas is virtually transparent to incoming solar radiation but absorbs outgoing terrestrial infra-red radiation – radiation that would otherwise escape to space and result in heat loss from the lower atmosphere. In general, through the mechanism of this so-called greenhouse effect, low levels of CO_2 in the atmosphere would be expected to lead to cooling, and high levels would be expected to produce a 'heat trap'. The same applies to levels of methane and nitrous oxide, which, molecule for molecule, are even more effective greenhouse gases than CO_2. Recently it has proved possible to retrieve CO_2 from gas bubbles preserved in layers of ice in deep-ice cores drilled from the polar regions. Analyses of changes in CO_2 concentrations in these cores have provided truly remarkable results and have demonstrated that CO_2 changes and climatic changes have progressed in approximate synchroneity over the last 160,000 years. Thus the last interglacial around 120,000 years ago was a time of high CO_2 levels, the last glacial maximum around 18,000 years ago of low CO_2 levels, and the early Holocene a time of very rapid rise in CO_2 levels. The reasons for the observed natural change in greenhouse gas concentrations are still the subject of active scientific research.

Once incoming radiation from the sun reaches the earth's surface it may be absorbed or reflected according to the nature of the surface, and in particular according to whether it is land or water, covered in vegetation or desert, and whether it is mantled by snow.

The effect of the received radiation on climate also depends on the distribution and altitude of land masses and oceans. These too are subject to change in a wide variety of ways: the plates that comprise the earth's crust are ever moving, mountain belts may grow or subside, and oceans and straits open and close. These processes shift areas into new latitudes, transform the world's wind belts and modify the climatically very important ocean currents.

In this discussion of causes it is also crucial to consider feedbacks. Such feedbacks are responses to the original forcing factors that act either to increase or intensify the original forcing

Plate 7.1
Air pollution in Cape Town, South Africa. The combustion of fossil fuels, including coal, to generate electricity and to power vehicles, is a major cause not only of local air pollution but also of the increase in greenhouse gas loadings in the atmosphere

(we call this positive feedback) or to decrease or reverse it (negative feedback). Clouds, ice and snow, and water vapour are three of the most important feedback mechanisms. An example of a positive feedback is the role of snow. Under cold conditions this falls rather than rain, changes the albedo (reflectivity) of the ground surface and causes further cooling of the air above it. Similarly, water vapour is a major greenhouse gas and a warmer climate produces more water vapour. This is because the rate of evaporation from the oceans and the water-holding capacity of the air would both increase as temperatures rise.

Finally, it may well be that the atmosphere and the oceans possess a degree of internal instability which furnishes a built-in mechanism of change so that some small and random change might, through the operation of positive feedbacks and the passage of thresholds, have extensive and long-term effects. Small triggers might have big consequences.

While humans are at present incapable of modifying some of these natural mechanisms of climate change – the output of solar radiation, the presence of fine interstellar matter, the earth's orbital variations, volcanic eruptions, mountain building and the overall pattern of land masses and oceans – there are some key areas where humans may be capable of making significant changes to global climates (plate 7.1). The most important categories of influence are in terms of the chemical composition of the atmosphere and the albedo of the earth's surface, though, as the following list shows, there are some others that also need to be considered.

Possible mechanisms
Gas emissions
   CO_2 – industrial and agricultural
   methane
   chlorofluorocarbons (CFCs)
   nitrous oxide
   krypton 85
   water vapour
   miscellaneous trace gases
Aerosol generation
Thermal pollution
Albedo change
   dust addition to ice caps
   deforestation
   overgrazing
Extension of irrigation
Alteration of ocean currents by constricting straits
Diversion of fresh waters into oceans

The greenhouse gases – carbon dioxide

The greenhouse effect occurs in the atmosphere because of the presence of certain gases that absorb infrared radiation (figure 7.2). Light and ultra-violet radiation from the sun are able to penetrate the atmosphere and warm the earth's surface. This energy is re-radiated as infrared radiation, which, because of its longer wavelength, is absorbed by certain substances such as water vapour, carbon dioxide and other trace gases. This causes the average temperature of the earth's surface and atmosphere to increase. Should the quantities of each substance in the atmosphere be increased, then the greenhouse effect will become enhanced (Hansen et al., 1981).

In reality the term 'greenhouse effect' is something of a misnomer. As Henderson-Sellers and Robinson (1986: 60) explain:

We know that a greenhouse maintains its higher internal temperature largely because the shelter it offers reduces the turbulent transfers of energy away from the surface rather than because of any radiative considerations. Thus while the greenhouse effect remains valid, and vital, for the atmosphere it might be better to think of the physical processes in terms of the 'leaky bucket' analogy . . . Here an increase in the amount of gas with absorption bands in the infrared part of the spectrum is represented by a decrease in the size of the hole in the bucket. The surface temperature, represented by the depth of the water in the bucket, rises as more absorbing gases enter the atmosphere.

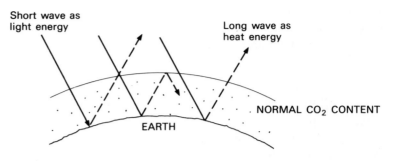

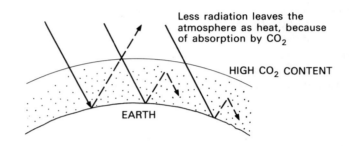

Figure 7.2
The greenhouse effect:
short-wave radiation from the
sun is absorbed by the earth's
surface which, in turn, radiates
heat at far longer wavelengths
because of its temperature of
around 280°K, compared with
6,000°K for the sun

Table 7.1 Sources of four principal greenhouse gases[1]

Gas	Natural sources	Human-derived sources
Carbon dioxide	Terrestrial biosphere Oceans	Fossil fuel combustion Cement production Land-use modification
Methane	Natural wetlands Termites Oceans and freshwater lakes	Fossil fuels (natural gas production, coal mines, petroleum industry, coal combustion) Enteric fermentation (e.g. cattle) Rice paddies Biomass burning Landfills Animal waste Domestic sewage
Nitrous oxide	Oceans Tropical soils (wet forests, dry savannas) Temperate soils (forests, grasslands)	Nitrogenous fertilizers Industrial sources Land-use modification (biomass burning, forest clearing) Cattle and feed lots
Chlorofluorocarbons[2]	Nil	Rigid and flexible foam Aerosol propellants Teflon polymers Industrial solvents

[1] Sources listed in order of decreasing magnitudes of emission except where otherwise indicated.
[2] Sources of chlorofluorocarbons not in order of decreasing magnitudes of emission.

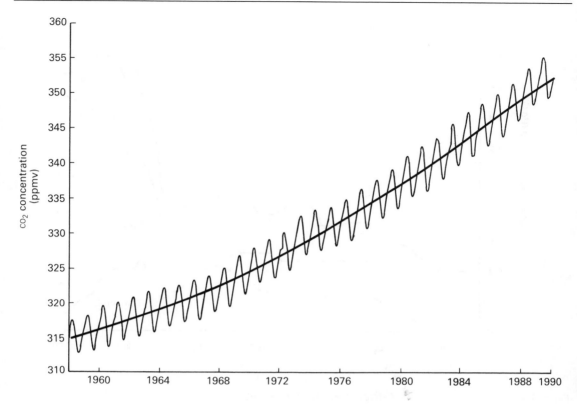

There are a number of ways in which humans have been enhancing the greenhouse effect, the most significant of which is to increase carbon dioxide levels in the atmosphere.

Since the beginning of the industrial revolution humans have been taking stored carbon out of the earth in the form of coal, petroleum and natural gas, and burning it to make carbon dioxide (CO_2), heat, water vapour and smaller amounts of sulphur dioxide (SO_2) and other gases. The pre-industrial level of carbon dioxide is a matter of some debate, but may have been as low as 260–270 ppm by volume (Wigley, 1983). The present level is over 350 ppmv, and the upward trend is evident in records from various parts of the world (figure 7.3). At the present rate it would reach 500 ppmv by the end of the twenty-first century. The prime cause of increased carbon dioxide emissions is fossil fuel combustion and cement production (*c.*5.5±0.5 GtC/yr in the 1980s), but with the release of carbon dioxide by changes in tropical land use (primarily deforestation) being a significant factor (*c.*1.6±1.0 GtC/yr). The amount of carbon derived from deforestation has increased greatly from about 0.4 GtC/yr in 1850 (figure 7.4) (Woodwell, 1992). The various relationships between land-use change and the build-up of greenhouse gases are reviewed by Adger and Brown (1994).

Figure 7.3
Mean monthly concentrations of atmospheric CO_2 at Mauna Loa, Hawaii. The yearly oscillation is explained mainly by the annual cycle of photosynthesis and respiration of plants in the northern hemisphere (based on data provided by the Scripps Institute of Oceanography)

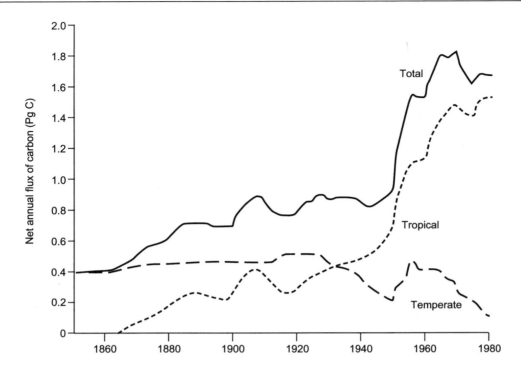

Figure 7.4
The net annual flux of carbon
from deforestation in tropical
and temperate zones globally,
1850–1980 (after Houghton
and Skole, 1990, figure 23.2,
and Woodwell, 1992,
figure 5.2)

Other greenhouse gases

In addition to carbon dioxide, it is probable that other gases
will contribute to the greenhouse effect. Individually their effects
may be minor, but as a group they may be major (Ramanathan,
1988). Indeed, molecule for molecule some of them may be
much more effective as greenhouse gases than CO_2, as the data
in table 7.2 show.

One of the more important of the trace gases is methane
(CH_4), which has a strong infrared absorption band at 7.66 μm.
Ice core studies and recent direct observations (figure 7.5b) suggest
that until the beginning of the eighteenth century background

Table 7.2 Radiative forcing relative to CO_2 per unit molecule change in the
atmosphere

GAS	Relative radiative forcing	Residence time in atmosphere (years)
CO_2	1	100
CH_4 (methane)	21	10
N_2O (nitrous oxide)	206	100–200
CFC-11	12,400	65
CFC-12	15,800	130

Source: extracted from Houghton et al., 1990, table 2.3, p. 53.

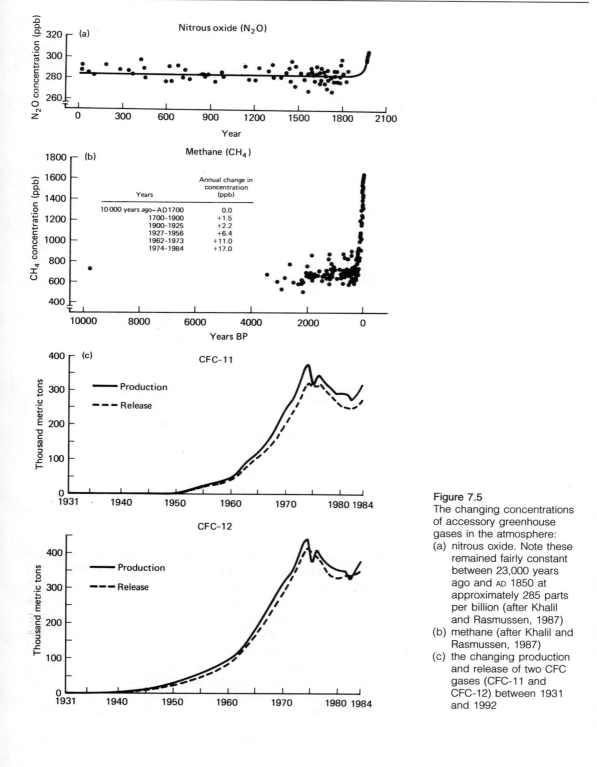

Figure 7.5
The changing concentrations of accessory greenhouse gases in the atmosphere:
(a) nitrous oxide. Note these remained fairly constant between 23,000 years ago and AD 1850 at approximately 285 parts per billion (after Khalil and Rasmussen, 1987)
(b) methane (after Khalil and Rasmussen, 1987)
(c) the changing production and release of two CFC gases (CFC-11 and CFC-12) between 1931 and 1992

Table 7.3 (a) Estimates of methane (CH_4) source strengths and sinks

Sources/sinks	Best estimate (10^6t a^{-1})	Range (10^6t a^{-1})
Sources		
Natural wetlands	115	100–200
Rice paddies	110	25–170
Enteric fermentation (animals)	80	65–100
Gas drilling, venting, transmission	45	25–50
Biomass burning	40	20–80
Termites	40	10–100
Landfills	40	20–70
Coal mining	35	19–50
Oceans	10	5–20
Fresh waters	5	1–25
CH_4 hydrate destabilization	5	0–100
Sinks		
Removal by soils	30	15–45
Reaction with OH	500	400–600
Atmospheric increase	44	40–48

(b) Estimates of nitrous oxide (N_2O) source strengths and sinks

Sources/sinks	Range (10^6t a^{-1})
Sources	
Oceans	1.4–2.6
Soils (tropical forests)	2.2–3.7
Soils (temperate forests)	0.7–1.5
Fossil fuel combustion	0.1–0.3
Biomass burning	0.02–0.2
Fertilizer (including ground water)	0.01–2.2
Sinks	
Removal by soils	Unknown
Photolysis in the stratosphere	7–13
Atmospheric increase	3–4.5

Source: data in UNEP, 1991.

levels were stable at around 600 parts per billion by volume (ppbv). They rose steadily between AD 1700 and 1900, and then increased still more rapidly, attaining levels that averaged 1,300 ppbv in the early 1950s and 1,600 ppbv by the mid-1980s (Khalil and Rasmussen, 1987) and over 1,700 ppbv in the 1990s. This increase of 2.5 times over background levels results primarily from increased rice cultivation in waterlogged paddy fields, the enteric fermentation produced in the growing numbers of flatulent domestic cattle, and the burning of oil and natural gas (Crutzen et al., 1986) (table 7.3).

Chlorofluorocarbons (CFCs), despite their relatively trace amounts in the atmosphere, have increased very markedly in terms of their emissions (figure 7.5c) and their concentrations in recent decades resulting from their use as refrigerants, foam makers, fire control agents and propellants in aerosol cans. They have a very strong greenhouse effect even in relatively small amounts. On the other hand, the ozone depletion they have caused in the stratosphere may counteract this effect, for stratospheric ozone depletion results in a decrease in radiative forcing (Houghton et al., 1992). Conversely, the build up of lower-level tropospheric ozone can contribute to the greenhouse effect.

Nitrous oxide (N_2O) is also no laughing matter, for it can contribute to the greenhouse effect, primarily by absorption of infrared at the 7.8 and 17 µm bands. Combustion of hydrocarbon fuels, the use of ammonia-based fertilizers, deforestation and biomass burning are among the processes that could lead to an increase in atmospheric N_2O levels (figure 7.5a and table 7.3b). Atmospheric N_2O concentrations have increased from around 275 ppbv in pre-industrial times to 311 ppbv in 1992.

Other trace gases that could play a greenhouse role include bromide compounds, carbon tetrafluoride, carbon tetrachloride and methyl chloride.

It is possible that the increase of global surface temperatures to be expected by 2050 as a result of these trace gases is 1.4–2.2°C, an increase that may be comparable to that anticipated from CO_2 itself (Macdonald, 1982; Bolin et al., 1986: xxi).

The continued role of greenhouse gases other than CO_2 in changing the climate is already not greatly less important than that of CO_2. If present trends continue, the combined concentrations of atmospheric CO_2 and other greenhouse gases would be radiatively equivalent to a doubling of CO_2 from pre-industrial levels possibly as early as the 2030s. The relative amounts of radiative forcing for different greenhouse gases since pre-industrial times are, according to the Intergovernmental Panel on Climate Change (1996), as follows (in watts per square metre):

CO_2	1.56 Wm^{-2}
CH_4	0.47 Wm^{-2}
N_2O	0.14 Wm^{-2}
CFCs and HCFCs	0.25 Wm^{-2}
tropospheric ozone	0.40 Wm^{-2}

Another important feature of the various greenhouse gases is their residence time in the atmosphere. CH_4 has a residence time in the atmosphere of about 10 years, the shortest of all the greenhouse gases. This means that if we could stop the enhanced emissions of that gas, its concentration in the atmosphere should fall to its natural level in a decade. By contrast N_2O (100–200 years) and CO_2 (c.100 years) have much longer residence times, so that even if we could control their sources immediately it would still take a very long time for them to fall to their natural levels.

Aerosols

Aerosols are finely divided solid or liquid particles dispersed in the atmosphere. In general terms it is believed that they affect climate because they intercept and scatter a small portion of incoming radiation from the sun, thus reducing the energy reaching the ground. This is called the 'direct effect' of aerosols. The direct effect increases with both the number and size of aerosols in the atmosphere. Aerosols also have an 'indirect effect'. This is because they are key elements in cloud formation. The number of cloud droplets that a cloud possesses is determined by the number of aerosols available on to which water vapour can condense. Clouds with large numbers of droplets reflect more sunlight back into space and so can also contribute to cooling. Aerosols do not, however, inevitably cause cooling (Weare et al., 1974). So, for example, Idso and Brazel (1978) and Brazel and Idso (1979) point to the two contrasting tendencies of dust: the backscattering effect producing cooling, and the thermal-blanketing effect causing warming. The second of these absorbs some of the earth's thermal radiation that would otherwise escape to space, and then re-radiates a portion of this back to the land surface, raising surface temperatures. They believe that natural dust from volcanic emissions tends to enter the stratosphere (where backscattering and cooling are the prime consequences), while anthropogenic dust more frequently occurs in the lower levels of the atmosphere, causing thermal blanketing and warming.

There is a variety of ways in which human activities have increased atmospheric aerobic loadings. One of these is industrial emission of smoke and dust particles (Davitya, 1969), though whether this is a matter of only local rather than regional or global significance is a matter for debate (Hoyt and Fröhlich, 1983).

Industrialization is not, however, the sole source of particles in the atmosphere, nor is a change in temperature the only possible consequence. Bryson and Barreis (1967), for example, argue that intensive agricultural exploitation of desert margins, such as in Rajasthan, India, would create a dust pall in the atmosphere by exposing larger areas of surface materials to deflation in dust storms. This dust pall, they believe, would so change atmospheric temperature that convection, and thus rainfall, would be reduced. Observations on dust levels over the Atlantic during the drought years of the late 1960s and early 1970s in the Sahel suggest that the degraded surfaces of that time led to a great (threefold) increase in atmospheric dust (Prospero and Nees, 1977). There is thus the possibility that human-induced desertization generates dust which could in turn

increase the degree of desertization by its effect on rainfall levels. Biomass burning of tropical savannas may also add large numbers of aerosols to the atmosphere.

The most catastrophic effects of anthropogenic aerosols in the atmosphere could be those resulting from a nuclear exchange between the great powers. Explosion, fire and wind might generate a great pall of smoke and dust in the atmosphere which would make the world dark and cold. It has been estimated that if the exchange reached a level of several thousand megatons, a 'nuclear winter' would be as low as −15° to −25°C (Turco et al., 1983), though more recent simulations by Schneider and Thompson (1988) suggest that some previous estimates may have been exaggerated. They suggest that in the northern hemisphere maximum average land surface summertime temperature depressions might be of the order of 5–15°C. The concept is discussed by Cotton and Piehlke (1995, Chapter 10).

Fears were also expressed that as a result of the severe smoke palls (plate 7.2) generated by the Gulf War in 1991 there might be severe climate impacts. Studies have suggested that because most of the smoke generated by the oil well fires stayed in the lower troposphere and had only a short residence time in the air, the effects were local (some cooling) rather than global, and that the operation of the monsoon was not affected to any

Plate 7.2
The Gulf War of 1991 led to the deliberate release and burning of oil in Kuwait. Fears were expressed at the time that smoke palls might have regional and global climate effects. In general, subsequent research has suggested that such fears may have been exaggerated

significant degree (Browning et al., 1991; Bakan et al., 1991). Furthermore, in the event the emissions of smoke particles were less than some forecasters had predicted, and they were also rather less black (Hobbs and Radke, 1992).

Warnings have been circulated (Boeck et al., 1975) about the possible role of a substance emitted from nuclear reactors, called Krypton 85. It is believed that its increasing presence will reduce the electrical resistance of the atmosphere between the oceans and the ionosphere. This in turn would affect the electrification of thunderclouds and, through that, precipitation levels.

Supersonic aircraft, both civil and military, discharge some water vapour into the stratosphere as contrails. At present, the water content of the stratosphere is low, as is the exchange of air between the lower stratosphere and other regions. Consequently, comparatively modest amounts of water vapour discharge by aircraft could have a significant effect on the natural balance. It has been calculated that 400 supersonic aircraft, making four flights per day, would place 150 million kilograms of water into the lower stratosphere. This could, over a period of years, double the existing water content of the atmosphere, thereby leading to a small rise in temperature of perhaps 0.6°C (Sawyer, 1971). However, statistics on the increase of cirrus levels as a result of stratospheric aircraft movements are still meagre (SCEP Report, in Matthews et al., 1971: 39) and their statistical significance is not always proven. Yet studies at Salt Lake City and Denver between 1949 and 1969 do show an upward trend in cirrus cloudiness which paralleled the increase in jet aircraft activity. The expected rapid development of civil aviation in future decades may make this an important element of anthropogenic atmospheric modification.

The greatest interest that climatologists currently have in aerosols as a cause of climatic change relates to sulphate aerosols (Charlson and Wigley, 1994). Over the world's oceans a major source of such aerosols is dimethylsulphide (DMS). These substances are produced by planktonic algae in sea water and then oxidize in the atmosphere to form sulphate aerosols. Because the albedo of clouds (and thus the earth's radiation budget) is sensitive to cloud-condensation nuclei density, any factor that controls planktonic algae may have an important impact on climate. The production of such plankton could be affected by water pollution in coastal areas or by global warming (Charlson et al., 1987). However, an even more important source of sulphate aerosols is the burning of fossil fuels and the subsequent emission of sulphur dioxide (SO_2) (Charlson et al., 1992), and these types of sulphate aerosol are concentrated over and downwind of major industrial regions. They have probably served to reduce the rate of global warming that has taken place in this century and may help to explain the cessation in global warming that took place in some regions between the 1940s and 1970s. Indeed, climate models that have predicted the amount of increase in global average temperature as a result of the rising concentrations of greenhouse gases have given a greater amount of temperature rise since the last century than has actually occurred. The newer climate models, which include the effect of these aerosols, produce predicted changes that have considerable similarity to the observed patterns of change (Taylor and Penner, 1994).

Vegetation and albedo change

Incoming radiation of all wavelengths is partly absorbed and partly reflected. Albedo is the term used to describe the proportion of energy reflected and hence is a measure of the ability of the surface to reflect radiation.

Land-use changes create differences in albedo which have important effects on the energy balance, and hence on the water balance, of an area. Tall rain forest may have an albedo as low as 9 per cent, while the albedo of a desert may be as high as 37 per cent (table 7.4).

There has been growing interest recently in the possible consequences of deforestation on climate through the effect of albedo change. Ground deprived of a vegetation cover as a result of deforestation and overgrazing (as in parts of the Sahel) has a very much higher albedo than ground covered in plants. This could affect temperature levels. Satellite imagery of the Sinai–Negev region of the Middle East shows an enormous difference

Table 7.4 Albedo values for different land-use types

Surface type	Location	Albedo (%)
Tall rain forest	Kenya	9
Lake	Israel	11.3
Peat and moss	England and Wales	12
Pine forest	Israel	12.3
Heather moorland	England and Wales	15
Evergreen scrub (maquis)	Israel	15.9
Bamboo forest	Kenya	16
Conifer plantation	England and Wales	16
Citrus orchard	Israel	16.8
Towns	England and Wales	17
Open oak forest	Israel	17.6
Deciduous woodland	England and Wales	18
Tea bushes	Kenya	20
Rough grass hillside	Israel	20.3
Agricultural grassland	England and Wales	24
Desert	Israel	37.3

Source: from miscellaneous data in Pereira, 1973, collated by author.

in image between the relatively dark Negev and the very bright Sinai–Gaza Strip area. This line coincides with the 1948–9 armistice line between Israel and Egypt and results from different land-use and population pressures. Otterman (1974) has suggested that this albedo change has produced temperature changes of the order of 5°C.

Charney et al. (1975) have argued that the increase in surface albedo, resulting from a decrease in plant cover, would lead to a decrease in the net incoming radiation, and an increase in the radiative cooling of the air. Consequently, they argue, the air would sink to maintain thermal equilibrium by adiabatic compression, and cumulus convection and its associated rainfall would be suppressed. A positive feedback mechanism would appear at this stage, for the lower rainfall would in turn adversely affect plants and lead to a further decrease in plant cover. However, this view is disputed by Ripley (1976), who suggests that Charney and his colleagues, while considering the impact of vegetation changes on albedo, have completely ignored the effect of vegetation on evapotranspiration. He points out that vegetated surfaces are usually cooler than bare ground since much of the absorbed solar energy is used to evaporate water, and concludes from this that protection from overgrazing and deforestation might, in contrast to Charney's views, be expected to lower surface temperatures and thereby reduce, rather than increase, convection and precipitation.

Removal of humid tropical rain forests has also been seen as a possible mechanism of anthropogenic climatic change through its effect on albedo. Potter et al. (1975) have proposed the following model for such change:

Deforestation
↓
Increased surface albedo
↓
Reduced surface absorption of solar energy
↓
Surface cooling
↓
Reduced evaporation and sensible heat flux
from the surface
↓
Reduced convective activity and rainfall
↓
Reduced release of latent heat, weakened Hadley circulation
and cooling in the mid and upper troposphere
↓
Increased tropical lapse rates
↓

Increased precipitation in the latitude bands
5–25°N and 5–25°S, and a decrease in the
equator–pole temperature gradient
↓
Reduced meridional transport of heat
and moisture out of equatorial regions
↓
Global cooling and decrease in precipitation
between 45–85°N and 40–60°S

However, more recent studies (Potter et al., 1981) have suggested that globally, over the past few thousand years, the climatic effects of albedo changes wrought by humans have been small and probably undetectable. Similarly, Henderson-Sellers and Gornitz (1984) sought to model the possible future effects of albedo changes produced by humans and also predicted that there would be but little alteration brought about by current levels of tropical deforestation.

On the other hand, Lean and Warrilow (1989) used a General Circulation Model (GCM) which showed greater changes than previous models and suggested that Amazon basin deforestation would, through the effects of changes in surface roughness and albedo, lead to reductions in both precipitation and evaporation. Likewise, a UK Meteorological Office GCM indicated that the deforestation of both Amazonia and Zaire would by changing surface albedo cause a decrease in precipitation levels (Mylne and Rowntree, 1992).

Budyko (1974) believes that the present extension of irrigation to about 0.4 per cent of the earth's surface (1.3 per cent of the land surface) would decrease the albedo, possibly on average by 10 per cent. The corresponding change of the albedo of the entire earth-atmosphere system would amount to about 0.03 per cent; enough, according to Budyko, to maintain the global mean temperature at a level nearly 0.1°C higher than it would otherwise be.

Forests, irrigation and climate

The replacement of forest with crops, in addition to leading to a change in surface albedo of the type just discussed, also changes some other factors that may have climatic significance, including surface roughness, leaf and stem areas and amount of evapotranspiration. Bonan (1997) has tried to model the climatic consequences of replacing the natural forests of the United States with crops and has argued that it would cause cooling of up to 2°C in the summer months over a wide region of the central United States. He suggests (p. 484) that 'land use practices that resulted in extensive deforestation in the Eastern United States, replacing forests with crop, have resulted in a significant climate change that is comparable to other well known anthropogenic climate forcings.'

The belief that forests can increase precipitation levels has a long history (Thornthwaite, 1956; Grove, 1997), and has been the basis of action programmes in many lands. For example, the American Timber Culture Act of 1873 was passed in the belief that if settlers were induced to plant trees on the Great Plains and prairies, precipitation would be increased sufficiently to eliminate the climatic hazards to agriculture. On the other hand, at much the same time, the view was expressed that 'rain follows the plough'. Aughey, working in Nebraska, for example, believed that, after the soil is 'broken', rain as it falls is absorbed by the soil 'like a huge sponge', and that the soil gives this absorbed moisture slowly back to the atmosphere by evaporation (cited by Thornthwaite, 1956: 569), and so increases the rainfall.

These two early and contradictory views illustrate the confusion that still surrounds this question today. Forests undoubtedly influence rates of evapotranspiration, the flow of streams, the level of groundwater and microclimates, but there is little reliable evidence to suggest either that regional rainfall is significantly increased by forest or that attempts to augment rainfall levels on desert margins by widespread planting of forest belts are likely to achieve much, for the aridity of deserts and their margins is controlled dominantly by the gross features of the general circulation, especially the subsiding air associated with the big high-pressure cells of the subtropics.

Although forests may not necessarily have a proven effect on regional or continental rainfall levels, they are far more effective than other vegetation types at trapping other kinds of precipitation, especially cloud, fog and mist. Hence deforestation or afforestation can affect water budgets through the degree to which they intercept non-rainfall precipitation. For example, in Hawaii, Norfolk Island pines (*Araucaria heterophylla*), in an area where the annual rainfall is 2,600 millimetres, condensed an additional 760 millimetres from heavy cloud. In the San Francisco area, Parsons (1960) recorded about 350 millimetres of drip from a pine tree on the Berkeley Hills during each of four rainless summers. In Japan this phenomenon has been used to good effect: trees are planted along the coast to intercept the inland drift of sea fogs.

There is one other land-use change that may result in measurable changes in precipitation; namely large-scale crop irrigation in semi-arid regions. The High Plains of the United States are normally covered with sparse grasses and have dry soils throughout the summer; evapotranspiration is then very low. In the last five decades irrigation has been developed throughout large parts of the area, greatly increasing summer evapotranspiration levels. Barnston and Schickendanz (1984) have produced strong statistical evidence of warm-season rainfall enhancement through irrigation in two parts of this area: one extending through Kansas, Nebraska and Colorado, and a second in the Texas Pandhandle. The largest absolute increase was in the latter area and, significantly, occurred in June, the wettest of the three heavily irrigated months. The effect appears to be especially important when stationary weather fronts occur, for this is a situation which allows for maximum interaction between the damp irrigated surface and the atmosphere. Hail storms and tornados are also significantly more prevalent than over non-irrigated regions (Nicholson, 1988).

The possible effects of water diversion schemes

The levels of the Aral and Caspian Seas in Central Asia have fallen, as have water tables all over the wide continental region. There have been proposals to divert some major rivers to help overcome these problems. However, this raises difficult questions 'because it appears to touch a peculiarly sensitive spot in the existing climatic regime of the northern hemisphere' (Lamb, 1977: 671). The low-salinity water which forms a 100–200 metre upper layer to the Arctic Ocean is in part caused by the input of fresh water from the large Russian and Siberian rivers. This low-salinity water is the medium in which the pack ice at present covering the polar ocean is formed. The tapping of any large proportion of this river flow might augment the area of salt water in the Arctic Ocean and thereby reduce the area of pack ice correspondingly. Temperatures over large areas might rise, which in turn might change the position and alignment of the main thermal gradients in the northern hemisphere and, with them, the jet stream and the development and steering of cyclonic activity. However, assessment of this particular climatic impact is still very largely speculative, and some numerical models indicate that the climate of the Arctic will not be drastically affected by river diversions (Semtner, 1984).

Lakes

It has often been implied that the presence of a large body of inland water must modify the climate around its shores, and therefore that artificial lakes have a significant effect on local or regional climates. Climatic changes produced by the construction of a reservoir are the result of a variety of factors (Vendrov, 1965): the creation of a body of water with a large heat capacity that reduces the continentality of the climate; the substitution of a water surface for a land surface and the rise of the groundwater level in the littoral zone supplying moisture to the evaporating surface (leading to a rise in wind velocity above the lake and in the littoral zone). Schemes have been put forward for augmenting desert rainfall by flooding desert basins in the Sahara, Kalahari and Middle East (see e.g. Schwarz, 1923). However, whether evaporation from lake surfaces can raise local precipitation levels is open to question, for precipitation depends more on atmospheric instability than upon the humidity content of the air. Moreover, most lakes are too small to affect the atmosphere materially in depth, so that their influence falls heavily under the sway of the regional circulation. In addition, one needs to remember that some of the world's driest deserts occur along coastlines. Thus a relatively small artificial lake would be even more impotent in creating rainfall (see Crowe, 1971: 443–50).

However, the climatic effect of artificial lakes is evident in other ways, notably in terms of a local reduction in frost hazard. In the case of the Rybinsk reservoir (*c*.4,500 square kilometres) in the CIS, it has been calculated that the climatic influence extends 10 kilometres from the lake and that the frost-free season has been extended by 5–15 days on average (D'Yakanov and Reteyum, 1965).

Urban climates

One consequence of the burning of fossil fuels is the production of heat. This is probably most important on the local scale, where it can be identified as the 'urban heat island'. On a broader scale, the amount of energy used by humans has been negligible compared both to the resources of solar energy and to the energy of photosynthesis of plants. In global terms the total amount of heat released by all human activity is roughly 0.01 per cent of the solar energy absorbed at the surface (Kellogg, 1978: 215). Such a small fraction would have a negligible effect on the overall heat balance of the earth.

It has been said that 'the city is the quintessence of man's capacity to inaugurate and control changes in his habitat' (Detwyler and Marcus, 1972). One way in which such control becomes evident is in a study of urban climates (Landsberg, 1981). Individual urban areas can at times, with respect to their weather, 'have similar impacts as a volcano, a desert, and as an irregular forest' (Changnon, 1973: 146). Some of the changes that can result are listed in table 7.5.

Compared with rural surfaces, city surfaces (table 7.5b) absorb significantly more solar radiation, because a higher proportion of the reflected radiation is retained by the high walls and

Table 7.5 (a) Average changes in climatic elements caused by cities

Element	Parameter	Urban compared with rural (–, less; +, more)
Radiation	On horizontal surface	–15%
	Ultraviolet	–30% (winter); –5% (summer)
Temperature	Annual mean	+0.7°C
	Winter maximum	+1.5°C
	Length of freeze-free season	+2 to 3 weeks (possible)
Wind speed	Annual mean	–20 to –30%
	Extreme gusts	–10 to –20%
	Frequency of calms	+5 to 20%
Relative	Annual mean	–6%
humidity	Seasonal mean	–2% (winter); –8% (summer)
Cloudiness	Cloud frequency + amount	+5 to 10%
	Fogs	+100% (winter); –30% (summer)
Precipitation	Amounts	+5% to 10%
	Days	+10%
	Snow days	–14%

(b) Effect of city surfaces

Phenomenon	Consequence
Heat production (the heat island)	Rainfall +
	Temperature +
Retention of reflected radiation by high walls and dark-coloured roofs	Temperature +
Surface roughness increase	Wind –
	Eddying +
Dust increase (the dust dome)	Fog +
	Rainfall + (?)

Source: H. Landsberg in Griffiths, 1976: 108.

dark-coloured roofs of the city streets. The concreted city surfaces have both great thermal capacity and conductivity, so that heat is stored during the day and released by night. By contrast, the plant cover of the countryside acts like an insulating blanket, so that rural areas tend to experience relatively lower temperatures by day and night, an effect enhanced by the evaporation and transpiration taking place. Another thermal change in cities, contributing to the development of the 'urban heat island', is the large amount of artificial heat produced by industrial, commercial and domestic users.

In general the highest temperature anomalies are associated with the densely built-up area near the city centre, and decrease markedly at the city perimeter. Observations in Hamilton, Ontario and Montreal, Quebec, suggested temperature changes of 3.8 and 4.0°C respectively per kilometre (Oke, 1978). Temperature differences also tend to be highest during the night. The form of the urban temperature effect has often been likened to

Table 7.6 Annual mean urban–rural temperature differences of cities

City	Temperature differences (°C)
Chicago, USA	0.6
Washington DC, USA	0.6
Los Angeles, USA	0.7
Paris, France	0.7
Moscow, Russia	0.7
Philadelphia, USA	0.8
Berlin, Germany	1.0
New York, USA	1.1
London, UK	1.3

Source: from data in J. T. Peterson, table 11.2, p. 136 and other sources, in Detwyler, 1971.

an 'island' protruding distinctly out of the cool 'sea' of the surrounding landscape. The rural–urban boundary exhibits a steep temperature gradient or 'cliff' to the urban heat island. Much of the rest of the urban area appears as a 'plateau' of warm air with a steady but weaker horizontal gradient of increasing temperature towards the city centre. The urban core may be a 'peak' where the urban maximum temperature is found. The difference between this value and the background rural temperature defines the *urban heat island intensity* ($T_{u-r}(\max)$) (Oke, 1978: 225).

Table 7.6 lists the average annual urban–rural temperature differences for several large cities. Values range from 0.6 to 1.3°C. The relationship between city size and urban–rural difference, however, is not necessarily linear; sizeable nocturnal temperature contrasts have been measured even in relatively small cities. Factors such as building density are at least as important as city size, and high wind velocities will tend to eliminate the heat island effect.

None the less, Oke (1978: 257) has found that there is some relation between heat-island intensity and city size. Using population as a surrogate of city size, $T_{u-r}(\max)$ is found to be proportional to the log of the population. Other interesting results of this study include the tendency for quite small centres to have a heat island, the observation that the maximum thermal modification is about 12°C, and the recognition of a difference in slope between the North American and the European relationships (figure 7.6). The explanation for this last result is not clear, but it may be related to the fact that population is a surrogate index of the central building density.

The relationships between maximum heat island intensity and urban population, the sky-view factor (a measure of building height and density) and the impermeable surface coverage of cities, are shown in figure 7.6. The heat island intensity increases with all three indices (Nakagawa, 1996).

(a)

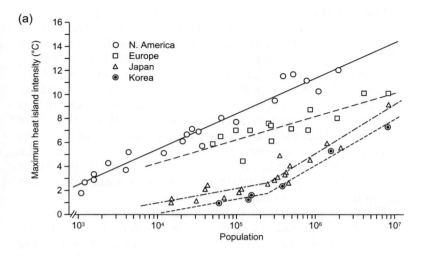

(b)

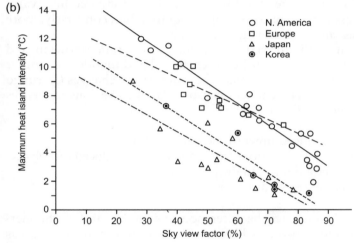

(c)

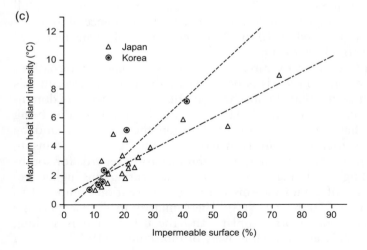

Figure 7.6
Relationships between the
maximum heat island intensity
and (a) urban population for
Japanese, Korean, North
American and European cities;
(b) the sky view factor for
Japanese, Korean, North
American and European cities;
(c) the ratio of impermeable
surface coverage for Japanese
and Korean cities (after
Nakagawa, 1996, figures 2, 3
and 4)

In many older towns and cities in Western Europe and North America, the process of 'counter-urbanization' has in recent years led to a decline in population, and it is worth considering whether this is reflected in a decline in the intensity of urban heat islands. One attempt to do this, in the context of London (Lee, 1992), revealed the perplexing finding that the heat-island intensity has decreased by day, but increased by night. The explanation that has been tentatively advanced to explain this is that there has been a decrease in the receipt of daytime solar radiation as a result of vehicular atmospheric pollution, whereas at night the presence of such pollution absorbs and re-emits significant amounts of outgoing terrestrial radiation, maintaining higher urban nocturnal minimum temperatures.

The existence of the urban heat island has a number of implications; city plants bud and bloom earlier; some birds are attracted to the thermally more favourable urban habitat; humans find added warmth stressful if the city is already situated in a warm area; less winter space-heating is required, but, conversely, more summer air-conditioning is necessary.

The urban-industrial effects on clouds, rain, snowfall and associated weather hazards such as hail and thunder are harder to measure and explain than the temperature changes (Darungo et al., 1978). The changes can be related to various influences (Changnon, 1973, p. 143):

thermally induced upward movement of air;
increased vertical motions from mechanically induced turbulence;
increased cloud and raindrop nuclei;
industrial increases in water vapour.

Table 7.7 illustrates the differences in summer rainfall, thunderstorms and hailstorms between various rural and urban areas in the United States. These data indicate that in cities rainfall increases ranged from 9 to 27 per cent, the incidence of thunderstorms increased by 10 to 42 per cent, and hailstorms increased by 67 to 430 per cent.

An interesting example of the effects of major conurbations on precipitation levels is provided by the London area. In this case it seems that the mechanical effect of the city was dominant in creating localized maxima of precipitation both by being a mechanical obstacle to air flow, on the one hand, and by causing frictional convergence of flow, on the other (Atkinson, 1975). A long-term analysis of thunderstorm records for southeast England is highly suggestive – indicating the higher frequencies of thunderstorms over the conurbation compared to elsewhere (Atkinson, 1968). The similarity in the morphology of the thunderstorm isopleth and the urban area is striking (figure 7.7a and b). Moreover, Brimblecombe (1977) shows a

Table 7.7 Areas of maximum increases (urban–rural difference) in summer rainfall and severe weather events for eight American cities

City	Rainfall		Thunderstorms		Hailstorms	
	%	Location*	%	Location*	%	Location*
St Louis	+15	B	+25	B	+276	C
Chicago	+17	C	+38	A,B,C	+246	C
Cleveland	+27	C	+42	A,B	+90	C
Indianapolis	0	–	0	–	0	–
Washington DC	+9	C	+36	A	+67	B
Houston	+9	A	+10	A,B	+430	B
New Orleans	+10	A	+27	A	+350	A,B
Tulsa	0	–	0	–	0	–

* A = within city perimeter
 B = 8–24 km downwind
 C = 24–64 km downwind.
Source: after Changnon, 1973, figure 1.5, p. 144.

steadily increasing thunderstorm frequency as the city has grown (figure 7.7c).

Similarly, the detailed Metromex investigation of St Louis in the United States (Changnon, 1978) shows that in the summer the city affects precipitation and other variables within a distance of 40 kilometres. Increases were found in various thunderstorm characteristics (about +10 to +115 per cent), hailstorm condition (+3 to +330 per cent), various heavy rainfall characteristics (+35 to +100 per cent) and strong gusts (+90 to +100 per cent).

Two main factors are involved in the effect that cities have on winds: the rougher surface they present in comparison with rural areas; and the frequently higher temperatures of the city fabric.

Buildings, especially those in cities with a highly differentiated skyline, exert a powerful frictional drag on air moving over and around them (Chandler, 1976). This creates turbulence, with characteristically rapid spatial and temporal changes in both direction and speed. The average speed of the winds is lower in built-up areas than over rural areas, but Chandler found that in London, when winds are light, speeds are greater in the inner city than outside, whereas the reverse relationship exists when winds are strong. The overall annual reduction of wind speed in central London is about 6 per cent, but for the higher-velocity winds (more than 1.5 metres per second) the reduction is more than doubled.

Studies in both Leicester and London, England, have shown that on calm, clear nights, when the urban heat-island effect is at its maximum, there is a surface inflow of cool air towards the zones of highest temperatures. These so-called 'country breezes'

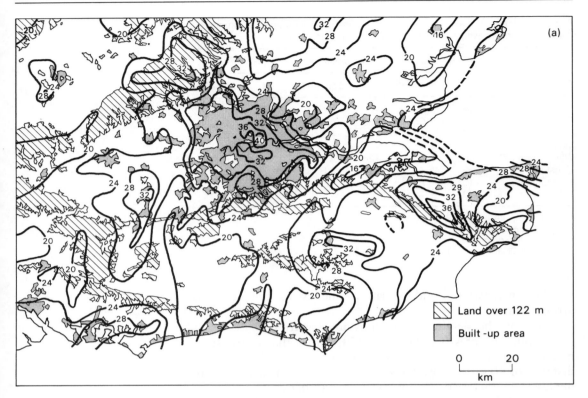

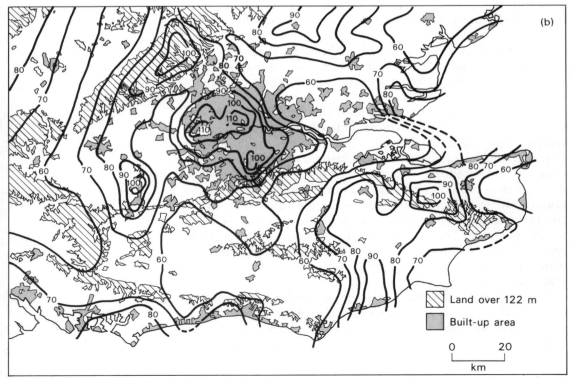

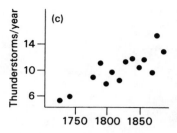

Figure 7.7
Thunder in south-east England:
(a) total thunder rain in south east England, 1951–60, expressed in inches (after Atkinson, 1968, figure 6)
(b) number of days with thunder overhead in south-east England, 1951–60 (after Atkinson, 1968, figure 5)
(c) thunderstorms per year in London (decadal means for whole year) (after Brimblecombe, 1977, figure 2)

have low velocities and become quickly decelerated by intense surface friction in the suburban areas. A practical implication of these breezes is that they transport pollution from the outer parts of an urban area into the city centre, accentuating the pollution problem during smogs.

Urban air pollution

The concentration of large numbers of people, factories, power stations and cars mean that large amounts of pollutants may be emitted into urban atmospheres. If weather conditions permit, the level of pollution may build up (plate 7.3). The nature of the pollutants (table 7.8) has changed as technologies have changed. For example, in the early phases of the industrial revolution in Britain the prime cause of air pollution in cities may have been the burning of coal, whereas now it may be vehicular emissions. Different cities may have very different levels of pollution, depending on factors such as the level of technology, size, wealth and anti-pollution legislation. Differences may also arise because of local topographic and climatic conditions. Photo-chemical smogs, for example, are a more serious threat in areas subjected to intense sunlight.

The variations in pollution levels between different cities are brought out in figure 7.8, which shows data for two types of pollution for a large range of city types. The data were prepared for the years 1980–4 by the Global Environmental Monitoring System of the United Nations Environmental Programme (UNEP). Figure 7.8a shows concentrations of total particulate matter. Most of this comes from the burning of poor-quality fuels. The shaded horizontal bar indicates the range of concentrations that UNEP considers a reasonable target for preserving human health. Note that the annual mean levels range from a low of about 35 µg per cubic metre to a high of about 800 µg per cubic metre: a range of about 25-fold! The higher values appear to be for rapidly growing cities in the developing countries. Some cities, however, such as Kuwait, may have unusually high values because of their susceptibility to dust storms from desert hinterlands.

Plate 7.3
In December 1952 the city
of London was affected by
severe smog. Visibility was
reduced and smog-masks
had to be worn out of doors.
Many people with weak chests
died. Since then, because of
legislation, the incidence of
smog has declined markedly

Figure 7.8 (*opposite*)
(a) The range of annual averages of total particulate matter concentrations measured at multiple sites within
    41 cities, 1980–1984. Each numbered bar represents a city, as follows: 1, Frankfurt; 2, Copenhagen; 3, Cali;
    4, Osaka; 5, Tokyo; 6, New York; 7, Vancouver; 8, Montreal; 9, Fairfield; 10, Chattanooga; 11, Medellin;
    12, Melbourne; 13, Toronto; 14, Craiova; 15, Houston; 16, Sydney; 17, Hamilton; 18, Helsinki; 19, Birmingham;
    20, Caracas; 21, Chicago; 22, Manila; 23, Lisbon; 24, Accra; 25, Bucharest; 26, Rio de Janeiro; 27, Zagreb;
    28, Kuala Lumpur; 29, Bombay; 30, Bangkok; 31, Illigan City; 32, Guangzhou; 33, Shanghai; 34, Jakarta;
    35, Tehran; 36, Calcutta; 37, Beijing; 38, New Delhi; 39, Xi'an; 40, Shenyang; 41, Kuwait City.
(b) The range of annual averages of sulphur dioxide concentrations measured at multiple sites within 54 cities,
    1980–1984. Each numbered bar represents a city, as follows: 1, Craiova; 2, Melbourne; 3, Auckland; 4, Cali;
    5, Tel Aviv; 6, Bucharest; 7, Vancouver; 8, Toronto; 9, Bangkok; 10, Chicago; 11, Houston; 12, Kuala Lumpur;
    13, Munich; 14, Helsinki; 15, Lisbon; 16, Sydney; 17, Christchurch; 18, Bombay; 19, Copenhagen;
    20, Amsterdam; 21, Hamilton; 22, Osaka; 23, Caracas; 24, Tokyo; 25, Wroclaw; 26, Athens; 27, Warsaw;
    28, New Delhi; 29, Montreal; 30, Medellin; 31, St Louis; 32, Dublin; 33, Hong Kong; 34, Shanghai; 35, New
    York; 36, London; 37, Calcutta; 38, Brussels; 39, Santiago; 40, Zagreb; 41, Frankfurt; 42, Glasgow;
    43, Guangzhou; 44, Manila; 45, Madrid; 46, Beijing; 47, Paris; 48, Xi'an; 49, São Paulo; 50, Rio de Janeiro;
    51, Seoul; 52, Tehran; 53, Shenyang; 54, Milan.
Source: Graedel and Crutzen (1993)

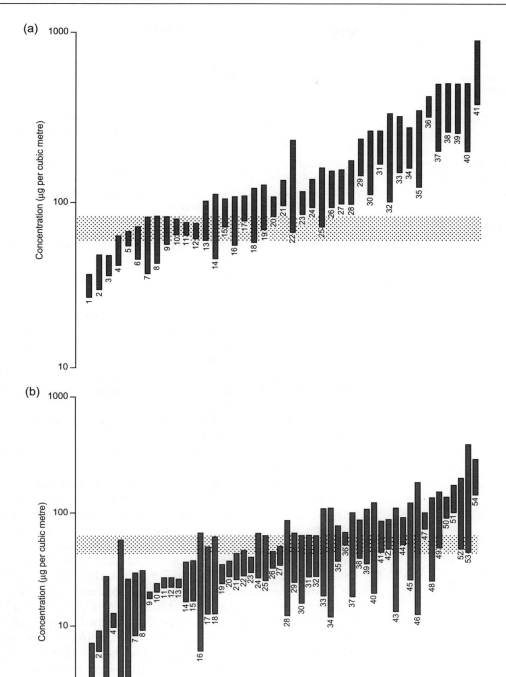

Table 7.8 Major urban pollutants

Type	Some consequences
Suspended particulate matter (characteristically 0.1–25 μm in diameter)	Fog, respiratory problems, carcinogens, soiling of buildings.
Sulphur dioxide (SO$_2$)	Respiratory problems, can cause asthma attacks. Damage to plants and lichens, corrosion of buildings and materials, production of haze and acid rain.
Photochemical oxidants: ozone and peroxyacetyl nitrate (PAN)	Headaches, eye irritation, coughs, chest discomfort, damage to materials (e.g. rubber), damage to crops and natural vegetation, smog.
Oxides of nitrogen (NOx)	Photochemical reactions, accelerated weathering of buildings, respiratory problems, production of acid rain and haze.
Carbon monoxide (CO)	Heart problems, headaches, fatigue, etc.
Toxic metals: lead	Poisoning, reduced educational attainments and increased behavioural difficulties in children.
Toxic chemicals: dioxins, etc.	Poisoning, cancers, etc.

The lower values tend to come from cities in the developed world (e.g. Western Europe, Japan and North America).

Figure 7.8b shows concentrations for sulphur dioxide. Much of this gas probably comes from the burning of high-sulphur coal. Once again, the horizontal shaded bar indicates the concentration range considered by UNEP to be a reasonable target for preserving human health.

These data indicate that the concentrations of sulphur dioxide can differ by as much as three times among different sites within the same urban areas and by as much as 30 times between different urban areas.

In some cities concentrations of pollutants have tended to fall over recent decades. This can result from changes in industrial technology or from legislative changes (e.g. clean air legislation, restrictions on car use, etc.). In many British cities, for example, legislation since the 1950s has reduced the burning of coal. As a consequence, fogs have become less frequent and the amount of sunshine has increased. Figure 7.9 shows the overall trends for the United Kingdom, and highlights the decreasing fog frequency and increasing sunshine levels (Musk, 1991). Similarly, lead concentration in the air in British cities has declined sharply following the introduction of unleaded petrol (Kirby, 1995). The concentrations of various pollutants have also been reduced in the Los Angeles area of California (figure 7.10). Here, carbon monoxide, non-methane hydrocarbon, nitrogen oxide and ozone concentrations have all fallen steadily over the period since the late 1960s (Lents and Kelly, 1993).

However, these three examples of improving trends come from developed countries. In many cities in poorer countries, pollution

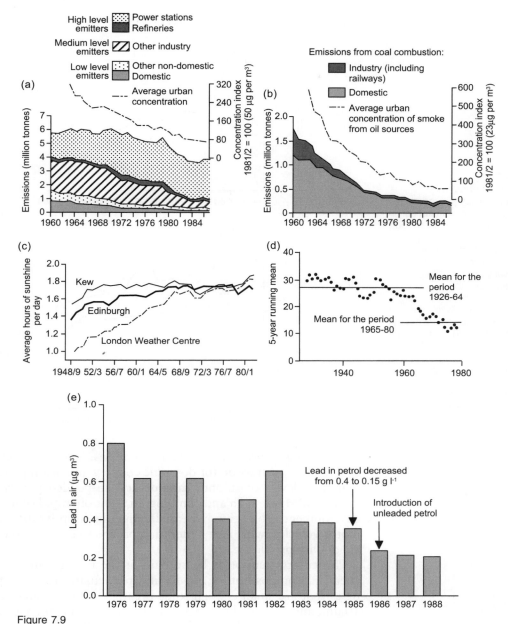

Figure 7.9
Trends in atmospheric quality in the United Kingdom:
(a) sulphur dioxide emissions from coal combustion and average urban concentrations;
(b) smoke emissions from coal combustion and average urban concentrations of oil smoke;
(c) increase in winter sunshine (10-year moving average) for London and Edinburgh city centres and for Kew, outer London
(d) annual fog frequency at 0900 GMT in Oxford, central England, 1926–80
(e) lead concentration (annual means) in UK sites (after Department of the Environment, data, Gomez and Smith, 1984, figure 3, and Kirby, 1995, figure 1)

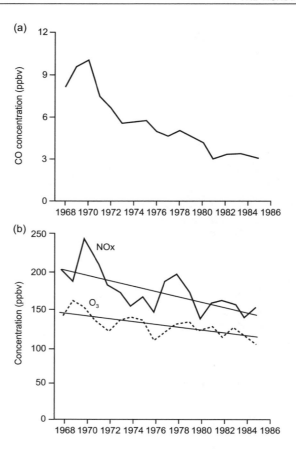

Figure 7.10
Air quality trends in Los
Angeles and its environs have
been measured continuously
and averaged over each hour.
The highest of the hourly
averages is then selected for
trend analysis
(a) downward trend in carbon
    monoxide (CO)
    concentrations; this trend
    is consistent with vehicular
    emission control measures
(b) trend for oxides of nitrogen
    (NOx) and ozone (O_3).
    Both are expressed in
    parts per billion by volume
    (ppbv)
(after Kuntesal and Chang,
1987 © Air Pollution Control
Association)

is increasing at present. In certain countries, heavy reliance on
coal, oil and even wood for domestic cooking and heating means
that their levels of sulphur dioxide and suspended particulate
matter (SPM) are high and climbing. In addition, rapid economic
development is bringing increased emissions from industry and
motor vehicles, which are generating progressively more serious
air-quality problems.

Particular attention is being paid at the present time to the
chemical composition of SPMs, and particularly to those particles that are small enough to be breathed in (i.e. smaller than
10 μm, and so often known as PM10s). Also of great concern in
terms of human health are elemental carbon (for example, from
diesel vehicles), polynuclear aromatic hydrocarbons (PAHs) and
toxic base metals (e.g. arsenic, lead, cadmium and mercury), in
part because of their possible role as carcinogens.

A major cause of urban air pollution is the development of
photochemical smog. The name originates from the fact that
most of the less desirable properties of such fog result from the
products of chemical reactions induced by sunlight. Unburned

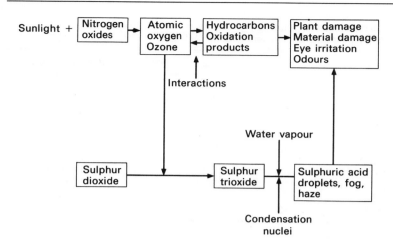

Figure 7.11
Possible reactions involving primary and secondary pollutants (after Haagen-Smit in Bryson and Kutzbach, 1968, figure 4. Reprinted by permission of the Association of American Geographers)

hydrocarbons play a major role in this type of smog formation and result from evaporation of solvents and fuels, as well as incomplete combustion of fossil fuels. In the presence of oxides of nitrogen, strong sunlight and stable meteorological conditions, complex chemical reactions occur, forming a family of peroxyacyl nitrates (sometimes collectively abbreviated to PANs).

Photochemical smog appears 'cleaner' than other kinds of fog in the sense that it does not contain the very large particles of soot that are so characteristic of smog derived from coal burning. However, the eye irritation and damage to plant leaves it causes make it unpleasant. Photochemical smog occurs particularly where there is large-scale combustion of petroleum products, as in car-dominated cities like Los Angeles. Its unpleasant properties include a high lead content; also a series of chemical reactions are triggered by sunlight (figure 7.11). For example, a photochemical decomposition of nitrogen dioxide into nitric oxide and atomic oxygen occurs, and the atomic oxygen can react with molecular oxygen to form ozone. Further ozone may be produced by the reaction of atomic oxygen with various hydrocarbons.

Photochemical smogs are not universal. Because sunlight is a crucial factor in their development they are most common in the tropics or during seasons of strong sunshine. Their especial notoriety in Los Angeles is due to a meteorological setting dominated at times by subtropical anti-cyclones with weak winds, clear skies and a subsidence inversion, combined with the general topographic situation and the high vehicle density (>1,500 vehicles per square kilometre). However, photochemical ozone pollution can, on certain summer days, reach appreciable levels even in the UK (Colbeck, 1988), especially in large cities such as London (table 7.9). Rigorous controls on vehicle emissions can

Table 7.9   Annual maximum hourly mean ozone concentrations at County
Hall, Westminster in Greater London, 1975–85

Year	Peak value (ppm)	Date of peak	No of days ≥ 0.080 ppm
1975	0.150	26 June	16
1976	0.212	27 June	26
1977	0.087	3 July	1
1978	0.103	29 July	2
1979	0.153	27 July	5
1980	0.116	27 August	2
1981	0.112	5 September	6
1982	0.091	3 August	1
1983	0.099	16 July	11
1984	0.088	6 July	3
1985	0.098	25 July	3

Source: Ball and Laxen, 1986.

greatly reduce the problem of high urban ozone concentrations
and this has been a major cause of the reduction in ozone levels
in Los Angeles over the last two decades, in spite of a growth in
that city's population and vehicle numbers (figure 7.12).

Air pollution: some further effects

This chapter has already made much reference to the ways in
which humans have changed the turbidity of the atmosphere
and the gases within it. However, the consequences of air pollution go further than either their direct impact on human health
or their impact on local, regional and global climates.

First of all, the atmosphere acts as a major channel for the
transfer of pollutants from one place to another, so that some
harmful substances have been transferred long distances from
their sources of emission. DDT is one example; lead is another.
Thus, from the start of the industrial revolution, the lead content
of the Greenland ice cap, although far removed from the source
of the pollutant (which is largely derived from either industrial
or auto-mobile emissions) rose very substantially (figure 7.13a).
The same applies to its sulphate content (figure 7.13b). An analysis of pond sediments from a remote part of North America
(Yosemite) indicates that lead levels were raised as a result of
human activities, being more than 20 times the natural levels.
The lead, which came in from atmospheric sources, showed a
fivefold elevation in the plants of the area and a fiftyfold elevation in the animals compared with natural levels (Shirahata et
al., 1980). Some estimates also compared the total quantities of
heavy metals that humans are releasing into the atmosphere
with emissions from natural sources (Nriagu, 1979). The increase was eighteenfold for lead, ninefold for cadmium, sevenfold for zinc and threefold for copper.

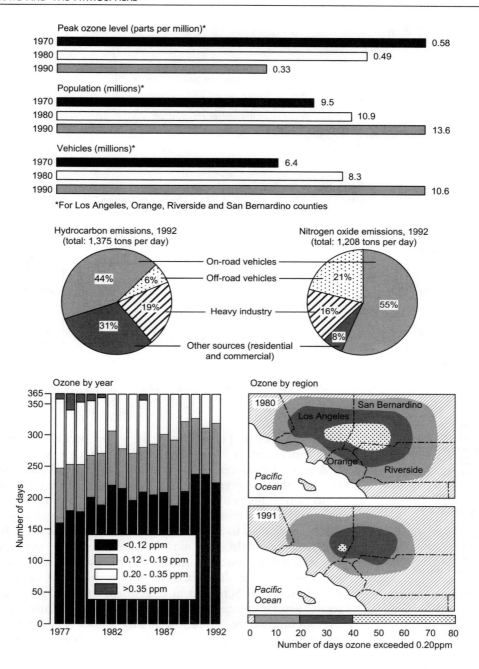

Figure 7.12
Air pollution in the Los Angeles area, 1970s to 1990s (after Lents and Kelly, 1993, p. 22)

There are, it must be stated, signs that as a result of pollution control regulations some of these trends are now being reversed. Boutron et al. (1991), for example, analysed ice and snow that has accumulated over Greenland in the previous two decades and found that lead concentrations had decreased by a factor of 7.5 since 1970. They attribute this to a curbing of the use of

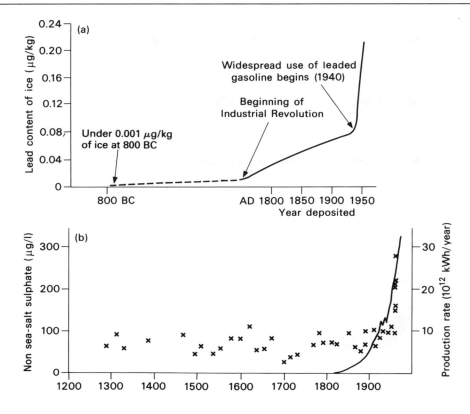

Figure 7.13
Trends in atmospheric quality:
(a) Lead content of the
    Greenland ice cap due
    to atmospheric fallout of
    the mineral on the snow
    surface. A dramatic upturn
    in worldwide atmospheric
    levels of lead occurred
    at the beginning of the
    industrial revolution in the
    nineteenth century and
    again after the more recent
    spread of the automobile
    (after Murozumi et al.,
    1969, p. 1247)
(b) The sulphate concentration
    on a sea-salt-free basis
    in north-west Greenland
    glacier ice samples as
    a function of year. The
    curve represents the world
    production of thermal
    energy from coal, lignite
    and crude oil (modified
    after Koide and Goldberg,
    1971, figure 1)

lead additives in petrol. Over the same period cadmium and zinc concentrations have decreased by a factor of 2.5.

A second example of the possible widespread and ramifying ecological consequences of atmospheric pollution is provided by 'acid rain' (Likens and Bormann, 1974). Acid rain is rain which has a pH of less than 5.65, this being the pH which is produced by carbonic acid in equilibrium with atmospheric CO_2. In many parts of the world, rain may be markedly more acid than this normal, natural background level. Snow and rain in the northeast United States have been known to have pH values as low as 2.1, while in Scotland in one storm the rain was the acidic equivalent of vinegar (pH 2.4). In the eastern United States the average annual precipitation acidity values tend to be around pH 4 (see figure 7.14), and the degree of acidulation appears to have increased between the 1950s and 1970s (see figure 7.15 for trends in Scandinavia).

It needs to be remembered that not all environmental acidification is caused by acid rain in the narrow sense. Acidity can reach the ground surface without the assistance of water droplets. This is as particulate matter and is termed 'dry deposition'. Furthermore, there are various types of 'wet precipitation' by

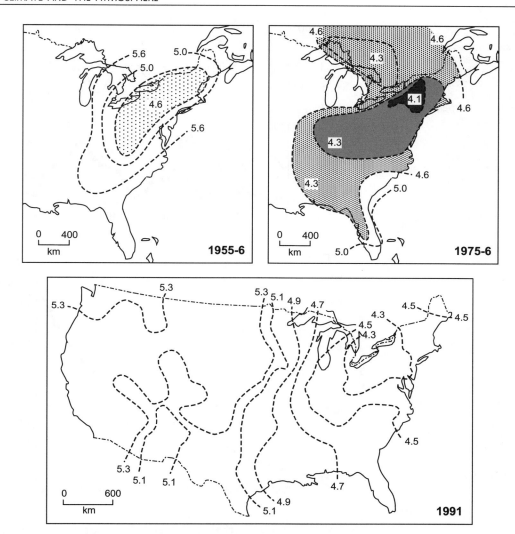

mist, hail, sleet or snow, in addition to rain itself. Thus some people prefer the term 'acid deposition' to 'acid rain'. The acidity of the precipitation in turn leads to greater acidity in rivers and lakes.

The causes of acid deposition are the increasing quantities of sulphur oxides (figure 7.16) and nitrogen oxides being emitted from fossil-fuel combustion. At present, roughly 60–70 per cent of the problem is due to sulphur oxide emissions, and the balance to the nitrogen oxides. However, in some regions, such as the west coast of the United States and Japan, the nitric acid contribution may well be of greater relative importance. Moreover, with the emissions of oxides of nitrogen appearing to increase at a faster rate than sulphur dioxide emissions, nitric acid will

Figure 7.14
Isopleths showing average pH for precipitation in North America. Note the low values for eastern North America and the relatively higher values to the west of the Mississippi. (Combined from Likens et al., 1979 and Graedel and Crutzen, 1995, with modifications)

Figure 7.15
Acidity of rain and snow for
four Scandinavian localities.
The acidity is expressed in
terms of the annual volume-
weighted concentration of
hydrogen ions in
microequivalents per litre
(a) Lista, Norway
(b) As, Norway
(c) Kise, Norway
(d) Kiruna, Sweden
(modified after Likens et al.,
1974, p. 45, in *Science*,
184, 1176–9, © 1979 by the
American Association for the
Advancement of Science)

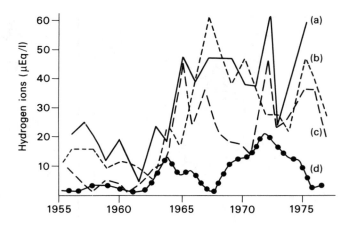

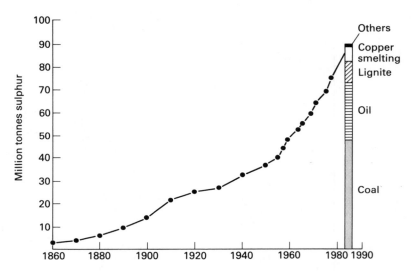

Figure 7.16
Global SO₂ emissions from
anthropogenic sources,
including the burning of coal,
lignite and oil, and copper
smelting

probably increase further in its relative importance. A possible
subsidiary reason for some of the observed increases in acidity
is that there may now be less calcium-rich dust in the atmosphere
over North America since the passing of the Dust Bowl years
(Stensland and Semonin, 1982).

Figure 7.17 shows how sulphate levels increased in European
precipitation between the 1950s and 1970s. Two main factors
contributed to the increasing seriousness of the problem at the
time. One was the replacement of coal by oil and natural gas.
The second, paradoxically, was a result of the implementation
of air pollution control measures (particularly increasing the
height of smoke stacks and installing particle precipitators). These
appear to have transformed a local 'soot problem' into a regional
'acid rain problem'. Coal burning produced a great deal of sulphate, but was largely neutralized by high calcium contents in the

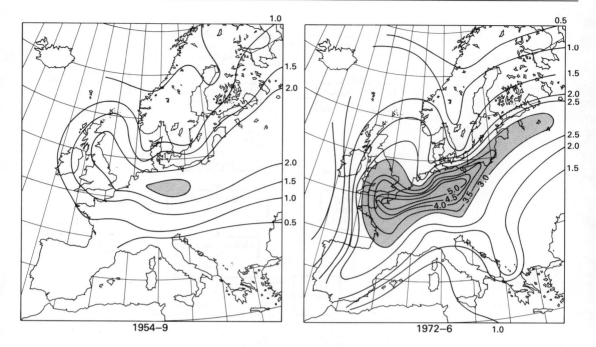

1954–9   1972–6

Figure 7.17
Annual mean concentration
of sulphate in precipitation in
Europe (mg S/l) (after Wallen
in Holdgate et al., 1982,
figure 2.3)

relatively unfiltered coal smoke emissions. Natural gas burning creates less sulphate but that which is produced is not neutralized. The new higher chimneys pump the smoke so high that it is dispersed over wide areas, whereas previously it returned to earth nearer the source.

Very long-term records of lake acidification can provide dramatic evidence for a recent magnification of the acid deposition problem. These are obtained by extracting cores from lake floors and analysing their diatom assemblages at different levels. The diatom assemblages reflect water acidity levels at the time they were living. Two studies serve to illustrate the trend. In southwest Sweden (Renberg and Hellberg, 1982) for most of postglacial time (that is, the last 12,500 years) the pH of the lakes appears to have decreased gradually from around 7.0 to about 6.0 as a result of natural ageing processes. However, over the last few decades, and especially since the 1950s, a further, more marked decrease occurred to present-day values of about 4.5. In Britain, the work of Battarbee and collaborators (1988a, b) shows that at sensitive sites pH values before around 1850 were close to 6.0 and that since then pH declines have varied between 0.5 and 1.5 units.

More direct monitoring of lake acidity also demonstrated that significant changes were taking place. Studies by Beamish et al. (1975) demonstrated that between 1961 and 1975 pH had declined by 0.13 units per year in George Lake, Canada, and a

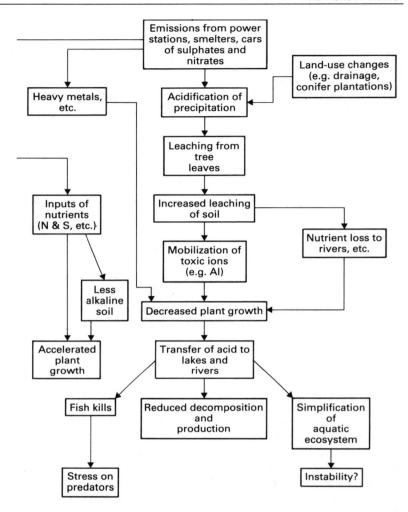

Figure 7.18
Pathways and effects of acid precipitation through different components of the ecosystem, showing some of the adverse and beneficial consequences

comparable picture emerges from Sweden, where Almer et al. (1974) found that the pH in some lakes had decreased by as much as 1.8 units since the 1930s.

The effects of acid rain (figure 7.18) are especially serious in areas underlain by highly siliceous types of bedrock (for example, granite, some gneisses, quartzite and quartz sandstone), such as the old shield areas of the Fenno-Scandian shield in Scandinavia and the Laurentide shield in Canada (Likens et al., 1979).

The ecological consequences of acid rain are still the subject of some debate. Krug and Frink (1983) have argued that acid rain only accelerates natural processes, and point out that the results of natural soil formation in humid climates include the leaching of nutrients, the release of aluminium ions and the acidification of soil and water. They also note that acidification by acid rain may be superimposed on longer-term acidification induced by changes in land use. Thus the regrowth of coniferous forests in what are now marginal agriculture areas, such as New England and highland Western Europe, can increase acidification of soils and water. Similarly, it is possible, though in general unproven (see Battarbee et al., 1985a), that in areas like western Scotland, a decline in upland agriculture and the regeneration of heathland could play a role in increasing soil and water acidification. Likewise Johnston et al. (1982) have suggested that acid rain can cause either a decrease or an increase in forest productivity, depending on local factors. For example, in soils where cation nutrients are abundant and sulphur or nitrogen are deficient, moderate inputs of acid rain are very likely to stimulate forest growth.

In general, however, it is the negative consequences of acid rain that have been stressed. One harmful effect is a change in soil character. The high concentration of hydrogen ions in acid rain causes accelerated leaching of essential nutrients, making them less available for plant use. Furthermore, the solubility of aluminium and heavy metal ions increases and, instead of being fixed in the soil's sorption complex, these toxic substances become available for plants or are transferred into lakes, where they become a major physiological stress for some aquatic organisms.

Fresh-water bodies with limited natural cations are poorly buffered and thus vulnerable to acid inputs. The acidification of thousands of lakes and rivers in southern Norway and Sweden during the past three decades has been attributed to acid rain, and this increased acidity has resulted in the decline of various species of fish, particularly trout and salmon. But fish are not the only aquatic organisms that may be affected. Fungi and moss may proliferate, organic matter may start to decompose less rapidly, and the number of green algae may be reduced.

Forest growth can also be affected by acid rain, though the evidence is not necessarily proven. Acid rain can damage foliage, increase susceptibility to pathogens, affect germination and reduce nutrient availability. However, since acid precipitation is only one of many environmental stresses, its impact may enhance, be enhanced by, or be swamped by other factors. For example, Blank (1985), in considering the fact that an estimated one-half of the total forest area of the former West Germany was showing signs of damage, referred to the possible role of ozone or of a run of hot, dry summers on tree health and growth. The whole question of forest decline is addressed in chapter 2.

Further useful information on the contentious issue of the effects of acid rain is contained in Likens and Butler (1981), Hutchinson and Havas (1980), Hornbeck (1981), Park (1987), Wellburn (1994) and Steinberg and Wright (1994).

The seriousness of acid rain caused by sulphur dioxide emissions in the western industrialized nations peaked in the

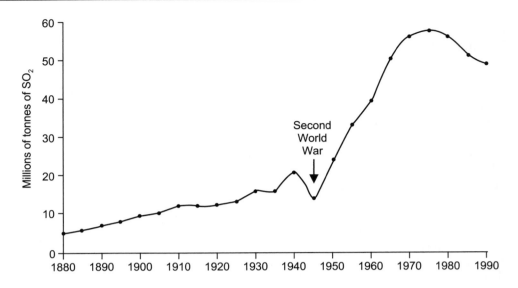

Figure 7.19
Trends in sulphate emissions in Europe 1880–1990, based on data in Mylona (1996)

Table 7.10 Emissions of sulphur and nitrogen oxides (as NO_2) in thousand tons a year for selected European countries in 1980 and 1993

County	Sulphur			Nitrogen oxides		
	1980	1993	1993 as % of 1980	1980	1993	1993 as % of 1980
Czech Republic	1,128	710	62.9	937	574	61.26
Denmark	226	78	34.51	274	264	96.35
Finland	292	60	20.55	264	253	95.83
France	1,669	568	34.03	1,823	1,519	83.32
Germany	3,743	1,948	52.04	2,440	2,904	84.42
Ireland	111	78	70.27	73	122	167.12
Italy	1,900	1,126	59.26	1,480	2,053	138.72
Netherlands	244	84	34.43	582	561	96.39
Norway	70	18	25.71	185	225	120.96
Poland	2,050	1,362	66.44	1,500	1,140	76.00
Spain	1,660	1,158	69.76	950	1,257	132.32
Sweden	254	50	19.69	424	399	94.10
UK	2,454	1,597	65.08	2,395	2,355	98.32
			47.28			103.47

mid-1970s or early 1980s (figure 7.19). Changes in industrial technology, in the nature of economic activity, and in legislation caused the output of SO_2 in Britain to decrease by 35 per cent between 1974 and 1990. Figures for Europe more generally are given in table 7.10. There are also increasing controls on the emission of NO_x in vehicle exhaust emissions, although emissions of nitrogen oxides in Europe have not declined. This means that the geography of acid rain may change, becoming less serious in the developed world, but increasing in locations like China, where economic development will continue to be fuelled by the burning of low-quality sulphur-rich coal in enormous quantities.

In those countries where acid rain remains or is increasing as a problem there are various methods available to reduce its damaging effects. One of these is to add powdered limestone to lakes to increase their pH values. However, the only really effective and practical long-term treatment is to curb the emission of the offending gases. This can be achieved in a variety of ways: by reducing the amount of fossil fuel combustion; by using less sulphur-rich fossil fuels; by using alternative energy sources that do not produce nitrate or sulphate gases (e.g. hydro-power or nuclear power); and by removing the pollutants before they reach the atmosphere. For example, after combustion at a power station, sulphur can be removed ('scrubbed') from flue gases by a process known as flue gas desulphurization (FGD), in which a mixture of limestone and water is sprayed into the flue gas, converting the sulphur dioxide (SO_2) into gypsum (calcium sulphate). NO_x in flue gas can be reduced by adding ammonia and passing it over a catalyst to produce nitrogen and water (a process called selective catalytic reduction or SCR). NO_x produced by cars can be reduced by fitting a catalytic converter.

Stratospheric ozone depletion

A very rapidly developing area of concern in pollution studies is the current status of stratospheric ozone levels. The atmosphere has a layer of relatively high concentration of ozone (O_3) at a height of about 16–18 kilometres in the polar latitudes and of about 25 kilometres in equatorial regions. This ozone layer is important because it absorbs incoming solar ultraviolet radiation, thus warming the stratosphere and creating a steep inversion of temperature at heights between about 15 and 50 kilometres. This in turn affects convective processes and atmospheric circulation, thereby influencing global weather and climate. However, the role of ozone in controlling receipts of ultraviolet radiation at the earth's surface has great ecological significance, because it modifies rates of photosynthesis. One class of organism that has been identified as being especially prone to the effects of increased ultraviolet radiation consequent upon ozone depletion are phytoplankton – aquatic plants that spend much of their time near the sea surface and are therefore exposed to such radiation. A reduction in their productivities would have potentially ramifying consequences, because these plants directly and indirectly provide the food for almost all fish.

Human activities appear to be causing ozone depletion in the stratosphere, most notably over the south polar regions, where an 'ozone hole' has been identified. Possible causes of ozone depletion are legion, and include various combustion products emitted from high-flying military and civil supersonic aircraft; nitrous oxide released from nitrogenous chemical fertilizers; and chlorofluorocarbons (CFCs) used in aerosol spray cans, refrigerant systems, and in the manufacture of foam fast-food containers.

However, in recent years the greatest attention has been focused on the role of CFCs, the production of which climbed greatly in the decade after the Second World War. These gases may diffuse upwards into the stratosphere where solar radiation causes them to become dissociated to yield chlorine atoms which react with and destroy the ozone. The process has been described thus by Titus and Seidel (1986: 4):

Because CFCs are very stable compounds, they do not break up in the lower atmosphere (known as the troposphere). Instead, they slowly migrate to the stratosphere, where ultraviolet radiation breaks them down, releasing chlorine.

Chlorine acts as a catalyst to destroy ozone; it promotes reactions that destroy ozone without being consumed. A chlorine (Cl) atom reacts with ozone (O_3) to form ClO and O_2. The ClO later reacts with another O_3 to form two molecules of O_2, which releases the Cl atoms. Thus two molecules of ozone are converted to three molecules of ordinary oxygen, and the chlorine is once again free to start the process. A single chlorine atom can destroy thousands of ozone molecules. Eventually, it returns to the troposphere, where it is rained out as hydrochloric acid.

The Antarctic ozone hole (plate 7.4) has been identified through satellite monitoring and by monitoring of atmospheric chemistry on the ground. The decrease in ozone levels at Halley Bay is shown in figure 7.20. The reasons why this zone of ozone depletion is so well developed over Antarctica include the very low temperatures of the polar winter, which seem to play a role in releasing chlorine atoms; the long sunlight hours of the polar summer, which promote photochemical processes; and the existence of a well-defined circulation vortex. This vortex is a region of very cold air surrounded by strong westerly winds, and air within this vortex is isolated from that at lower latitudes, permitting chemical reactions to be contained rather than more widely diffused. No such clearly defined vortex exists in the northern hemisphere, though ozone depletion does seem to have occurred in the Arctic as well (Proffitt et al., 1990). Furthermore, observations in the last few years indicate that the Antarctic ozone hole is spreading over wider areas and persisting longer into the Antarctic summer. It is also possible that the situation could be worsened by emissions of volcanic ash into the atmosphere (as from Mount Pinatubo), for these can also cause chemical reactions that lead to ozone depletion (Mintzer and Miller, 1992).

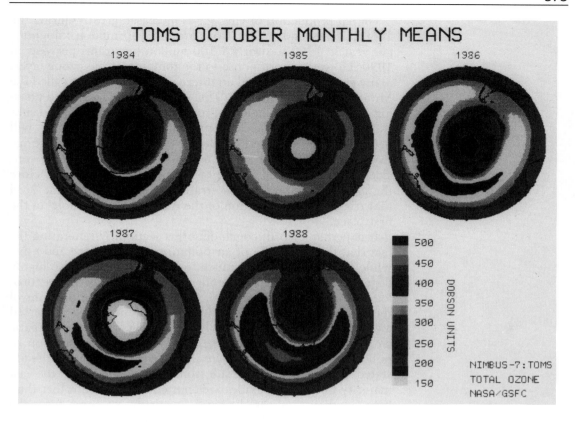

Plate 7.4
In the 1980s ground observations and satellite monitoring of atmospheric ozone levels indicated that a 'hole' had developed in the stratospheric ozone layer above Antarctica. These Nimbus satellite images show the ozone concentrations (in Dobson units) for the month of October between 1984 and 1988

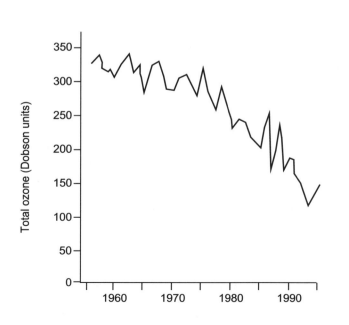

Figure 7.20
Total ozone during October over Halley Bay, Antarctica (based on data provided by the British Antarctic Survey)

Global production of CFC gases increased greatly during the 1960s, 1970s and 1980s, from around 180 million kilograms per year in 1960 to nearly 1,100 million kilograms per year in 1990. However, in response to the thinning of the ozone layer, many governments signed an international agreement called the Montreal Protocol in 1987. This pledged them to a rapid phasing out of CFCs and halons. Production has since dropped substantially. However, because of their stability, these gases will persist in the atmosphere for decades or even centuries to come. Even with the most stringent controls that are now being considered, it will be the middle of the twenty-first century before the chlorine content of the stratosphere falls below the level that triggered the formation of the Antarctic 'ozone hole' in the first place.

It is perhaps worth concluding this section by reiterating the fact that the variability of air pollution in both space and time, while undoubtedly owing much to human action, is also highly dependent on natural conditions, and in particular on meteorological conditions (Thompson, 1978). Climatic factors control the efficiency with which effluents are diluted and dispersed away from primary sources, while local relief and microclimatic circumstances modify the interaction between the emission of pollutants and the atmosphere. Moreover, as we have already stated, not all pollutants result from human endeavour: the vast majority of nitrogen oxides are emitted by bacteria rather than by car exhausts, most chlorides in rain water come from the oceans, and a considerable proportion of atmospheric dust is injected into the air by such mechanisms as volcanic eruptions, and by dust storms removing fine silt from glacial outwash fans or desert basins.

Deliberate climatic modification

It has long been a human desire to modify weather and climate, but it is only since the Second World War and the development of high-altitude observations of clouds that serious attempts have been made to modify such phenomena as rainfall, hailstorms and hurricanes (Cotton and Piehlke, 1995).

The most fruitful human attempts to augment natural precipitation have been through cloud seeding. Rainmaking experiments of this type are based on three main assumptions (Chorley and More, 1967: 159).

1   Either the presence of ice crystals in a supercooled cloud is necessary to release snow and rain or the presence of comparatively large water droplets is necessary to initiate the coalescence process.

2   Some clouds precipitate inefficiently or not at all, because these components are naturally deficient.

3   The deficiency can be remedied by seeding the clouds artificially, either with solid carbon dioxide (dry ice) or silver iodide, to produce crystals, or by introducing water droplets or large hygroscopic nuclei (e.g. salt).

These methods of seeding are not universally productive and in many countries expenditure on such procedures has been reduced. Under conditions of orographic lift and in thunderstorm cells, when nuclei are insufficient to generate rain by natural means, some augmentation may be attained, especially if cloud temperatures are of the order of -10 to $-15°C$. The increase of precipitation gained under favourable conditions may be of the order of 10–20 per cent in any one storm. In lower latitudes, where cloud-top temperatures frequently remain above 0°C, silver-iodide or dry-ice seeding is not applicable. Therefore alternative methods have been introduced whereby small water droplets of 50 millimetres diameter are sprayed into the lower layers of deep clouds, so that the growth of cloud particles will be stimulated by coalescence. Other techniques of warm cloud seeding include the feeding of hygroscopic particles into the lower air layers near the updraught of a growing cumulus cloud (Breuer, 1980).

The results of the many experiments now carried out on cloud seeding are still controversial, very largely because we have an imperfect understanding of the physical processes involved. This means that the evidence has to be evaluated on a statistical rather than a scientific basis so, although precipitation may occur after many seeding trials, it is difficult to decide to what extent artificial stimulation and augmentation is responsible. It also needs to be remembered that this form of planned weather modification applies to small areas for short periods. As yet, no means exist to change precipitation appreciably over large areas on a sustained basis.

Other types of deliberate climatic modification have also been attempted (Hess, 1974). For example, it has been thought that the production of many more hailstone embryos by silver-iodide seeding will yield smaller hailstones which would both be less damaging and more likely to melt before reaching the ground (plate 7.5). Some results in Russia have been encouraging, but one cannot exclude the possibility that seeding may sometimes even increase hail damage (Atlas, 1977). Similar experiments have been conducted in lightning suppression. The concept here is to produce in a thundercloud, again by silver-iodide seeding, an abnormal abundance of ice crystals that would act as added corona points and thus relieve the electrical potential gradient by corona discharge before a lightning strike could develop (Panel on Weather and Climate Modification, 1966: 4–8).

Plate 7.5
The damage caused by
large hailstones can be
considerable. These examples,
the size of a baseball, fell near
Neligh, Nebraska, in June
1950. There is, therefore,
a considerable incentive to
explore ways of reducing
their impact through climatic
modification experiments

Hurricane modification is perhaps the most desirable aim of those seeking to suppress severe storms, because of the extremely favourable benefit-to-cost ratio of the work. The principle once again is that of introducing freezing nuclei into the ring of clouds around the hurricane centre to trigger the release of the latent heat of fusion in the eye-wall cloud system which, in turn, diminishes the maximum horizontal temperature gradients in the storm, causing a hydrostatic lowering of the surface temperature. This eventually should lead to a weakening of the damaging winds (Smith, 1975: 212). A 15 per cent reduction in maximum winds is theoretically possible, but as yet work is largely at an experimental stage. A 30 per cent reduction in maximum winds was claimed following seeding of Hurricane Debbie in 1969, but US attempts to seed hurricanes were discontinued in the 1970s following the development of new computer models of the effects of seeding on hurricanes. The models suggested that although maximum winds might be reduced by 10–15 per cent on average, seeding may increase the winds just outside the region of maximum winds by 10–15 per cent, may either increase or decrease the maximum storm surge, and may or may not affect the direction of the storm (Sorkin, 1982). Seeding of hurricanes is unlikely to recommence until such uncertainties can be resolved.

Much research in Russia has been devoted to the possibility of removing the Arctic Sea ice and to ascertain the effects of such action on the climate of northern areas (see Lamb, 1977: 660). One proposal was to dam the Bering Strait, thereby blocking off water flow from the Pacific. The assumption was that more, and warmer, Atlantic water would be drawn into the central Arctic, improving temperature conditions in that area. Critics have pointed to the possibility of adverse changes in temperatures elsewhere, together with undesirable change in precipitation character and amount.

Fog dispersal, vital for airport operation, is another aim of weather modification. Seeding experiments have shown that fog consisting of supercooled droplets can be cleared by using liquid propane or dry ice. In very cold fogs this seeding method causes rapid transformation of water droplets into ice particles. Warm fogs with temperatures above freezing point occur more frequently than supercooled fogs in mid-latitudes and are more difficult to disperse. Some success has been achieved using sodium chloride and other hygroscopic particles as seeding agents, but the most effective method is to evaporate the fog. The French have developed the 'turboclair' system in which jet engines are installed alongside the runway at major airports and the engines produce short bursts of heat to evaporate the fog and improve visibility as an aircraft approaches (Hess, 1974).

In regions of high temperature, dark soils then become overheated, and the resultant high evapotranspiration rates lead to moisture deficiencies. Applications of white powders (India and Israel) or of aluminium foils (Hungary) increase the reflection from the soil surface and reduce the rate at which insulation is absorbed. Temperatures of the soil surface and subsurface are lowered (by as much as 10°C), and soil moisture is conserved (by as much as 50 per cent).

The planting of windbreaks is an even more important attempt by humans to modify local climate deliberately. Shelter belts have been in use for centuries in many windswept areas of the globe, both to protect soils from blowing and to protect the plants from the direct effects of high velocity winds. The size and effectiveness of the protection depends on their height, density, shape and frequency. However, the belts may have consequences for microclimate beyond those for which they were planted. Evaporation rates are curtailed; snow is arrested, and its melting waters are available for the fields; but the temperatures may become more extreme in the stagnant space in the lee of the belt, creating an increase in frost danger.

Traditional farmers in many societies have been aware of the virtues of microclimatic management (Wilken, 1972). They manage shade by employing layered cropping systems or by covering the plant and soil with mulches; they may deliberately try to modify albedo conditions. Tibetan farmers, for example, reportedly throw dark rocks on to snow-covered fields to promote late spring melting; and in the Paris area of France some very dense stone walls were constructed to absorb and radiate heat.

Conclusion

Changes in the composition of the earth's atmosphere as a result of human emissions of trace gases, and changes in the nature of land cover, have caused great concern in recent years. Global warming, ozone depletion and acid rain have become central issues in the study of environmental change. Although most attention is often paid to climatic change resulting from greenhouse gases, there is a whole series of other mechanisms which have the potential to cause climatic change. Most notably, this chapter has pointed to the importance of other changes in atmospheric composition and properties, whether these are caused by aerosol generation or albedo change.

However, the greenhouse effect and global warming may prove to have great significance for the environment and for human activities. Huge uncertainties remain about the speed, degree, direction and spatial patterning of potential change. None the less, if the earth warms up by a couple of degrees over the next hundred or so years, the impacts, some negative and some positive, are unlikely to be trivial. This is something that will form a focus of the next chapter.

For many people, especially in cities, the immediate climatic environment has already been changed. Urban climates are different in many ways from those of their rural surroundings. The quality of the air in many cities has been transformed by a range of pollutants, but under certain circumstances clean air legislation and other measures can cause rapid and often remarkable improvements in this area.

The same is true of two major pollution issues – ozone depletion and acid deposition. Both processes have serious environmental consequences and their effects may remain with us for many years; but both can be slowed down or even reversed by regulating the production and output of the offending gases.

The Future

8

Man's advent has not been a mere solitary fact, nor have the alterations which he has effected been confined solely to the relations that subsist between himself and nature. He has set in motion a series of changes which have reacted on each other in countless circles, both through the organic and the inorganic world. Nor are they confined to the past; they still go on; and, as years roll away, they must produce new modifications and reactions, the stream of change ever widening, carrying with it man himself, from whom it took its rise, and who is yet in no small degree involved in the very revolutions which he originates.

Sir Archibald Geikie (1901: 425)

Introduction

Much of this book has in one sense been historical, attempting to outline the various changes that have been wrought by humans on the face of the earth over the last few million years as levels of population and the nature of technological progress have changed. None the less, from time to time reference has been made to the future and to the uncertainties that surround it. Given the rapid expansion in human numbers and the increasing sophistication and pervasiveness of certain technological developments, the pace of change over the next decades is likely to become such that the human impact will be greatly magnified.

There is now an unprecedented interest in the environmental future, and few people are unaware of the threat and potential dangers posed by nuclear pollution, species extinction, acid rain, the carbon dioxide problem and desertification. It is for this reason that various attempts have been made to formulate informed predictions of likely changes as a guide to governments and administrators. One of the most notable of these was the report prepared and first published in 1980 as *Global 2000*, an American compilation for the President of the United States (Council on Environmental Quality and the Department of State, 1982). Some of its findings on such matters as the extinctions of species that may result from deforestation in the moist tropics have already been referred to. The report was based on an enormous amount of informed research by a large number of often distinguished scientists. Although it was careful not to be excessively negative, and weighed up various alternative scenarios, its basic conclusions were fundamentally pessimistic (p. 1):

If present trends continue, the world in 2000 will be more crowded, more polluted, less stable ecologically, and more vulnerable to disruption than the world we live in now. Serious stresses involving population, resources and environment are clearly visible ahead. Despite greater material output, the world's people will be poorer in many ways than they are today. For hundreds of millions of the desperately poor, the outlook for food and other necessities of life will be no better. For many it will be worse. Barring revolutionary advances in technology, life for most people on earth will be more precarious in 2000 than it is now – unless the nations of the world act decisively to alter current trends.

In 1984, however, Simon and Kahn edited a volume entitled *The resourceful earth: a response to Global 2000*, which took a very much more optimistic view, and which included contributions from a variety of distinguished scientists, including some eminent geographers. Note the alternative view of the future which they proposed (pp. 1–2):

If present trends continue, the world in 2000 will be *less crowded* (though more populated), *less polluted*, *more stable ecologically* and *less vulnerable to resource-supply disruption* than the world we live in now . . . The world's people will be *richer* in most ways than they are today . . . The outlook for food and other necessities of life will be *better* . . . life for most people on earth will be *less precarious economically than it* is now. [emphasis in original]

Since then Simon (1995, 1996) has produced optimistic volumes. There have also been vigorous attacks on the alleged extreme views of environmentalists (e.g. Bailey, 1995). At the same time, however, other scientists have come together to give cogent views of the significance of such issues as the greenhouse effect (e.g. Houghton et al., 1996).

There are many possible reasons why there should be such a wide disparity in attitudes to the likely future of the earth. Among these are a paucity of background data on the present and past state of the earth, the problems of extrapolating data into the future by curve-fitting procedures, the crudeness and assumptions of so many models, the complexity of interactions in the environment, uncertainty about the nature of technological developments, and possible political constraints on the use of fossil fuels and nuclear power. Inevitably, also, differences in opinion may result from ideological and political differences between authors. However, whatever the standpoint of the investigator, it is unquestionable that our fundamental knowledge is still far too slim for us to build reliable prognoses of the future on it, and that geographers and practitioners in allied disciplines need to seek more information.

The climatic future

Given the multitude of forcing factors and feedbacks, and the complex links between air, water, land and the organic world, an immediate reaction to the question 'Can we predict future climates?' would be 'no'. Certainly, in a whole variety of ways our predictive skills are both blunt and naïve; but the implications of even quite modest climate changes are so profound for so many aspects of human existence that we have to grapple with the problem and develop our capabilities. This is precisely what that great co-operative endeavour, the Intergovernmental Panel on Climate Change (IPCC), has been doing since it was established in 1988 (Houghton et al., 1996; Watson et al., 1996).

The main means of projecting future climates are climate models called General Circulation Models (GCMs). They are based upon physical laws that describe the atmospheric and oceanic dynamics and physics, and upon empirical relationships, and their depiction as mathematical equations. These equations are solved numerically using large computers. The main uncertainties in simulations using GCMs arise from problems of adequately representing clouds and their radiative properties, the coupling between the atmosphere and the oceans, and the detailed processes that operate at the land surface. Like all models they involve a range of assumptions and they remain relatively coarse in scale. It is not surprising, perhaps, that different GCMs give quite different projections.

Assessment of the future scale and trend of climatic change is full of imponderables. This has been well stated by Warrick (1988: 223–4):

Prediction of the rate at which atmospheric CO_2 will accumulate in the atmosphere depends largely on assumptions regarding future fossil fuel consumption and on knowledge of the global carbon cycle. Added to CO_2 are changes in the atmospheric concentrations of other 'greenhouse' gases – methane, chlorofluorocarbons, nitrous oxide, ozone – which in aggregate, are becoming just as important relatively speaking, as CO_2. Taking into account the projected concentrations of these gases (with the associated uncertainties), there could possibly be an 'equivalent' doubling of CO_2 by as early as the 2030s . . . But changes in climate will not follow step-by-step. Owing to the thermal inertia of the oceans, changes in the global surface temperature will probably lag behind by several decades. Thus the question of 'when' is clouded.

The question of 'where' is even more problematic. Climatic changes will vary in direction and magnitude from one agricultural region to the next. Unfortunately, GCMs (General Circulation Models) cannot yet provide reliable predictions of climatic change at regional scales.

Future energy scenarios are notoriously variable, and hence there are huge uncertainties with respect to fossil fuel consumption. Major problems include the prediction of economic growth in different economies and the indeterminate nature of future energy choice (as, for example, between nuclear and conventional power generation) (Bach, 1988).

In addition, there are major limitations in our ability to predict what will happen to the CO_2 released by the burning of fossil fuels and by deforestation and soil humus destruction. Major uncertainties surround the role of oceans as sinks for CO_2, the role of oceanic biota in regulating global carbon dynamics (see Clark et al., 1982), and the role of different terrestrial biomes in extracting CO_2 from the atmosphere.

A major problem lies in the reliability of GCMs, many of which have been over-simple and have many debatable assumptions. Dickinson (1986: 275–6) has urged caution:

The GCMs have tended to use submodels that are considerably over-simplified . . . Oversimplified ocean and sea-ice models, together with questionable descriptions for ice and snow albedos, suggest that ice-albedo radiative feedbacks in GCMs are of doubtful accuracy. Likewise, the models used for land surface water, snow budgets, and evapotranspiration have been too simple to allow quantitative evaluation of the effects of realistic changes in land surface processes on climate.

It is for these sorts of reasons that GCM scenarios for the future of climate have been so variable (Schlesinger and Mitchell, 1985).

The real world is more complex than the models that are designed to represent it, and the ocean–land–cryosphere–biosphere is governed by numerous feedback mechanisms which respond to changes in atmospheric composition and temperature. Macdonald (1982) lists the following for consideration:

1 *Temperature-infrared radiation feedback (negative)*
   As the temperature rises, so does the outgoing infrared radiation.
2 *Water-greenhouse feedback (positive)*
   As temperature rises, the amount of water vapour increases since relative humidity is dependent on temperature. An increase in water vapour content increases the infrared blanket of the atmosphere and this causes a further temperature increase.

3 *Ice and snow cover albedo feedback (positive)*
   As temperature rises snow cover and ice will melt, decreasing the earth's surface albedo and thereby increasing the absorption of sunlight.
4 *Cloud cover feedback (positive)*
   As temperature rises, cloud cover becomes greater. This traps infrared radiation and so the temperature rises.
5 *Cloud cover feedback (negative)*
   As temperature rises producing more cloud, the atmospheric albedo increases, and this tends to reflect incoming solar radiation and so to lower the temperature (see Idso, 1982).
6 *Carbon dioxide biofeedback (negative)*
   Increasing the carbon dioxide concentration in the atmosphere will increase the growth of the biosphere, which will fix carbon via photosynthesis and so tend to reduce the atmospheric level.
7 *Iceberg formation feedback (negative and positive)*
   As temperature rises, ice sheets will break up, producing, at least initially, large numbers of icebergs which will increase the earth's albedo until all the ice melts.
8 *Temperature-convection feedback (negative)*
   As temperature is increased at the ground, convection will increase, carrying sensible heat away from the ground and cooling it.
9 *Evaporation–precipitation feedback (negative and positive)*
   As temperature increases the intensity of the hydrological cycle will increase. Increased evaporation will carry off the additional temperature rise in latent heat, while increased precipitation will be working to return that same energy.
10 *Sea-ice ocean current feedback (positive)*
   The bottom-water producing currents are driven in the Antarctic by the formation of sea-ice in the Weddell Sea. An increase in temperature will inhibit or stop sea-ice formation and remove the pump which drives these currents. As a result the cold bottom waters will not be replenished and an increase in ocean temperature will occur, leading to a loss of CO_2 from the oceans and a further increase in temperature.
11 *Biomass-albedo feedback (positive)*
   Stimulation of new growth as a response to increased temperatures will cause a decrease in albedo which will stimulate a temperature rise.

A potential problem in forecasting future climates is that change may take place not solely in a gradual and progressive manner, but by stepwise changes across thresholds. Positive feedbacks may be involved. A notable example of this type of possibility is Flohn's (1982) contention that a global warming of 4°C will cause a sudden warming of central Arctic sea-ice. Because, in his view, the Antarctic ice will be less radically and less quickly affected, a globally asymmetric cryosphere will result, which will thereby cause a further profound change in the global general atmospheric circulation. Similarly, Bell (1982) has suggested that a marked warming in high latitudes could trigger the release of large quantities of methane, currently trapped in permafrost, subsea permafrost, and sea-ice, thereby adding significantly to the greenhouse effect (see also Nisbet, 1990).

A comparable position has been taken by Woodwell (1990: 5), who believes:

The speed of the warming may be determined less by human activities, unless rapid steps are taken to stop it, than by the warming itself, the climatically caused destruction of forests, and the oxidation of organic matter in soils and bogs. As the warming progresses and trees die at the warmer and drier margins of their distributions, organic matter from plants and soils will be released as carbon into the atmosphere, speeding the warming.

A second method of climate prediction is the climate analogue scenario, which is based on the use of a suitably defined ensemble of warm years from the recent instrumental record, which is then compared with either the long-term mean or a similarly defined cold-year ensemble (Webb and Wigley, 1985). This was the approach adopted by Wigley et al. (1980), who sought to predict the pattern of future temperature and precipitation changes by comparing the five warmest years in the period 1925–74 with the five coldest. As they recognized, however, they were dealing with a temperature shift of only 0.6°C, so the patterns obtained would not necessarily be representative of an even warmer climatic future. Furthermore, as Webb and Wigley (1985: 242) point out, 'The use of past warm periods as potential analogs of a world with high CO_2 concentrations is based on the assumption that, given similar boundary conditions (as represented by the state of the oceans, the land surface and the cryosphere), the general circulation of the lower atmosphere will respond in a similar way, even with differing forcing mechanisms.' Such an assumption may not be valid.

A third approach to climatic prediction is the study of selected past warm periods. Flohn and Dansgard (1984), for example, examined the Early Medieval optimum, the Holocene optimum (at around 6,000 years ago), the last interglacial (c.120,000 years ago), and the Late Tertiary. This method holds considerable promise, though there are major limitations imposed by data gaps and dating uncertainties for many parts of the world. Furthermore, boundary conditions may have been different in the Pliocene of the Late Tertiary, largely because of the effect that the Himalayan uplift has had since that time on the location of the jet streams. Even the status of the Holocene 'altithermal' or 'hypsithermal' can be questioned, and Webb and Wigley (1985: 252) believe that

More data (and derived temperature estimates) are needed before it can be demonstrated that the global mean temperature was higher at 6000 BP than it is today. The current data suggest that the global mean temperature at 6000 BP was probably within 1°C of today's temperature.

Schneider (1984: 190) has also expressed doubts about using the altithermal as an analogue, stating that:

Unless we can be sure that the Altithermal warm period was caused by an increase in CO_2 (or some equivalent annual, global scale external forcing) we cannot argue convincingly that a global warming induced by burning fossil fuels would result in similar regional climatic shifts – assuming, of course, we could know reliably what Altithermal seasonal and regional climatic patterns actually were.

The patterning of change

Most models indicate that the increase in global surface temperature for a doubling of greenhouse gases will be around 1.5–4.5°C. In 1995, the IPCC gave a 'best estimate' for an increase in global mean surface air temperature relative to 1990 of about 2°C by 2100 (Houghton et al., 1996). However, the warming will not occur in a regular and standard manner all over the world. There will be major differences between different regions (figure 8.1). Maximum warming is predicted to occur in high northern latitudes in late summer and autumn associated with reduced sea-ice and snow cover. Temperature rises in that zone could be greater than 6°C.

Simulations of the pattern of precipitation change (figure 8.2) are rather more complex and show marked differences between different GCMs. However, more recent models 'produce enhanced precipitation in high latitudes and the tropics throughout the year, and in mid-latitudes in winter' (Houghton et al., 1990: 145). Changes in the dry tropics are generally small, with both increases and decreases, though most models simulate an enhancement of precipitation associated with a strengthening of the south-west monsoon in Asia.

One of the few reasonably comprehensive attempts to predict future precipitation changes by combining the evidence from palaeoclimates, modern analogue data and GCMs is provided by Budyko and Izrael (1991). They argue on the basis of palaeoclimatic data that under conditions of warmth comparable to

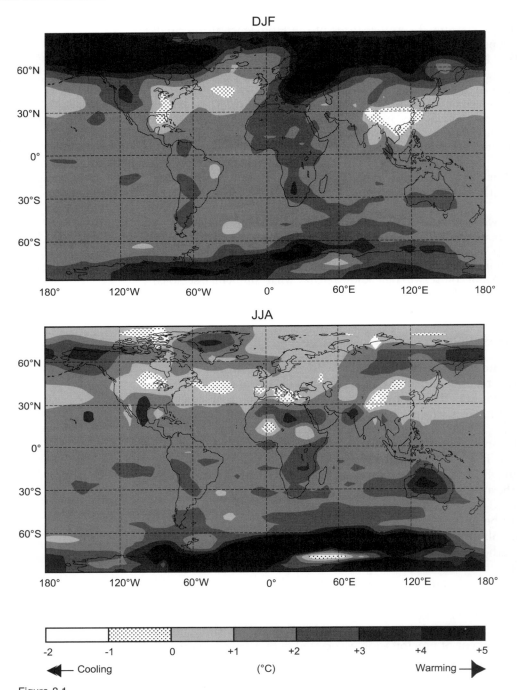

Figure 8.1
Seasonal change in surface temperature from 1880–1989 to 2040–2049 in simulations with aerosol effects included. DJF = December, January, February. JJA = June, July, August (modified after IPCC, 1995, figure 6.10, p. 307)

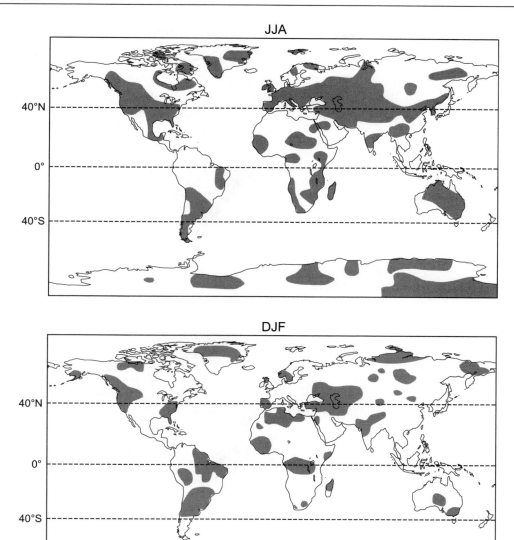

JJA

DJF

those that existed in the Pliocene (figure 8.3) precipitation would increase (by up to 30 centimetres per year) over most of Eurasia and the Sahara. On the other hand, under conditions comparable to those of the Holocene Atlantic Optimum, with temperatures up to 1°C warmer than the present, there is a large zone (between 50° and 30°N) where precipitation levels would decline (by up to 20 centimetres per year in central North America). There would, however, be improved moisture conditions in the subtropical regions (between 10° and 20°N) and in higher

Figure 8.2
Seasonal changes in precipitation for a doubling of atmospheric CO_2. Shaded areas are those parts of the land surface which will show either no change or a reduction (modified after IPCC, 1995, figure 6.11, p. 308)

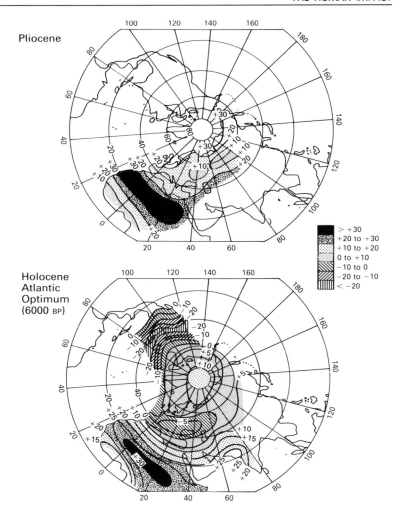

Figure 8.3
Deviations in annual
precipitation means (cm) for
two past warm phases (from
Budyko and Izrael, 1991)

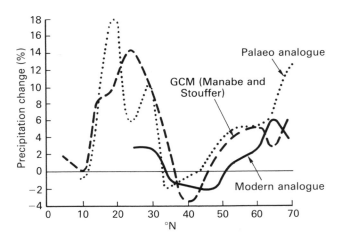

Figure 8.4
Relative changes in mean
latitudinal precipitation on the
continents of the northern
hemisphere with 1°C higher
mean surface air temperature
(after Budyko and Izrael, 1991,
figure 1.5)

latitudes (more than 60°N). This pattern of changes is attributed to a northward shift both of the intertropical convergence zone (ITCZ) and of westerly cyclone tracks. In sum, Budyko and Izrael believe that under conditions of marked warming both the mid-latitude and lower-latitude arid zones of the northern hemisphere will be wetter, whereas under conditions of less marked warming (i.e. by around 1°C) areas like the Sahara and Thar will become moister, but areas like the High Plains of the United States or the steppes of the CIS will become drier (figure 8.4).

Potential consequences of global warming

If global warming and associated changes in rainfall and soil moisture amounts occur, then the effects on natural environments may be substantial and rapid. These impacts have been reviewed by the IPCC (Watson et al., 1996). The following are some of the thought-provoking comments made in their report.
    On forests:

A substantial fraction (a global average of one-third, varying by region from one-seventh to two-thirds) of the existing forested area of the world will undergo major changes in broad vegetation types . . . climate change is expected to occur at a rapid rate relative to the speed at which forest species grow, reproduce and establish themselves . . . the species composition of forests is likely to change; entire forest types may disappear.

On deserts and desertification:

Deserts are likely to become more extreme – in that, with few exceptions, they are projected to become hotter but not significantly wetter.

On the cryosphere:

Between one-third and one-half of existing mountain glacier mass could disappear over the next 100 years.

On mountain regions:

The projected decrease in the extent of mountain glaciers, permafrost and snow cover caused by a warmer climate will affect hydrologic systems, soil stability and related socio-economic systems . . . Recreational industries – of increasing economic importance to many regions – are also likely to be disrupted.

On coastal systems:

Climate change and a rise in sea level or changes in storm surges could result in the erosion of shores and associated habitat, increased salinity of estuaries and freshwater aquifers, altered tidal ranges in rivers and bays, a change in the pattern of chemical and microbiological contamination in coastal areas, and increased coastal flooding. Some coastal ecosystems are particularly at risk, including saltwater marshes, mangrove ecosystems, coastal wetlands, coral reefs, coral atolls, and river deltas.

Some landscape types will be highly sensitive to global warming. This may be the case because they are located in zones where it is forecast that climate will change to an above average degree. This applies, for instance, in the high latitudes of North America and Eurasia, where the degree of warming may be three or four times greater than the presumed global average. It may also be the case for some critical areas where particularly substantial changes in rainfall may result from global warming. For example, various methods of climatic prediction produce scenarios in which the American High Plains will become considerably drier. Other landscapes will be highly sensitive because certain landscape-forming processes are very closely controlled by climatic conditions. If such landscapes are close to a particular climatic threshold then quite modest amounts of climate change can switch them from one state to another.

The rest of this chapter will discuss a selection of themes: changes in the frequency of tropical cyclones; changes in the cryosphere; changes in the biosphere; changes in the hydrosphere; changes in the aeolian environment; and changes in the coastal environment.

Changes in tropical cyclones

Tropical cyclones (hurricanes) are highly important geomorphological agents, in addition to being notable natural hazards, and are closely related in their places of origin to sea temperature conditions. They develop only where sea-surface temperatures (SSTs) are in excess of 26.5°C. Moreover, their frequency over the last century has changed in response to changes in temperature, and there is even evidence that their frequency was reduced during the Little Ice Age (see Spencer and Douglas, 1985). It is, therefore, possible that as the oceans warm up, so the geographical spread and frequency of hurricanes will increase. Furthermore, it is also likely that the intensity of these storms will be magnified. Emanuel (1987) used a GCM which predicted that with a doubling of present atmospheric concentrations of CO_2 there will be an increase of 40–50 per cent in the destructive potential of hurricanes. More recently Krutson et al. (1998) simulated hurricane activity for an SST warming of 2.2°C and found that this yielded hurricanes that were more intense by 3–7 metres per second for wind speed, an increase of 5–12 per cent.

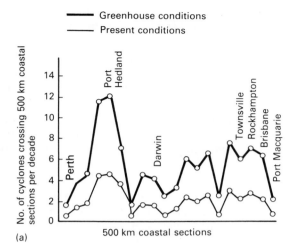

(a)

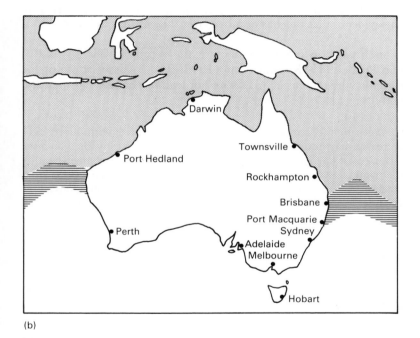

(b)

Figure 8.5
(a) The present frequency of
    cyclones crossing 500 km
    long sections of the
    Australian coast, and an
    estimate of the frequency
    under conditions with a
    2°C rise in temperature
(b) The area where February
    sea surface temperatures
    around Australia are
    currently greater than
    27°C (stippled) and the
    additional area with such
    temperatures with a 2°C
    rise in temperature
    (hatched)
(modified after Henderson-
Sellers and Blong, 1989,
figures 4.12, 4.14)

An increase in hurricane intensity and frequency would have
numerous geomorphological consequences in low latitudes,
including accentuated river flooding and coastal surges, severe
coast erosion, accelerated land erosion and siltation, and the
killing of corals (because of fresh-water and siltation effects)
(De Sylva, 1986).

Figure 8.5 indicates one scenario of the likely latitudinal change
in the extent of warm, cyclone-generating sea water in the Australian region, using as a working threshold for cyclone genesis

a summer (February) SST of 27°C. Although cyclones do occur to the south of this line, they are considerably more frequent to the north of it. Under greenhouse conditions it is probable that on the margins of the Great Sandy Desert near Port Hedland the number of cyclones crossing the coast will approximately triple from around four per decade to twelve per decade (Henderson-Sellers and Blong, 1989).

However, it needs to be pointed out that the IPCC (Houghton et al., 1990: xxv) was somewhat equivocal on the question of the extent to which warming would stimulate cyclone activity, stating:

Although the areas of sea having temperatures over this critical value (26.5°C) will increase as the globe warms, the critical temperature itself may increase in a warmer world. Although the theoretical maximum intensity is expected to increase with temperature, climate models give no consistent indication whether tropical storms will increase or decrease in frequency or intensity as climate changes; neither is there any evidence that this has occurred over the past few decades.

Such caution is also evident in the recent review by Walsh and Pittock (1998); and Gray (1993: 96) discusses some of the reasons why increased SSTs may not lead to changes in hurricane activity:

Changes in the atmospheric circulation associated with future SST increases may not be the same as the historical variations on which the estimates of hurricane frequency are based. Since these changes modify the tropical storms' environment, they could prevent the storms from reaching the potential increased strength estimated from thermodynamic considerations. For example, an increase in vertical shear of the zonal wind over the tropical Atlantic would tend to decrease the frequency of tropical storm formation, thereby moderating an increase in frequency from higher SSTs, while a decrease in vertical shear would tend to have the opposite effect.

The cryosphere

Of considerable importance in terms of both climatic feedbacks and the nature of future sea-level changes is the response of the Arctic and Antarctic ice masses to a global warming.

With respect to the Arctic ice, the situation is controversial and unclear. On the one hand there is the view of Budyko (1982: 233) who believes that 'if the mean global temperature rises to 2–3° higher than in preindustrial time, it will be enough for the many-year-ice in the Arctic to disappear.' By contrast, Flohn (1982) believes a temperature increase of 4–5°C would cause an ice-free Arctic Ocean, creating conditions broadly analogous to those of the Pliocene. These views are disputed by Parkinson and Kellogg (1979), who contend that while a 5°C atmosphere temperature rise would cause Arctic pack-ice to disappear in August and September, it would reappear during the winter months. Bentley (1984) argues that in all probability no less than a 10°C temperature increase would be required to remove Arctic sea-ice in summer.

Equal uncertainties surround the future of the West Antarctic ice-sheet. One extreme view was postulated by Mercer (1978), who believed that predicted increases in temperature at 80°S would start a 'catastrophic' deglaciation of the area, leading to a sudden 5 metre rise in sea-level. A less extreme view has been put forward by Thomas et al. (1979). They recognize that higher temperatures will weaken the ice-sheets by thinning them, enhancing lines of weakness, and promoting calving, but they contend that deglaciation would be rapid rather than catastrophic, the whole process taking 400 years or so. Robin (1986) takes a broadly intermediate position. He contends (p. 355) that:

A catastrophic collapse of the West Antarctic ice sheet is not imminent, but better oceanographic knowledge is required before we can assess whether a global temperature rise of 3.5°C might start such a collapse by the end of the next century. Even then it is likely to take at least 200 years to raise sea level by another 5 metres.

Predictions of the rate of ice sheet response to global warming are highly problematic because they involve the use of models with many assumptions or unknowns. Many factors are involved, including the rate at which accumulation occurs. This is crucial in the context of Antarctica, where warmer temperatures could cause more snow deposition in an environment which is currently extremely arid. Such increased accumulation might offset increased rates of ablation (Gregory and Oelemans, 1998). Budd (1991) has calculated that increases of around 4°C in temperatures in the Antarctic would lead to a 30 per cent increase in net accumulation. Also important, however, are varying snow surface albedos, variability in rate of calving of outlet glaciers, the protective and buttressing role of ice shelves, and the geometry and slope of the ice sheet (Huybrechts et al., 1990). The rapidity with which ice sheets and glaciers decay (and hence sea level may rise) depends greatly on the terminal environment and certain topographic thresholds (Sugden, 1991).

Alpine glaciers are highly responsive to climatic change, as is made evident by their frequent and rapid fluctuations during the course of the Holocene (see Grove, 1988). Substantial neoglacial advances and retreats have been caused by quite limited fluctuations in temperature and/or precipitation. The historical picture of glacial response to such fluctuations is, however,

Table 8.1 Calculated contemporary and future areas of permafrost in the northern hemisphere (10^6 km^2)

Zone	Contemporary	2050	% change
Continuous permafrost	11.7	8.5	−27
Discontinuous permafrost	5.6	5.0	−11
Sporadic permafrost	8.1	7.9	−2
Total	25.4	21.4	−16

Source: based on Nelson and Anisimov, 1993.

complex, and this behoves us to exercise caution in predicting the future picture of response to just one control of glacial state, i.e. warming. Changes in precipitation (which control rate of accumulation) and cloudiness (which can control rate of ablation) will also be significant factors in determining glacier equilibrium, while topographic controls, the effects of size and varying propensity to surging will create local complexities of response.

Bearing such caveats in mind, it is none the less highly likely that most alpine glaciers will show increasing rates of retreat in a warmer world, and given the rates of retreat experienced in many areas in response to the warming episode of the first decades of the twentieth century, it is probable that many glaciers will disappear altogether. For example, in New Zealand Chinn (1988) calculates that if temperatures rise by 3.6–6.3°C, snow lines will migrate vertically by 300–500 metres, and around 1,000 out of the country's 3,000 glaciers will disappear.

As we have seen, under most models of climatic change the degree of temperature increase will be greatest in high latitudes. With a doubling of atmospheric CO_2 concentration a high latitude temperature rise of perhaps even as much as 10°C might be expected (Bell, 1982). Given that the existence and distribution of permafrost are so closely related to temperature, it is probable that major changes will occur (Anisimov and Nelson, 1996). Barry (1985) estimates that an average northward displacement of the southern permafrost boundary by 150 ± 50 kilometres would be expected for each 1°C warming so that a total *maximum* displacement of between 1,000 and 2,000 kilometres is possible. The IPCC (Watson et al., 1996: 253) suggests a 15 per cent shrinkage in total permafrost area by 2050 (table 8.1). Figure 8.6 shows predictions of permafrost displacement for Canada and Siberia.

However, major uncertainty surrounds the rate at which permafrost degradation will occur. As it is probably a slow process, permafrost will continue to exist for an extended time in areas of *continuous* permafrost. In areas of discontinuous or sporadic permafrost the rate will vary greatly depending on local material conductivity, snow cover and vegetation. Changes in

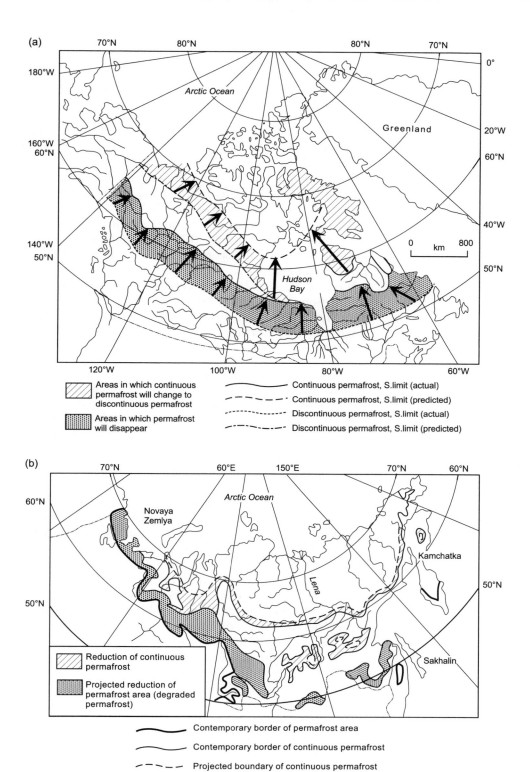

(a)

180°W

70°N   80°N   80°N   70°N   0°

Arctic Ocean

Greenland   20°W

160°W
60°N   60°N

40°W

140°W
50°N   0   km   800

50°N

Hudson
Bay

120°W   100°W   80°W   60°W

Areas in which continuous permafrost will change to discontinuous permafrost

Areas in which permafrost will disappear

——————— Continuous permafrost, S.limit (actual)
– – – – – – Continuous permafrost, S.limit (predicted)
· · · · · · · · · · · Discontinuous permafrost, S.limit (actual)
–·–·–·–·– Discontinuous permafrost, S.limit (predicted)

(b)

70°N   60°E   150°E   70°N   60°N

60°N

Arctic Ocean

Novaya
Zemlya

Kamchatka

50°N

50°N

Lena

Reduction of continuous permafrost

Projected reduction of permafrost area (degraded permafrost)

Sakhalin

——————— Contemporary border of permafrost area

——————— Contemporary border of continuous permafrost

– – – – – Projected boundary of continuous permafrost

Figure 8.6
Predicted changes in permafrost: (a) in Canada as a result of a surface temperature increase of 4°C; (b) in Siberia as a result of a surface temperature increase of 2°C (modified from Atmospheric Environment Service, Canada, in French, 1996, figure 17.5 and Anisimov, 1989)

vegetation type and in snow cover in a warmer world may modify the direct consequences of warmer surface temperatures (Bøer et al., 1990).

There is historical evidence that permafrost can degrade relatively rapidly. For example, during the warm optimum of the Holocene (c.6,000 years ago) the southern limit of discontinuous permafrost in the Russian Arctic was up to 600 kilometres north of its present position (Koster, 1994). Similarly, during the warming phase of recent decades, researches have demonstrated that along the Mackenzie Highway in Canada, between 1962 and 1988, the southern fringe of the discontinuous zone has moved north by about 120 kilometres in response to an increase over the same period of 1°C in mean annual temperature (Kwong and Tau, 1994).

Woo et al. (1992) make certain predictions based on the assumption that a greenhouse warming of 4–5°C causes a spatially uniform increase in surface temperature of the same magnitude over northern Canada:

Permafrost in over half of what is now the discontinuous zone could be eliminated. The retreat will be accelerated if warming is accompanied by increased snowfall, particularly during the autumn freeze-up period. The boundary between continuous and discontinuous permafrost may shift northward by hundreds of kilometres, but because of its links to the treeline, its ultimate position and the time to reach it are speculative. It is possible that a warmer climate could ultimately eliminate continuous permafrost from the whole of the mainland of North America, restricting its presence only to the Arctic Archipelago.

Smith (1993) provides a good review of likely changes in the Canadian tundra. Where the permafrost is ice-rich, or contains massive ground ice segregations, subsidence and settling due to thawing will occur, inducing a thermokarst relief which can alter drainage patterns and stream courses. Coastal retreat will also gather momentum as permafrost degrades in coastal lowlands, and large areas may be inundated as depressions are lowered to below current sea levels because of thaw settlement. River banks and lake and reservoir shorelines might also become amenable to faster rates of erosion, while slopes could become less stable and the active zone become thicker.

The biosphere

Climate change is likely to have a whole suite of biological consequences (Gates, 1992). Indeed, some models suggest that wholesale changes in biomes will occur. Using the GCM developed by Manabe and Stouffer (1980) for a doubling of CO_2 levels, and mapping the present distribution of ecosystem types

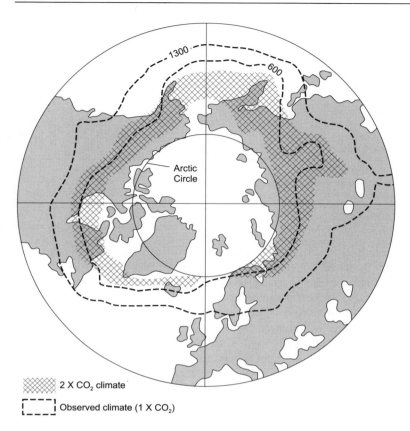

2 X CO₂ climate

Observed climate (1 X CO₂)

Figure 8.7
The northern and southern
boundaries of the boreal forest
are approximately defined by
the 600 and 1300 growing
degree-day isopleths. These
are shown in their current
positions and in the positions
they would occupy under a
warming associated with a
doubling of CO_2 levels (after
Kauppi and Posch, 1988)

in relation to contemporary temperature conditions, Emmanuel et al. (1985) found *inter alia* that the following changes would take place: boreal forests would contract from their present position of comprising 23 per cent of total world forest cover to less than 15 per cent; grasslands would increase from 17.7 per cent of all world vegetation types to 28.9 per cent; deserts would increase from 20.6 to 23.8 per cent; and forests would decline from 58.4 to 47.4 per cent. Potential changes in the geographical extent of boreal forests in the northern hemisphere are especially striking. Kauppi and Posch (1988), using a Goddard Institute of Space Sciences GCM, modelled the northern and southern boundaries of the boreal forest in response to a climate warming associated with a doubling of atmospheric carbon dioxide levels (figure 8.7). Note how the southern boundary moves from the southern tip of Scandinavia to the northernmost portion.

More recently, Smith et al. (1992) have attempted to model the response of Holdridge's life zones to global warming and associated precipitation changes as predicted by a range of different GCMs. These are summarized for some major biomes in

Table 8.2   Changes in areal coverage ($km^2 \times 10^3$) of major biomes as a result of the climatic conditions predicted for a $\times$ 2 CO_2 world by various General Circulation Models

Biome	Current	OSU	GFDL	GISS	UKMO
Tundra	939	−302	−5.5	−314	−573
Desert	3,699	−619	−630	−962	−980
Grassland	1,923	380	969	694	810
Dry forest	1,816	4	608	487	1,296
Mesic forest	5,172	561	−402	120	−519

OSU   = Oregon State University
GFDL = Geophysical Fluid Dynamic Laboratory
GISS = Goddard Institute for Space Studies
UKMO = UK Meteorological Office.
Source: from Smith et al., 1992, table 3.

table 8.2. All the GCMs predict conditions that would lead to a very marked contraction in the tundra and desert biomes, and an increase in the areas of grassland and dry forests. There is, however, some disagreement as to what will happen to mesic forests as a whole, though within this broad class the humid tropical rainfall element will show an expansion.

Theoretically a rise of 1°C in mean temperature could cause a poleward shift of vegetation zones of about 200 kilometres (Ozenda and Borel, 1990). However, uncertainties surround the question of how fast plant species would be able to move to and to settle new habitats suitable to the changed climatic conditions. Postglacial vegetation migration rates appear to have been in the range of a few tens of kilometres per *century*. For a warming of 2–3°C forest bioclimates could shift northwards about 4–6° of latitude in a century, indicating the need for a migration rate of some tens of kilometres per *decade*. Furthermore, migration could be hampered because of natural barriers, zones of cultivation, etc.

Zabinski and Davis (1989) modelled the potential changes in the range of certain tree species in eastern North America, using a GISS GCM and a doubling of atmospheric carbon dioxide levels. The differences between their present ranges and their predicted ranges of each species is very large (figure 8.8).

Altitudinal changes in vegetation zones will be of considerable significance. In general, Peters (1988) believes that with a 3°C temperature change vegetation belts will move about 500 metres in altitude. One consequence of this would be the probable elimination of the Douglas fir (*Pseudotsuga taxifolia*) from the lowlands of California and Oregon, because rising temperatures would preclude the seasonal chilling this species requires for seed germination and shoot growth.

Vegetation will also probably be changed by variations in the role of certain extreme events that cause habitat disturbance,

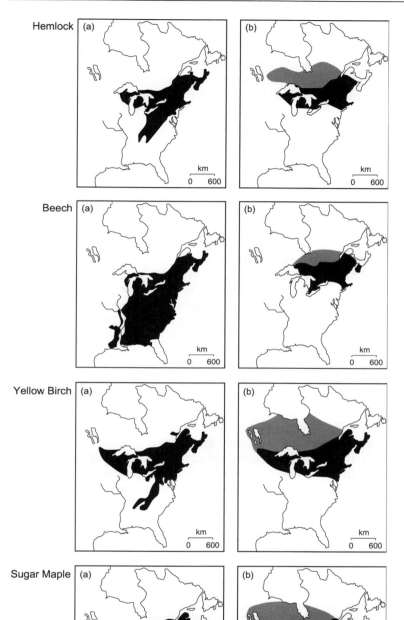

Hemlock

Beech

Yellow Birch

Sugar Maple

Figure 8.8
Present and future range of four tree species in eastern North America: (a) present range; (b) range in AD 2090 under the GISS GCM $2 \times CO_2$ scenario. The black area is the projected occupied range considering a rate of migration of 100 km per century. The grey area is the potential projected range with climate change (after Zabinski and Davis, 1989)

including fire (exacerbated by drought), wind storms, hurricanes and coastal flooding (Overpeck et al., 1990). Other possible non-climatic effects of elevated CO_2 levels on vegetation include changes in photosynthesis, stomatal closure and carbon fertilization (Idso, 1983), though these are still matters of controversy (Solomon and West, 1986). However, especially at high altitudes, it is possible that elevated CO_2 levels would have potentially significant effects on tree growth, causing growth enhancement (La Marche et al., 1984).

The hydrosphere

Changes in temperatures, precipitation quantities, and the timing and form of precipitation would have highly important hydrological consequences (Jones et al., 1996). In areas affected by snowfall today, the changes may be especially marked. In a warmer world, there would be a tendency for a marked decrease to occur in the proportion of winter precipitation that falls as snow. Furthermore, there would be an earlier and shorter spring snowmelt. The first of these two effects would cause greater winter rainfall and hence winter runoff, since less overall precipitation would enter snowpacks to be held over until spring snowmelt. The second effect would intensify spring runoff, leading to additional adverse consequences for both summer runoff level and for spring and summer soil moisture levels (Gleick, 1986). Table 8.3, taken from Gleick's work, indicates likely changes in runoff for two temperature and five precipitation scenarios in such regions, and demonstrates clearly the anticipated direction and seasonality of runoff changes.

On the other hand, in high-latitude tundra environments warmer winters may cause more snow to fall, thereby creating increased runoff levels in the summer months (Barry, 1985). Indeed, Budyko (1982: 242) predicts that because of increased precipitation (perhaps by as much as 500–600 millimetres per

Table 8.3 The effects of hypothetical temperature and precipitation scenarios on runoff (after Gleick, 1986)

Temperature change	Precipitation change (%)				
	−20	−10	0	+10	+20
(a) Winter (December, January, February)					
+2°C	−24	−9	+8	+25	+44
+4°C	−4	+14	+34	+54	+75
(b) Summer (June, July, August)					
+2°C	−42	−32	−22	−12	−1
+4°C	−73	−68	−62	−55	−49

Table 8.4 Approximate % decrease in runoff for a 2°C increase in temperature

Initial temperature (°C)	Precipitation (mm y⁻¹)					
	200	300	400	500	600	700
−2	26	20	19	17	17	14
0	30	23	23	19	17	16
2	39	30	24	19	17	16
4	47	35	25	20	17	16
6	100	35	30	21	17	16
8		53	31	22	20	16
10		100	34	22	22	16
12			47	32	22	19
14			100	38	23	19

Source: data in Revelle and Waggoner, 1983.

year in the tundra zone), runoff in Russia north of 58–60°N will increase by a factor of 2 to 3. However, runoff in cold northern regions will not merely be a response to changes in the amount and seasonality of precipitation and evapotranspiration. As permafrost thaws, groundwater recharge may increase and surface runoff decrease. This tendency will be supplemented as the water storage capacity of thickening active layers increases. As permafrost degrades in the discontinuous permafrost zone, subterranean flow conduits may be reopened, causing spring discharge to occur, especially in karstic (limestone) areas. The greater amount of groundwater flow may lead to more extensive river ice formation in the Arctic. This could result in more serious flooding during the spring/early summer break-up (Woo et al., 1992).

No less significant runoff changes may be anticipated for semi-arid environments, such as the south-west United States. For example, Revelle and Waggoner (1983) suggest that in the event of there being a 2°C rise in temperature and a 10 per cent reduction in precipitation, water supplies would be diminished by 76 per cent in the Rio Grande region and by 40 per cent in the Upper Colorado. Table 8.4 demonstrates that a 2°C rise in temperature would be most serious for water supplies and runoff in those regions where the mean annual precipitation is less than about 400 millimetres per year. There is reason to believe that because of increased temperatures and rainfall reductions the severity of droughts in the High Plains could be more severe than in the Dust Bowl years of the 1930s (Rosenzweig and Hillel, 1993) (figure 8.9).

Projected summer dryness in areas such as the North American Great Plains (Manabe et al., 1981) may be accentuated by a positive feedback process involving changes in cloud cover. Manabe and Wetherald (1986: 627) suggested that:

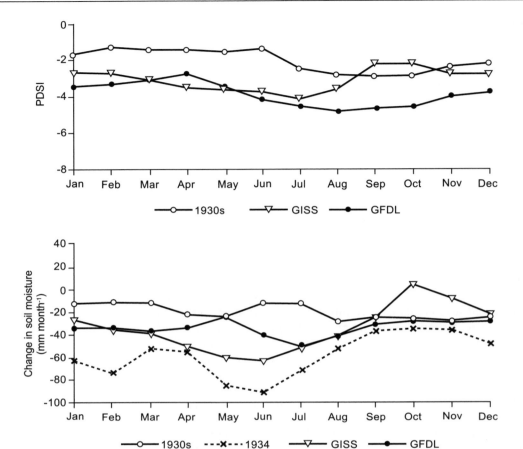

Figure 8.9
A comparison of drought conditions in a warmer world as predicted by the GISS and GFDL GCM-based scenarios with those that occurred in Dodge City, Kansas, in the Dust Bowl years of the 1930s:
(a) The Palmer Drought Severity Index (PDSI) relative to 1951–80 in Dodge City, Kansas. The PDSI is a widely used index for evaluating drought conditions in the USA and is based on deviations from normal precipitation and temperature conditions
(b) Soil moisture anomalies calculated by the PDSI method
(modified from Rosenzweig and Hillel, 1993)

When soil moisture is reduced, a large fraction of radiative energy absorbed by the continental surface is ventilated through the upward flux of sensible heat rather than through evaporation. Accordingly, the temperature of the continental surface and the overlying layer increases, resulting in the general reduction in relative humidity and precipitation in the lower troposphere of the model. Accompanying the reduction in relative humidity is a reduction in total cloud amount, causing an increase in solar energy reaching the continental surface. Thus the radiation energy absorbed by the continental surface also increases, raising the rate of potential evaporation.

Substantial hydrological changes may also occur in the catchments of the Great Lakes of North America because of changes in precipitation, snowmelt timing and evapotranspirational loss. Models produced by Croley (1990) and Hartmann (1990) suggest that for a doubling of CO_2 levels there may be a 23–51 per cent reduction in net basin supplies to all the Great Lakes, and that as a result lake levels will fall at rates ranging from 13 millimetres per decade (for Lake Superior) to 93 millimetres per decade (for Lake Ontario).

The general response of rivers to global warming is reviewed by Arnell (1996), who is particularly informative on the British situation. A useful study for Australia, showing the difference in response between wet tropical catchments (increases of up to 25 per cent in annual runoff) and those in the south (decreases of up to 35 per cent) for the year 2030 is Chiew et al. (1995).

The aeolian environment

Changes in precipitation and evapotranspiration rates will probably have a marked impact on the aeolian environment. Rates of deflation, sand and dust entrainment and dune formation are closely related to soil moisture conditions and the extent of vegetation cover. Areas that are marginal in terms of their stability with respect to aeolian processes will be particularly susceptible, and this has been made evident, for example, through recent studies of the semi-arid portions of the United States (e.g. the High Plains). Repeatedly through the Holocene they have flipped from states of vegetated stability to states of drought-induced surface instability. It is only with the recent availability of thermoluminescent and optical dates that this sensitivity to quite minor perturbations has become evident (e.g. Stokes and Gaylord, 1993).

Muhs and Maat (1993) have used the output from GCMs, combined with a dune mobility index (which incorporates wind strength and the ratio of mean annual precipitation to potential evapotranspiration), to show that sand dunes and sand sheets on the Great Plains are likely to become reactivated over a significant part of the region, particularly if the frequencies of wind speeds above the threshold velocity were to increase by even a moderate amount.

For another part of the United States, Washington State, Stetler and Gaylord (1996) have suggested that with a 4°C warming vegetation would be greatly reduced and that as a consequence sand dune mobility would increase by over 400 per cent.

The same considerations apply to dust storm generation. For example, the GCM from the Geophysical Fluids Dynamics Laboratory (GFDL) in Princeton, New Jersey, when applied to Nebraska and Kansas (Smith and Tirpak, 1990), shows that the pattern of temperature increase and precipitation decrease in a world with a doubling of atmospheric CO_2 levels will be broadly comparable to those experienced in the devastating Dust Bowl years of the 1930s. In some years during that period, climatic stations recorded well over 100 dust storm events in the year (Goudie and Middleton, 1992). Such a view is supported by Wheaton (1990), who has modelled the effects of climate change on dust storm generation in Canada.

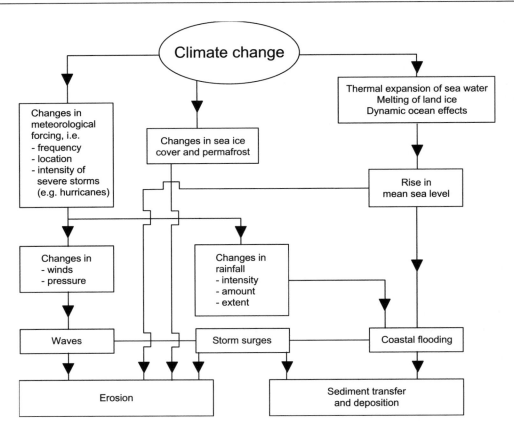

Coastal environments

Climate change can have a whole suite of impacts (Eisma, 1995) on coastal environments (figure 8.10). Among the most important geomorphological consequences of a global warming would be a worldwide rise in sea level. This would occur as a result of two distinct processes: the thermal expansion of the upper layers of the oceans, and the melting of land ice (US Department of Energy, 1985; Titus, 1986).

The degree of sea-level rise could be moderated by reservoir construction. Newman and Fairbridge (1986) have calculated that between 1957 and 1982 human intervention stored as much as 0.75 millimetres per year of sea-level rise potential in reservoirs and irrigation projects. Conversely, Sahagian et al. (1994) have argued that human actions have accelerated the rate of sea-level rise. They believe that groundwater withdrawal, surface water diversion and land-use changes (e.g. draining of wetlands, and reduction in soil moisture levels following land-use changes) may have contributed to at least 30 per cent of observed twentieth-century sea-level rise.

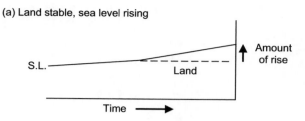

(a) Land stable, sea level rising

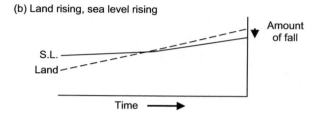

(b) Land rising, sea level rising

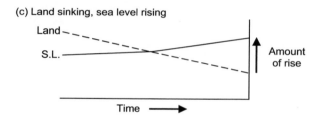

(c) Land sinking, sea level rising

Figure 8.11
The effects of global sea-level rise will be either compounded or mitigated according to whether the local environment is one that is stable, rising or sinking

The degree of tidal inundation may show local variations. Goemans (1986), for example, has pointed out that the tidal system of the North Sea is rather complex, with several amphidromic points (points of zero tidal range). These points might move as a result of sea-level change and could therefore lead to changes in local tidal amplitudes and of the relative high-water level.

The effects of global sea-level rise will be compounded in those areas that suffer from local subsidence as a result of local tectonic movements, isotatic adjustments and fluid abstraction. Areas where land is rising because of isotasy (e.g. Fennoscandia or the Canadian Shield) or because of tectonic uplift (e.g. much of the Pacific coast of the Americas) will be less at risk than subsiding regions (e.g. the deltas of the Mississippi and Nile rivers) (figure 8.11).

The anticipated rise of sea-level over the next century is a subject of contention, largely because of uncertainties about the behaviour of Antarctic ice (Oppenheimer, 1998). In the 1980s there were expectations that sea level could rise by over 3.5 metres by 2100. Now, however, there is a tendency to view

Plate 8.1
Bangladesh already suffers from severe flooding. This is a problem that may increase in coming decades should sea level rise or if tropical storms become more frequent and intense (as is predicted by some General Circulation Models)

such values as excessive, and the IPCC (Houghton et al., 1990: 279) concluded 'that a rise of more than 1 metre over the next century is unlikely'. None the less, this is a rate three to six times faster than that experienced over the last 100 years (1–2 millimetres per year). The Panel also recognized that there were still large uncertainties associated with the future contributions of Antarctica and Greenland to sea-level rise. In 1995, the best estimate for sea-level change by 2100 compared to the present was a rise of 38–55 centimetres (Houghton et al., 1996).

The results of sea-level rise in sensitive areas (e.g. low-lying coasts, deltas, marshes) will be marked. Broadus et al. (1986), for example, calculate that were the sea level to rise by just 1 metre in 100 years, 12–15 per cent of Egypt's arable land would be lost and 16 per cent of the population would have to be relocated. With a 3 metre rise the figures would be a 20 per cent loss of arable land and a need to relocate 21 per cent of the population. In Bangladesh (plate 8.1), a 1 metre rise would inundate 11.5 per cent of the total land area of the state and affect 9 per cent of the population directly, while a 3 metre rise would inundate 29 per cent of the land area and affect 21 per cent of the population (figure 8.12). Many of the world's major conurbations might be flooded in whole or in part, and sewers and drains rendered inoperative (Kuo, 1986). It needs to be remembered, however, that deltas will not be affected solely by

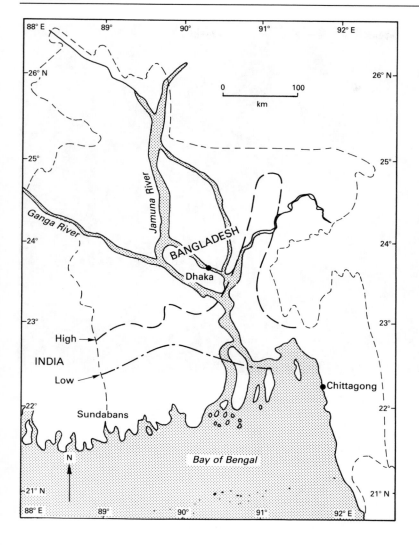

Figure 8.12
Projected areas of flooding as
a result of sea-level change in
Bangladesh, for two scenarios
(low = 1 metre and high = 3
metres) (modified after
Broadus et al., 1986, figure 7)

sea-level changes. The delta lands of Bangladesh, for example,
receive very heavy sediment loads from the rivers that feed them,
so that it is the relative powers of accretion and inundation that
will be crucial (Milliman et al., 1989).

Salt marshes, including the mangrove swamps of the tropics
(plate 8.2), are potentially highly vulnerable in the face of sea-
level rise, particularly in those circumstances where sea defences
and other barriers prevent the landward migration of marshes
as sea level rises. However, salt marshes are highly dynamic
features and in some situations may well be able to cope, even
with quite rapid rises of sea level (table 8.5) (Reed, 1995).

Reed (1990) suggests that salt marshes in riverine settings
may receive sufficient inputs of sediment that they are able to
accrete rapidly enough to keep pace with projected rises of sea

Plate 8.2
Mangrove swamp in the
Seychelles, Indian Ocean.
Such swamps are valuable
wetland resources. They are
already under great human
pressures in many areas,
and may also be susceptible
to the effects of any future
sea-level rise

Table 8.5 Salt marsh vulnerability

Less sensitive
Areas of high sediment input
Areas of high tidal range (high sediment transport potential)
Areas with effective organic accumulation

More sensitive
Areas of subsidence
Areas of low sediment input (e.g. cyclically abandoned delta areas)
Mangroves (longer life cycle, therefore slower response)
Constraint by sea walls, etc. (nowhere to go)
Microtidal areas (rise in sea-level represents a larger proportion of total tidal range)
Reef settings (lack of allogenic sediment)

level. Areas of high tidal range, such as the marshes of the
Severn Estuary in England/Wales, are also areas of high sediment transport potential and may thus be less vulnerable to sea-
level rise. Likewise, some vegetation associations, e.g. *Spartina*
swards, may be relatively more effective than others at encouraging accretion, and organic matter accumulation may itself be
significant in promoting vertical build-up of some marsh surfaces. For marshes that are dependent upon inorganic sediment
accretion, increased storm activity and beach erosion which
might be associated with the greenhouse effect could conceiv-
ably mobilize sufficient sediments in coastal areas to increase
their sediment supply.

Marsh areas that may be highly prone to sea-level rise include areas of deltaic sedimentation where, because of sediment movement controls (e.g. reservoir construction) or cyclic changes in the locations of centres of deposition, rates of sediment supply are low. Such areas may also be areas with high rates of subsidence. A classic example of this is portions of the Mississippi Delta. Park et al. (1986) undertook a survey of coastal wetlands in the United States and suggested that sea-level change could, by 2100, lead to a loss of between 22 and 56 per cent of the 1975 wetland area, according to the degree of sea-level rise that takes place.

Ray et al. (1992) have examined the past and future response of salt marshes in Virginia in the United States to changes in sea level, and have been particularly interested in the response of mid-lagoon marshes. Between 6,000 and 2,000 years ago, when the rate of sea level rise was about 3 millimetres per year, the lagoons were primarily open water environments with little evidence of mid-lagoon marshes. After AD 1200, land subsidence in the area slowed down, and sea level rise was only about 1.3 millimetres per year. This permitted mid-lagoon marshes to expand and flourish. Such marshes in general lack an inorganic sediment supply so that their upward growth rate approximates only about 1.5 millimetres per year. Present rates of sea-level rise exceed that figure (they are about 2 millimetres per year) and cartographic analysis shows a 16 per cent loss of marsh between 1852 and 1968. As sea-level rise accelerates as a result of global warming, almost all mid-lagoon marsh will be lost.

Rises in sea level will increase nearshore water depths and thereby modify wave refraction patterns. This means that wave energy amounts will also change at different points along a particular shoreline. Pethick (1993) maintains that this could be significant for the classic Scolt Head Island salt marshes of Norfolk, which are at present within a low to medium wave energy zone. After a 1.0 metre rise in sea level these marshes will experience high wave energy because of the migration of wave foci. Pethick remarks (p. 166):

The result will be to force the longshore migration of salt marsh and mudflat systems over distances of up to ten kilometres in 50 years – a yearly migration rate of 200 metres. It is doubtful whether salt marsh vegetation could survive in such a transitory environment – although intertidal mudflat organisms may be competent to do so – and a reduction or total loss of these open coast wetlands may result.

One particular type of marsh that may be affected by anthropogenically accelerated sea-level rise is the mangrove swamp. As with other types of marsh the exact response will depend on the local setting, sources and rates of sediment supply, and the rate of sea-level rise itself. However, mangroves may respond rather differently from other marshes in that they are composed of relatively long-lived trees and shrubs, which means that the speed of zonation change will be less (Woodroffe, 1990).

The degree of disruption is likely to be greatest in microtidal areas, where any rise in sea level represents a larger proportion of the total tidal range than in macrotidal areas. The setting of mangrove swamps will be very important in determining how they respond. River-dominated systems with large allochthonous sediment supply will have faster rates of shoreline progradation and deltaic plain accretion and so may be able to keep pace with relatively rapid rates of sea-level rise. By contrast in reef settings, in which sedimentation is primarily autochthonous, mangrove surfaces are less likely to be able to keep up with sea-level rises.

This is the view of Ellison and Stoddart (1990), who argued that low island mangrove ecosystems (mangals) have in the past been able to keep up with a sea-level rise of up to 8–9 centimetres per century, but that at rates over 12 centimetres per century they had not been able to persist. On this basis they concluded (p. 161):

the predicted possible rates of greenhouse-induced sea-level rise of 100–200 cm/100 years make it inevitable that most mangals will collapse as viable coastal ecosystems. This implies that mangrove ecosystems of low islands will be more vulnerable to rising sea-level than those of high islands and continental shores.

The ability of mangrove propagules to take root and become established in intertidal areas subjected to a higher mean sea-level is in part dependent on species. In general, the large propagule species (for example, *Rhizophora* spp.) can become established in rather deeper water than can the smaller propagule species (for example, *Avicennia* spp.). The latter has aerial roots which project only vertically above tidal muds for short distances (Snedaker, 1993).

Mangrove colonization and migration would also be influenced by salinity conditions so that any speculations above mangrove response to sea-level rise must also incorporate allowance for changes in rainfall and fresh-water runoff.

In arid areas, such as the Middle East, extensive tracts of coastline are fringed by low-level salt plains called *sabkhas*. These features are generally regarded as equilibrium forms that are produced by depositional processes (e.g. alluvial siltation, aeolian inputs, evaporite formation, faecal pellet deposition) and plantation processes (e.g. wind erosion and storm surge effects). They tend to occur at or about high tide level. Because of the range of depositional processes involved in their development they might be able to adjust to a rising sea level, but quantitative data on present and past rates of accretion are sparse. A large proportion

Table 8.6 Some global influences on state of coral reefs

Increase in sea surface temperatures
Will cause stress (bleaching etc.) in some areas
Will stimulate growth in some areas

Increase in storm frequency and intensity
Will build up islands by throwing up coral debris
Will erode reefs
Will cause change in species composition

Increase in sea levels
Will stimulate reef growth (if slow)
Will cause inundation and cause corals to give up (if fast)
Or if corals are stressed (e.g. by pollution, increased temperatures, ultraviolet
    effects, etc.)

---

of the industrial and urban infrastructure of the United Arab
Emirates is located on or in close proximity to *sabkhas*.

Another potentially susceptible environment is the coral reef
(table 8.6) (Stoddart, 1990; Spencer, 1995). In the 1980s there
were widespread fears that if rates of sea-level rise were high
(perhaps 2–3 metres or more by 2100) they would submerge
whole atoll systems (e.g. the Maldives). However, with the reduced expectations for the degree of sea-level rise that may occur,
there has arisen a belief that coral reefs may survive and even
prosper with moderate rates of sea-level rise. Kinsey and Hopley
(1991), for instance, believe that reef growth will be stimulated
by the rising sea levels of a warmer world and that reef production could double in the next 100 years from around 900 million
to 1,800 million tonnes per year. On the other hand, Buddemeier
and Smith (1988) took a rather pessimistic view of the response
of reefs to a rate of sea-level rise of 15 millimetres per year over
the next century, believing that sustained maxima of reef growth
were generally less than 10 millimetres per year.

However, reef accretion is not the sole response of reefs to
sea-level rises, for reef tops are frequently surmounted by small
islands (cays and motus) composed of clastic debris. Such islands
might be very susceptible to sea-level rise. On the other hand,
were warmer seas to produce more storms, then the deposition
of large amounts of very coarse debris could in some circumstances lead to their enhanced development.

Increased sea-surface temperatures could have deleterious consequences for corals which are near their thermal maximum of
tolerance, and increased temperatures in recent years have been
identified as one potential or actual cause of a phenomenon
called coral bleaching. This involves the loss of symbiotic
zooxanthellae or a decrease in the photosynthetic pigmentation
in these zooxanthellae. This in turn has an adverse effect on the
coral host since these photosynthetic symbionts typically supply
about two-thirds of the coral's nutrients and also facilitate

calcification. Those corals thus stressed by a rise in temperature or by various types of pollution, might well find it more difficult to cope with rapidly rising sea levels than would healthy corals. Moreover, it is possible that increased ultraviolet radiation due to ozone depletion could aggravate bleaching and mortality caused by global warming. Various studies have addressed the issue of coral bleaching (e.g. Brown, 1990), and some have demonstrated an increasing incidence of bleaching events during the warm years of the 1980s and 1990s (e.g. Glynn, 1996).

Indeed, Goreau and Hayes (1994) have produced maps of global coral bleaching episodes between 1983 and 1991 and have related them to maps of sea surface temperatures over that period. They find that areas of severe bleaching are related to what they describe as ocean 'hot spots' where marked positive temperature anomalies exist. They argue that coral reefs are ecosystems that may be uniquely prone to the effects of global warming:

If global warming continues almost all ecosystems can be replaced by migration of species from lower latitudes, except for the warmest ecosystems. These have no source of immigrants already adapted to warmer conditions. Their species must evolve new environmental tolerances if their descendants are to survive, a much slower process than migrations.

Bird (1986: 84) has summarized the general effects of accelerated coastal submergence as follows:

- On cliffed coasts submergence is likely to accelerate coastline recession, except on outcrops of hard rock formations, where the high and low tide lines will simply move up the cliff face. Existing shore platforms and abrasion ramps will disappear beneath the sea.
- The shores of deltas and coastal plains will retreat, except where they are maintained by coastal sedimentation.
- Beaches will be narrowed, and beach erosion will become much more extensive and severe than it is now.
- Inlets, embayments and estuaries will be enlarged and deepened, and increasing salinity penetration will cause a regression of coastal ecosystems: where possible, mangrove and salt marsh communities will move back into terrain presently occupied by freshwater vegetation.
- Coastal lagoons will also become larger and deeper, but the enclosing barriers may transgress landward on to them. If the barriers are submerged, or destroyed by erosion, the lagoons will become coastal inlets or embayments.
- Low-lying areas on coastal plains, such as *sabkhas* (saline depressions now subject to occasional marine flooding) on arid coasts, will be flooded to form permanent lagoons.

Plate 8.3
Increasing numbers of humans inhabit erodible, low-lying coastal areas. Such areas may become increasingly subject to coast erosion and inundation if sea level rises. The fragile barrier island systems of the mid-Atlantic coast of the USA, shown here after a severe winter storm on 7 March 1962, are good examples of sensitive and susceptible systems

Figure 8.13
The so-called Bruun Rule whereby a rise in sea-level causes beach erosion. If the sea level (s) rises by a unit amount so will the offshore bottom (s'). The sand necessary to raise the bottom (area b') can be supplied by waves eroding the upper part of the beach (area b). R is the amount of shoreline recession and W is the 'active' portion of the profile participating in the adjustment (from Committee on Engineering Implications of Changes in Relative Mean Sea Level, 1987, p. 54, figure 5-3)

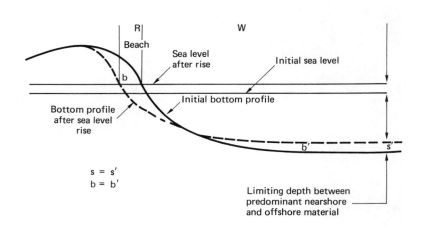

- Upward growth of coral and associated organisms will be stimulated on fringing biogenic reefs, keeping pace with the marine transgression or lagging somewhat behind it.
- Erosion, structural damage, and marine flooding caused by storm surges or tsunamis will intensify because of the greater heights of the waves arriving through deepening coastal waters.
- Water tables will rise in coastal regions, and soil and water salinity will be augmented.

Sandy beaches are held to be sensitive because of the so-called Bruun rule. The Bruun rule (Bruun, 1962) predicts future rates of coastal erosion in response to rising sea-level. Bruun envisaged a profile of equilibrium (figure 8.13) in which the volume of material removed during shoreline retreat is transferred on to the adjacent shoreface/inner shelf, thus maintaining the original bottom profile and near-shore shallow conditions. With a rise in sea level, additional sediment has to be added to the below water portion of the beach profile. One source of such material is beach erosion, and estimates of beach erosion of about 100 metres for every 1 metre rise in sea-level have been postulated (Committee on Engineering Implications of Changes in Relative Mean Sea Level, 1987). However, although the concept is intuitively appealing, it is also difficult to confirm or quantify without precise bathymetric surveys and integration of complex near-shore profiles over a long period of time. Moreover, an appreciable time lag may occur in shoreline response, which is highly dependent upon local storm frequency. Furthermore, the model is essentially a two-dimensional one in which the role of longshore sediment movement is not considered. It is also assumed that no substantial offshore leakage of sediment occurs. Accurate determination of sediments budgets in three dimensions is still replete with problems. This is an area where much work is required.

Whatever the problems of modelling, however, sandy beaches will tend to disappear from locations where they are already narrow and backed by high ground or swamp and marsh, but will probably tend to persist where they can retreat across wide beach ridge plains. This has particular implications for some major seaside resorts.

Future uncertainty

In almost all attempts to predict the future major uncertainties arise. One can provide various explanations for this. First, environmental change is the result of very complex interactions between several closely coupled non-linear systems. The complexity creates problems for both modelling and comprehension, while the non-linearity means that the dimensions of a response are not by any means necessarily directly proportional to the size of the stimulus that promotes change. Secondly, prediction of environmental change depends on models and many models are imperfect because of the gross assumptions they employ (see the discussion of General Circulation Models for an exemplification of this problem). Thirdly, nature always has surprises in store (e.g. various types of extreme event or catastrophe) and these will not cease in the future. They may work to counteract or increase the consequences of various human actions. Fourthly, there are bound to be factors that we have ignored, which turn out to be highly important. For example, who several decades ago could have predicted the role of CFCs in increasing the ozone hole? Fifthly, identification of future trends requires some knowledge of background or natural levels. Often we lack the necessary long-term data to enable us to ascertain whether observed trends have happened before and whether they are or are not cyclical. Sixthly, some key issues are less easy to predict than others and therefore preclude accurate prediction of phenomena that depend on them. For example, without a clear idea of precipitation patterns in a warmer world it is well-nigh impossible to predict the response of rivers and biota. Seventhly, we may often be able to predict that change will probably occur, but we find it much less easy to predict the speed of response. Ice caps and glaciers may well melt in a warmer world, but how fast? Finally, there are problems of definition. Without clear definitions of phenomena, measurement is difficult, 'results' can be meaningless and trends or changes difficult to identify. We find it difficult to define such phenomena as desertification and deforestation.

Thus there is considerable uncertainty surrounding many of the potential effects of global climatic change and many examples of this uncertainty have been given in this chapter (e.g. with respect to coral reefs, tropical cyclone frequency and the future of the great ice caps). However, such uncertainty should not engender complacency: the uncertainties may be large, but the risks are huge. Some of the potential change may not only cause great changes to the physical landscape but also create a series of environmental hazards and costs that have grave human implications.

Future responses

There is now very great interest in how we might adapt to global warming, should it occur. Such adaptations would be necessary if we could not limit emissions of greenhouse gases sufficiently to rule out the possibility of significant warming.

It is often said that there are two types of adaptation that may be necessary. The first of these is *reactive adaptation*, whereby we respond to climatic change after it occurs. The second is *anticipatory adaptation*, in which we take steps in advance of climatic change to minimize any potentially negative effects or to increase our ability to adapt to changes rapidly and inexpensively.

Reactive adaptation may well be feasible and effective. In many parts of the world we may well be able to adapt to the most likely ways in which the climate may change. For example, we could substitute heat- and drought-resistant crops for those whose yields are reduced. Infrastructure is generally replaced on a much faster timescale than climatic change; so it could be adapted to changes in climate. It can also be argued in favour of reactive adaptation that it does not involve prematurely spending money in advance of uncertain changes.

On the other hand, one can argue that rapid climate change, or significant increases in the intensity and frequency of extreme events such as floods, storms or droughts, could make reactive adaptations difficult and could pose immediate problems for large numbers of people. Equally, some policies would have significant benefits even under current environmental conditions and would be valuable from a cost-benefit perspective even if no climatic change took place. These types of anticipatory policies are often called 'no regrets policies' (table 8.7) because they will succeed whether or not climatic change takes place, meaning that policymakers should never have to regret their adoption. 'No regrets' policies may, none the less, be expensive. However, rather than adapting to climatic change, there is much to be said for reducing the emissions of greenhouse gases. The IPCC (Watson et al., 1996) summarized some of the options: implementing low-cost measures, such as energy conservation and

Table 8.7 Examples of 'no-regrets' policies: In response to possible global warming

Policy area and measures	Benefits
Coastal zone management	
Wetland preservation and migrations	Maintains healthy wetlands which are more likely to have higher value than artificially created replacements. Maintains existing coastal fisheries that are difficult to relocate
Integrated development of coastal datasets	Integrated data allow formation of comprehensive planning and identification of regions most likely to be affected by physical or social changes. Allows effects of changes to be examined beyond the local or regional scale
Improved development of coastal models	Improved modelling allows more accurate evaluation of how coastal systems respond to climate change and also to other shocks
Land-use planning	Sensible land-use planning, such as the use of land setbacks to control shoreline development, better preserves the landscape and also minimizes the concerns of beach erosion from any cause
Water resources	
Conservation	Reducing demand can increase excess supply, giving more safety margin for future droughts. Using efficient technologies such as drip irrigation reduces demand to some extent. Preserving some flexibility of demand is useful as less valuable uses allow reduced demand during droughts.
Market allocation	Market-based allocation allows water to be diverted to its most efficient uses, in contrast with non-market mechanisms that can result in wasteful uses. Market allocations are able to respond more rapidly to changing supply conditions, and also tend to lower demand, conserving water
Pollution control	Improving water quality by improving the quality of incoming emissions provides greater water quality safety margins during droughts and makes water supply systems less vulnerable to declines in quality because of climate change
River basin planning	Comprehensive planning across a river basin can allow for imposition of cost-effective solutions to water quality and water supply problems. Planning can also help cope with population growth and changes in supply and demand from many causes, including climate change
Drought contingency planning	Plans for short-term measures to adapt to droughts. These measures would help offset droughts of known or greater intensity and duration
Human health	
Weather/health watch warning systems	Warning systems to notify people of heat stress conditions or other dangerous weather situations will allow people to take necessary precautions. This can reduce heat stress and other types of fatalities both now and if heat waves become more severe
Improved public health and pest management procedures	Many diseases which will spread if climate changes are curable or controllable, and efforts in these areas will raise the quality of human life both now and if climate change occurs
Improved surveillance systems	More and better data on the incidence and spread of diseases are necessary to better determine the future patterns of infection and disease spread. This information is helpful under any scenario
Ecosystems	
Protect biodiversity and nature	Biodiversity protection maintains ecological diversity, and richness preserves variety in genotypes for medical and other research. A more diverse gene pool provides more candidates for successful adaption to climate change. One possibility is to preserve endangered species outside of their natural habitat, such as in zoos

Table 8.7 *(cont'd)*

Policy area and measures	Benefits
Protect and enhance migration corridors	Such policies help maintain an ecosystem and animal and tree species diversity. Corridors and buffer zones around current reserve areas that include different altitudes and ecosystems are more likely to withstand climate change by increasing the likelihood of successful animal and tree migration
Watershed protection	Forest cover provides watershed protection, including protection from bank erosion, siltation, and soil loss. All of these functions are extremely valuable whether climate changes occur or not
Agriculture	
Irrigation efficiency	Many improvements are possible and efficient from a cost–benefit standpoint. Improvements allow greater flexibility to future change by reducing water consumption without reducing crop yields
Development of new crop types	Development of more and better heat- and drought-resistant crops will help alleviate current and future world food demand by enabling production in marginal areas to expand. Improvements will be critical, as world population continues to increase, with or without climate change

Source: After Smith et al. (1995), table 3.

energy efficiency; phasing out existing distortionary policies, such as some fossil-fuel subsidies which increase emissions; switching from more to less carbon-intensive fuels or to carbon-free fuels; enhancing or expanding greenhouse gas sinks or reservoirs, such as forests; and reducing methane and nitrous oxide emissions from industrial processes, landfills, agriculture, fossil-fuel extraction and transport.

One can argue that the central challenge for policymakers in coming decades will be to find ways of allowing the global economy to grow at a moderate rate while at the same time maintaining or enhancing the protection of wilderness, the prevention of pollution and the sustenance of ecological resources. We cannot be sure that we will find policies that enable this to happen. Governments and societies will inevitably need to make difficult trade-offs between economic growth and environmental protection. We cannot envisage a situation where there is indefinite growth in the human population and indefinite growth in the consumption of resources. We need to ensure, to use Sir Crispin Tickell's phrase (Tickell, 1993), that humans are not 'a suicidal success'.

Conclusion

9

The power of non-industrial and pre-industrial civilizations

Thus far we have looked at the impact that human societies have had on the different components of the physical environment: vegetation, animals, soils, water, landforms and climate. It has become apparent that Marsh was correct over a century ago to express his cogently argued views of the importance of human agency in environmental change. Since his time the impact that humans have had on the environment has increased, as has our awareness of this impact. However, it is worth making the point here that, although much of the concern expressed about the undesirable effects humans have tends to focus on the role played by sophisticated industrial societies, this should not blind us to the fact that many highly significant environmental changes were and are being achieved by non-industrial societies. In recent years it has become apparent that fire, in particular, enabled early societies to alter vegetation substantially, so that plant assemblages that were once thought to be natural climatic climaxes may in reality be in part anthropogenic fire climaxes. This applies to many areas of both savanna and mid-latitude grassland (see p. 62). Such alteration of natural vegetation has been shown to pre-date the arrival of European settlers in the Americas (Denevan, 1992), New Zealand and elsewhere, and the effects of fire may have been compounded by the use of the stone axe and by the grazing effects of domestic animals. In turn, the removal and modification of vegetation would have led to adjustment in fauna. It is also apparent that soil erosion resulting from vegetation removal has a long history and that it was regarded as a threat by the classical authors. Recent studies (see p. 80) tend to suggest that some of the major environmental changes in highland Britain and similar parts of Western Europe that were once explained by climatic changes can be more effectively explained by the activities of Mesolithic and Neolithic peoples. This applies, for example, to the decline in the numbers of certain plants in the pollen record and to the development of peat bogs and podzolization (see p. 175). Even soil salinization may have started at an early date because of the adoption of irrigation practices in arid areas, and its effects on crop yields were noted in Iraq more than 4,000 years ago (see p. 168). Similarly (see p. 142), there is an increasing body of evidence that the hunting practices of early human groups may have caused great changes in the world's megafauna as early as 11,000 years ago.

In spite of the increasing pace of world industrialization and urbanization, it is ploughing and pastoralism which are responsible for many of our most serious environmental problems and which are still causing some of our most widespread changes in the landscape. Thus soil erosion brought about by agriculture is, it can be argued, a more serious pollutant of the world's waters than is industry: many of the habitat changes which so effect wild animals are brought about through agricultural expansion (see p. 130); and soil salinization and desertification can be regarded as two of the most serious problems facing the human race. Land-use and land cover changes, such as the conversion of forests to fields, may be as effective in causing anthropogenic changes in climate as the more celebrated burning of fossil fuels and emission of industrial aerosols into the atmosphere. The liberation of CO_2 in the atmosphere through agricultural expansion, changes in surface albedo values and the production of dust, are all major ways in which agriculture may modify world climates.

The proliferation of impacts

A further point we can make is that, with developments in technology, the number of ways in which humans are affecting the environment is proliferating. It is these recent changes, because of the uncertainty which surrounds them and the limited amount of experience we have of their potential effects, which have caused greatest concern. Thus it is only since the Second World War, for example, that humans have had nuclear reactors for electricity generation, that they have used powerful pesticides such as DDT, and that they have sent supersonic aircraft into the stratosphere. Likewise, it is only since around the turn of the century that the world's oil resources have been extensively exploited, that chemical fertilizers have become widely used, and that the internal combustion engine has revolutionized the scale and speed of transport and communications.

Above all, however, the complexity, frequency and magnitude of impacts is increasing, partly because of steeply rising population levels and partly because of a general increase in per

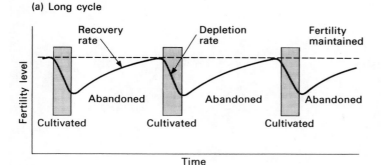

(a) Long cycle

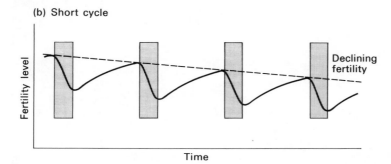

(b) Short cycle

Figure 9.1
Land rotation and population
density. The relationship of soil
fertility cycles to cycles of
slash-and-burn agriculture:
(a) fertility levels are
    maintained under the long
    cycles characteristic of
    low-density populations
(b) fertility levels are declining
    under the shorter cycles
    characteristic of increasing
    population density. Notice
    that in both diagrams the
    curves of both depletion
    and recovery have the
    same slope (after Haggett,
    1979, figure 8.4)

capita consumption. Thus some traditional methods of land use such as shifting agriculture (see p. 58) and nomadism, which have been thought to sustain some sort of environmental equilibrium, seem to break down and to cause environmental deterioration when population pressures exceed a particular threshold.

This is illustrated for shifting agriculture systems by figure 9.1, which shows the relationship of soil fertility levels to cycles of slash-and-burn agriculture. Fertility can be maintained (figure 9.1a) under the long cycles characteristic of low-density populations. However, as population levels increase, the cycles necessarily become shorter, and soil fertility levels are not maintained, thereby imposing greater stresses on the land (figure 9.1b).

At the other end of the spectrum, increasing incomes, leisure and ease of communication have generated a stronger demand for recreation and tourism in the developed nations (plate 9.1). These have created additional environmental problems (see p. 103), especially in coastal and mountain areas (see e.g. Bayfield, 1973). Some of the environmental consequences of recreation, which are reviewed at length by Liddle (1997), can be listed as follows:

1   desecration of cave formations by speleologists;
2   trampling by human feet leading to soil compaction;

Plate 9.1
The impact of recreation pressures is well displayed at a prehistoric hill-fort, Badbury Rings, Dorset, England. Pedestrians and motorcyclists have caused severe erosion of the ramparts

3   nutrient additions at campsites by people and their pets;
4   decrease in soil temperatures because of snow compaction by snowmobiles;
5   footpath erosion and off-road vehicle erosion;
6   dune reactivation by trampling;
7   vegetation change due to trampling and collecting;
8   creating of new habitats by cutting trails and clearing campsites;
9   pollution of lakes and inland waterways by gasoline discharge from outboard motors and by human waste;
10 creation of game reserves and protection of ancient domestic breeds;
11 disturbance of wildlife by proximity of persons and by hunting, fishing and shooting;
12 conservation of woodland for pheasant shooting.

Likewise, it is apparent when considering the range of possible impacts of one major type of industrial development that they are significant. As table 9.1 indicates, the exploitation of an oil-field and all the activities that it involves (for example, pipelines,

Table 9.1 Qualitative environmental impacts of mineral industries with particular reference to an oilfield

| Facility | Direction of the impact and reaction of the environment | | | |
	Land	Air	Water	Biocenosis
Well	Alienation of land surface Extraction of oil associated gas, ground water Pollution by crude oil, refined products, drilling mud Disturbance of internal structure of soil and subsoil Destruction of soil	Pollution by associated gas and volatile hydrocarbons, products of combustion	Withdrawal of surface water and ground water Pollution by crude oil and refined products, salination of fresh water Disturbance of water balance of both subsurface and surface waters	Pollution by crude oil and refined products Disturbance and destruction over a limited surface area
Pipeline	Alienation of land Accidental oil spills Disturbance of landforms and internal structure of soil and subsoil	Pollution by volatile hydrocarbons	Disturbance and destruction over a limited surface area	
Motor roads	Alienation of land Pollution by oil products Disturbance of landforms and internal structure of soil and subsoil	Pollution by combustion products, volatile hydrocarbons, sulphur dioxide, nitrogen oxides	Pollution by combustion products Disturbances and destruction over limited surface area	
Collection point	Alienation of land Pollution by crude oil and refined products (spills) Disturbance of internal structure of soil and subsoil	Pollution by volatile hydrocarbons	Disturbance and destruction over limited surface area	

Source: Denisova, 1977, table 2, p. 650.

new roads, refineries, drilling, etc.) have a wide range of likely effects on land, air, water and organisms.

Conversely, if one takes one ecosystem type as an example – the coral reef – one can see the diversity of stresses to which they are now being exposed (figure 9.2) as a result of a whole range of different human activities.

A very substantial amount of change has been achieved in recent decades. Table 9.2, based on the work of Kates et al. (1991), attempts to make quantitative comparisons of the human impact on ten 'component indicators of the biosphere'. For each component they defined total net change clearly induced by humans to be 0 per cent for 10,000 years ago and 100 per cent for 1985. They then estimated dates by which each component had reached successive quartiles (i.e. 5, 50 and 75 per cent) of its 1985 total change. They believe that about half of the components have changed more in the single generation since 1950 than in the whole of human history before that date.

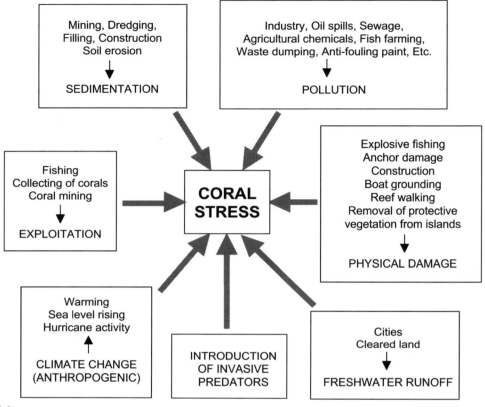

Figure 9.2
Some causes of anthropogenic
stress on coral reef ecosystems

Table 9.2 Chronologies of human-induced transformations

(a) Quartiles of change from 10,000 BC to mid 1980s

	Dates of quartiles		
Form of transformation	25%	50%	75%
Deforested area	1700	1850	1915
Terrestrial vertebrate diversity	1790	1880	1910
Water withdrawals	1925	1955	1975
Population size	1850	1950	1970
Carbon releases	1815	1920	1960
Sulphur releases	1940	1960	1970
Phosphorus releases	1955	1975	1980
Nitrogen releases	1970	1975	1980
Lead releases	1920	1950	1965
Carbon tetrachloride production	1950	1960	1970

Table 9.2 *(cont'd)*

(b) % change by the time of Marsh and Princeton symposium

	% change	
Form of transformation	*1860*	*1950*
Deforested area	50	90
Terrestrial vertebrate diversity	25–50	75–100
Water withdrawals	15	40
Population size	30	50
Carbon releases	30	65
Sulphur releases	5	40
Phosphorus releases	<1	20
Nitrogen releases	<1	5
Lead releases	5	50
Carbon tetrachloride production	0	25

Source: from Kates et al., 1991, table 1.3.

Changes are reversible

It is evident that while humans have imposed many undesirable and often unexpected changes on the environment, they often have the capacity to modify the rate of such changes or to reverse them. There are cases where this is not possible: once soil has been eroded from an area it cannot be restored; once a plant or animal has become extinct it cannot be brought back; and once a laterite iron pan has become established it is difficult to destroy.

However, through the work of George Perkins Marsh and others, people became aware that many of the changes that had been set in train needed to be reversed or reduced in degree. Sometimes this has simply involved discontinuing a practice which has proved undesirable (such as the cavalier use of DDT), or replacing it with another which is less detrimental in its effect. Often, however, specific measures have been taken which have involved deliberate decisions of management and conservation. Denson (1970), for example, outlines a sophisticated six-stage model for wildlife conservation:

1 immediate physical protection from humans and from changes in the environment;
2 educational efforts to awaken the public to the need for protection and to gain acceptance of protective measures;
3 life-history studies of the species to determine their habitat requirements and the causes of their population decline;
4 dispersion of the stock to prevent loss of the species by disease or by a chance event such as fire;
5 captive breeding of the species to assure higher survival of young, to aid research and to reduce the chances of catastrophic loss;

6   habitat restoration or rehabilitation when this is necessary before introducing the species.

Many conservation measures have been successful, while others have created as many problems as they were intended to solve. This applies, for example, to certain schemes for the reduction of coast erosion.

The concern with preservation and conservation has been long-standing, with many important landmarks. Interest has grown dramatically in recent years. Lowe (1983) has identified four stages in the history of British Nature Conservation:

1   the natural history/humanitarium period   (1830–90)
2   the preservation period   (1870–1940)
3   the scientific period   (1910–70)
4   the popular/political period   (1960–present)

The first of these stages was rooted in a strong enthusiasm for natural history, and the crusade against cruelty to animals. Although many Victorian naturalists were avid collectors, numerous clubs were established to study nature and some of them sought to preserve species to make them available for observation. As we shall see, certain acts were introduced at this time to protect birds. During the preservationist period, there was the formation of a spate of societies devoted to preserving open land and its associated wildlife (for example, the National Trust, 1894; and the Council for the Preservation of Rural England, 1926). There was a growing sense of vulnerability of wildlife and landscapes to urban and industrial expansion, and geographers like Vaughan Cornish (see Goudie, 1972b) campaigned for the creation of national parks and the preservation of scenery, made possible through the National Parks and Access to Countryside Act of 1949. From the First World War onwards ecological research developed, and there arose an increasing understanding of ecological relationships. Scientists pressed for the regulation of habitats and species, and the Nature Conservancy was established in 1949. In the 1960s and the years that followed popular interest in conservation and widespread media attention first developed. This was partly generated by pollution incidents (such as the wrecks of the *Torrey Canyon* and *Amoco Cadiz*), and a gathering sense of impending environmental doom. Ecology became a political issue in various European nations, including the UK. In many countries major developments in land use, construction and industrialization now have to be preceded by the production of an Environmental Impact Assessment.

Thus in some countries, and in connection with particular species, conservation and protection have had a long and sometimes successful impact. In Britain, for example, the Wild Birds

Protection Act dates back to 1880, and the Sea Birds Protection Act even further, to 1869. The various acts have been modified and augmented over the years to outlaw egg-collecting, pole-trapping, plumage importation and the capture or possession of a range of species. The effectiveness of the different acts can be measured in real terms. Over the last 60 years no species of British breeding birds have been lost due to lack of protection – the only major loss has been the Kentish plover which was in any case on the edge of its range. Perhaps more importantly, several species have successfully recolonized Britain, the most celebrated being the avocet and the osprey. Today both are firmly established, together with other species lost in the nineteenth century: the black-tailed godwit, the goshawk and the bittern. Also as a result of protection the red kites of Wales have not only survived but increased in number, and the peregrine falcon maintains its largest numbers in Europe outside Spain.

One further ground which gives some basis for hope that humans may soon be reconciled with the environment is that there are some signs of a widespread shift in public attitudes to nature and the environment. These changing social values, combined with scientific facts, influence political action. This point of view, which acts as an antidote for some of the more pessimistic views of the world's future, was elegantly presented by Ashby (1978). He contended that the rudiments of a healthy environmental ethic are developing, and explained (pp. 84–5):

Its premise is that respect for nature is more moral than lack of respect for nature. Its logic is to put the Teesdale Sandwort . . . into the same category of value as a piece of Ming porcelain, the Yosemite Valley in the same category as Chartres Cathedral: a Suffolk landscape in the same category as a painting of the landscape by Constable. Its justification for preserving these and similar things is that they are unique, or irreplaceable, or simply part of the fabric of nature, just as Chartres and the painting by Constable are part of the fabric of civilisation; also that we do not understand how they have acquired their durability and what all the consequences would be if we destroy them.

Although there may be considerable controversy surrounding the precise criteria that can be used to select and manage sites that are particularly worthy of conservation (Goldsmith, 1983), there are none the less many motives behind the increasing desire to protect species and landscapes. These can be listed under the following general headings:

1 *The ethical.* It is asserted that wild species have a right to coexist with us on our planet, and that we have no right to exterminate them. Nature, it is maintained, is not simply there for humans to transform and modify as they please for their own utilitarian ends.

2   *The scientific.* We know very little about our surrounding environments, including, for example, the rich insect faunas of the tropical rain forest; therefore such environments should be preserved for future scientific study.

3   *The aesthetic.* Plants and animals, together with landscapes, may be beautiful and so enrich the life of humans.

4   *The need to maintain genetic diversity.* By protecting species we maintain the species diversity upon which future plant- and animal-breeding work will depend. Once genes have been lost (see chapter 2, section on 'The change in genetic and species diversity') they cannot be replaced.

5   *Environmental stability.* It is argued that in general the more diverse an ecosystem is, the more checks and balances there are to maintain stability. Thus environments that have been greatly simplified by humans may be inherently unstable, and prone to disease, etc.

6   *Recreational.* Preserved habitats and landscapes have enormous recreational value, and in the case of some game reserves and national parks may have economic value as well (for example, the safari industry of East Africa).

7   *Economic.* Many of the species in the world are still little known, and there is the possibility that we have great storehouses of plants and animals, which, when knowledge improves, may become useful economic resources.

8   *Future generations.* One of the prime arguments for conservation, whether of beautiful countryside, rare species, soil or mineral resources, is that future generations (and possibly ourselves later in life) will require them, and may think badly of a generation that has squandered them.

9   *Unintended impacts.* As we have seen so often in this book, profligate or unwise actions can lead to side-effects and consequences that may be disadvantageous to humans.

Some of these arguments (for example, 4, 5, 6, 7 and 9) are more utilitarian than others, and some may be subject to doubt – it could, for example, be argued that future generations will have technology to use new resources and may not need some of those we regard as essential – but overall they provide a broadly based platform for the conservation ideal (Myers, 1979).

The susceptibility to change

Ecosystems respond in different ways to the human impact, and some are more vulnerable to human perturbation than others (Kasperson et al., 1995). It has often been thought, for example, that complex ecosystems are more stable than simple ones. Thus in Clements' Theory of Succession the tendency towards community stabilization was ascribed in part to an increasing level of integration of community functions. As Goodman (1975: 238) has expressed it:

In general the predisposition to expect greater stability of complex systems was probably a combined legacy of eighteenth century theories of political economics, aesthetically and perhaps religiously motivated attraction to the belief that the wondrous variety of nature must have some purpose in an orderly work, and ageless folkwisdom regarding eggs and baskets.

Indeed, as Murdoch (1975) has pointed out, it makes good intuitive sense that a system with many links, or 'multiple fail-safes', is more stable than one with few links or feedback loops. As an example, if a type of herbivore is attacked by several predatory species, the loss of any one of these species will be less likely to allow the herbivore to erupt or explode in numbers than if only one predator species were present and that single predator type disappeared.

Various other arguments have been marshalled to support the idea that great diversity and complexity affords greater ability to minimize the magnitude, duration and irreversibility of changes brought about by some external perturbation such as humans (Noy-Meir, 1974). It has been stated that natural systems, which are generally more diverse than artificial systems such as crops or laboratory populations, are also more stable. Likewise, the tropical rain forest has been thought of as more diverse and more stable than less complex temperate communities, while simple Arctic ecosystems or oceanic islands have always appeared highly vulnerable to disturbance brought about by anthropogenic plant and animal introductions (see p. 87).

However, considerable doubt has been expressed as to whether the classic concept of the causal linkage between diversity/complexity and stability is entirely valid (see e.g. Hurd et al., 1971). Murdoch (1975) indicates that there is no convincing field evidence that diverse natural communities are generally more stable than simple ones. He cites various papers which show that fluctuations of microtine rodents (lemmings, field voles, etc.) are as violent in relatively complex temperate zone ecosystems as they are in the less complex Arctic zone ecosystems. This is supported by Goodman (1975: 239), who writes:

As for the apparent stability of tropical biota, that could well be an illusion attributable to insufficient study of bewilderingly complex assemblages in which many species are so poorly represented in samples of feasible size that even considerable fluctuations might go undetected. Indeed, there are countervailing anecdotes regarding ecological instability in the tropics, such as recent reports on an insect virtually defoliating the wild Brazil-nut trees in Bolivia and of monkeys succumbing in large numbers to epidemics.

He goes on to add: 'There is growing awareness of the surprising susceptibility of the rain forest ecosystems to man-made perturbation.' This is a point of view supported by May (1979) and discussed by Hill (1975). Hill points out that a very high species diversity is frequently associated with areas which have relatively constant physical environmental conditions over the course of a year and a series of years. The rain forest may be constructed to be such an environment, and one where this constancy has allowed the presence of many specialized species, each pursuing a narrow range of activities. It has been argued that because of the high degree of specialization, the indigenous species have a limited ability to recover from major stresses caused by human intervention.

Goodman (1975) has also queried the sufficiency of the argument in its reference to the apparent instability of island ecosystems, suggesting that islands, being evolutionary backwaters and dead-ends, may accumulate species that are especially susceptible to competitive or exploitative displacement. In this case, lack of diversity may not necessarily be the sole or prime cause of instability.

The apparent instability of agricultural compared to natural communities is also often attributed to lack of diversity (see p. 103), and indeed modern agriculture does involve significant ecosystem simplification. However, such instability as there is may not, once again, necessarily result from simplification. Other factors could promote instability: agricultural communities are disrupted, even destroyed more frequently and more massively as part of the cultivation process than those natural systems we tend to think of as stable; the component species of natural systems are co-evolved (co-adapted), and this is not usually true of agricultural communities. As Murdoch (1975: 799) suggests, it may be that:

natural systems are more stable than crop systems because their interacting species have had a long shared evolutionary history. In contrast with these natural communities the dominant plant species of a crop system is thrust into an often alien landscape . . . the crops have undergone radical selection in breeding programmes, often losing their genetic defence mechanisms.

Thus the idea that complex natural ecosystems will be less susceptible to human interference and that simple artificial ecosystems will inevitably be unstable are not necessarily tenable. None the less, it is apparent that there are differences in susceptibility between different ecosystem types, and these differences may result from factors other than the degree of diversity and complexity (Cairns and Dickson, 1977).

Some systems tend to be *vulnerable*. Lakes, for example, are natural traps and sumps and are thus more vulnerable to the effect of disadvantageous inputs than are rivers (which are continually receiving new inputs) or oceans (which are so much larger). Other systems display the property of *elasticity* – the ability to recover from damage. This may be because nearby epicentres exist to provide organisms to reinvade a damaged system. Small, isolated systems will often tend to possess low elasticity (see p. 152). Two of the most important properties, however, are *resilience* (being a measure of the number of times a system can recover after stress), and *inertia* (the ability to resist displacement of structural and functional characteristics).

Two systems which display resilience and inertia are deserts and estuaries. In both cases their indigenous organisms are highly accustomed to variable environmental conditions. Thus most desert fauna and flora evolved in an environment where the normal pattern is one of more or less random alternations of short favourable periods and long stress periods. They have pre-adapted resilience (Noy-Meir, 1974) so that they can tolerate extreme conditions, have the ability for rapid recovery, have various delay and trigger mechanisms (in the case of plants) and have flexible and opportunistic eating habits (in the case of beasts). Estuaries, on the other hand, though the subject of increasing human pressures, also display some resilience. The vigour of their water circulation continuously and endogenously renews the supply of water, food, larvae, etc.; this aids recovery. Also, many species have biological characteristics which provide special advantages in estuarine survival (Cronin, 1967). These characteristics usually protect the species against the natural violence of estuaries and are often helpful in resisting external forces such as humans.

Human influence or nature?

From many of the examples given in this book it is apparent that in many cases of environmental change it is impossible to state, without risk of contradiction, that people rather than nature are responsible. Most systems are complex and human agency is but one component of them, so that many human actions can lead to end-products which are intrinsically similar to those that may be produced by natural forces. It is a case of equifinality, whereby different processes can lead to basically similar results. Humans are not always responsible for some of the changes with which they are credited. This book has given many examples of this problem and a selection is presented in table 9.3. Deciphering the cause is often a ticklish problem, given the intricate interdependence of different components of ecosystems, the frequency and complexity of environmental changes, and the varying relaxation times that different ecosystem

Table 9.3   Human influence or nature? Some examples, with page references to this book where applicable

| Change | Causes | | Page reference |
	Natural	Anthropogenic	
Late Pleistocene animal extinction	Climate	Hunting	142
Death of savanna trees	Soil salinization through climatically induced ground-water rise	Over-grazing	–
Desertification in semi-arid areas	Climatic change	Over-grazing, etc.	67
Holocene peat-bog development in highland Britain	Climatic change and progressive soil deterioration	Deforestation and ploughing	175
Holocene elm and linden decline	Climatic change	Feeding and stalling of animals	80
Tree encroachment into alpine pastures in USA	Temperature amelioration	Cessation of burning	–
Gully development	Climatic change	Land-use change	288
Early twentieth-century climatic warming	Changes in solar emission and volcanic activity	CO_2-generated greenhouse effect	328
Increasing coast recession	Rising sea-level	Disruption of sediment supply	311
Increasing coastal flood risk	Rising sea-level, natural subsidence	Pumping of aquifers creating subsidence	284
Increasing river flood intensity	Higher intensity rainfall	Creation of drainage ditches	225
Ground collapse	Karstic process	Dewatering by over-pumping	280
Forest decline	Drought	Air, soil and water pollution	94

components may have when subject to a new impulse. This problem plainly does not apply to the same extent to changes which have been brought about deliberately and knowingly by humans, but it does apply to the many cases where humans may have initiated change inadvertently and unintentionally.

This fundamental difficulty means that environmental impact statements of any kind are extremely difficult to make. As we have seen, humans have been living on the earth and modifying it in different degrees for several millions of years, so that it is problematical to reconstruct any picture of the environment before human intervention. We seldom have any clear baseline against which to measure changes brought about by human society. Moreover, even without human interference, the environment would be in a perpetual state of flux on a great many different time-scales (Goudie, 1992). In addition, as Wall and Wright (1977: 3) have demonstrated, there are spatial and temporal discontinuities between cause and effect. For example, erosion in one locality may lead to deposition in another, while destruction of key elements of an animal's habitat may lead to population declines throughout its range. Likewise, in a time context, a considerable interval may elapse before the full implications of an activity are apparent. Also, because of the complex interaction between different components of different environmental systems and subsystems it is almost impossible to measure total environmental impact. For example, changes in soil may lead to changes in vegetation, which in turn may trigger changes in water quality and in animal populations. Primary impacts give rise to a myriad successive repercussions throughout ecosystems which may be impracticable to trace and monitor. Quantitative cause-and-effect relationships can seldom be established.

Into the unknown

During the 1980s and 1990s the full significance of possible future environmental changes has become apparent, and national governments and international institutions have begun to ponder whether the world is entering a spasm of unparalleled humanly induced modification. Our models and predictions are still highly inadequate, and there are great ranges in some of the values we give for such crucial changes as sea-level rise and global climatic warming, but the balance of scientific argument favours the view that change will occur and that change will be substantial. Some of the changes may be advantageous for humans or for particular ecosystems; others will be extremely disadvantageous. But all change, if it is rapid and of a great magnitude, is likely to create uncertainties and instabilities. The study of future events will not only become a major concern for the environmental sciences but will also become a major concern for economists, sociologists, lawyers and political scientists. George Perkins Marsh was a lawyer and politician, but it is only now, over a century since he wrote *Man and nature*, that the wisdom, perspicacity and prescience of his ideas have begun to be given the praise and attention they deserve.

Guide to Reading

CHAPTER ONE

Freedman, B., 1995, *Environmental ecology* (2nd edn). San Diego: Academic Press. An enormously impressive and wide-ranging study with a strong ecological emphasis.

Goudie, A. S. (ed.), 1997, *The human impact reader: readings and case studies*. Oxford: Blackwell. A collection of key papers on many of the themes discussed in this book.

Goudie, A. S. and Viles, H., 1997, *The earth transformed*. Oxford: Blackwell. An introductory treatment of the human impact, with many case studies.

Mannion, A. M., 1995, *Agriculture and environmental change*. London: Wiley. A new and comprehensive study of the important role that agriculture plays in land transformation.

Meyer, W. B., 1996, *Human impact on the earth*. Cambridge: Cambridge University Press. A good point of entry to the literature that brims over with thought-provoking epigrams.

Middleton, N. J., 1999, *The global casino* (2nd edn). London: Edward Arnold. An introductory text, by a geographer, which is well illustrated and clearly written.

Pickering, K. T. and Owen, L. A., 1997, *Global environmental issues* (2nd edn). London: Routledge. A well-illustrated survey.

Ponting, C., 1991, *A green history of the world*. London: Penguin. An engaging and informative treatment of how humans have transformed the earth through time.

Simmons, I. G., 1996, *Changing the face of the earth: culture, environment and history* (2nd edn). Oxford: Blackwell. A characteristically amusing and perceptive review of many facets of the role of humans in transforming the earth, from an essentially historical perspective.

State of the Environment Advisory Council, 1996, *Australia: state of the environment 1996*. Collingwood, Victoria, Australia: CSIRO Publishing. A large compendium which discusses major environmental issues in the context of Australia.

Turner, B. L. II (ed.), 1990, *The earth as transformed by human action*. Cambridge: Cambridge University Press. A great analysis of global and regional changes over the past 300 years.

CHAPTER TWO

Crutzen, P. J. and Goldammer, J. G., 1993, *Fire in the environment.* Chichester: Wiley. A particularly useful study of fire's importance.

Grainger, A., 1990, *The threatening desert: controlling desertification.* London: Earthscan. A very readable and wide-ranging review of desertification.

Grainger, A., 1992, *Controlling tropical deforestation.* London: Earthscan. An up-to-date introduction with a global perspective.

Holzner, W., Werger, M. J. A., Werger, I. and Ikusima, I. (eds), 1983, *Man's impact on vegetation.* Hague: Junk. A wide-ranging edited work with examples from many parts of the world.

Meyer, W. B. and Turner, B. L. II (eds), 1994, *Changes in land use and land cover: a global perspective.* Cambridge: Cambridge University Press. An excellent edited survey of human transformation of the biosphere.

Simmons, I. G., 1979, *Biogeography: natural and cultural.* London: Edward Arnold. A useful summary of natural and human impacts on ecosystems.

CHAPTER THREE

Clutton-Brock, J., 1987, *A natural history of domesticated mammals.* London: British Museum/Cambridge: Cambridge University Press. A global survey of the domestication of a wide range of animals from asses to zebu cattle.

Drake, J. A. et al. (eds), 1989, *Biological invasions: a global perspective.* Chichester: Wiley. A collection of essays that consider both animal and plant invaders.

Martin, P. S. and Klein, R. G., 1984, *Pleistocene extinctions.* Tucson: University of Arizona Press. An enormous survey of whether or not late Pleistocene extinctions were caused by humans.

Wilson, E. O., 1992, *The diversity of life.* Cambridge, MA: Harvard/Belknap. A beautifully written and highly readable discussion of biodiversity.

CHAPTER FOUR

Johnson, D. L. and Lewis, L. A., 1995, *Land degradation: creation and destruction.* Oxford: Blackwell. A broadly based study of intentional and unintentional causes of many aspects of land degradation.

McTainsh, G. and Boughton, W. C., 1993, *Land degradation processes in Australia.* Melbourne: Longman Cheshire. An Australian perspective on soil modification.

Morgan, R. P. C., 1995, *Soil erosion and conservation* (2nd edn). Harlow: Longman. A revised edition of a fundamental work.

Russell, J. S. and Isbell, F. R. (eds), 1986, *Australian soils: the human impact.* St Lucia: University of Queensland Press. A detailed collection of studies on land degradation in Australia.

CHAPTER FIVE

Clark, R. B., 1997, *Marine pollution* (4th edn). Oxford: Clarendon Press. The standard work on this important topic.

Gleick, P. H. (ed.), 1993, *Water in crisis: a guide to the world's freshwater resources*. New York: Oxford University Press. A compendium of information on trends in water quality and consumption.

Petts, G. E., 1985, *Impounded rivers: perspectives for ecological management*. Chichester: Wiley. A treatment of the many consequences of dam construction.

CHAPTER SIX

Brookes, A., 1988, *Channelised rivers*. Chichester: Wiley. An advanced research monograph with broad scope.

Goudie, A. S., 1993, Human influence in geomorphology. *Geomorphology*, 7, 37–59. A general review, with a concern with the future, in a major journal.

Nir, D., 1983, *Man, a geomorphological agent: an introduction to anthropic geomorphology*. Jerusalem: Keter. A general survey that was ahead of its time.

CHAPTER SEVEN

Elsom, D., 1996, *Smog alert: managing urban air quality*. London: Earthscan. A very readable and informative guide.

Graedel, T. E. and Crutzen, P. J., 1993, *Atmospheric change: an earth system perspective*. San Francisco: Freeman. A well-illustrated study of changes, past, present and future.

Houghton, J. T., 1997, *Global warming: the complete briefing* (2nd edn). Cambridge: Cambridge University Press. An introduction by a leading scientist.

Houghton, J. T., Meira Filho, L. G., Callander, B. A., Harris, N., Kaltenberg, A. and Maskell, K. (eds), 1996, *Climate change 1995: the science of the climate change*. Cambridge: Cambridge University Press. A report from the Intergovernmental Panel on Climate Change on the causes and trends of climate change.

Kemp, D. D., 1994, *Global environmental issues: a climatological approach* (2nd edn). London: Routledge. An introductory survey.

Williams, M. A. J. and Balling, R. C., 1996, *Interactions of desertification and climate*. London: Edward Arnold. A study of land cover changes and their effects.

CHAPTER EIGHT

Arnell, N., 1996, *Global warming, river flows and water resources*. Chichester: Wiley. A study of potential future hydrological changes.

Eisma, D. (ed.), 1995, *Climate change: impact on coastal habitation*. Boca Raton, FL: Lewis. A study of the effects of climate change and sea-level rise.

Gates, D. M., 1993, *Climate change and its biological consequences*. Sunderland, MA: Sinauer Associates. A textbook that considers some of the effects of climate change on the biosphere.

Leggett, J., 1996, *Climate change and the financial sector*. Munich: Gerling Akademie Verlag. A study of why climate change may have a severe impact on economies and business.

CHAPTER NINE

Liddle, M., 1997, *Recreation ecology*. London: Chapman & Hall. An excellent review of the multiple ways in which recreation and tourism have an impact on the environment.

O'Riordan, T. (ed.), 1995, *Environmental science for environmental management*. Harlow: Longman Scientific. A multi-author guide to managing environmental change.

Roberts, N. (ed.), 1994, *The changing global environment*. Oxford: Blackwell. A good collection of case studies with a wide perspective.

Simon, J. L., 1996, *The ultimate resource 2*. Princeton: Princeton University Press. A view that the state of the world is improving.

—— (ed.), 1995, *The state of humanity*. Oxford: Blackwell. A multi-author work with an optimistic message.

Turner, B. L. II (ed.), 1990, *The earth as transformed by human action*. Cambridge: Cambridge University Press. A study of change in the last 300 years.

References

Abler, R., Adams, J. S. and Gould, P., 1971, *Spatial organisation: the geographer's view of the world*. Englewood Cliffs: Prentice Hall.

Abul-Atta, A. A., 1978, *Egypt and the Nile after the construction of the High Aswan Dam*, Cairo: Ministry of Irrigation and Land Reclamation.

Ackerman, W. C., Harmeston, R. H. and Sinclair, R. A., 1970, Some long-term trends in water quality of rivers and lakes. *Eos*, 51, 516–22.

Adger, W. N. and Brown, K., 1994, *Land use and the causes of global warming*, Chichester: Wiley.

Ågren, C. and Elvingson, P., 1996, *Still with us: the acidification of the environment is still going on*. Göteborg: Swedish NGO Secretariat on Acid Rain.

Ahlgren, I. F., 1974, The effect of fire on soil organisms. In I. I. Kozlowski and C. C. Ahlgren (eds), *Fire and ecosystems*. New York: Academic Press, 47–72.

Alabaster, J. S., 1972, Suspended solids and fisheries. *Proceedings of the Royal Society*, 180B, 395–406.

Al-Ibrahim, A. A., 1991, Excessive use of ground-water resources in Saudi Arabia: impacts and policy options. *Ambio*, 20, 34–7.

Allchin, B., Goudie, A. S. and Hegde, K. T. M., 1977, *The prehistory and palaeogeography of the Great Indian Desert*. London: Academic Press.

Allen, J. C. and Barnes, D. F., 1985, The causes of deforestation in developing countries. *Annals of the Association of American Geographers*, 75, 163–84.

Allen, J., Golson, J. and Jones, R. (eds), 1977, *Sunda and Sahul. Prehistoric studies in southeast Asia, Melanesia and Australia*. London: Academic Press.

Almer, B., Dickson, W., Ekstrom, C., Hornstrom, E. and Miller, U., 1974, Effects of acidification on Finnish lakes. *Ambio*, 3, 30–6.

Andel, T. H. van, Zangger, E. and Demitrack, A., 1990, Land use and soil erosion in prehistoric and historical Greece. *Journal of Field Archaeology*, 17, 379–96.

Anderson, D. M., 1994, Red tides. *Scientific American*, 271 (2), 52–8.

Anderson, J. M. and Spencer, T., 1991, Carbon, nutrient and water balances of tropical rainforest ecosystems subject to disturbance. *MAB Digest*, 7.

Anisimov, O. A., 1989, Changing climate and permafrost distribution in the Soviet Arctic. *Physical Geography* 10, 282–93.

Anisimov, O. A. and Nelson, F. E., 1996, Permafrost distribution in the northern hemisphere under scenarios of climatic change. *Global and Planetary Change*, 14, 59–72.

Arber, M. A., 1946, Dust-storms in the Fenland around Ely. *Geography*, 31, 23–6.

Arnell, N., 1996, *Global warming, river flows and water resources.* Chichester: Wiley.

Aron, W. I. and Smith, S. H., 1971, Ship canals and aquatic ecosystems. *Science*, 174, 13–20.

Ashby, E., 1978, *Reconciling man with the environment.* London: Oxford University Press.

Atkinson, B. W., 1968, A preliminary examination of the possible effect of London's urban area on the distribution of thunder rainfall, 1951–60. *Transactions of the Institute of British Geographers*, 44, 97–118.

——, 1975, The mechanical effect of an urban area on convective precipitation. Occasional paper 3, Department of Geography, Queen Mary College, University of London.

Atlas, P., 1977, The paradox of hail suppression. *Science*, 195, 139–45.

Aubertin, G. M. and Patric, J. H., 1974, Water quality after clearcutting a small watershed in West Virginia. *Journal of environmental quality*, 3, 243–9.

Aubréville, A., 1949, *Climats, forêts et desertification de L'Afrique tropicale.* Paris: Societé d'Edition Géographiques Maritimes et Coloniales.

Bach, W., 1979, Short-term climatic alterations caused by human activities: status and outlook. *Progress in physical geography*, 3, 55–83.

Bach, W., 1988, Modelling and climatic response to greenhouse gases. In S. Gregory (ed.), *Recent climatic change: a regional approach.* London: Bellhaven Press, 7–19.

Bailey, R. (ed.), 1995, *The true state of the planet.* New York: Free Press.

Bakan, S., Chlono, A., Cubasch, U., Feichter, J., Graf, H., Grassl, H., Hasselman, K., Kirchner, I., Latif, M., Roeckner, E., Samsen, R., Schlese, U., Schrivener, D., Schult, I., Sielman, F. and Welks, W., 1991, Climate response to smoke from the burning oil wells in Kuwait. *Nature*, 351, 367–71.

Ball, D. F., 1975, discussion in J. G. Evans, S. Limbrey and H. Cleere (eds), The effect of man on the landscape: the Highland zone. *Council for British Archaeology Research Report*, 11, 26.

Ball, D. J. and Laxen, D. P. H., 1986, *Ambient ozone concentrations in Greater London 1975 to 1985.* London: London Scientific Services.

Ballantyne, C. K., 1991, Late Holocene erosion in upland Britain: climatic deterioration or human influence? *The Holocene*, 1, 81–5.

Balling, R. C. and Wells, S. G., 1990, Historical rainfall patterns and arroyo activity within the Zuni river drainage basin, New Mexico. *Annals of the Association of American Geographers*, 80, 603–17.

Banks, H. O. and Richter, R. C., 1953, Sea-water intrusion into groundwater basins bordering the California coast and inland bays. *Transactions of the American Geophysical Union*, 34, 575–82.

Barber, K. E., 1981, *Peat stratigraphy and climatic change*. Rotterdam: Balkema.

Barends, F. B. J., Brouwer, F. J. J. and Schröder, F. H., 1995, *Land subsidence*, IAHS Publication 234. Wallingford, Oxon: International Association of Hydrological Sciences.

Bari, M. A. and Schofield, N. J., 1992, Lowering of a shallow, saline water table by extensive eucalypt reforestation. *Journal of Hydrology*, 133, 273–91.

Barnston, A. G. and Schickendanz, P. T., 1984, The effect of irrigation on warm season precipitation in the southern Great Plains. *Journal of climate and applied meteorology*, 23, 865–88.

Barrass, R., 1974, *Biology, food and people*. London: Hodder & Stoughton.

Barry, R. G., 1985, The cryosphere and climate change. In M. C. Mac-Cracken and F. M. Luther (eds), *Detecting the climatic effects of increasing carbon dioxide*. Washington DC: US Dept of Energy, 111–48.

Bartlett, H. H., 1956, Fire, primitive agriculture, and grazing in the tropics. In W. L. Thomas (ed.), *Man's role in changing the face of the earth*. Chicago: University of Chicago Press, 692–720.

Barton, B. A., 1977, Short-term effect of highway construction on the limnology of a small stream in southern Ontario. *Freshwater biology*, 7, 99–108.

Barton, M. E., 1987, The sunken lanes of southern England: engineering geological considerations. *Geological Society Engineering Geology Special Publication*, 4, 411–18.

Bates, M., 1956, Man as an agent in the spread of organisms. In W. L. Thomas (ed.), *Man's role in changing the face of the earth*. Chicago: University of Chicago Press, 788–804.

Battarbee, R. W., 1977, Observations in the recent history of Lough Neagh and its drainage basin. *Philosophical transactions of the Royal Society of London*, 281B, 303–45.

Battarbee, R. W., Flower, R. J., Stevenson, A. C. and Rippey, B., 1985a, Lake acidification in Galloway: a palaeoecological test of competing hypotheses. *Nature*, 314, 350–2.

Battarbee, R. W., Appleby, P. G., Odel, K. and Flower, R. J., 1985b, ^{210}Pb dating of Scottish Lake sediments, afforestation and accelerated soil erosion. *Earth surface processes and landforms*, 10, 137–42.

Battarbee, R. W., Flower, R. J., Stevenson, A. C., Jones, V. J., Harriman, R. and Appleby, P. G., 1988a, Diatom and chemical evidence for reversibility of acidification of Scottish lochs. *Nature*, 332, 530–2.

Battarbee, R. W. and 15 collaborators, 1988b, *Lake acidification in the United Kingdom 1800–1986*. London: Ensis.

Battistini, R. and Verin, P., 1972, Man and the environment in Madagascar. *Monographiae biologicae*, 21, 331–7.

Baver, L. D., Gardner, W. H. and Gardner, W. R., 1972, *Soil physics* (4th edn). New York: Wiley.

Baxter, R. M., 1977, Environmental effects of dams and impoundments. *Annual Review of ecology and systematics*, 8, 255–83.

Bayfield, N. G., 1973, Human pressures on soils in mountain areas. *Welsh soils discussion group report*, 14, 36–49.

——, 1979, Recovery of four heath communities on Cairngorm, Scotland, from disturbance by trampling. *Biological conservation*, 15, 165–79.

Beamish, R. J., Lockhart, W. L., Van Loon, J. C. and Harvey, H. H., 1975, Long-term acidification of a lake and resulting effects on fishes. *Ambio*, 4, 98–102.

Beaumont, P., 1978, Man's impact on river systems: a world-wide view. *Area*, 10, 38–41.

Beckinsale, R. P., 1969, Human responses to river regimes. In R. J. Chorley (ed.), *Water, earth and man*. London: Methuen, 487–509.

—— , 1972, The effect upon river channels of sudden changes in sediment load. *Acta geographica debrecina*, 10, 181–6.

Begon, M., Harper, J. L. and Townshend, C. R., 1986, *Ecology: individuals, populations and communities*. Oxford: Blackwell Scientific.

Behre, K.-E. (ed.), 1986, *Anthropogenic indicators in pollen diagrams*. Rotterdam: Balkema.

Bell, J. N. B. and Cox, R. A., 1975, Atmospheric ozone and plant damage in the United Kingdom. *Environmental pollution*, 8, 163–70.

Bell, M. L., 1982, The effect of land-use and climate on valley sedimentation. In A. F. Harding (ed.), *Climatic change in later prehistory*. Edinburgh: Edinburgh University Press, 127–42.

Bell, M. and Walker, M. J. C., 1992, *Late Quaternary environmental change*. Harlow: Longman Scientific and Technical.

Bell, M. L. and Nur, A., 1978, Strength changes due to reservoir-induced pore pressure and stresses and application to Lake Oroville. *Journal of geophysical research*, 83, 4469–83.

Bell, P. R., 1982, Methane hydrate and the carbon dioxide question. In W. G. Clark (ed.), *Carbon dioxide review 1982*. Oxford: Oxford University Press, 401–5.

Bendell, J. F., 1974, Effects of fire on birds and mammals. In T. T. Kozlowski and C. C. Ahlgren (eds), *Fire and ecosystems*. New York: Academic Press, 73–138.

Bender, B., 1975, *Farming in prehistory from hunter-gatherer to food-producer*. London: Baker.

Bennett, C. F., 1968, Human influences in the zoogeography of Panama. *Ibero-Americana*, 51.

Bennett, H. H., 1938, *Soil conservation*. New York: McGraw-Hill.

Bentley, C. R., 1984, Some aspects of the cryosphere and its role in climatic change. *Geophysical monograph*, 29, 207–20.

Bernabo, J. C. and Webb, T., III, 1977, Changing patterns in the Holocene pollen record of northeastern North America: mapped summary. *Quaternary research*, 8, 64–96.

Berry, B. J. L., 1974, *Land use, urban form and environmental quality*. Chicago: Department of Geography Research Series.

Bidwell, O. W. and Hole, F. D., 1965, Man as a factor of soil formation. *Soil science*, 99, 65–72.

Biglane, K. E. and Lafleur, R. A., 1967, Notes on estuarine pollution with emphasis on the Louisiana Gulf Coast. *Publication 83, American association for the advancement of science*, 690–2.

Binford, M. W., Brenner, M., Whitmore, T. J., Higuera-Grundy, A., Deevey, E. S. and Leyden, B., 1987, Ecosystems, palaeoecology, and human disturbance in subtropical and tropical America. *Quaternary Science Review*, 6, 115–28.

Bird, E. C. F., 1979, Coastal processes. In K. J. Gregory and D. E. Walling (eds), *Man and environmental processes*. Folkestone: Dawson, 82–101.

Bird, E. C. F., 1985, *Coastline changes. A global view*. Chichester: Wiley.
—— , 1986, Potential effects of sea level rise on the coasts of Australia, Africa and Asia. In J. G. Titus (ed.), *Effects of changes in stratospheric ozone and global climate*. Washington DC: UNEP/USEPA, 83–98.
—— , 1996, *Beach management*. Chichester: Wiley.
Birkeland, P. W. and Larson, E. E., 1978, *Putnam's geology*. New York: Oxford University Press.
Birks, H. J. B., 1986, Late-Quaternary biotic changes in terrestrial and lacustrine environments, with particular reference to north-west Europe. In B. E. Berglund (ed.), *Handbook of Holocene palaeo-ecology and palaeohydrology*. Chichester: Wiley, 3–65.
—— , 1988, Long-term ecological change in the British uplands. *British ecological society special publication*, 7, 37–56.
Blackburn, W. H. and Tueller, P. T., 1970, Pinjon and juniper invasion in Black sagebrush communities in east-central Nevada. *Ecology*, 51, 841–8.
Blackburn, W. H., Knight, R. W. and Schuster, J. L., 1983, Saltcedar influence on sedimentation in the Brazos River. *Journal of soil and water conservation*, 37, 298–301.
Blainey, G., 1975, *Triumph of the nomads. A history of ancient Australia*. Melbourne: Macmillan.
Blank, I. W., 1985, A new type of forest decline in Germany. *Nature*, 314, 311–14.
Blumer, M., 1972, Submarine seeps: are they a major source of open ocean oil pollution? *Science*, 176, 1257–8.
Böckh, A., 1973, Consequences of uncontrolled human activities in the Valencia lake basin. In M. T. Farvar and J. P. Milton (eds), *The careless technology*. London: Tom Stacey, 301–17.
Bockman, O. C., Kaarstad, O., Lie, O. H. and Richards, I., 1990, *Agriculture and fertilizers*. Oslo: Norsk Hydro.
Boeck, W. L., Shaw, D. T. and Vonnegut, B., 1975, Possible consequences of global dispersions of Krypton 85. *Bulletin of the American Meteorological Society*, 56, 527.
Bøer, E. M., Koster, E. and Lundberg, E., 1990, Greenhouse impact in Fennoscandia: preliminary findings of a European workshop on the effects of climatic change. *Ambio*, 19, 2–10.
Bolin, B., Doos, B. R., Jager, J. and Warrick, R. A. (eds), 1986, *The greenhouse effect, climatic change and ecosystems*. Chichester: Wiley.
Bonan, G. B., 1997, Effects of land use on the climate of the United States. *Climatic Change*, 37, 449–86.
Boorman, L. A., 1977, Sand-dunes. In R. S. K. Barnes (ed.), *The coastline*. London: Wiley, 161–97.
Booysen, P. de V. and Tainton, N. M. (eds), 1984, *Ecological effects of fire in South African ecosystems*. Berlin: Springer-Verlag.
Bormann, F. H., Likens, G. E., Fisher, D. W. and Pierce, R. S., 1968, Nutrient loss accelerated by clear cutting of a forest ecosystem. *Science*, 159, 882–4.
Bourne, W. R. P., 1970, Oil pollution and bird conservation. *Biological conservation*, 2, 300–2.
Boussingault, J. B., 1845, *Rural economy* (2nd edn). London: Baillière.
Boutron, C. F., Görlach, U., Candelone, J.-P., Bolshov, M. A. and Delmas, R. J., 1991, Decrease in anthropogenic lead, cadmium and zinc in Greenland snows since the late 1960s. *Nature*, 353, 153–6.

Brampton, A. H., 1998, Cliff conservation and protection: methods and practices to resolve conflicts. In J. Hooke (ed.), *Coastal defence and earth science conservation*. London: Geological Society, 21–31.

Brassington, F. C. and Rushton, K. R., 1987, A rising water table in central Liverpool. *Quarterly Journal of Engineering Geology*, 20, 151–8.

Bravard, J-P. and Petts, G. E., 1996, Human impacts on fluvial hydrosystems. In G. E. Petts and C. Amoros (eds), *Fluvial hydrosystems*. London: Chapman & Hall, 242–62.

Brazel, A. J. and Idso, S. B., 1979, Thermal effects of dust on climate. *Annals of the Association of American Geographers*, 69, 432–7.

Breuer, G., 1980, *Weather modification: prospects and problems*. Cambridge: Cambridge University Press.

Bridges, E. M., 1978, Interaction of soil and mankind in Britain. *Journal of soil science*, 29, 125–39.

Brimblecombe, P., 1977, London air pollution 1500–1900. *Atmospheric environment*, 11, 1157–62.

—— , 1987, *The big smoke*. London: Methuen.

Broadus, J., Milliman, J., Edwards, S., Aubrey, D. and Gable, F., 1986, Rising sea level and damming of rivers: possible effects in Egypt and Bangladesh. In J. G. Titus (ed.), *Effects of changes in stratospheric ozone and global climate*. Washington DC: UNEP/USEPA, 165–89.

Broecker, W. S., Takahashi, T., Simpson, H. J. and Peng, T. H., 1979, Fate of fossil fuel carbon dioxide and the global carbon budget. *Science*, 206, 409–18.

Brookes, A., 1985, River channelization: traditional engineering methods, physical consequences, and alternative practices. *Progress in physical geography*, 9, 44–73.

—— , 1987, The distribution and management of channelized streams in Denmark. *Regulated rivers*, 1, 3–16.

—— , 1988, *Channelised rivers*. Chichester: Wiley.

Brooks, A. P. and Brierly, G. J., 1997, Geomorphic responses of lower Bega River to catchment disturbance, 1851–1926. *Geomorphology*, 18, 291–304.

Brown, A. A. and Davis, K. P., 1973, *Forest fire control and its use* (2nd edn). New York: McGraw-Hill.

Brown, B. E., 1990, Coral bleaching. *Coral Reefs*, 8, 153–232.

Brown, E. H., 1970, Man shapes the earth. *Geographical journal*, 136, 74–85.

Brown, J. H., 1989, Patterns, modes and extents of invasions by vertebrates. In J. A. Drake (ed.), *Biological invasions: a global perspective*. Chichester: Wiley, 85–109.

Brown, L. R., Flavin, C. and Kane, H., 1992, *Vital signs*. London: Earthscan.

Brown, S. and Lugo, A. E., 1990, Tropical secondary forests. *Journal of Tropical Ecology*, 6, 1–32.

Browning, K. A., Allah, R. J., Ballard, B. P., Barnes, R. T. H., Bennetts, D. A., Maryon, R. H., Mason, P. J., McKenna, D., Mitchell, J. F. B., Senior, C. A., Slingo, A. and Smith, F. B., 1991, Environmental effects from burning oil wells in Kuwait. *Nature*, 351, 363–7.

Bruijnzeel, L. A., 1990, *Hydrology of moist tropical forests and effects of conversion: a state of knowledge review*. Amsterdam: Free University for UNESCO International Hydrological Programme.

Brunhes, J., 1920, *Human geography*. London: Harrap.

Bruun, P., 1962, Sea level rise as a cause of shore erosion. *American Society of Civil Engineers Proceedings: Journal of Waterways and Harbors Division*, 88, 117–30.

Bryan, G. W., 1979, Bioaccumulation of marine pollutants. *Philosophical transactions of the Royal Society*, 286B, 483–505.

Bryan, K., 1928, Historic evidence of changes in the channel of Rio Puerco, a tributary of the Rio Grande in New Mexico. *Journal of geology*, 36, 265–82.

Bryson, R. A. and Barreis, D. A., 1967, Possibility of major climatic modifications and their implications: northwest India, a case for study. *Bulletin of the American Meteorological Society*, 48, 136–42.

Bryson, R. A. and Kutzbach, J. E., 1968, *Air pollution*. Commission on college geography resource paper 2. Washington DC: Association of American Geographers.

Budd, W. F., 1991, Antarctica and global change. *Climatic change*, 19, 271–99.

Buddemeier, R. W. and Smith, S. V., 1988, Coral reef growth in an era of rapidly rising sea level: predictions and suggestions for long-term research. *Coral Reefs*, 7, 51–6.

Budyko, M. I., 1974, *Climate and life*. New York: Academic Press.

—— , 1982, *The earth's climate: past and future*. New York: Academic Press.

Budyko, M. I. and Izrael, Y. A., 1991, *Anthropogenic climatic change*. Tuscon: University of Arizona Press.

Bulmer, S., 1982, Human ecology and cultural variation in prehistoric New Guinea. *Monographiae biologicae* (New Guinea), 42, 169–206.

Burcham, L. T., 1970, Ecological significance of alien plants in California grasslands. *Proceedings of the Association of American Geographers*, 2, 36–9.

Burke, R., 1972, Stormwater runoff. In R. T. Oglesby, C. A. Carlson and J. A. McCann (eds), *River ecology and man*. New York: Academic Press, 727–33.

Burney, A. D., 1993, Recent animal extinctions: recipes for disaster. *American Scientist*, 81, 530–41.

Burrin, P. J., 1985, Holocene alluviation in southeast England and some implications for palaeohydrological studies. *Earth surface processes and landforms*, 10, 257–71.

Burt, T. P., Donohoe, M. A. and Vann, A. R., 1983, The effect of forestry drainage operations on upland sediment yields: the results of a stormbased study. *Earth surface processes and landforms*, 8, 339–46.

Burt, T. P. and Haycock, N. E., 1992, Catchment planning and the nitrate issue: a UK perspective. *Progress in Physical Geography*, 16, 379–404.

Busack, S. D. and Bury, R. B., 1974, Some effects of off-road vehicles and sheep grazing on lizard population in the Mojave Desert. *Biological conservation*, 6, 179–83.

Bush, M. B., 1988, Early Mesolithic disturbance: a force on the landscape. *Journal of Archaeological Science*, 15, 453–62.

Bussing, C., 1972, The impact of feedlots. In D. D. MacPhail (ed.), *The High Plains: problem of semiarid environments*. Fort William: Colorado State University, 78–86.

Butzer, K. W., 1972, *Environment and archaeology – an ecological approach to prehistory*. London: Methuen.

—— , 1974, Accelerated soil erosion: a problem of man–land relationships. In I. R. Manners and M. W. Mikesell (eds), *Perspectives on environments*. Washington DC: Association of American Geographers.

—— , 1976, *Early hydraulic civilization in Egypt*. Chicago: University of Chicago Press.

—— , 1977, Environment, culture and human evolution. *American scientist*, 65, 572–84.

—— , 1980, Adaptation to global environmental change. *Professional geographer*, 32, 269–78.

Cadle, R. D., 1971, The chemistry of smog. In W. H. Matthews, W. W. Kellogg and G. D. Robinson (eds), *Man's impact on the climate*. Cambridge, Mass.: MIT Press, 339–59.

Cairns, J. and Dickson, D. K., 1977, Recovery of streams from spills of hazardous materials. In J. Cairns, K. L. Dickson and E. E. Herricks (eds), *Recovery and restoration of damaged ecosystems*. Charlottesville: University Press of Virginia, 24–42.

Carbon Dioxide Assessment Committee, 1983, *Changing climate*. Washington: National Academic Press.

Carlson, C. W., 1978, Research in ARS related to soil structure. In W. W. Emerson, R. D. Bond and A. R. Dexter (eds), *Modification of soil structure*. Chichester: Wiley, 279–84.

Carrara, P. E. and Carroll, T. R., 1979, The determination of erosion rates from exposed tree roots in the Piceance Basin, Colorado. *Earth surface processes*, 4, 407–17.

Carter, D. L., 1975, Problems of salinity in agriculture. *Ecological studies*, 15, 26–35.

Carter, L. J., 1977, Soil erosion: the problem persists despite the billions spent on it. *Science*, 196, 409–11.

Carter, R. W. G., 1988, *Coastal environments*. London: Academic Press.

Casey, H. and Clarke, R. T., 1979, Statistical analysis of nitrate concentrations from the River Frome (Dorset) for the period 1965–76. *Freshwater biology*, 9, 91–7.

Cathcart, R. B., 1983, Mediterranean Basin – Sahara reclamation. *Speculations in science and technology*, 6, 150–2.

Challinor, D., 1968, Alteration of surface soil characteristics by four tree species. *Ecology*, 49, 286–90.

Chambers, F. M., Dresser, P. Q. and Smith, A. G., 1979, Radiocarbon dating evidence on the impact of atmospheric pollution on upland peats. *Nature*, 282, 829–31.

Champion, T., Gramble, C., Shennan, S. and Whittle, A., 1984, *Prehistoric Europe*. London: Academic Press.

Chancellor, W. J., 1977, Compaction of soil by agricultural equipment. *Division of agricultural sciences, University of California bulletin*, 1881.

Chandler, T. J., 1965, *The climate of London*. London: Hutchinson.

—— , 1976, The climate of towns. In T. J. Chandler and S. Gregory (eds), *The climate of the British Isles*. London: Longman, 307–29.

Changnon, S. A., 1968, The La Porte weather anomaly – fact or fiction? *Bulletin of the American Meteorological Society*, 49, 4–11.

Changnon, S. A., 1973, Atmospheric alterations from man-made biospheric changes. In W. R. D. Sewell (ed.), *Modifying the weather: a social assessment*. Adelaide: University of Victoria, 135–84.

——, 1978, Urban effects on severe local storms at St Louis. *Journal of applied meteorology*, 17, 578–86.

Charlson, R. J., Lovelock, J. E., Andreae, M. O. and Warren, S. G., 1987, Oceanic phytoplankton, atmospheric sulphur, cloud albedo and climate. *Nature*, 326, 655–61.

Charlson, R. J., Schwartz, S. E., Hales, J. M., Cess, R. D., Coakley, J. A., Hansen, J. E. and Hoffmann, D. J., 1992, Climate forcing by anthropogenic aerosols. *Science*, 255, 423–30.

Charlson, R. J. and Wigley, T. M. L., 1994, Sulfate aerosol and climatic change. *Scientific American*, 270 (2), 28–35.

Charney, J., Stone, P. H. and Quirk, W. J., 1975, Drought in the Sahara: a bio-geophysical feedback mechanism. *Science*, 187, 434–5.

Charnock, A., 1983, A new course for the Nile. *New scientist*, October, 285–8.

Chester, D. K. and James, P. A., 1991, Holocene alluviation in the Algarve, southern Portugal: the case for an anthropogenic cause. *Journal of Archaeological Science*, 18, 73–87.

Chesters, G. and Konrad, J. F., 1971, Effects of pesticide usage on water quality. *Bio science*, 21, 565–9.

Chi, S. C. and Reilinger, R. E., 1984, Geodetic evidence for subsidence due to groundwater withdrawal in many parts of the United States of America. *Journal of Hydrology*, 67, 155–82.

Chiew, F. H. S., Wetton, P. H., McMahon, T. A. and Pittock, A. B., 1995, Simulation of the impacts of climate change on runoff and soil moisture in Australian catchments. *Journal of Hydrology*, 167, 121–47.

Child, B. A., 1985, Bush encroachment and rangeland deterioration. Paper presented at a symposium on 'The deterioration of natural resources in Africa', School of Geography, University of Oxford, 1 July, 1985.

Childe, V. G., 1936, *Man makes himself*. London: Watts.

Chinn, T. J., 1988, Glaciers and snowlines. In Ministry for the Environment (ed.), *Climate Change: the New Zealand response*. Wellington: Ministry for the Environment, 238–40.

Chorley, R. J. and More, R. J., 1967, The interaction of precipitation and man. In R. J. Chorley (ed.), *Water, earth and man*. London: Methuen, 157–66.

Clark, G., 1962, *World prehistory*. Cambridge: Cambridge University Press.

——, 1977a, Domestication and social evolution. In J. Hutchinson, J. G. G. Clark, E. M. Jope and R. Riley (eds), *The early history of agriculture*. Oxford: Oxford University Press, 5–11.

——, 1977b, *World prehistory in new perspective*. Cambridge: Cambridge University Press.

Clark, J. R., 1977, *Coastal ecosystem management*. New York: Wiley.

Clark, J. S., Cachier, H., Goldammer, J. G. and Stocks, B. (eds), 1997, *Sediment records of biomass burning and global change*. Berlin: Springer Verlag.

Clark, R. B., 1997, *Marine pollution* (4th edn). Oxford: Clarendon Press.

Clark, R. R., 1963, *Grimes Graves*. London: Her Majesty's Stationery Office.

Clark, W. C. (ed.), 1982, *Carbon dioxide review 1982*. Oxford: Clarendon Press.

Clark, W. C. and 7 co-workers, 1982, The carbon dioxide question: perspectives for 1982. In W. C. Clark (ed.), *Carbon dioxide review 1982*. Oxford: Oxford University Press, 3–44.

Clarke, R. T. and McCulloch, J. S. G., 1979, The effect of land use on the hydrology of small upland catchments. In G. E. Hollis (ed.), *Man's impact on the hydrological cycle in the United Kingdom*. Norwich: Geo Abstracts, 71–8.

Clutton-Brock, J., 1987, *A natural history of domesticated mammals*. London: British Museum/Cambridge: Cambridge University Press.

Coates, D. R. (ed.), 1976, *Geomorphology and engineering*. Stroudsburg: Dowden, Hutchinson and Ross.

—— , 1977, Landslide perspective. *Reviews in engineering geology*, 3, 3–28.

—— , 1983, Large-scale land subsidence. In R. Gardner and H. Scoging (eds), *Mega-geomorphology*. Oxford: Oxford University Press, 212–34.

Cochrane, R., 1977, The impact of man on the natural biota. In A. G. Anderson (ed.), *New Zealand in maps*. Section 14. London: Hodder & Stoughton.

Coe, M., 1981, Body size and the extinction of the Pleistocene megafauna. *Palaeoecology of Africa*, 13, 139–45.

—— , 1982, The bigger they are . . . *Oryx*, 16, 225–8.

Coffey, M., 1978, The dust storms. *Natural history* (New York), 87, 72–83.

Colbeck, I., 1988, Photochemical ozone pollution in Britain. *Science progress*, 72, 207–26.

Cole, M. M., 1963, Vegetation and geomorphology in northern Rhodesia: an aspect of the distribution of the Savanna of Central Africa. *Geographical journal*, 129, 290–310.

Cole, M. M. and Smith, R. F., 1984, Vegetation as an indicator of environmental pollution. *Transactions of the Institute of British Geographers*, 9, 477–93.

Cole, S., 1970, *The Neolithic revolution* (5th edn). London: British Museum (Natural History).

Collier, C. R., 1964, Influences of strip mining on the hydrological environment of parts of Beaver Creek Basin, Kentucky, 1955–59. *United States geological survey professional paper*, 472-B.

Committee on the Atmosphere and the Biosphere, 1981, *Atmosphere–biosphere interactions: towards a better understanding of the ecological consequences of fossil fuel combustion*. Washington DC: National Academy Press.

Committee on Engineering Implications of Changes in Relative Mean Sea Level, 1987, *Responding to changes in sea level*. Washington DC: National Academy Press.

Conacher, A. J., 1979, Water quality and forests in Southwestern Australia: review and evaluation. *Australian geographer*, 14, 150–9.

Conacher, A. J. and Conacher, J., 1995, *Rural land degradation in Australia*. Melbourne: Oxford University Press.

Conacher, A. J. and Sala, M. (eds), 1998, *Land degradation in Mediterranean environments of the world*. Chichester: Wiley.

Conway, G. R. and Pretty, J. N., 1991, *Unwelcome harvest: agriculture and pollution*. London: Earthscan.

Conway, V. M., 1954, Stratigraphy and pollen analysis of southern Pennine blanket peats. *Journal of Geology*, 42, 117–47.

Cooke, G. W., 1977, Waste of fertilizers. *Philosophical transactions of the Royal Society of London*, 281B, 231–41.

Cooke, R. U. and Doornkamp, J. C., 1990, *Geomorphology in environmental management* (2nd edn). Oxford: Clarendon Press.

Cooke, R. U. and Reeves, R. W., 1976, *Arroyos and environmental change in the American south-west*. Oxford: Clarendon Press.

Cooke, R. U., Brunsden, D., Doornkamp, J. C. and Jones, D. K. C., 1982, *Urban geomorphology in drylands*. Oxford: Oxford University Press.

Coones, P. and Patten, J. H. C., 1986, *The landscape of England and Wales*. Harmondsworth: Penguin Books.

Cooper, C. F., 1961, The ecology of fire. *Scientific American*, 204, 4, 150–60.

Corlett, R. T., 1995, Tropical secondary forests. *Progress in Physical Geography*, 19, 159–72.

Costa, J. E., 1975, Effects of agriculture on erosion and sedimentation in the Piedmont province, Maryland. *Bulletin of the Geological Society of America*, 86, 1281–6.

Cotton, W. R. and Piehlke, R. A., 1995, *Human impacts on weather and climate*. Cambridge: Cambridge University Press.

Council on Environmental Quality and the Department of State, 1982, *The global 2000 report for the President*. Harmondsworth: Penguin Books.

Coupland, R. T. (ed.), 1979, *Grassland ecosystems of the world: analysis of grasslands and their uses*. Cambridge: Cambridge University Press.

Cowell, E. B., 1976, Oil pollution of the sea. In R. Johnson (ed.), *Marine pollution*. London: Academic Press, 353–401.

Crisp, D. T., 1977, Some physical and chemical effects of the Cow Green (Upper Teesdale) impoundment. *Freshwater biology*, 7, 109–20.

Critchley, W. R. S., Reij, C. and Willcocks, T. J., 1994, Indigenous soil and water conservation: a review of the state of knowledge and prospects for building on traditions. *Land Degradation and Rehabilitation*, 5, 293–314.

Croley, T. E., 1990, Laurentian Great Lakes double-CO_2 climate change hydrological impacts. *Climatic Change*, 17, 27–47.

Cronin, L. E., 1967, The role of man in estuarine processes. *Publication 83, American Association for the Advancement of Science*, 667–89.

Cronk, Q. C. B. and Fuller, J. L., 1995, *Plant invaders*. London: Chapman & Hall.

Crowe, P. R., 1971, *Concepts in climatology*. London: Longman.

Crozier, M. J., Marx, S. L. and Grant, I. J., 1978, Impact of off-road recreational vehicles on soil and vegetation. *Proceedings of the 9th New Zealand Geography Conference*, Dunedin, 76–9.

Crutzen, P. J., Aselmann, I. and Sepler, W., 1986, Methane production by domestic animals, wild ruminants, other herbivores, fauna and humans. *Tellus*, 38B, 271–84.

Crutzen, P. J. and Goldammer, J. G. (eds), 1993, *Fire in the environment*. Chichester: Wiley.

Cumberland, K. B., 1961, Man in nature in New Zealand. *New Zealand geographer*, 17, 137–54.

Currey, D. T., 1977, The role of applied geomorphology in irrigation and groundwater studies. In J. R. Hails (ed.), *Applied geomorphology*. Amsterdam: Elsevier, 51–83.

Custodio, E., Iribar, V., Manzano, B. and Galofre, M., 1986, Evolution of sea water chemistry in the Llobregat Delta, Barcelona, Spain. In *Proceedings of the 9th Salt Water Chemistry Meeting, Delft*.

Cwynar, L. C., 1978, Recent history of fire and vegetation from laminated sediment of Greenleaf Lake, Algonquin Park, Ontario. *Canadian journal of botany*, 56, 10–21.

Daniel, T. C., McGuire, P. E., Stoffel, D. and Millfe, B., 1979, Sediment and nutrient yield from residential construction sites. *Journal of environmental quality*, 8, 304–8.

Darby, H. C., 1956, The clearing of the woodland in Europe. In W. L. Thomas (ed.), *Man's role in changing the face of the Earth*. Chicago: University of Chicago Press, 183–216.

Darling, F. F., 1956, Man's ecological dominance through domesticated animals on wild lands. In W. L. Thomas (ed.), *Man's role in changing the face of the Earth*. Chicago: University of Chicago Press, 778–87.

Darungo, F. P., Allee, P. H. and Weickmann, H. K., 1978, Snowfall induced by a power plant plume. *Geophysical research letters*, 5, 515–17.

Daubenmire, R., 1968, Ecology of fire in grassland. *Advances in ecological research*, 5, 209–66.

Davis, B. N. K., 1976, Wildlife, urbanisation and industry. *Biological conservation*, 10, 249–91.

Davitaya, F. F., 1969, Atmospheric dust content as a factor affecting glaciation and climatic change. *Annals of the Association of American Geographers*, 59, 552–60.

Deadman, A., 1984, Recent history of *Spartina* in northwest England and in North Wales and its possible future development. In P. Doody (ed.), *Spartina anglica in Great Britain*. Shrewsbury: Nature Conservancy Council, 22–4.

Denevan, M. W., 1992, The pristine myth: the landscapes of the Americas in 1492. *Annals of the Association of American Geographers*, 82, 369–85.

Denisova, T. B., 1977, The environmental impact of mineral industries. *Soviet geography*, 18, 646–59.

Denson, E. P., 1970, The trumpeter swan, *Olor Buccinator;* a conservation success and its lessons. *Biological conservation*, 2, 251–6.

Department of Environment, 1984, *Digest of environmental pollution and water statistics for 1983*. London: HMSO.

——, 1989, *Digest of environmental pollution and water statistics for 1988*. London: HMSO.

De Sylva, D., 1986, Increased storms and estuarine salinity and other ecological impacts of the greenhouse effect. In J. G. Titus (ed.), *Effects of changes in stratospheric ozone and global climate*, vol. 4, *Sea level rise*. Washington DC: UNEP/USEPA, 153–64.

Detwyler, T. R. (ed.), 1971, *Man's impact on environment*. New York: McGraw-Hill.

Detwyler, T. R. and Marcus, M. G., 1972, *Urbanisation and environment: the physical geography of the city*. Belmont: Duxbury Press.

Di Castri, F., 1989, History of biological invasions with special emphasis on the old world. In J. A. Drake (ed.), *Biological invasions: a global perspective*. Chichester: Wiley, 1–30.

Dickinson, R. E., 1986, Impact of human activities on climate – a frame-work. In W. C. Clark and R. E. Munn (eds), *Sustainable development of the biosphere*. Cambridge: Cambridge University Press, 252–89.

Dicks, B., 1977, Changes in the vegetation of an oiled Southampton Water salt marsh. In J. Cairns, K. L. Dickson and E. E. Herricks (eds), *Recovery and restoration of damaged ecosystems*. Charlottesville: University of Virginia, 72–101.

Dimbleby, G. W., 1974, The legacy of prehistoric man. In A. Warren and F. B. Goldsmith (eds), *Conservation in practice*. London: Wiley, 179–89.

Dobson, M. C., 1991, De-icing salt damage to trees and shrubs. *Forestry Commission Bulletin*, 101.

Dodd, A. P., 1959, The biological control of the prickly pear in Australia. In A. Keast, R. L. Crocker and C. S. Christian (eds), *Biogeography and ecology in Australia*. The Hague: Junk, 565–77.

Doerr, A. and Guernsely, L., 1956, Man as a geomorphological agent: the example of coal mining. *Annals of the Association of American Geographers*, 46, 197–210.

Dolan, R., Godfrey, P. J. and Odum, W. E., 1973, Man's impact on the barrier islands of North Carolina. *American scientist*, 61, 152–62.

Donkin, R. A., 1979, Agricultural terracing in the Aboriginal New World. *Viking Fund publications in anthropology*, 561.

Doody, P. (ed.), 1984, *Spartina anglica in Great Britain*. Shrewsbury: Nature Conservancy Council.

Doughty, R. W., 1974, The human predator: a survey. In I. R. Manners and M. V. Mikesell (eds), *Perspectives on environment*. Washington DC: Association of American Geographers, 152–80.

—— , 1978, The English sparrow in the American landscape: a paradox in nineteenth century wildlife conservation. Research paper 19, School of Geography, University of Oxford.

Douglas, I., 1969, The efficiency of humid tropical denudation systems. *Transactions of the Institute of British Geographers*, 46, 1–6.

—— , 1983, *The urban environment*. London: Arnold.

Down, C. G. and Stocks, J., 1977, *Environmental impact of mining*. London: Applied Science Publishers.

Drake, J. A. et al. (eds), 1989, *Biological invasions: a global perspective*. Chichester: Wiley.

Dregne, H. E., 1986, Desertification of arid lands. In F. El-Baz and M. H. A. Hassan (eds), *Physics of Desertification*. Dordrecht: Nijhoff, 4–34.

Dregne, H. E. and Tucker, C. J., 1988, Desert encroachment. *Desertification Control Bulletin*, 16, 16–19.

Drennan, D. S. H., 1979, Agricultural consequences of groundwater development in England. In G. E. Hollis (ed.), *Man's impact on the hydrological cycle in the United Kingdom*. Norwich: Geo Abstracts, 31–8.

Driscoll, R., 1983, The influence of vegetation on the swelling and shrinkage of clay soils in Britain. *Géotechnique*, 33, 293–326.

Dunne, T. and Leopold, L. B., 1978, *Water in environmental planning*. San Francisco: Freeman.

D'Yakanov, K. N. and Reteyum, A. Y., 1965, The local climate of the Rybinsk reservoir. *Soviet geography*, 6, 40–53.

Eden, M. J., 1974, Palaeoclimatic influences and the development of savanna in southern Venezuela. *Journal of biogeography*, 1, 95–109.

Edington, J. M. and Edington, M. A., 1977, *Ecology and environmental planning*. London: Chapman & Hall.

Edlin, H. L., 1976, The Culbin sands. In J. Leniham and W. W. Fletcher (eds), *Reclamation*. Glasgow: Blackie, 1–31.

Edmonson, W. T., 1975, Fresh water pollution. In W. W. Murdoch (ed.), *Environment*. Sunderland: Sinauer Associates, 251–71.

Edwards, A. M. C., 1975, Long term changes in the water quality for agricultural catchments. In R. D. Hey and T. D. Daniels (eds), *Science technology and environmental management*. Farnborough: Saxon House, 111–22.

Edwards, K. J., 1985, The anthropogenic factor in vegetational history. In K. J. Edwards and W. P. Warren (eds), *The Quaternary history of Ireland*. London: Academic Press, 187–200.

Ehrenfeld, D. W., 1972, *Conserving life on earth*. New York: Oxford University Press.

Ehrlich, P. R. and Ehrlich, A. H., 1970, *Population, resources, environment: issues in human ecology*. San Francisco: Freeman.

—— , 1982, *Extinction*. London: Gollancz.

Ehrlich, P. R., Ehrlich, A. H. and Holdren, J. P., 1977, *Ecoscience: population, resources, environment*. San Francisco: Freeman.

Eisma, D. (ed.), 1995, *Climate change: impact on coastal habitation*. Boca Raton (FL): Lewis.

Ellenberg, H., 1979, Man's influence on tropical mountain ecosystems in South America. *Journal of ecology*, 67, 401–16.

Ellis, J. B., 1975, Urban stormwater pollution. *Middlesex Polytechnic research report*, 1.

Ellison, J. C. and Stoddart, D. R., 1990, Mangrove ecosystem collapse during predicted sea level rise: Holocene analogues and implications. *Journal of Coastal Research*, 7, 151–65.

Elsom, D., 1987, *Atmospheric pollution* (revised reprint 1989). Oxford and Cambridge, Mass.: Basil Blackwell.

—— , 1992, *Atmospheric pollution* (2nd edn). Oxford: Blackwell.

—— , 1996, *Smog alert: managing urban air quality*. London: Earthscan.

Elton, C. S., 1958, *The ecology of invasions by plants and animals*. London: Methuen.

Emanuel, K. A., 1987, The dependence of hurricane intensity on climate. *Nature*, 326, 483–5.

Emmanuel, W. R., Shugart, H. H. and Stevenson, M. P., 1985, Climatic change and the broad-scale distribution of terrestrial ecosystem complexes. *Climatic change*, 7, 29–43.

Engelhardt, F. R. (ed.), 1985, *Petroleum effects in the Arctic environment*. London: Elsevier Applied Science.

Evans, D. M., 1966, Man-made earthquakes in Denver. *Geotimes*, 10, 11–18.

Evans, J. G., Limbrey, S. and Cleere, H. (eds), 1975, *The effect of man on the landscape: the Highland zone*. Council for British Archaeology research report 11.

Evans, R. and Northcliffe, S., 1978, Soil erosion in North Norfolk. *Journal of agricultural science*, 90, 185–92.

Evenari, M., Shanan, L. and Tadmor, N. H., 1971, Runoff agriculture in the Negev Desert of Israel. In W. G. McGinnies, R. J. Goldman and P. Paylore (eds), *Food, fiber and the arid lands*. Tucson: University of Arizona Press, 312–22.

Fairhead, J. and Leach, M., 1996, *Misreading the African landscape: society and ecology in a forest-savanna mosaic*. Cambridge: Cambridge University Press.

—— , 1998, *Reframing deforestation*. London: Routledge.

Feare, C. J., 1978, The decline of booby (Sulidae) population in the Western Indian Ocean. *Biological conservation*, 14, 295–305.

Ferrians, O. J., Kachadoorian, R. and Green, G. W., 1969, Permafrost and related engineering problems in Alaska. *United States geological survey professional paper*, 678.

Fillenham, L. F., 1963, Holme Fen Post. *Geographical journal*, 129, 502–3.

Fisher, J., Simon, N. and Vincent, J., 1969, *The red book – wildlife in danger*. London: Collins.

Fitzpatrick, J., 1994, *A continent transformed: human impact on the natural vegetation of Australia*. Melbourne: Oxford University Press.

Flenley, J. R., 1979, *The equatorial rain forest: a geological history*. London: Butterworth.

Flenley, J. R., King, A. S. M., Jackson, J., Chew, C., Teller, J. and Prentice, M. E., 1991, The late Quaternary vegetational and climatic history of Easter Island. *Journal of Quaternary Science*, 6, 85–115.

Flohn, H., 1982, Climate change and an ice-free Arctic Ocean. In W. C. Clark (ed.), *Carbon dioxide review 1982*. Oxford: Oxford University Press, 145–79.

Flohn, H. and Dansgaard, W., 1984, Selected climates from the past and their relevance to possible future climate. In H. Flohn and R. Fantechi (eds), *The climate of Europe: past, present and future*. Dordrecht: Reidel, 198–268.

Food and Agricultural Organization of the United Nations, 1980, *1979 Production Yearbook*. Rome: FAO.

Foster, I. D. L., Dearing, J. A. and Appleby, R. G., 1986, Historical trends in catchment sediment yields: a case study in reconstruction from lake-sediment records in Warwickshire, UK. *Hydrological science journal*, 31, 427–43.

Foster, T., 1976, *Bushfire*. Sydney: Reed.

Fox, H. L., 1976, The urbanizing river: a case study in the Maryland Piedmont. In D. R. Coates (ed.), *Geomorphology and engineering*, Stroudsburg: Dowden, Hutchinson and Ross, 245–71.

Frankel, O. H., 1984, Genetic diversity, ecosystem conservation and evolutionary responsibility. In F. D. Castri, F. W. G. Baker and M. Hadley (eds), *Ecology in practice*, vol. I. Dublin: Tycooly, 4315–27.

Freedman, B., 1995, *Environmental ecology* (2nd edn). San Diego: Academic Press.

Freeland, W. J., 1990, Large herbivorous mammals: exotic species in northern Australia. *Journal of Biogeography*, 17, 445–9.

French, H. M., 1976, *The periglacial environment*. London: Longman.

——, 1996, *The periglacial environment* (2nd edn). Harlow: Longman.

French, P. W., 1997, *Coastal and estuarine management*. London: Routledge.

Frenkel, R. E., 1970, Ruderal vegetation along some California roadsides. *University of California publications in geography*, 20.

Fuller, R., Hill, D. and Tucker, G., 1991, Feeding the birds down on the farm: perspectives from Britain. *Ambio*, 20(6), 232–7.

Gade, D. W., 1976, Naturalization of plant aliens: the volunteer orange in Paraguay. *Journal of Biogeography*, 3, 269–79.

Galay, V. J., 1983, Causes of river bed degradation. *Water resources research*, 19, 5, 1057–90.

Gameson, A. L. H. and Wheeler, A., 1977, Restoration and recovery of the Thames Estuary. In J. Cairns, K. L. Dickson and E. E. Herricks (eds), *Recovery and restoration of damaged ecosystems*. Charlottesville: University Press of Virginia, 92–101.

Gates, D. M., 1993, *Climate change and its biological consequences*. Sunderland, MA: Sinauer.

Geertz, C., 1963, *Agricultural involution: the process of ecological change in Indonesia*. Berkeley: University of California Press.

Geikie, A., 1901, *The scenery of Scotland viewed in connection with its physical geology* (3rd edn). London: Macmillan.

Gerasimov, I. P., 1976, Problems of natural environment transformation in Soviet constructive geography. *Progress in geography*, 9, 75–99.

Gerasimov, I. P., Armand, D. L. and Yefron, K. M., 1971, *Natural resources of the Soviet Union: their use and renewal*. San Francisco: Freeman.

GESAMP (IMO/FAO/UNESCO/WMO/IAEA/UN/UNEP Joint Group of Experts on the Scientific Aspects of Marine Pollution), 1990, *The state of the marine environment*. UNEP Regional Seas Reports and Studies, 115, Nairobi.

Ghassemi, F., Jakeman, A. and Nix, H. A., 1995, *Salinisation of land and water resources*. Wallingford: CAB International.

Gifford, G. F. and Hawkins, R. H., 1978, Hydrological impact of grazing on infiltration: a critical review. *Water resources research*, 14, 305–13.

Gilbert, G. K., 1917, Hydraulic mining debris in the Sierra Nevada. *United States geological survey professional paper*, 105.

Gilbert, O. L., 1970, Further studies on the effect of sulphur dioxide on lichens and bryophytes. *New phytologist*, 69, 605–27.

——, 1975, Effects of air pollution on landscape and land-use around Norwegian aluminium smelters. *Environmental pollution*, 8, 113–21.

Gill, T. E., 1996, Eolian sediments generated by anthropogenic disturbance of playas: human impacts on the geomorphic system and geomorphic impacts on the human system. *Geomorphology*, 17, 207–28.

Gillespie, R., Horton, D. R., Ladd, P., Macumber, P. G., Rich, I. H., Thorne, R. and Wright, R. V. S., 1978, Lancefield Swamp and the extinction of the Australian megafauna. *Science*, 200, 1044–8.

Gillon, D., 1983, The fire problem in tropical savannas. In F. Bourlière (ed.), *Tropical savannas*. Oxford: Elsevier Scientific, 617–41.

Gimingham, C. H., 1981, Conservation: European Heathlands. In R. L. Spect (ed.), *Heathlands and related shrublands*. Amsterdam: Elsevier Scientific, 249–59.

Gimingham, C. H. and de Smidt, I. T., 1983, Heaths and natural and semi-natural vegetation. In W. Holzner, M. J. A. Werger and I. Ikusima (eds), *Man's impact on vegetation*. Hague: Junk, 185–99.

Glacken, C. J., 1956, Changing ideas of the habitable world. In W. L. Thomas (ed.), *Man's role in changing the face of the earth*. Chicago: University of Chicago Press, 70–92.

——, 1963, The growing second world within the world of nature. In F. R. Fosberg (ed.), *Man's place in the island ecosystem*. Honolulu: Bishop Museum Press, 75–100.

——, 1967, *Traces on the Rhodian shore: nature and culture in western thought from ancient times to the end of the eighteenth century*. Berkeley: University of California Press.

Gleick, P. H., 1986, Regional water resources and global climatic change. In J. G. Titus (ed.), *Effects of changes in stratospheric ozone and global climate*, vol. 3, *Climatic change*. Washington DC: UNEP/USEPA, 217–49.

—— (ed.), 1993, *Water in crisis: a guide to the world's freshwater resources*. New York: Oxford University Press.

Glynn, P., 1996, Coral reef bleaching: facts, hypotheses and implications. *Global Change Biology*, 2, 495–510.

Godbole, N. N., 1972, Theories on the origin of salt lakes in Rajasthan, India. *Proceedings of the 24th International Geological Congress*, Section 10, 354–7.

Goemans, T., 1986, The sea also rises: the ongoing dialogue of the Dutch with the sea. In J. G. Titus (ed.), *Effects of changes in stratospheric ozone and global climate*, vol. 4, *Sea level rise*. Washington DC: UNEP/USEPA, 47–56.

Goldberg, E. D., Hodge, V., Koide, M., Griffin, J., Gamble, E., Bicker, O. P., Metisoff, G., Holdren, G. R. and Brown, R., 1978, A pollution history of Chesapeake Bay. *Geochimica et cosmochimica acta*, 42, 1413–25.

Goldsmith, F. B., 1983, Evaluating nature. In A. Warren and F. B. Goldsmith (eds), *Conservation in perspective*. Chichester: Wiley, 233–46.

Goldsmith, V., 1979, Coastal dunes. In R. A. Davis (ed.), *Coastal sedimentary environments*. New York: Springer-Verlag.

Gomez, B. and Smith, C. G., 1984, Atmospheric pollution and fog frequency in Oxford, 1926–1980. *Weather*, 39, 379–84.

Gong, Zi-Tong, 1983, Pedogenesis of paddy soil and its significance in soil classification. *Soil science*, 135, 5–10.

Goodman, D., 1975, The theory of diversity–stability relationships in ecology. *Quarterly review of biology*, 50, 237–66.

Goreau, T. J. and Hayes, R. L., 1994, Coral bleaching and ocean 'hot spots'. *Ambio*, 23, 176–80.

Gorman, M., 1979, *Island ecology*. London: Chapman & Hall.

Gottschalk, L. C., 1945, Effects of soil erosion on navigation in Upper Chesapeake Bay. *Geographical review*, 35, 219–38.

Goudie, A. S., 1972a, The concept of post-glacial progressive desiccation. Research paper 4, School of Geography, University of Oxford.

——, 1927b, Vaughan Cornish: geographer. *Transactions of the Institute of British Geographers*, 55, 1–16.

——, 1973, *Duricrusts of tropical and subtropical landscapes*. Oxford: Clarendon Press.

——, 1977, Sodium sulphate weathering and the disintegration of Mohenjo-Daro, Pakistan. *Earth surface processes*, 2, 75–86.

——, 1983, Dust storms in space and time. *Progress in physical geography*, 7, 502–30.

—— (ed.), 1990, *Techniques for desert reclamation*. Chichester: Wiley.

——, 1992, *Environmental change*, 3rd edn. Oxford: Clarendon Press.

——, 1993, Human influence in geomorphology. *Geomorphology*, 7, 37–59.

——, 1995, *The changing earth: rates of geomorphological processes*. Oxford: Blackwell.

——, 1997, *The human impact reader: readings and case studies*. Oxford: Blackwell.

Goudie, A. S. and Middleton, N. S., 1992, The changing frequency of dust storms through time. *Climate Change*, 20, 197–225.

Goudie, A. S. and Viles, H. A., 1997, *Salt weathering hazards*. Chichester: Wiley.

——, 1997, *The earth transformed*. Oxford: Blackwell.

Goudie, A. S., Viles, H. A. and Pentecost, A., 1993, The late-holocene tufa decline in Europe. *The Holocene*, 3, 181–6.

Goudie, A. S. and Wilkinson, J. C., 1977, *The warm desert environment*. Cambridge: Cambridge University Press.

Gourou, P., 1961, *The tropical world* (3rd edn). London: Longman.

Gowlett, J. A. J., Harris, J. W. K., Walton, D. and Wood, B. A., 1981, Early archaeological sites, hominid remains and traces of fire from Chesowanja, Kenya. *Nature*, 284, 125–9.

Graedel, T. E. and Crutzen, P. J., 1993, *Atmospheric change: an earth system perspective*. San Francisco: Freeman.

——, 1995, *Atmosphere, climate and change*. New York: Scientific American Library.

Graetz, D., 1994, Grasslands. In W. B. Meyer and B. L. Turner II (eds), *Changes in land use and land cover: a global perspective*. Cambridge: Cambridge University Press, 125–47.

Graf, W. K., 1977, Network characteristics in suburbanizing streams. *Water resources research*, 13, 459–63.

Graf, W. L., 1988, *Fluvial processes in dryland rivers*. Berlin: Springer-Verlag.

Grainger, A., 1990, *The threatening desert: controlling desertification*. London: Earthscan.

——, 1992, *Controlling tropical deforestation*. London: Earthscan.

Gray, R., 1993, Regional meteorology and hurricanes. In G. A. Maul (ed.), *Climatic change in the Intra-Americas Sea*. London: Edward Arnold, 87–99.

Grayson, D. K., 1977, Pleistocene avifaunas and the overkill hypothesis. *Science*, 195, 691–3.

——, 1988, Perspectives on the archaeology of the first Americans. In R. C. Carlisle (ed.), *Americans before Columbus: Ice Age origins*. Pittsburgh: University of Pittsburgh, 107–23.

Green, F. H. W., 1976, Recent changes in land use and treatment. *Geographical journal*, 142, 12–26.

Green, F. H. W., 1978, Field drainage in Europe. *Geographical journal*, 144, 171–4.

Green, H. S. et al., 1981, Pontnewydd cave in Wales – a new Middle Pleistocene hominid site. *Nature*, 294, 707–13.

Green, R. C., 1975, Adaptation and change in Maori culture. *Monographiae biologicae* (New Guinea), 27, 591–661.

Greenland, D. J., 1977, Soil drainage by intensive arable cultivation: temporary or permanent. *Philosophical transactions of the Royal Society of London*, 281B, 193–208.

Greenland, D. J. and Lal, R., 1977, *Soil conservation and management in the humid tropics*. Chichester: Wiley.

Gregory, J. M. and Oelemans, J., 1998, Simulated future sea-level rise due to glacier melt based on regionally and seasonally resolved temperature changes. *Nature*, 391, 474–6.

Gregory, K. J., 1985a, The impact of river channelization. *Geographical journal*, 151, 53–74.

—— , 1985b, *The nature of physical geography*. London: Arnold.

Gregory, K. J. and Walling, D., 1973, *Drainage basin form and process: a geomorphological approach*. London: Arnold.

—— , 1979, *Man and environmental processes*. Folkestone: Dawson.

Grieve, A. M., 1987, Salinity and waterlogging in the Murray–Darling basin. *Search*, 18, 72–4.

Griffiths, J. F., 1976, *Applied climatology, an introduction* (2nd edn). Oxford: Oxford University Press.

Grigg, D., 1970, *The harsh lands*. London: Macmillan.

Gross, M. G., 1972, Geological aspects of waste solids and marine waste deposits, New York Metropolitan region. *Bulletin of the Geological Society of America*, 83, 3163–76.

Grove, J. M., 1988, *The Little Ice Age*. London: Routledge.

Grove, R. H., 1983, *The future for forestry*. Cambridge: British Association of Nature Conservationists.

—— , 1990, The origins of environmentalism. *Nature*, 345, 11–14.

—— , 1997, *Ecology, climate and empire: colonialism and global environmental history, 1400–1940*. Cambridge: White Horse Press.

Grover, H. D. and Musick, H. B., 1990, Shrubland encroachment in southern New Mexico, USA: an analysis of desertification processes in the American southwest. *Climatic Change*, 17, 305–30.

Guidon, N. and Delibrias, G., 1986, Carbon 14 dates point to man in the Americas 32,000 years ago. *Nature*, 321, 769–71.

Guilday, J. E., 1967, Differential extinction during late Pleistocene and recent time. In P. S. Martin and H. E. Wright (eds), *Pleistocene extinctions*. New Haven: Yale University Press, 121–40.

Haggett, P., 1979, *Geography: a modern synthesis* (3rd edn). London: Prentice Hall.

Haigh, M. J., 1978, Evolution of slopes on artificial landforms – Blaenavon, UK. Research paper 183, Department of Geography, University of Chicago.

Hails, J. R. (ed.), 1977, *Applied geomorphology*. Amsterdam: Elsevier.

Hanes, T. L., 1971, Succession after fire in the chaparral of Southern California. *Ecological monographs*, 41, 27–52.

Hannah, L., Lohse, D., Hutchinson, C., Carr, L. and Lankerani, A., 1994, A preliminary inventory of human disturbance of world ecosystems. *Ambio*, 23, 246–50.

Hansen, J., Johnson, D., Lacis, A., Lebedeff, S., Lee, P., Rims, D. and Russell, G., 1981, Climatic impact of increasing atmospheric carbon dioxide. *Science*, 213, 957–66.

Hansen, J., Fung, I., Lacis, A., Rind, D., Lebedeff, S., Ruedy, R. and Russell, G., 1988, Global climate changes as forecast by Goddard Institute for Space Studies three dimensional model. *Journal of geophysical research*, 93, 9341–64.

Happ, S. C., 1944, Effect of sedimentation on floods in the Kickapoo Valley, Wisconsin. *Journal of geology*, 52, 53–68.

Harlan, J. R., 1975a, Our vanishing genetic resources. *Science*, 188, 617–22.

—— , 1975b, *Crops and man*. Madison: American Society of Agronomy.

—— , 1976, The plants and animals that nourish man. *Scientific American*, 235, 3, 88–97.

Harris, D. R., 1966, Recent plant invasions in the arid and semi-arid southwest of the United States. *Annals of the Association of American Geographers*, 56, 408–22.

—— (ed.), 1980, *Human ecology in savanna environments*. London: Academic Press.

—— (ed.), 1996, *The origins and spread of agriculture and pastoralism in Eurasia*. London: UCL Press.

Hartmann, H. C., 1990, Climate change impacts on Laurentian Great Lakes levels. *Climatic Change*, 17, 49–67.

Harvey, A. M. and Renwick, W. H., 1987, Holocene alluvial fan and terrace formation in the Bowland Fells, Northwest England. *Earth surface processes and landforms*, 12, 249–57.

Harvey, A. M., Oldfield, F., Baron, A. F. and Pearson, G. W., 1981, Dating of post-glacial landforms in the central Howgills. *Earth surface processes and landforms*, 6, 401–12.

Hawksworth, D. L., 1990, The long-term effects of air pollutants on lichen communities in Europe and North America. In G. M. Woodwell (ed.), *The earth in transition: patterns and processes of biotic impoverishment*. Cambridge: Cambridge University Press, 45–64.

Hay, J., 1973, Salt cedar and salinity on the Upper Rio Grande. In M. T. Farvar and J. P. Milton (eds), *The careless technology*. London: Tom Stacey, 288–300.

Haynes, C. V., 1991, Geoarchaeological and palaeohydrological evidence for a Clovis-age drought in North America and its bearing on extinction. *Quaternary research*, 35, 438–50.

Heathwaite, A. L., Johnes, P. J. and Peters, N. E., 1996, Trends in nutrients. *Hydrological Processes*, 10, 263–93.

Heinselman, M. L. and Wright, H. E., 1973, The ecological role of fire in natural conifer forests of western and northern America. *Quaternary research*, 3, 317–482.

Helldén, U., 1985, Land degradation and land productivity monitoring – needs for an integrated approach. In A. Hjört (ed.), *Land management and survival*. Uppsala: Scandinavian Institute of African Studies, 77–87.

Helliwell, D. R., 1974, The value of vegetation for conservation. II: M1 motorway area. *Journal of environmental management*, 2, 75–8.

Henderson-Sellers, A. and Blong, R., 1989, *The greenhouse effect: living in a warmer Australia*. Kensington, NSW: New South Wales University Press.

Henderson-Sellers, A. and Gornitz, V., 1984, Possible climatic impacts of land cover transformation with particular emphasis on tropical deforestation. *Climatic change*, 6, 231–57.

Henderson-Sellers, A. and Robinson, P. J., 1986, *Contemporary climatology*. London: Longman.

Hess, W. N. (ed.), 1974, *Weather and climate modification*. New York: Wiley.

Hewlett, J. D., Post, H. E. and Doss, R., 1984, Effect of clear-cut silviculture on dissolved ion export and water yield in the Piedmont. *Water resources research*, 20, 7, 1030–8.

Heywood, V. H., 1989, Patterns, extents and modes of invasions by terrestrial plants. In J. A. Drake (ed.), *Biological invasions: a global perspective*. Chichester: Wiley, 31–55.

Hickey, J. J. and Anderson, O. W., 1968, Chlorinated hydrocarbons and eggshell changes in raptorial and fish-eating birds. *Science*, 162, 271–2.

Hill, A. R., 1975, Ecosystems stability in relation to stresses caused by human activities. *Canadian geographer*, 19, 206–20.

Hillel, D., 1971, Artificial inducement of runoff as a potential source of water in arid lands. In W. G. McGinnies, B. J. H. Goldman and P. Paylore (eds), *Food, fiber and the arid lands*. Tucson: University of Arizona Press, 324–30.

Hills, T. L., 1965, Savannas: a review of a major research problem in tropical geography. *Canadian geographer*, 9, 216–28.

Hobbs, P. V. and Radke, L. F., 1992, Airborne studies of the smoke from the Kuwait oil fires. *Science*, 256, 987–91.

Holdgate, M. W., 1979, *A perspective on environmental pollution*. Cambridge: Cambridge University Press.

Holdgate, M. W. and Wace, N. M., 1961, The influence of man on the floras and faunas of southern islands. *Polar record*, 10, 473–93.

Holdgate, M. W., Kassas, M. and White, G. F., 1982, *The world environment 1972–1982*. Dublin: Tycooly.

Hollis, G. E., 1975, The effects of urbanization on floods of different recurrence interval. *Water resources research*, 11, 431–5.

—— , 1978, The falling levels of the Caspian and Aral Seas. *Geographical journal*, 144, 62–80.

—— , 1988, Rain, roads, roofs and runoff: hydrology in cities. *Geography*, 73, 9–18.

Hollis, G. E. and Luckett, J. K., 1976, The response of natural river channels to urbanization: two case studies from Southeast England. *Journal of hydrology*, 30, 351–63.

Holtz, W. G., 1983, The influence of vegetation on the swelling and shrinking of clays in the United States of America. *Géotechnique*, 33, 159–63.

Holzer, T. L., 1979, Faulting caused by groundwater extraction in South-central Arizona. *Journal of geophysical research*, 84, 603–12.

Holzner, W., Werger, M. J. A., Werger, I. and Ikusima, I. (eds), 1983, *Man's impact on vegetation*. The Hague: Junk.

Hopkins, B., 1965, Observations on savanna burning in the Olikemeji forest reserve, Nigeria. *Journal of applied ecology*, 2, 367–81.

Hornbeck, J. W., 1981, Acid rain: facts and fallacies. *Journal of forestry*, 79, 438–43.

Hotes, F. L. and Pearson, E. A., 1977, Effects of irrigation on water quality. In E. B. Worthington (ed.), *Arid land irrigation in developing countries: environmental problems and effects.* Oxford: Pergamon, 127–58.

Houghton, J. T., 1997, *Global warming: the complete briefing* (2nd edn). Cambridge: Cambridge University Press.

Houghton, J. T., Jenkins, G. J. and Ephraums, J. J., 1990, *Climate change: the IPCC Scientific Assessment.* Cambridge: Cambridge University Press.

Houghton, J. T., Callander, B. A. and Varney, S. K. (eds), 1992, *Climate change 1992: the supplementary report of the IPCC scientific assessment.* Cambridge: Cambridge University Press.

Houghton, J. T., Meira Filho, L. G., Callander, B. A., Harris, N., Kaltenberg, A. and Maskell, K. (eds), 1996, *Climate change 1995: the science of the climate change.* Cambridge: Cambridge University Press.

Houghton, R. H. and Skole, D. L., 1990, Carbon. In B. L. Turner II (ed.), *The earth transformed by human action.* Cambridge: Cambridge University Press, 393–408.

Howard, K. W. F. and Beck, P. J., 1993, Hydrochemical implications of groundwater contamination by road de-icing chemicals. *Journal of Contaminant Hydrology*, 12, 245–68.

Howe, G. M., Slaymaker, H. O. and Harding, D. M., 1966, Flood hazard in mid-Wales. *Nature*, 212, 584–5.

—— , 1967, Some aspects of the flood hydrology of the upper catchments of the Severn and Wye. *Transactions of the Institute of British Geographers*, 41, 33–58.

Hoyt, D. V. and Fröhlich, C., 1983, Atmospheric transmissions at Davos, Switzerland, 1909–1979. *Climatic change*, 5, 61–71.

Huang, W., Clochon, R., Gu, Y., Larick, R., Fang, Q., Schwartz, H., Yonge, C., de Vos, J. and Rink, W., 1995, Early *homo* and associated artefacts from Asia. *Nature*, 378, 275–8.

Hudson, B. J., 1979, Coastal land reclamation with special reference to Hong Kong. *Reclamation review*, 2, 3–16.

Hudson, N., 1971, *Soil conservation.* London: Batsford.

—— , 1987, Soil and water conservation in semi-arid areas. *FAO soils bulletin*, 55.

Hughes, M. K., Lepp, N. W. and Phipps, D. A., 1980, Aerial heavy metal pollution and terrestrial ecosystems. *Advances in ecological research*, 11, 217–327.

Hughes, R. J., Sullivan, M. E. and Yok, D., 1991, Human-induced erosion in a highlands catchment in Papua New Guinea: the prehistoric and contemporary records. *Zeitschrift für Geomorphologie, Supplement-band*, 83, 227–39.

Hughes, T. P., 1994, Catastrophes, phase shifts, and large-scale degradation of a Caribbean coral reef. *Science*, 265, 1547–51.

Hull, S. K. and Gibbs, J. N., 1991, Ash dieback: a survey of non-woodland trees. *Forestry Commission Bulletin*, 93, 32.

Huntington, E., 1914, *The climatic factor as illustrated in arid America.* Carnegie Institution of Washington Publication, 192.

Hurd, L. E., Mellinger, M. W., Wold, L. L. and McNaughton, S. J., 1971, Stability and diversity at three trophic levels in terrestrial successional ecosystems. *Science*, 173, 1134–6.

Husar, R. B. and Husar, J. D., 1991, Sulfur. In B. L. Turner, W. C. Clark, R. W. Kates, J. F. Richards, J. T. Matthews and W. B. Meyer (eds), *The earth as transformed by human action.* Cambridge: Cambridge University Press, 409–21.

Hutchinson, G. E., 1973, Eutrophication. *American scientist*, 61, 269–79.

Hutchinson, G. L. and Mosier, A. R., 1979, Nitrous oxide emissions from an irrigated cornfield. *Science*, 205, 1125–7.

Hutchinson, T. C. and Havas, M., 1980, *Effects of acid precipitation on terrestrial ecosystems.* New York: Plenum.

Huybrechts, P., Letreguilly, A. and Rech, N., 1990, The Greenland ice sheet and greenhouse warming. *Palaeogeography, Palaeoclimatology, Palaeoecology*, 89, 399–412.

Idso, S. B., 1982, *Carbon dioxide: friend or foe?* Tempe: IBR Press.

——, 1983, Carbon dioxide and global temperature: what the data show. *Journal of environmental quality*, 12, 159–63.

——, 1989, *Carbon dioxide and global change: earth in transition.* Tempe, Arizona: IBR Press.

Idso, S. B. and Brazel, A. J., 1978, Climatological effects of atmospheric particulate pollution. *Nature*, 274, 781–2.

Ikawa-Smith, F., 1980, Current issues in Japanese archaeology. *American scientist*, 68, 134–45.

Illies, J., 1974, *Introduction to zoogeography.* London: Macmillan.

Imeson, A. C., 1971, Heather burning and soil erosion on the North Yorkshire Moors. *Journal of applied ecology*, 8, 537–41.

Innes, J. L., 1983, Lichenometric dating of debris-flow deposits in the Scottish Highlands. *Earth surface processes and landforms*, 8, 579–88.

——, 1987, *Air pollution and forestry.* Forestry Commission bulletin, 70.

——, 1992, Forest decline. *Progress in Physical Geography*, 16, 1–64.

Innes, J. L. and Boswell, R. C., 1990, Monitoring of forest condition in Great Britain 1989. *Forestry Commission Bulletin*, 94, 57.

Institute of Hydrology, 1991, *Institute of Hydrology Report 1990–91.* Wallingford: Institute of Hydrology.

Irving, W. M., 1985, Context and chronology of early man in the Americas. *Annual review of anthropology*, 14, 529–55.

Isaac, E., 1970, *Geography of domestication.* Englewood Cliffs: Prentice Hall.

Isachenko, A. G., 1974, On the so-called anthropogenic landscapes. *Soviet geography*, 15, 467–75.

——, 1975, Landscape as a subject of human impact. *Soviet geography*, 16, 631–43.

Isaksen, I. S. A. (ed.), 1988, *Tropospheric ozone: regional and global scale interactions.* Dordrecht: Reidel.

Ives, J. D. and Messerli, B., 1989, *The Himalayan dilemma: reconciling development and conservation.* London: Routledge.

Jacks, G. V. and Whyte, R. O., 1939, *The rape of the earth: a world survey of soil erosion.* London: Faber & Faber.

Jacobs, J., 1969, *The economy of cities.* New York: Random House.

——, 1975, Diversity, stability and maturity in ecosystems influenced by human activities. In W. H. Van Dobben and R. H. Lowe-McConnell (eds), *Unifying concepts in ecology.* The Hague: Junk, 187–207.

Jacobsen, T. and Adams, R. M., 1958, Salt and silt in ancient Meso-potamian agriculture. *Science*, 128, 1251–8.

Jarman, M. R., 1977, Early animal husbandry. In J. Hutchinson, J. G. G. Clark, E. M. Jope and R. Riley (eds), *The early history of agriculture*. Oxford: Oxford University Press, 85–97.

Jarvis, P. H., 1979, The ecology of plant and animal introductions. *Progress in physical geography*, 3, 187–214.

Jeffries, M., 1997, *Biodiversity and conservation*. London: Routledge.

Jennings, J. N., 1952, *The origin of the Broads*. Royal Geographical Society research series, 2.

——, 1966, Man as a geological agent. *Australian journal of science*, 28, 150–6.

Jenny, H., 1941, *Factors of soil formation*. New York: McGraw-Hill.

Jickells, T. D., Carpenter, R. and Liss, P. S., 1991, Marine environ-ment. In B. L. Turner, W. C. Clark, R. W. Kates, J. F. Richards, J. T. Matthews and W. B. Meyer (eds), *The earth as transformed by human action*. Cambridge: Cambridge University Press, 313–34.

Joern, A. and Keeler, K. H. (eds), 1995, *The changing prairie*. New York: Oxford University Press.

Johannessen, C. L., 1963, Savannas of interior Honduras. *Ibero-Americana*, 46.

Johnson, A. I. (ed.), 1991, *Land subsidence*. International Association of Hydrological Sciences Publication 200.

Johnson, D. L. and Lewis, L. A., 1995, *Land degradation: creation and destruction*. Oxford: Blackwell.

Johnson, N. M., 1979, Acid rain: neutralization within the Hubbard Brook ecosystem and regional implications. *Science*, 204, 497–9.

Johnston, D. W., 1974, Decline of DDT residues in migratory song-birds. *Science*, 186, 841–2.

Johnston, D. W., Turner, J. and Kelly, J. M., 1982, The effects of acid rain on forest nutrient status. *Water resources research*, 18, 448–61.

Jones, D. K. C., 1983, Human occupance and the physical environ-ment. In R. J. Johnston and J. C. Doornkamp (eds), *The changing geography of the United Kingdom*. London: Methuen, 327–61.

Jones, J. A. A., Liu, C., Woo, M.-K. and Kung, H.-T. (eds), 1996, *Regional hydrological response to climate change*. Dordrecht: Kluwer.

Jones, P. D., Raper, S. C. B., Bradley, R. S., Diaz, H. F., Kelly, P. M. and Wigley, T. M. L., 1986, Northern hemisphere surface air temperature variations, 1851–1984. *Journal of climate and applied meteorology*, 25, 161–79.

Jones, R., Benson-Evans, K. and Chambers, F. M., 1985, Human influence upon sedimentation in Llangorse Lake, Wales. *Earth sur-face processes and landforms*, 10, 227–35.

Judd, W. R., 1974, Seismic effects of reservoir impounding. *Engineering geology*, 8, 1–212.

Judson, S., 1968, Erosion rates near Rome, Italy. *Science*, 160, 1444–5.

Kadomura, H., 1983, Some aspects of large-scale land transformation due to urbanization and agricultural development in recent Japan. *Advances in space research*, 2, 8, 169–78.

——, 1994, Climatic change, droughts, desertification and land degradation in the Sudano-Sahelian region: a historico-geographical perspective. In H. Kadomura (ed.), *Savannization processes in tropical*

Africa II. Tokyo: Department of Geography, Tokyo Metropolitan University, 203–28.

Karnes, L. B., 1971, Reclamation of wet and overflow lands. In G.-H. Smith (ed.), *Conservation of natural resources*. New York: Wiley, 241–55.

Kasperson, V. X., Kasperson, R. E. and Turner, B. L. II, 1995, *Regions at risk: comparisons of threatened environments*. Tokyo: United Nations University Press.

Kates, R. W., Turner, B. L. I. and Clark, W. C., 1990, The great transformation. In B. L. Turner, W. C. Clark, R. W. Kates, J. F. Richards, J. T. Matthews and W. B. Meyer (eds), *The earth as transformed by human action*. Cambridge: Cambridge University Press, 1–17.

Kauppi, P. and Posch, M., 1988, A case study of the effects of CO_2-induced climatic warming on forest growth and the forest sector. In M. L. Parry, T. R. Carter and N. T. Konijn (eds), *The impact of climatic variations in agriculture*, vol. 1. Dordrecht: Kluwer, 183–95.

Keefer, D. K., de France, S. D., Mosely, M. E., Richardson, J. B., Satterlee, D. R. and Day-Lewis, A., 1998, Early maritime economy and El Niño events at Quebrada Tachuay, Peru. *Science*, 281, 1833–1935.

Keller, E. A., 1976, Channelisation: environmental, geomorphic and engineering aspects. In D. R. Coates (ed.), *Geomorphology and engineering*. Stroudsburg: Dowden, Hutchinson and Ross, 115–40.

Kellman, M., 1975, Evidence for late glacial age fire in a tropical montane savanna. *Journal of biogeography*, 2, 57–63.

Kellogg, W. W., 1977, *Effects of human activities on global climate*. WHO technical note, 156.

——, 1978, Global influence of mankind on the climate. In J. Gribbin (ed.), *Climatic change*. London: Cambridge University Press, 205–27.

——, 1982, Precipitation trends on a warmer earth. In R. A. Peck and J. R. Hummel (eds), *Interpretation of climate and photochemical models, ozone and temperature measurements*. New York: American Institute of Physics, 35–46.

Kemp, D. D., 1994, *Global environmental issues: a climatological approach* (2nd edn). London: Routledge.

Kent, M., 1982, Plant growth problems in colliery spoil reclamation. *Applied geography*, 2, 83–107.

Khalil, M. A. K. and Rasmussen, R. A., 1987, Atmospheric methane: trends over the last 10,000 years. *Atmospheric environment*, 21, 2445–52.

Kiersch, G. A., 1965, The Vaiont reservoir disaster. *Mineral information service*, 18, 129–38.

Kilgore, B. M. and Taylor, D., 1979, Fire history of a sequoia–mixed conifer forest. *Ecology*, 60, 129–42.

King, C. A. M., 1974, Coasts. In R. U. Cooke and J. C. Doornkamp (eds), *Geomorphology in environmental management*. Oxford: Clarendon Press, 188–222.

——, 1975, *Introduction to physical and biological oceanography*. London: Edward Arnold.

Kinsey, D. W. and Hopley, D., 1991, The significance of coral reefs as global carbon sinks – response to greenhouse. *Palaeogeography, Palaeoclimatology, Palaeoecology*, 89, 363–77.

Kirby, C., 1995, Urban air pollution. *Geography*, 80, 375–92.

Kirch, P. V., 1982, Advances in Polynesian prehistory: three decades in review. *Advances in world archaeology*, 2, 52–102.

Kirkpatrick, J., 1994, *A continent transformed: human impact on the natural vegetation of Australia*. Melbourne: Oxford University Press.

Kittredge, J. H., 1948, *Forest influences*. New York: McGraw-Hill.

Klein, R. G., 1983, The stone age prehistory of southern Africa. *Annual review of anthropology*, 12, 25–48.

Knox, J. C., 1977, Human impacts on Wisconsin stream channels. *Annals of the Association of American Geographers*, 67, 323–42.

——, 1987, Historical valley floor sedimentation in the Upper Mississippi Valley. *Annals of the Association of American Geographers*, 77, 224–44.

Kohen, J., 1995, *Aboriginal environmental impacts*. Sydney: University of New South Wales Press.

Koide, M. and Goldberg, E. D., 1971, Atmospheric and fossil fuel combustion. *Journal of geophysical research*, 76, 6589–96.

Komar, P. D., 1976, *Beach processes and sedimentation*. Englewood Cliffs: Prentice Hall.

Komarov, B., 1978, *The destruction of nature in the Soviet Union*. London: Pluto Press.

Koster, E. A., 1994, Global warming and periglacial landscapes. In N. Roberts (ed.), *The changing global environment*. Oxford: Blackwell, 150–72.

Kotlyakov, V. M., 1991, The Aral Sea basin: a critical environmental zone. *Moscow Environment*, 33(1), 4–9, 36–8.

Kovda, V. A., 1980, *Land aridization and drought control*. Boulder: Westview Press.

Kramer, R., van Schaik, C. and Johnson, J., 1997, *Last stand: protected areas and the defense of tropical biodiversity*. New York: Oxford University Press.

Krantz, G. S., 1970, Human activities and megafaunal extinctions. *American scientist*, 58, 164–70.

Krug, E. C. and Frink, C. R., 1983, Acid rain on acid soil: a new perspective. *Science*, 221, 520–5.

Krutson, T. R., Tuleya, R. E. and Kurihara, Y., 1998, Simulated increase of hurricane intensities in a CO_2 warmed world. *Science*, 279, 1018–20.

Kühlmann, D. H., 1988, The sensitivity of coral reefs to environmental pollution. *Ambio*, 17, 13–21.

Kuo, C., 1986, Flooding in Taipeh, Taiwan and coastal drainage. In J. G. Titus (ed.), *Effects of changes in stratospheric ozone and global climate*. Washington DC: UNEP/USEPA, 37–46.

Kwong, Y. T. J. and Tau, T. Y., 1994, Northward migration of permafrost along the Mackenzie Highway and climatic warming. *Climatic Change*, 26, 399–419.

La Marche, V. C., Graybill, D. A., Fritts, H. C. and Rose, M. R., 1984, Increasing atmospheric carbon dioxide: tree ring evidence for growth enhancement in natural vegetation. *Science*, 225, 1019–21.

Labadz, J. C., Burt, T. P. and Potter, A. W. L., 1991, Sediment yield and delivery in the blanket peat moorlands of the southern Pennines. *Earth Surface Processes and Landforms*, 16, 255–71.

Lamb, H. H., 1977, *Climate: present, past and future. 2: Climatic history and the future.* London: Methuen.

Lambert, J. H., Jennings, J. N., Smith, C. T., Green, C. and Hutchinson, J. N., 1970, *The making of the Broads: a reconsideration of their origin in the light of new evidence.* Royal Geographical Society research series, 3.

Lamprey, H., 1975, The integrated project on arid lands. *Nature and resources*, 14, 2–11.

Landes, K. K., 1973, Mother nature as an oil polluter. *Bulletin of the American Association of Petroleum Geologists*, 57, 637–41.

Landsberg, H. E., 1970, Man-made climatic changes. *Science*, 170, 1265–8.

—— , 1981, *The urban climate.* New York: Academic Press.

Langford, T. E., 1972, A comparative assessment of thermal effects in some British and North American rivers. In R. T. Oglesby, C. A. Carlson and J. A. McCann (eds), *River ecology and man.* New York: Academic Press, 318–51.

Langford, T. E. L., 1990, *Ecological effects of thermal discharges.* London: Elsevier Applied Science.

Lanly, J. P. and Clement, J., 1979, Present and future natural forest and plantation areas in the tropics. *Unasylva*, 31, 12–20.

Lanly, J. P., Singh, K. D. and Janz, K., 1991, FAO's 1990 reassessment of tropical forest cover. *Nature and Resources*, 27, 21–6.

Laporte, L. F., 1975, *Encounter with the earth.* San Francisco: Canfield Press.

Larick, R. and Ciochon, R. L., 1996, The African emergence and early Asian dispersals of the genus Homo. *American Scientist*, 84, 538–51.

La Roe, E. T., 1977, Dredging – ecological impacts. In J. R. Clarke (ed.), *Coastal ecosystem management.* New York: Wiley, 610–14.

Larson, F., 1940, The role of bison in maintaining the short grass plains. *Ecology*, 21, 113–21.

Lawson, D. E., 1986, Response of permafrost terrain to disturbance: a synthesis of observations from northern Alaska, USA. *Arctic and Alpine research*, 18, 1–17.

Le Houérou, H. N., 1977, Biological recovery versus desertization. *Economic geography*, 63, 413–20.

Lean, J. and Warrilow, D. A., 1989, Simulation of the regional climatic impact of Amazon deforestation. *Nature*, 342, 126–33.

Lee, D. O., 1992, Urban warming – an analysis of recent trends in London's heat island. *Weather*, 47, 50–6.

Leggett, J., 1996, *Climate change and the financial sector.* Munich: Gerling Akademie.

Lemon, P. C., 1968, Effects of fire on an African plateau grassland. *Ecology*, 49, 316–22.

Lently, A. D., 1994, Agriculture and wildlife: ecological implications of subsurface irrigation drainage. *Journal of Arid Environments*, 28, 85–94.

Lents, J. M. and Kelly, W. J., 1993, Clearing the air in Los Angeles. *Scientific American*, October, 18–25.

Leopold, L. B., 1951, Rainfall frequency: an aspect of climatic variation. *Transactions of the American Geophysics Union*, 32, 347–57.

Leopold, L. B., Wolman, M. G. and Miller, J. P., 1964, *Fluvial processes in geomorphology.* San Francisco: Freeman.

Lewin, J., Bradley, S. B. and Macklin, M. G., 1983, Historical valley alluviation in mid-Wales. *Geological journal*, 18, 331–50.

Liddle, M., 1997, *Recreation ecology*. London: Chapman & Hall.

Likens, G. E. and Bormann, F. H., 1974, Acid rain: a serious regional environmental problem. *Science*, 184, 1176–9.

Likens, G. E. and Butler, T. J., 1981, Recent acidification of precipitation in North America. *Atmospheric environment*, 15, 1103–9.

Likens, G. E., Wright, R. F., Galloway, J. N. and Butler, T. J., 1979, Acid rain. *Scientific American*, 241, 4, 39–47.

Lloyd, J. W., 1986, A review of aridity and groundwater. *Hydrological processes*, 1, 63–78.

Lockwood, J. G., 1979, *Causes of climate*. London: Arnold.

Lowe, P. D., 1983, Values and institutions in the history of British nature conservation. In A. Warren and F. B. Goldsmith (eds), *Conservation in perspective*. Chichester: Wiley, 329–52.

Lowe-McConnell, R. H., 1975, Freshwater life on the move. *Geographical magazine*, 47, 768–75.

Lugo, A. E., Cintron, G. and Goenaga, C., 1981, Mangrove ecosystems under stress. In G. W. Barrett and R. Rosenberg (eds), *Stress effects on natural ecosystems*. Chichester: John Wiley, 129–53.

Lugo, A. E., 1988, Estimating reductions in the diversity of tropical forest species. In E. O. Wilson (ed.), *Biodiversity*. Washington DC: National Academy Press.

Luke, R. H., 1962, *Bush fire control in Australia*. Melbourne: Hodder & Stoughton.

Lund, J. W. G., 1972, Eutrophication. *Proceedings of the Royal Society of London*, 180B, 371–82.

Lyell, C., 1835, *Principles of geology* (4th edn), vol. III. London: Murray (12th edn 1875).

Lynch, J. A., Rishel, G. B. and Corbett, E. S., 1984, Thermal alteration of streams draining clearcut watersheds: quantification and biological implications. *Hydrobiologia*, 111, 161–9.

Mabbutt, J. A., 1985, Desertification of the world's rangelands. *Desertification Control Bulletin*, 12, 1–11.

Macdonald, G. J., 1982, *The long-term impacts of increasing atmospheric carbon dioxide levels*. Cambridge, Mass.: Ballinger.

Macfarlane, M. J., 1976, *Laterite and landscape*. London: Academic Press.

Macklin, M. G. and Lewin, J., 1986, Terraced fills of Pleistocene and Holocene age in the Rheidol Valley, Wales. *Journal of Quaternary Science*, 1, 21–34.

Macklin, M. G., Passmore, D. G., Stevenson, A. C., Colwey, A. C., Edwards, D. N. and O'Brien, C. F., 1991, Holocene alluviation and land-use change on Callaly Moor, Northumberland, England. *Journal of Quaternary Science*, 6, 225–32.

McKnight, T. L., 1959, The feral horse in Anglo-America. *Geographical review*, 49, 506–25.

——, 1971, Australia's buffalo dilemma. *Annals of the Association of American Geographers*, 61, 759–73.

McLennan, S. M., 1993, Weathering and global denudation. *Journal of Geology*, 101, 295–303.

McTainsh, G. and Boughton, W. C., 1993, *Land degradation processes in Australia*. Melbourne: Longman Cheshire.

Mader, H. J., 1984, Animal habitat isolation by roads and agricultural fields. *Biological conservation*, 29, 81–96.

Magilligan, F. J., 1985, Historical floodplain sedimentation in the Galena River basin, Wisconsin and Illinois. *Annals of Association of American Geographers*, 75, 583–94.

Maignien, R., 1966, *A review of research on laterite*. UNESCO, Natural resources research, 4.

Mainguet, M., 1995, *L'homme et la sécheresse*. Paris: Masson.

Makkaveyev, N. I., 1972, The impact of water engineering projects on geomorphic processes in stream valleys. *Soviet geography*, 13, 387–93.

Maltby, E., 1986, *Waterlogged wealth. Why waste the world's wet places?* London: Earthscan.

Manabe, S. and Stouffer, R. J., 1980, Sensitivity of a global climate model to an increase of CO_2 concentration in the atmosphere. *Journal of atmospheric science*, 37, 99–118.

Manabe, S. and Wetherald, R. T., 1986, Reduction in summer soil wetness by an increase in atmospheric carbon dioxide. *Science*, 232, 626–8.

Manabe, S., Wetherald, R. T. and Stouffer, R. J., 1981, Summer dryness due to an increase of atmospheric CO_2 concentration. *Climate change*, 3, 347–86.

Mandel, S., 1977, The overexploitation of groundwater resources in dry regions. In Y. Munklak and S. F. Singer (eds), *Arid zone development: potentialities and problems*. Cambridge, Mass.: Ballinger, 31–52.

Manners, I. R., 1978, Agricultural activities and environmental stress. In K. A. Hammond (ed.), *Sourcebook of the environment*. Chicago: University of Chicago Press, 263–94.

Manners, I. R. and Mikesell, M. W. (eds), 1974, *Perspectives on environment*. Washington DC: Association of American Geographers.

Mannion, A. M., 1991, *Global environmental change*. Harlow: Longman.

—— , 1992, Acidification and eutrophication. In A. M. Mannion and S. R. Bowlby (eds), *Environmental issues in the 1990s*. Chichester: Wiley, 177–95.

—— , 1995, *Agriculture and environmental change*. Chichester: Wiley.

—— , 1997, *Global environmental change* (2nd edn). Harlow: Longman.

Manshard, W., 1974, *Tropical agriculture*. London: Longman.

Mark, A. F. and McSweeney, G. D., 1990, Patterns of impoverishment in natural communities: case studies in forest ecosystems – New Zealand. In G. M. Woodwell (ed.), *The earth in transition: patterns and processes of biotic impoverishment*. Cambridge: Cambridge University Press, 151–76.

Marker, M. E., 1967, The Dee estuary: its progressive silting and salt marsh development. *Transactions of the Institute of British Geographers*, 41, 65–71.

Marks, P. L. and Bormann, F. H., 1972, Revegetation following forest cutting: mechanisms for return to steady-state nutrient cycling. *Science*, 176, 914–15.

Marquiss, M., Newton, I. and Ratcliffe, D. A., 1978, The decline of the raven, *Corvus corax*, in relation to afforestation in southern Scotland and northern England. *Journal of applied ecology*, 15, 129–44.

Marsh, G. P., 1864, *Man and nature*. New York: Scribner.

——, 1965, *Man and nature*, edited by D. Lowenthal. Cambridge, Mass.: Belknap Press.

Marshall, L. G., 1984, Who killed Cock Robin? An investigation of the extinction controversy. In P. S. Martin and R. G. Klein (eds), *Quaternary extinctions*. Tucson: University of Arizona Press.

Martens, L. A., 1968, Flood inundation and effects of urbanization in Metropolitan Charlotte, North Carolina. *United States geological survey water supply paper*, 1591-C.

Martin, P. S., 1967, Prehistoric overkill. In P. S. Martin and H. E. Wright (eds), *Pleistocene extinctions*. New Haven: Yale University Press, 75–120.

——, 1974, Palaeolithic players on the American stage: man's impact on the late Pleistocene megafauna. In J. D. Ives and R. G. Barry (eds), *Arctic and alpine environments*. London: Methuen.

——, 1982, The pattern and meaning of Holarctic mammoth extinction. In D. M. Hopkins, J. V. Matthews, C. S. Schweger and S. B. Young (eds), *Paleoecology of Beringia*. New York: Academic Press, 399–408.

Martin, P. S. and Klein, R. G., 1984, *Pleistocene extinctions*. Tucson: University of Arizona Press.

Martin, P. S. and Wright, H. E. (eds), 1967, *Pleistocene extinctions*. New Haven: Yale University Press, 75–120.

Martinez, J. D., 1971, Environmental significance of salt. *Bulletin of the American Association of Petroleum Geologists*, 55, 810–25.

Marx, J. L., 1975, Air pollution: effects on plants. *Science*, 187, 731–3.

Mather, A. S., 1983, Land deterioration in upland Britain. *Progress in physical geography*, 7, 2, 210–28.

Matley, I. M., 1966, The Marxist approach to the geographical environment. *Annals of the Association of American Geographers*, 56, 97–111.

Matthews, W. H., Kellogg, W. W. and Robinson, G. D. (eds), 1971, *Man's impact on the climate*. Cambridge, Mass.: MIT Press.

Maugh, T. H., 1979, The Dead Sea is alive and well . . . *Science*, 205, 178.

May, R. M., 1979, Fluctuations in abundance of tropical insects. *Nature*, 278, 505–7.

May, T., 1991, Südspanische matorrales als Kulturofolge-vegetation. *Geoökodynamik* 12, 87–107.

Mead, W. R., 1954, Ridge and furrow in Buckinghamshire. *Geographical journal*, 120, 34–42.

Meade, R. H., 1991, Reservoirs and earthquakes. *Engineering Geology*, 30, 245–62.

——, 1996, River-sediment input to major deltas. In J. D. Milliman and B. V. Haq (eds), *Sea-level rise and coastal subsidence*. Dordrecht: Kluwer, 63–85.

Meade, R. H. and Parker, R. S., 1985, Sediment in rivers in the United States. *United States Geological Survey Water Supply Paper*, 2276, 49–60.

Meade, R. H. and Trimble, S. W., 1974, Changes in sediment loads in rivers of the Atlantic drainage of the United States since 1900. *Publication of the International Association of Hydrological Science*, 113, 99–104.

Meadows, M. E. and Linder, H. P., 1993, A palaeoecological perspective on the origin of Afromontane grasslands. *Journal of Biogeography*, 20, 345–55.

Mee, L. D., 1992, The Black Sea in crisis: a need for concerted international action. *Ambio*, 21, 278–86.

Mellanby, K., 1967, *Pesticides and pollution*. London: Fontana.

Mendelssohn, H., 1973, Ecological effects of chemical control of rodents and jackals in Israel. In H. T. Farvar and J. P. Milton (eds), *The careless technology*. London: Tom Stacey, 527–44.

Mercer, D. E. and Hamilton, L. S., 1984, Mangrove ecosystems: some economic and natural benefits. *Nature and resources*, 20, 14–19.

Mercer, J. H., 1978, West Antarctic ice sheet and CO_2 greenhouse effect: a threat of disaster. *Nature*, 271, 321–5.

Merryfield, D. L. and Moore, P. D., 1971, Prehistoric human activity and blanket peat initiation on Exmoor. *Nature*, 250, 439–41.

Meybeck, M., 1979, Concentration des eaux fluviales en éléments majeurs et apports en solution aux océans. *Revue de géographie physique et géologie dynamique*, 21a, 215–46.

Meyer, W. B., 1996, *Human impact on the earth*. Cambridge: Cambridge University Press.

Meyer, W. B. and Turner, B. L. II (eds), 1994, *Changes in land use and land cover: a global perspective*. Cambridge: Cambridge University Press.

Micklin, P. P., 1972, Dimensions of the Caspian Sea problem. *Soviet geography*, 13, 589–603.

Middleton, N. J., 1999, *The global casino* (2nd edn). London: Arnold.

Middleton, N. J. and Thomas, D. S. G., 1997, *World atlas of desertification* (2nd edn). London: Edward Arnold.

Mieck, I., 1990, Reflections on a typology of historical pollution: complementary conceptions. In P. Brimblecombe and C. Pfister (eds), *The silent countdown*. Berlin: Springer Verlag, 73–80.

Mikesell, M. W., 1969, The deforestation of Mount Lebanon. *Geographical review*, 59, 1–28.

Miller, R. S. and Botkin, D. B., 1974, Endangered species: models and predictions. *American scientist*, 62, 172–81.

Milliman, J. D., 1990, Fluvial sediment in coastal seas: flux and fate. *Nature and Resources*, 26, 12–22.

Milliman, J. D., Broadus, J. M. and Gable, F., 1989, Environmental and economic impacts of rising sea level and subsiding deltas: the Nile and Bengal examples. *Ambio*, 18, 340–5.

Milliman, J. D., Qin, Y. S., Ren, M. E. and Yoshiki Saita, 1987, Man's influence on erosion and transport of sediment by Asian rivers: the Yellow River (Huanghe) example. *Journal of geology*, 95, 751–62.

Milne, W. G., 1976, Induced seismicity. *Engineering geology*, 10, 83–388.

Ministry of Agriculture, Fisheries and Food, 1976, *Agriculture and water quality*. London: Her Majesty's Stationery Office.

Mintzer, I. M. and Miller, A. S., 1992, Stratospheric ozone depletion: can we save the sky? In *Green Globe Yearbook 1992*. Oxford: Oxford University Press, 83–91.

Mitchell, J. F. B., 1983, The seasonal response of a general circulation model to changes in CO_2 and sea temperatures. *Quarterly journal of the Royal Meteorological Society*, 109, 113–52.

Mitchell, J. F. B., Wilson, C. A. and Cunnington, W. M., 1987, On CO_2 climate sensitivity and model dependence of results. *Quarterly journal of the Royal Meteorological Society*, 113, 293–322.

Montgomery, D. R., 1997, What's best on the banks? *Nature*, 388, 328–9.

Mooney, H. A. and Parsons, D. J., 1973, Structure and function of the California Chaparral – an example from San Dimas. *Ecological studies*, 7, 83–112.

Moore, D. M., 1983, Human impact on island vegetation. In W. Holzner, M. J. A. Werger and I. Ikusima (eds), *Man's impact on vegetation*. The Hague: Junk, 237–48.

Moore, N. W., Hooper, M. D. and Davis, B. N. K., 1967, Hedges, I. Introduction and reconnaissance studies. *Journal of applied ecology*, 4, 201–20.

Moore, P. D., 1973, Origin of blanket mires. *Nature*, 256, 267–9.

—— , 1986, Unravelling human effects. *Nature*, 321, 204.

Moore, R. M., 1959, Ecological observations on plant communities grazed by sheep in Australia. In A. Keast, R. L. Crocker and C. S. Christian (eds), *Biogeography and ecology in Australia*. The Hague: Junk, 500–13.

Moore, T. R., 1979, Land use and erosion in the Machakos Hills. *Annals of the Association of American Geographers*, 69, 419–31.

Morgan, G. S. and Woods, C. A., 1986, Extinction and the zoogeography of West Indian land mammals. *Biological journal of the Linnean Society*, 28, 167–203.

Morgan, R. P. C., 1977, *Soil erosion in the United Kingdom: field studies in the Silsoe area, 1973–75*. National College of Agricultural Engineering, occasional paper, 4.

—— , 1979, *Soil erosion*. London: Longman.

—— , 1995, *Soil erosion and conservation* (2nd edn). Harlow: Longman.

Morgan, W. B. and Moss, R. P., 1965, Savanna and forest in Western Nigeria. *Africa*, 35, 286–93.

Mote, V. L. and Zumbrunne, C., 1977, Anthropogenic environmental alteration of the Sea of Azov. *Soviet geography*, 18, 744–59.

Moyle, P. B., 1976, Fish introductions in California: history and impact on native fishes. *Biological conservation*, 9, 101–18.

Mrowka, J. P., 1974, Man's impact on stream regimen and quality. In I. R. Manners and M. W. Mikesell (eds), *Perspectives on environment*, Washington DC: Association of American Geographers.

Mudge, G. P., 1983, The incidence and significance of ingested lead pellet poisoning in British wildfowl. *Biological conservation*, 27, 333–72.

Muhly, J. D., 1997, Artifacts of the Neolithic, Bronze and Iron Ages. In E. M. Myers (ed.), *The Oxford encyclopaedia of archaeology in the Near East*. New York: Oxford University Press, vol. 4, 5–15.

Muhs, D. R. and Matat, P. B., 1993, The potential response of eolian sands to greenhouse warming and precipitation reduction on the Great Plains of the United States. *Journal of Arid Environments*, 25, 351–61.

Munn, R. E., 1996, Global change: both a scientific and a political issue. In R. E. Munn, J. W. M. La Rivière and N. van Lookeren Campagne (eds), *Policy making in an era of global environmental change*. Dordrecht: Kluwer, 1–15.

Murdoch, W. W., 1975, Diversity, complexity, stability and pest control. *Journal of applied ecology*, 12, 795–807.

Murozumi, M., Chow, T. J. and Paterson, C., 1969, Chemical concentrations of pollutant lead aerosols, terrestrial dusts and sea salt in Greenland and Antarctic snow strata. *Geochimica et cosmoschimica acta*, 33, 1247–94.

Murton, R. K., 1971, *Man and birds*. London: Collins.

Musk, L. F., 1991, The fog hazard. In A. H. Perry and L. J. Symons (eds), *Highway meteorology*. London: Spon, 91–130.

Myers, N., 1979, *The sinking ark: a new look at the problem of disappearing species*. Oxford: Pergamon Press.

——, 1983, Conversion rates in tropical moist forests. In F. B. Golley (ed.), *Tropical rain forest ecosytems*. Amsterdam: Elsevier Scientific, 289–300.

——, 1984, *The primary source: tropical forests and our future*. New York: Norton.

——, 1988, *Natural resource systems and human exploitation systems: physiobiotic and ecological linkages*. World Bank policy planning and research staff, environment department working paper, 12.

——, 1990, The biodiversity challenge: expanded hot-spots analysis. *The Environmentalist*, 10 (4), 243–56.

——, 1992, Future operational monitoring of tropical forests: an alert strategy. In J. P. Mallingreau, R. da Cunha and C. Justice (eds), *Proceedings World Forest Watch Conference*. Sao Jose dos Campos, Brazil, 9–14.

Mylne, M. F. and Rowntree, P. R., 1992, Modelling the effects of albedo change associated with tropical deforestation. *Climatic Change*, 21, 317–43.

Mylona, S., 1996, Sulphur dioxide emissions in Europe, 1880–1991 and their effect on sulphur concentrations and depositions. *Tellus*, 48B, 662–89.

Nakagawa, K., 1996, Recent trends of urban climatological studies in Japan, with special emphasis on the thermal environments of urban areas. *Geographical Review of Japan*, ser. B, 69, 206–24.

Nakano, T. and Matsuda, I., 1976, A note on land subsidence in Japan. *Geographical reports of Tokyo Metropolitan University*, 11, 147–62.

National Academy of Sciences, 1972, *The earth and human affairs*. San Francisco: Camfield Press.

Nature Conservancy Council, 1977, *Nature conservation and agriculture*. London: Her Majesty's Stationery Office.

——, 1984, *Nature conservation in Great Britain*. Shrewsbury: Nature Conservancy Council.

Nesterov, A. I., 1974, The development of anthropogenic landscape science at the Geography Faculty of Voronezh University. *Soviet geography*, 15, 463–6.

Newman, J. R., 1979, Effects of industrial pollution on wildlife. *Biological conservation*, 15, 181–90.

Newman, W. S. and Fairbridge, R. W., 1986, The management of sea-level rise. *Nature*, 320, 319–21.

Newton, J. G., 1976, Induced and natural sinkholes in Alabama: continuing problem along highway corridors. In F. R. Zwanig (ed.), *Subsidence over mines and caverns*. Washington DC: National Academy of Sciences, 9–16.

Nicholson, S. E., 1978, Climatic variations in the Sahel and other African regions during the past five centuries. *Journal of arid environments*, 1, 3–24.

——, 1988, Land surface atmosphere interaction: physical processes and surface changes and their impact. *Progress in physical geography*, 12, 36–65.

Nicod, J., 1986, Facteurs physico-chimiques de l'accumulation des formations travertineuses. *Mediterranée*, 10, 161–4.

Nihlgård, B. J., 1997, Forest decline and environmental stress. In D. Brune, D. V. Chapman, M. D. Gwynne and J. M. Pacyna (eds), *The global environment*. Weinheim: VCH.

Nikonov, A. A., 1977, Contemporary technogenic movements of the Earth's crust. *International geology review*, 19, 1245–58.

Nir, D., 1983, *Man, a geomorphological agent: an introduction to anthropic geomorphology*. Jerusalem: Keter.

Nisbet, E., 1990, Climate change and methane. *Nature*, 347, 23.

Nordstrom, K. F., 1994, Beaches and dunes of human-altered coasts. *Progress in Physical Geography*, 18, 497–516.

Nossin, J. J., 1972, Landsliding in the Crati basin, Calabria, Italy. *Geologie en Mijnbouw*, 51, 591–607.

Nowlis, J. S., Roberts, C. M., Smith, A. H. and Siirila, 1997, Human-enhanced impacts of a tropical storm on nearshore coral reefs. *Ambio*, 26, 515–21.

Noy-Meir, I., 1974, Stability in arid ecosystems and effects of men on it. *Proceedings of the 12th International Congress of Ecology*, Wageningen, 220–5.

Nriagu, J. O., 1979, Global inventory of natural and anthropogenic emissions of trace metals in the atmosphere. *Nature*, 279, 409–11.

Nriagu, J. O. and Pacyna, J. M., 1988, Quantitative assessment of worldwide contamination of air, water and soils by trace metals. *Nature*, 337, 134–9.

Nunn, P. D., 1991, *Human and natural impacts on Pacific island environments*. Occasional paper of the East–West Environment and Policy Institute, 13. Honolulu.

Nutalaya, P. and Ran, J. L., 1981, Bangkok: the sinking metropolis. *Episodes*, 4, 3–8.

Nye, P. H. and Greenland, D. J., 1964, Changes in the soil after clearing tropical forest. *Plant and Soil*, 21, 101–12.

Oberle, M., 1969, Forest fires: suppression policy has its ecological drawbacks. *Science*, 165, 568–71.

OECD (Organisation for Economic Co-operation and Development), 1991, *The state of the environment*. Paris: OECD.

Oke, T. R., 1978, *Boundary layer climates*. London: Methuen.

Oppenheimer, M., 1998, Global warming and the stability of the West Antarctic ice sheet. *Nature*, 393, 325–32.

O'Riordan, T. (ed.), 1995, *Environmental science for environmental management*. Harlow: Longman Scientific.

O'Sullivan, P. E., Coard, M. A. and Pickering, D. A., 1982, The use of laminated lake sediments in the estimation and calibration of erosion rates. *Publication of the International Association of Hydrological Science*, 137, 385–96.

Otterman, J., 1974, Baring high albedo soils by overgrazing: a hypothesised desertification mechanism. *Science*, 186, 531–3.

Overpeck, J. T., Rind, D. and Goldberg, R., 1990, Climate-induced changes in forest disturbance and vegetation. *Nature*, 343, 51–3.

Oxley, D. J., Fenton, M. B. and Carmody, G. R., 1974, The effects of roads on populations of small mammals. *Journal of applied ecology*, 11, 51–9.

Ozenda, P. and Borel, J. L., 1990, The possible responses of vegetation to a global climatic change. In M. M. Boer and R. S. de Groot (eds), *Landscape – ecological impact of climatic change*. Amsterdam: IOS Press, 221–49.

Page, H., 1982, Some notes on the geomorphological and vegetational history of the saltings at Brean. *Somerset Archaeology and Natural History*, 120–5.

Page, M. J. and Trustrum, N. A., 1997, A late Holocene lake sediment record of the erosion response to land use change in a steepland catchment, New Zealand. *Zeitschrift für Geomorphologie*, 41, 369–92.

Pakiser, L. C., Eaton, J. P., Healy, J. H. and Raleigh, C. B., 1969, Earthquake prediction and control. *Science*, 166, 1467–74.

Panel on Weather and Climate Modification, 1966, *Weather and climate modification problems and prospects*, publication 1350. Washington DC: National Academy of Sciences.

Panofsky, H., 1976, Man's impact on climate. In T. A. Ferrar (ed.), *The urban costs of climate modification*. New York: Wiley.

Park, C. C., 1977, Man-induced changes in stream channel capacity. In K. J. Gregory (ed.), *River channel change*. Chichester: Wiley, 121–44.

—— , 1987, *Acid rain: rhetoric and reality*. London: Methuen.

Park, R. A., Armentano, T. V. and Cloonan, C. L., 1986, Predicting the effects of sea level rise on coastal wetlands. In J. G. Titus (ed.), *Effects of changes in stratospheric ozone and global climate*, vol. 4, *Sea level rise*. Washington DC: UNEP/USEPA, 129–52.

Parkinson, C. L. and Kellogg, W. W., 1979, Arctic sea ice decay simulated for a CO_2-induced temperature rise. *Climatic change*, 2, 149–62.

Parsons, J. J., 1960, Fog drip from coastal stratus. *Weather*, 15, 58.

Passmore, J., 1974, *Man's responsibility for nature*. London: Duckworth.

Pawson, E., 1978, *The early industrial revolution: Britain in the eighteenth century*. London: Batsford.

Peck, A. J., 1975, Effects of land use on salt distribution in the soil. *Ecological studies*, 15, 77–90.

—— , 1978, Salinization of non-irrigated soils and associated streams: a review. *Australian journal of soil research*, 16, 157–68.

—— , 1983, Response of groundwaters to clearing in Western Australia. *Papers, international conference on groundwater and man*, 2, 327–35.

Peet, R. K., Glenn-Lewin, D. C. and Wolf, J. W., 1983, Prediction of man's impact on plant species diversity. In W. Holzner, M. J. A. Werger and I. Ikusima (eds), *Man's impact on vegetation*. The Hague: Junk, 41–54.

Peglar, S. M. and Birks, H. J. B., 1993, The mid-Holocene *Ulmus* fall at Diss Mere, Norfolk, south-east England – disease and human impact? *Vegetation History and Archeology*, 2, 61–8.

Peierls, B. L., Caraco, N. F., Pace, M. L. and Cole, J. J., 1991, Human influence on river nitrogen. *Nature*, 350, 386.

Pennington, W., 1974, *The history of British vegetation* (2nd edn). London: English University Press.

——, 1981, Records of a lake's life in time: the sediments. *Hydrobiologia*, 79, 197–219.

Pereira, H. C., 1973, *Land use and water resources in temperate and tropical climates*. Cambridge: Cambridge University Press.

Perla, R., 1978, Artificial release of avalanches in North America. *Arctic and Alpine research*, 10, 235–40.

Pernetta, J., 1995, What is global change? *Global Change Newsletter*, 21, 1–3.

Peters, R. L., 1988, The effect of global climatic change on natural communities. In E. O. Wilson (ed.), *Biodiversity*. Washington DC: National Academy Press, 450–61.

Petersen, K. K., 1981, *Oil shale, the environmental challenges*. Golden, Colorado: Colorado School of Mines.

Pethick, J., 1993, Shoreline adjustments and coastal management: physical and biological processes under accelerated sea level rise. *Geographical Journal*, 159, 162–8.

Petts, G. E., 1979, Complex response of river channel morphology subsequent to reservoir construction. *Progress in physical geography*, 3, 329–62.

——, 1985, *Impounded rivers: perspectives for ecological management*. Chichester: Wiley.

Petts, G. E. and Lewin, J., 1979, Physical effects of reservoirs on river systems. In G. E. Hollis (ed.), *Man's impact on the hydrological cycle in the United Kingdom*. Norwich: Geobooks, 79–91.

Pickering, K. T. and Owen, L. A., 1997, *Global environmental issues* (2nd edn). London: Routledge.

Pimentel, D., 1976, Land degradation: effects on food and energy resources. *Science*, 194, 149–55.

Pimentel, D., Harvey, C., Resosuddarmo, P., Sinclair, K., Kurz, D., McNair, M., Crist, S., Shpritz, L., Fitton, L., Saffouri, R. and Blair, R., 1995, Environmental and economic costs of soil erosion and conservation benefits. *Science*, 267, 1117–22.

Pitt, D., 1979, Throwing light on a black secret. *New scientist*, 81, 1022–5.

Pluhowski, E. J., 1970, Urbanization and its effects on the temperature of the streams on Long Island, New York. *United States geological survey professional paper*, 627-D.

Pollard, E. and Miller, A., 1968, Wind erosion in the East Anglian Fens. *Weather*, 23, 414–17.

Ponting, C., 1991, *A green history of the world*. London: Penguin.

Poore, M. E. D., 1976, The values of tropical moist forest ecosystems. *Unasylva*, 28, 127–43.

Pope, J. C., 1970, Plaggen soils in the Netherlands. *Geoderma*, 4, 229–55.

Porter, E., 1978, *Water management in England and Wales*. Cambridge: Cambridge University Press.

Potter, G. L., Ellsaesser, H. W., MacCracken, M. C. and Luther, F. M., 1975, Possible climatic impact of tropical deforestation. *Nature*, 258, 697–8.

Potter, G. L., Ellsaesser, H. W., MacCracken, M. C. and Ellis, J. C., 1981, Albedo change by man: test of climatic effects. *Nature*, 291, 47–9.

Powell, M., 1985, Salt, seed and yields in Sumerian agriculture: a critique of progressive salinization. *Zeitschrift für Assyrologie und Vorderasiatische Archaologie*, 75, 7–38.

Preston, A., 1973, Heavy metals in British waters. *Nature*, 242, 95–7.

Price, M. F., 1989, Global change: defining the ill-defined. *Environment*, 31 (8), 18–20, 42–4.

Price, M. and Reed, D. W., 1989, The influence of mains leakage and urban drainage on groundwater levels beneath conurbations in the United Kingdom. *Proceedings Institution of Civil Engineers*, 86(I), 31–9.

Prince, H. C., 1959, Parkland in the Chilterns. *Geographical review*, 49, 18–31.

——, 1962, Pits and ponds in Norfolk. *Erdkunde*, 16, 10–31.

——, 1964, The origin of pits and depressions in Norfolk. *Geography*, 49, 15–32.

——, 1979, Marl pits or dolines of the Dorset Chalklands? *Transactions of the Institute of British Geographers*, new series, 4, 116–17.

Proffitt, M. H., Margitan, J. J., Kelly, K. K., Loewenstein, M., Podolske, J. R. and Chan, K. R., 1990, Ozone loss in the Arctic polar vortex inferred from high-altitude aircraft measurements. *Nature*, 347, 31–3.

Prokopovich, N. P., 1972, Land subsidence and population growth. *24th International Geological Congress Proceedings*, 13, 44–54.

Prospero, J. M. and Nees, R. T., 1977, Dust concentration in the atmosphere of the equatorial North Atlantic; possible relationship to Sahelian drought. *Science*, 196, 1196–8.

Pyne, S. J., 1982, *Fire in America – a cultural history of wildland and rural fire*. Princeton: Princeton University Press.

Rackham, O., 1980, *Ancient woodland*. London: Arnold.

Radley, J., 1962, Peat erosion on the high moors of Derbyshire and west Yorkshire. *East Midlands Geographer*, 3 (1), 40–50.

Radley, J. and Sims, C., 1967, Wind erosion in East Yorkshire. *Nature*, 216, 20–2.

Raison, R. J., 1979, Modification of the soil environment by vegetation fires with particular reference to nitrogen transformation: a review. *Plant and Soil*, 51, 73–108.

Raleigh, C. B., Healy, J. H. and Bredehoeft, J. D., 1976, An experiment in earthquake control at Rangeley, Colorado. *Science*, 191, 1230–7.

Ramanathan, V., 1988, The greenhouse theory of climate change: a test by an inadvertent global experiment. *Science*, 240, 293–9.

Ranwell, D. S., 1964, *Spartina* salt marshes in southern England, II: rate and seasonal pattern of sediment accretion. *Journal of Ecology*, 52, 79–94.

Ranwell, D. S. and Boar, R., 1986, *Coast dune management guide*. Monks Wood: Institute of Terrestrial Ecology.

Rapp, A., 1974, *A review of desertization in Africa – water, vegetation and man*. Secretariat for International Ecology, Stockholm, report no. 1.

Rapp, A., Murray-Rust, D. H., Christansson, C. and Berry, L., 1972, Soil erosion and sedimentation in four catchments near Dodoma, Tanzania. *Geografiska annaler*, 54A, 255–318.

Rapp, A., Le Houérou, H. N. and Lundholm, B., 1976, Can desert encroachment be stopped? *Ecological bulletin*, 24.

Rasid, H., 1979, The effects of regime regulation by the Gardiner Dam on downstream geomorphic processes in the South Saskatchewan River. *Canadian geographer*, 23, 140–58.

Rasool, S. I. and Schneider, S. H., 1971, Atmospheric carbon dioxide and aerosols: effects of large increases on global climate. *Science*, 173, 138–41.

Ratcliffe, D. A., 1974, Ecological effects of mineral exploitation in the United Kingdom and their significance to nature conservation. *Proceedings of the Royal Society of London*, 339A, 355–72.

Ray, C., Hayden, B. P., Bulger, A. J. and McCormick-Ray, G., 1992, Effects of global warming on the biodiversity of coastal-marine zones. In R. L. Peters and T. E. Lovejoy (eds), *Global warming and biological diversity*. New Haven: Yale University Press, 91–102.

Reck, R. A, 1975, Aerosols and polar temperature changes. *Science*, 188, 728–30.

Reclus, E., 1871, *The earth* (2 vols). London: Chapman & Hall.

—— , 1873, *The ocean, atmosphere and life*. New York: Harper and Brothers.

Reed, C. A., 1970, Extinction of mammalian megafauna in the old world late Quaternary. *Bioscience*, 20, 284–8.

Reed, D. J., 1990, The impact of sea level rise on coastal salt marshes. *Progress in Physical Geography*, 14, 465–81.

Reed, D. J., 1995, The response of coastal marshes to sea level rise: survival or submergence? *Earth Surface Processes and Landforms*, 20, 39–48.

Reed, L. A., 1980, Suspended-sediment discharge, in five streams near Harrisburg, Pennsylvania, before, during, and after highway construction. *United States geological survey water supply paper*, 2072.

Reij, C., Scoones, I. and Toulmin, C. (eds), 1996, *Sustaining the soil: indigenous soil and water conservation in Africa*. London: Earthscan.

Renberg, I. and Hellberg, T., 1982, The pH history of lakes in SW Sweden, as calculated from the subfossil diatom flora of the sediments. *Ambio*, 11, 30–3.

Revelle, R. R. and Waggoner, P. E., 1983, Effect of a carbon dioxide-induced climatic change on water supplies in the western United States. In Carbon Dioxide Assessment Committee, *Changing climate*. Washington DC: National Academy Press, 419–32.

Rhoades, J. D., 1990, Soil salinity – causes and controls. In A. S. Goudie (ed.), *Techniques for Desert Reclamation*. Chichester: Wiley, 109–34.

Richards, J. F., 1991, Land transformation. In B. L. Turner, W. C. Clark, R. W. Kates, J. F. Richards, J. T. Matthews and W. B. Meyer (eds), *The earth as transformed by human action*. Cambridge: Cambridge University Press, 163–78.

Richards, K. S. and Wood, R., 1977, Urbanization, water redistribution, and their effect on channel progresses. In K. J. Gregory (ed.), *River channel changes*. Chichester: Wiley, 369–88.

Richards, P. W., 1952, *The tropical rainforest*. Cambridge: Cambridge University Press.

Richardson, J. A., 1976, Pit heap into pasture. In J. Lenihan and W. W. Fletcher (eds), *Reclamation*. Glasgow: Blackie, 60–93.

Richardson, S. J. and Smith, J., 1977, Peat wastage in the East Anglian Fens. *Journal of soil science*, 28, 485–9.

Richter, D. O. and Babbar, L. I., 1991, Soil diversity in the tropics. *Advances in Ecological Research*, 21, 315–89.

Rinderer, T. E., Oldroyd, R. P. and Sheppard, W. S., 1993, Africanized bees in the US. *Scientific American*, 269 (6), 52–8.

Ripley, E. A., 1976, Drought in the Sahara: insufficient geophysical feedback? *Science*, 191, 100.

Ritchie, J. C., 1972, Sediment, fish and fish habitat. *Journal of soil and water conservation*, 27, 124–5.

Robb, G. A. and Robinson, J. D. F., 1995, Acid drainage from mines. *Geographical Journal*, 161, 47–54.

Roberts, N., 1998, *The Holocene: an environmental history* (2nd edn). Oxford: Basil Blackwell.

Roberts, N. (ed.), 1994, *The changing global environment*. Oxford: Blackwell.

Robin, G. de Q., 1986, Changing the sea level. In B. Bolin et al. (eds), *The greenhouse effect, climatic change and ecosystems*. Chichester: Wiley, 322–59.

Robinson, M., 1979, The effects of pre-afforestation ditching upon the water and sediment yields of a small upland catchment. Working paper 252, School of Geography, University of Leeds.

—— , 1990, *Impact of improved land drainage on river flows*. Institute of Hydrology, Wallingford, Report 113.

Robinson, M. A. and Lambrick, G. H., 1984, Holocene alluviation and hydrology in the Upper Thames Basin. *Nature*, 308, 809–14.

Rodda, J. C., Downing, R. A. and Law, F. M., 1976, *Systematic hydrology*. London: Newnes-Butterworth.

Romme, W. H. and Despain, D. G., 1989, The Yellowstone fires. *Scientific American*, 261, 21–9.

Roots, C., 1976, *Animal invaders*. Newton Abbot: David & Charles.

Rose, R., 1970, Lichens as pollution indicators. *Your environment*, 5.

Rosenzweig, C. and Hillel, D., 1993, The dust bowl of the 1930s: analog of greenhouse effect in the Great Plains? *Journal of Environmental Quality*, 22, 9–22.

Rosepiler, M. J. and Reilinger, R., 1977, Land subsidence due to water withdrawal in the vicinity of Pecos, Texas. *Engineering geology*, 11, 295–304.

Routson, R. C., Wildung, R. E. and Bean, R. M., 1979, A review of the environmental impact of ground disposal of oil shale wastes. *Journal of environmental quality*, 8, 14–19.

Royal Commission on Environmental Pollution, 1972, *Pollution in some British estuaries and coastal waters*. London: Her Majesty's Stationery Office.

Royal Society Study Group, 1983, *The nitrogen cycle of the United Kingdom*. London: The Royal Society.

Rozanov, B. G., Targulian, V. and Orlov, D. S., 1991, Soils. In B. L. Turner, W. C. Clark, R. W. Kates, J. F. Richards, J. T. Matthews and W. B. Meyer (eds), *The earth as transformed by human action*. Cambridge: Cambridge University Press, 203–14.

Russell, E. J., 1961, *The world of the soil*. London: Fontana.

Russell, J. S. and Isbell, R. F. (eds), 1986, *Australian soils: the human impact*. St Lucia: University of Queensland Press.

Ryder, M. L., 1966, The exploitation of animals by man. *Advancement of science*, 23, 9–18.

Ryding, S. O. and Rast, R. W., 1989, *The control of eutrophication of lakes and reservoirs*. Paris: UNESCO.

Sabadell, J. E., Risley, E. M., Jorgensen, H. T. and Thornton, B. S., 1982, *Desertification in the United States: Status and Issues*. Bureau of Land Management, Department of the Interior.

Sahagian, D. L., Schwartz, F. W. and Jacobs, D. K., 1994, Direct anthropogenic contribution to sea level rise in the twentieth century. *Nature*, 367, 54–7.

Sanchez, P. A. and Buol, S. W., 1975, Soils of the tropics and the world food crisis. *Science*, 188, 598–603.

Sanders, W. M., 1972, Nutrients. In R. T. Oglesby, C. A. Carlson and J. A. McCann (eds), *River ecology and man*. New York: Academic Press, 389–415.

Sarmiento, G. and Monasterio, M., 1975, A critical consideration of the environmental conditions associated with the occurrence of savanna ecosystems in tropical America. *Ecological studies*, 11, 233–50.

Sarre, P., 1978, The diffusion of Dutch elm disease. *Area*, 10, 81–5.

Sauer, C. O., 1938, Destructive exploitation in modern colonial expansion. *International Geographical Congress, Amsterdam*, vol. III, sect. IIIC, 494–9.

——, 1952, *Agricultural origins and dispersals*, Isaiah Bowman lecture series, 2. New York: American Geographical Society.

——, 1969, *Seeds, spades, hearths and herds*. Cambridge, Mass.: MIT Press.

Savage, M., 1991, Structural dynamics of a southwestern pine forest under chronic human influence. *Annals of the Association of American Geographers*, 81, 271–89.

Savini, J. and Kammerer, J. C., 1961, Urban growth and the water regime. *United States geological survey water supply paper*, 1591A.

Sawyer, J. S., 1971, Possible effects of human activity on world climate. *Weather*, 26, 252–62.

Schimper, A. F. W., 1903, *Plant-geography upon a physiological basis*. Oxford: Clarendon Press.

Schindler, D. W., 1974, Eutrophication and recovery in experimental lakes: implications for lake management. *Science*, 184, 897–9.

Schlesinger, M. E. and Mitchell, J. F. B., 1985, Model projections of equilibrium response to increased carbon dioxide. In M. C. MacCracken and F. M. Luther (eds), *Projecting the climatic effects of increasing carbon dioxide*. Washington DC: US Dept of Energy, 81–147.

Schmid, J. A., 1974, The environmental impact of urbanization. In I. R. Manners and M. W. Mikesell (eds), *Perspectives on environment*. Washington DC: Association of American Geographers.

——, 1975, Urban vegetation. Research paper, Department of Geography, University of Chicago, 161.

Schmieder, O., 1927a, The Pampa – a natural or culturally induced grassland? *University of California publications in geography*, 2, 255–70.

——, 1927b, Alteration of the Argentine Pampa in the colonial period. *University of California publications in geography*, 2, 303–21.

Schneider, S. H., 1984, On the empirical verification of model-predicted CO_2-induced climatic effects. *Geophysical monograph*, 29, 187–201.

Schneider, S. H. and Thompson, S. L., 1988, Simulating the effects of nuclear war. *Nature*, 333, 221–7.

Schumm, S. A., 1977, *The fluvial system*. New York: Wiley.

Schumm, S. A., Harvey, M. D. and Watson, C. C., 1984, *Incised channels: morphology, dynamics and control*. Littleton, Colorado: Water Resources Publications.

Schwarz, E. H. L., 1923, *The Kalahari or Thirsland redemption*. Cape Town and Oxford: Oxford University Press.

Schwarz, H. E., Emel, J., Dickens, W. J., Rogers, P. and Thompson, J., 1991, Water quality and flows. In B. L. Turner, W. C. Clark, R. W. Kates, J. F. Richards, J. T. Matthews and W. B. Meyer (eds), *The earth as transformed by human action*. Cambridge: Cambridge University Press, 253–70.

Scott, D. F., 1997, The contrasting effects of wildfire and clear felling on the hydrology of a small catchment. *Hydrological Processes*, 11, 543–55.

Scott, G. J., 1977, The role of fire in the creation and maintenance of savanna in the Montana of Peru. *Journal of biogeography*, 4, 143–67.

Scott, W. S. and Wylie, N. P., 1980, The environmental effects of snow dumping: a literature review. *Journal of Environmental Management*, 10, 219–40.

Sears, P. B., 1957, Man the newcomer: the living landscape and a new tenant. In L. H. Russwurm and E. Sommerville (eds), *Man's natural environment, a system approach*. North Scituate: Duxbury, 43–55.

Segall, P., 1989, Earthquakes triggered by fluid extraction. *Geology*, 17, 942–6.

Selby, M. J., 1979, Slopes and weathering. In K. J. Gregory and D. E. Walling (eds), *Man and environmental processes*. Folkestone: Dawson, 105–22.

Semaw, S., Renne, P., Harris, J. W. K., Feibel, C. S., Bernov, R. L., Fesseha, N. and Mowbray, K., 1997, 2.5-million-year-old stone tools from Gona, Ethiopia. *Nature*, 385, 333–6.

Semtner, A. J., 1984, The climate response of the Arctic Ocean to Soviet river diversions. *Climatic change*, 6, 109–30.

Shaler, N. S., 1912, *Man and the earth*. New York: Duffield.

Sheail, J., 1971, *Rabbits and their history*. Newton Abbot: David & Charles.

Sheffield, A. T., Healy, T. R. and McGlone, M. S., 1995, Infilling rates of a steepland catchment estuary, Whangamata, New Zealand. *Journal of Coastal Research*, 11 (4), 1294–1308.

Sherlock, R. L., 1922, *Man as a geological agent*. London: Witherby.

Sherratt, A., 1981, Plough and pastoralism: aspects of the secondary products revolution. In I. Hodder, G. Isaac and N. Hammond (eds), *Pattern of the past*. Cambridge: Cambridge University Press, 261–305.

Sherratt, A., 1997, Climatic cycles and behaviour revolutions: the emergence of modern humans and the beginning of farming. *Antiquity*, 71, 271–87.

Shiklomanov, I. A., 1985, Large scale water transfers. In J. C. Rodda (ed.), *Facets of hydrology II*. Chichester: Wiley, 345–87.

Shirahata, H., Elias, R. W., Patterson, C. C. and Koide, M., 1980, Chronological variations in concentrations and isotopic composition of anthropogenic atmospheric lead in sediments of a remote sub-alpine pond. *Geochimica et cosmochimica acta*, 44, 149–62.

Simmonds, N. W., 1976, *Evolution of crop plants*. London: Longman.

Simmons, I., 1979, *Biogeography: natural and cultural*. London: Arnold.

——, 1989, *Changing the face of the earth*. Oxford: Basil Blackwell.

——, 1993, *Environmental history: a concise introduction*. Oxford: Blackwell.

——, 1996, *Changing the face of the earth: culture, environment and history*, 2nd edn. Oxford: Blackwell.

Simon, J. L., 1981, *The ultimate resource*. Princeton: Princeton University Press.

——, 1996, *The ultimate resource 2*. Princeton: Princeton University Press.

—— (ed.), 1995, *The state of humanity*. Oxford: Blackwell.

Simon, J. L. and Kahn, H., 1984, *The resourceful earth – a response to global 2000*. Oxford: Basil Blackwell.

Simoons, F. J., 1974, Contemporary research themes in the cultural geography of domesticated animals. *Geophysical review*, 64, 557–76.

Sinclair, A. R. E. and Fryxell, J. M., 1985, The Sahel of Africa: ecology of a disaster. *Canadian journal of zoology*, 63, 987–94.

Smith, C. T., 1978, *An historical geography of western Europe before 1800* (2nd edn). London: Longman.

Smith, J. B. and Tirpak, D. A. (eds), 1990, *The potential effects of global climate change on the United States*. New York: Hemisphere.

Smith, J. B., Carmichael, J. J. and Titus, J. G., 1995, Adaptation policy. In K. M. Strzepak and J. B. Smith (eds), *As climate changes: international impacts and implications*. Cambridge: Cambridge University Press, 201–10.

Smith, J. E. (ed.), 1968, *Torrey Canyon pollution and marine life*. Cambridge: Cambridge University Press.

Smith, K., 1975, *Principles of applied climatology*. London: McGraw-Hill.

Smith, L. B. and Hadley, C. H., 1926. The Japanese beetle. *Department Circular, US Department of Agriculture*, 363, 1–66.

Smith, M. W., 1993, Climate change and permafrost. In H. M. French and O. Slaymaker (eds), *Canada's cold environments*. Montreal and Kingston: McGill-Queens University Press, 292–311.

Smith, N., 1976, *Man and water*. London: Davies.

Smith, T. M., Shugart, H. H., Bonan, G. B. and Smith, J. B., 1992, Modelling the potential response of vegetation to global climate change. *Advances in Ecological Research*, 22, 93–116.

Smith, W. H., 1974, Air pollution – effects on the structure and function of the temperate forest ecosystem. *Environmental pollution*, 6, 111–29.

Snedaker, S. C., 1993, Impact on mangroves. In G. Maul (ed.), *Climatic change in the Intra-Americas Sea*. London: Edward Arnold, 282–305.

Snelgrove, A. K., 1967, *Geohydrology of the Indus River, West Pakistan*. Hyderabad: Sind University Press.

So, C. L., 1971, Mass movements associated with the rainstorm of 1966 in Hong Kong. *Transactions of the Institute of British Geographers*, 53, 55–65.

Solomon, A. M. and West, D. C., 1986, Atmospheric carbon dioxide change: agent of future forest growth or decline. In J. G. Titus (ed.), *Effects of changes in stratospheric ozone and global climate*, vol. 3, *Climatic change*. Washington DC: UNEP/USEPA, 23–8.

Somerville, M., 1858, *Physical geography* (4th edn). London: Murray.

Sopper, W. E., 1975, Effects of timber harvesting and related management practices on water quality in forested watersheds. *Journal of environmental quality*, 4, 24–9.

Sorkin, A. J., 1982, *Economic aspects of natural hazards*. Lexington: Lexington Books.

Southwick, C. H., 1976, *Ecology and the quality of our environment* (2nd edn). New York: Van Nostrand.

Spanier, E. and Galil, B. S., 1991, Lessepsian migration: a continuous biogeographical process. *Endeavour*, 15, 102–6.

Sparks, B. W. and West, R. G., 1972, *The Ice Age in Britain*. London: Methuen.

Spate, O. H. K. and Learmonth, A. T. A., 1967, *India and Pakistan*. London: Methuen.

Speight, M. C. D., 1973, Oudoor recreation and its ecological effects. A bibliography and review. Discussion papers in conservation, 4, University College, London.

Spencer, J. E. and Hale, G. A., 1961, The origin, nature and distribution of agricultural terracing. *Pacific viewpoint*, 2, 1–40.

Spencer, J. E. and Thomas, W. L., 1978, *Introducing cultural geography* (2nd edn). New York: Wiley.

Spencer, T., 1995, Potentialities, uncertainties and complexities in the response of coral reefs to future sea level rise. *Earth Surface Processes and Landforms*, 20, 49–64.

Spencer, T. and Douglas, I., 1985, The significance of environmental change: diversity, disturbance and tropical ecosystems. In I. Douglas and T. Spencer (eds), *Environmental change and tropical geomorphology*. London: Allen & Unwin, 13–33.

Sperling, C. H. B., Goudie, A. S., Stoddart, D. R. and Poole, G. C., 1979, Origin of the Dorset dolines. *Transactions of the Institute of British Geographers*, new series, 4, 121–4.

Speth, W. W., 1977, Carl Ortwin Sauer on destructive exploitation. *Biological Conservation*, 11, 145–60.

Stanley, D. J., 1996, Nile delta: extreme case of sediment entrapment on a delta plain and consequent coastal land loss. *Marine Geology*, 129, 189–95.

State of the Environment Advisory Council, 1996, *Australia: state of the environment 1996*. Collingwood, Victoria, Aus.: CSIRO Publishing.

Steadman, D. W., Stafford, T. W., Donahue, D. J. and Jull, A. J. T., 1991, Chronology of Holocene vertebrate extinction in the Galápagos Islands. *Quaternary Research*, 36, 126–33.

Steinberg, C. E. W. and Wright, R. F. (eds), 1994, *Acidification of freshwater ecosystems: implications for the future*. Chichester: Wiley.

Stensland, G. J. and Semonin, R. G., 1982, Another interpretation of the pH trend in the United States. *Bulletin of the American Meteorological Society*, 63, 11, 1277–84.

Stephens, J. C., 1956, Subsidence of organic soils in the Florida Everglades. *Proceedings of the Soil Science Society of America*, 20, 77–80.

Sternberg, H. O'R., 1968, Man and environmental change in South America. *Monographiae biologicae*, 18, 413–45.

Stetler, L. L. and Gaylord, D. R., 1996, Evaluating eolian-climate interactions using a regional climate model from Hanford, Washington (USA). *Geomorphology*, 17, 99–113.

Stevenson, A. C., Jones, V. J. and Battarbee, R. W., 1990, The cause of peat erosion: a palaeolimnological approach. *New Phytologist*, 114, 727–35.

Stewart, O. C., 1956, Fire as the first great force employed by man. In W. L. Thomas (ed.), *Man's role in changing the face of the earth*. Chicago: University of Chicago Press, 115–33.

Stiles, D., 1995, An overview of desertification as dryland degradation. In D. Stiles (ed.), *Social aspects of sustainable dryland management*. Chichester: Wiley.

Stocking, M., 1984, Erosion and soil productivity: a review. *FAO soil conservation programme land and water development division consultants' working paper*, 1.

Stoddart, D. R., 1968, Catastrophic human interference with coral atoll ecosystems. *Geography*, 53, 25–40.

——, 1971, Coral reefs and islands and catastrophic storms. In J. A. Steers (ed.), *Applied coastal geomorphology*. London: Macmillan, 154–97.

——, 1990, Coral reefs and islands and predicted sea-level rise. *Progress in Physical Geography*, 14, 521–36.

Stokes, S. and Gaylord, D., 1993, Geochronology of Clear Creek Dune Field. A test of the performance of optical dating. *Quaternary Research*, 39, 274–81.

Stott, P. A., 1978, Tropical rain forest in recent ecological thought: The reassessment of a non-renewable resource. *Progress in Physical Geography*, 2, 80–98.

Strahler, A. N. and Strahler, A. H., 1973, *Environmental geoscience: interaction between natural systems and man*. Santa Barbara: Hamilton.

Strandberg, C. H., 1971, Water pollution. In G. H. Smith (ed.), *Conservation of natural resources* (4th edn). New York: Wiley, 189–219.

Sturrock, F. and Cathie, J., 1980, Farm modernisation and the countryside. Occasional paper 12, Department of Land Economy, University of Cambridge.

Sugden, D. E., 1991, The stepped response of ice sheets to climatic change. In C. Harris and B. Stonehouse (eds), *Antarctica and global climatic change*. London: Bellhaven Press, 107–14.

Swank, W. T. and Douglass, J. E., 1974, Streamflow greatly reduced by converting deciduous hardwood stands to pine. *Science*, 18, 857–9.

Swanston, D. N. and Swanson, F. J., 1976, Timber harvesting, mass erosion and steepland forest geomorphology in the Pacific north-west. In D. R. Coates (ed.), *Geomorphology and engineering*. Stroudberg: Dowden, Hutchinson & Ross, 199–221.

Swift, J., 1974, *The other Eden. A new approach to man, nature and society*. London: Dent.

Swift, L. W. and Messer, J. B., 1971, Forest cuttings raise temperatures of small streams in the southern Appalachians. *Journal of soil and water conservation*, 26, 111–16.

Swift, M. J. and Sanchez, P. A., 1984, Biological management of tropical soil fertility for sustained productivity. *Nature and resources*, 20, 2–10.

Swisher, C. C., Curtis, G. H., Jacob, T., Getty, A. G. and Suprijo, A., 1994, Age of the earliest known hominids in Java, Indonesia. *Science*, 263, 1118–21.

Tajikistan Academy of Sciences, 1975, *Induced seismicity of the Nurek reservoir*. Dushanbe: Tajik Academy.

Tallis, J. H., 1965, Studies on southern Pennine peats, IV: evidence of recent erosion. *Journal of Ecology*, 53, 509–20.

Tallis, J. H., 1985, Erosion of blanket peat in the southern Pennines: new light on an old problem. In R. H. Johnson (ed.), *The geomorphology of north-west England*. Manchester: Manchester University Press, 313–36.

Taylor, C., 1975, *Fields in the English landscape*. London: Dent.

Taylor, J. A., 1985, Bracken encroachment rates in Britain. *Soil use and management*, 1, 53–6.

Taylor, K. E. and Penner, J. E., 1994, Response of the climatic system to atmospheric aerosols and greenhouse gases. *Nature*, 369, 734–7.

Terborgh, J., 1992, *Diversity and the tropical rain forest*. New York: Freeman.

Thirgood, J. V., 1981, *Man and the Mediterranean forest – a history of resource depletion*. London: Academic Press.

Thomas, D. S. G. and Middleton, N. J., 1994, *Desertification: exploding the myth*. Chichester: Wiley.

Thomas, R. H., Sanderson, T. J. O. and Rose, K. E., 1979, Effect of climatic warming on the West Antarctic ice sheet. *Nature*, 277, 355–8.

Thomas, W. L. (ed.), 1956, *Man's role in changing the face of the Earth*. Chicago: University of Chicago Press.

Thompson, J. R., 1970, Soil erosion in the Detroit metropolitan area. *Journal of Soil and Water Conservation*, 25, 8–10.

Thompson, R., 1978, Atmospheric contamination – a review of the air pollution problem. *Geographical Papers, Department of Geography, University of Reading*, 68.

Thornthwaite, C. W., 1956, Modification of the rural microclimates. In W. L. Thomas (ed.), *Man's role in changing the face of the Earth*. Chicago: University of Chicago Press, 567–83.

Tickell, C., 1993, The human species: a suicidal success. *Geographical Journal*, 159, 215–26.

Tiffen, M., Mortimore, M. and Gichuki, F. N., 1994, *More people, less erosion: environmental recovery in Kenya*. London: Wiley.

Titus, J. G. (ed.), 1986, *Effects of changes in stratospheric ozone and global climate*, vol. 4, *Sea level rise*. Washington DC: UNEP/USEPA.

Titus, J. G. and Seidel, S., 1986, Overview of the effects of changing the atmosphere. In J. G. Titus (ed.), *Effects of changes in stratospheric ozone and global climate*. Washington DC: UNEP/USEPA, 3–19.

Tivy, J., 1971, *Biogeography. A study of plants in the ecosphere*. Edinburgh: Oliver and Boyd.

Tivy, J. and O'Hare, G., 1981, *Human impact on the ecosystem*. Edinburgh: Oliver and Boyd.

Todd, D. K., 1959, *Groundwater hydrology*. New York: Wiley.

——, 1964, Groundwater. In Ven Te Chow (ed.), *Handbook of applied hydrology*, sect. 13. New York: McGraw-Hill.

Tolba, M. K. and El-Kholy, O. A., 1992, *The world environment, 1972–1992*. London: UNEP/Chapman & Hall.

Tomaselli, R., 1977, Degradation of the Mediterranean maquis. *UNESCO, MAB technical note*, 2, 33–72.

Tomlinson, T. E., 1970, Trends in nitrate concentration in English rivers in relation to fertilizer use. *Water treatment and examination*, 19, 277–89.

Trimble, S. W., 1974, *Man-induced soil erosion on the southern Piedmont*. Soil Conservation Society of America.

——, 1976, Modern stream and valley sedimentation in the Driftless Area, Wisconsin, USA. *23rd International Geographical Congress*, sect. 1, 228–31.

——, 1988, The impact of organisms on overall erosion rates within catchments in temperate regions. In H. A. Viles (ed.), *Biogeomorphology*. Oxford: Basil Blackwell, 83–142.

——, 1997a, Contribution of stream channel erosion to sediment yield from an urbanizing watershed. *Science*, 278, 1442–4.

——, 1997b, Stream channel erosion and change resulting from riparian forests. *Geology*, 25, 467–9.

Trimble, S. W. and Lund, S. W., 1982, Soil conservation and the reduction of erosion and sedimentation in the Coon Creek Basin, Wisconsin. *US geological survey professional paper*, 1234.

Trimble, S. W. and Mendel, A. C., 1995, The cow as a geomorphic agent: a critical review. *Geomorphology*, 13, 233–53.

Troels-Smith, J., 1956, Neolithic period in Switzerland and Denmark. *Science*, 124, 876–9.

Tubbs, C., 1984, *Spartina* on the south coast: an introduction. In P. Doody (ed.), *Spartina anglica in Great Britain*. Shrewsbury: Nature Conservancy Council, 3–4.

Turco, R. P., Toon, O. B., Ackermann, T. P., Pollack, J. B. and Sagan, C., 1983, Nuclear winter: global consequences of multiple nuclear explosions. *Science*, 222, 1283–92.

Turner, B. L. II (ed.), 1990, *The earth as transformed by human action*. Cambridge: Cambridge University Press.

Turner, B. L., Kasperson, R. E., Meyer, W. B., Dow, K. M., Golding, D., Kasperson, J. X., Mitchell, R. C. and Ratick, S. J., 1990, Two types of global environmental change: definitional and spatial-scale issues in their human dimensions. *Global Environmental Change*, 1, 14–22.

Turner, I. M., 1996, Species loss in fragments of tropical rain forest: a review of the evidence. *Journal of Applied Ecology*, 33, 200–9.

Tyldesley, J. A. and Bahn, P. G., 1983, The use of plants in the European palaeolithic: a review of the evidence. *Quaternary Science Reviews*, 2, 53–81.

UNEP, 1989, *Environmental data report*. Oxford and Cambridge, Mass.: Basil Blackwell.

——, 1991, *United Nations Environment Programme environmental data report*, 3rd edn. Oxford: Basil Blackwell.

Uri, N. D. and Lewis, J. A., 1998, The dynamics of soil erosion in US agriculture. *Science of the Total Environment*, 218, 45–58.

US Bureau of Entomology and Plant Quarantine, 1941, *Insect pest survey bulletin*, 21, 801–2.

US Department of Energy, 1985, *Glaciers, ice sheets, and sea level: effect of a CO_2-induced climatic change*. Washington DC: US Dept of Energy.

Usher, M. B., 1973, *Biological management and conservation*. London: Chapman & Hall.

Vale, T. R., 1974, Sagebrush conversion projects: an element of contemporary environmental change in the western United States. *Biological conservation*, 6, 272–84.

Vale, T. R. and Vale, G. R., 1976, Suburban bird population in west-central California. *Journal of biogeography*, 3, 157–65.

Vandermeulen, J. H. and Hrudey, S. E. (eds), 1987, *Oil in freshwater: chemistry, biology, countermeasure technology*. New York: Pergamon.

Vankat, J. L., 1977, Fire and man in Sequoia National Park. *Annals of the Association of American Geographers*, 67, 17–27.

Vaudour, J., 1986, Travertins holocènes et pression anthropique. *Méditerranée*, 10, 168–73.

Veblen, T. T. and Stewart, G. H., 1982, The effects of introduced wild animals on New Zealand forests. *Annals of the Association of American Geographers*, 72, 372–97.

Vendrov, S. L., 1965, A forecast of changes in natural conditions in the northern Ob'basin in case of construction of the lower Ob'Hydro Project. *Soviet geography*, 6, 3–18.

Vice, R. B., Guy, H. P. and Ferguson, G. E., 1969, Sediment movement in an area of suburban highway construction, Scott Run Basin, Fairfax, County, Virginia, 1961–64. *United States geological survey water supply paper*, 1591-E.

Viessman, W., Knapp, J. W., Lewis, G. L. and Harbaugh, T. E., 1977, *Introduction to hydrology* (2nd edn). New York: IEP.

Viets, F. G., 1971, Water quality in relation to farm use of fertilizer. *Bioscience*, 21, 460–7.

Viles, H. A. and Spencer, T., 1995, *Coastal problems: geomorphology, ecology and society at the coast*. London: Edward Arnold.

Vine, H., 1968, Developments in the study of soils and shifting agriculture in tropical Africa. In R. P. Moss (ed.), *The soil resources of tropical Africa*. Cambridge: Cambridge University Press, 89–119.

Vita-Finzi, C., 1969, *The Mediterranean valleys*. Cambridge: Cambridge University Press.

Vitousek, P. M., Gosz, J. R., Gruer, C. C., Melillo, J. M., Reiners, W. A. and Todd, R. L., 1979, Nitrate losses from disturbed ecosystems. *Science*, 204, 469–73.

Vogl, R. J., 1974, Effects of fires on grasslands. In T. T. Kozlowski and C. C. Ahlgren (eds), *Fire and ecosystems*. New York: Academic Press, 139–94.

——, 1977, Fire: a destructive menace or a rational process. In J. Cairns, K. L. Dickson and E. E. Herricks (eds), *Recovery and restoration of damaged ecosystems*. Charlottesville: University Press of Virginia, 261–89.

Von Broembsen, S. L., 1989, Invasions of natural ecosystems by plant pathogens. In J. A. Drake (ed.), *Biological invasions: a global perspective*. Chichester: Wiley, 77–83.

Wagner, R. H., 1974, *Environment and man*. New York: Norton.

Waithaka, J. M., 1996, Elephants: a keystone species. In T. R. Mc-Clanahan and T. P. Young (eds), *East African ecosystems and their conservation*. New York: Oxford University Press, 284–5.

Waldichuk, M., 1979, Review of the problems. *Philosophical transactions of the Royal Society*, 286B, 399–429.

Walker, H. J., 1988, *Artificial structures and shorelines*. Dordrecht: Kluwer.

Walker, H. J., Coleman, J. M., Roberts, H. H. and Tye, R. S., 1987, Wetland loss in Louisiana. *Geografiska annaler*, 69A, 189–200.

Wall, G. and Wright, C., 1977, The environmental impact of outdoor recreation. *Department of Geography publication series, University of Waterloo, Canada*, 11.

Walling, D. E. and Gregory, K. J., 1970, The measurement of the effects of building construction on drainage basin dynamics. *Journal of hydrology*, 11, 129–44.

Walling, D. E. and Quine, T. A., 1991, *Recent rates of soil loss from areas of arable cultivation in the UK*. IAHS Publication 203, 123–31.

Wallwork, K. L., 1956, Subsidence in the mid-Cheshire industrial area. *Geographical journal*, 122, 40–53.

—— , 1960, Some problems of subsidence and land use in the mid-Cheshire industrial area. *Geographical journal*, 126, 191–9.

—— , 1974, *Derelict land*. Newton Abbot: David & Charles.

Walsh, K. and Pittock, B., 1998, Potential changes in tropical storms, hurricanes, and extreme rainfall events as a result of climate change. *Climate Change*, 39, 199–213.

Walsh, R. P., Hudson, R. N. and Howells, K. A., 1982, Changes in the magnitude-frequency of flooding and heavy rainfalls in the Swansea Valley since 1875. *Cambria*, 9, 2, 36–60.

Wang, W. C., Yung, Y. L., Lacis, A. A., Mo, T. and Hanson, J. E., 1976, Greenhouse effects due to man-made perturbations of trace gases. *Science*, 194, 685–90.

Ward, R. C., 1978, *Floods – a geographical perspective*. London: Macmillan.

Ward, S. D., 1979, Limestone pavements – a biologist's view. *Earth science conservation*, 16, 16–18.

Warren, A. and Maizels, J. K., 1976, *Ecological change and desertification*. London: University College.

—— , 1977, Ecological change and desertification. In United Nations, *Desertification: its causes and consequences*. Oxford: Pergamon, 171–260.

Warrick, R. A., 1988, Carbon dioxide, climatic change and agriculture. *Geographical journal*, 154, 221–33.

Watson, A., 1976, The origin and distribution of closed depressions in south-west Lancashire and north-west Cheshire. Unpublished BA dissertation, University of Oxford.

Watson, R. T., Zinyowera, M. C. and Moss, R. H. (eds), 1996, *Climate change 1995. Impacts, adaptations and mitigation of climate change: scientific-technical analyses*. Cambridge: Cambridge University Press.

Weare, B. C., Temkin, R. L. and Snell, C. M., 1974, Aerosols and climate: some further considerations. *Science*, 186, 827–8.

Weaver, H., 1974, Effects of fire on temperate forests: western United States. In T. T. Kozlowski and C. C. Ahlgren (eds), *Fire and ecosystems*. New York: Academic Press, 279–319.

Weaver, J. E., 1954, *North American prairie*. Lincoln: Johnsen.

Webb, T. and Wigley, T. M. L., 1985, What past climates can indicate about a warmer world. In M. C. MacCracken and F. M. Luther (eds), *Projecting the climatic effects of increasing carbon dioxide*. Washington DC: US Dept of Energy, 237–57.

Weber, P., 1993, Reviving coral reefs. In L. R. Brown (ed.), *State of the world 1993*. London: Earthscan, 42–60.

Wein, R. W. and Maclean, D. A. (eds), 1983, *The role of fire in northern circumpolar ecosystems*. Chichester: Wiley.

Weisrock, A., 1986, Variations climatiques et periodes de sedimentation carbonatée a l'Holocène-l'age des depôts. *Mediterrannée*, 10, 165–7.

Wellburn, A., 1988, *Air pollution and acid rain: the biological impact*. London: Longman.

—— , 1994, *Air pollution and climatic change: the biological impact* (2nd edn). New York: Longman.

Wells, N. A. and B. Andriamihaja, 1993, The initiation and growth of gullies in Madagascar: are humans to blame? *Geomorphology*, 8, 1–46.

Wells, P. V., 1965, Scarp woodlands, transported grass soils and concept of grassland climate in the Great Plains region. *Science*, 148, 246–9.

Wertine, T. A., 1973, Pyrotechnology: man's first industrial uses of fire. *American scientist*, 61, 670–82.

Westhoff, V., 1983, Man's attitude towards vegetation. In W. Holzner, M. J. A. Werger and I. Ikusima (eds), *Man's impact on vegetation*. The Hague: Junk, 7–24.

Westing, A. and Pfeiffer, E. W., 1972, The cratering of Indochina. *Scientific American*, 226, 5, 21–9.

Wheaton, E. E., 1990, Frequency and severity of drought and dust storms. *Canadian Journal of Agricultural Economics*, 38, 695–700.

Whitaker, J. R., 1940, World view of destruction and conservation of natural resources. *Annals of the Association of American Geographers*, 30, 143–62.

Whitmore, T. M., Turner, B. L., Johnson, D. L., Kates, R. W. and Gottschang, T. R., 1990, Long term population change. In B. L. Turner, W. C. Clark, R. W. Kates, J. T. Matthews and W. B. Meyer (eds), *The Earth as transformed by human action*. Cambridge: Cambridge University Press, 26–39.

Whitney, G. G., 1994, *From coastal wilderness to fruited plain*. Cambridge: Cambridge University Press.

Whittaker, E., 1961, Temperatures in heath fires. *Journal of ecology*, 49, 709–15.

Whitten, A. J., Damanik, S. J., Anwar, J. and Nazaruddin, H., 1987, *The ecology of Sumatra*. Yogyukarta: Gadjah Mada University Press.

Whyte, A. V. I., 1977, Guidelines for field studies in environmental perception. *UNESCO, MAB technical note 5*.

Wigley, T. M. L., 1983, The pre-industrial carbon dioxide level. *Climatic change*, 5, 315–20.

Wigley, T. M. L., Jones, P. D. and Kelly, P. M., 1980, Scenario for a warm, high-CO_2 world. *Nature*, 283, 17–21.

Wilken, G. C., 1972, Microclimate management by traditional farmers. *Geographical review*, 62, 544–60.

Wilkinson, W. B. and Brassington, F. C., 1991, Rising groundwater levels – an international problem. In R. A. Downing and W. B. Wilkinson (eds), *Applied groundwater hydrology – a British perspective*. Oxford: Clarendon Press, 35–53.

Williams, G. P., 1978, The case of the shrinking channels – the North Platte and Platte rivers in Nebraska. *United States geological survey circular*, 781.

Williams, M., 1970, *The draining of the Somerset levels*. Cambridge: Cambridge University Press.

—— , 1988, The death and rebirth of the American forest: clearing and reversion in the United States, 1900–1980. In J. F. Richards and R. P. Tucker (eds), *World deforestation in the twentieth century*. Durham, NC and London: Duke University Press, 211–29.

—— , 1989, *Americans and their Forests*, Cambridge: Cambridge University Press.

—— , 1990, *Wetlands: a threatened landscape*. Oxford: Basil Blackwell.

—— , 1994, Forests and tree cover. In W. B. Meyer and B. L. Turner II (eds), *Changes in land use and land cover: a global perspective*. Cambridge: Cambridge University Press.

Williams, M. A. J. and Balling, R. C., 1996, *Interactions of desertification and climate*. London: Arnold.

Williams, P. W. (ed.), 1993, Karst terrains: environmental changes and human impact. *Catena Supplement*, 25.

Williams, R. S. and Moore, J. G., 1973, Iceland chills lava flow. *Geotimes*, 18, 14–17.

Williamson, D. R., 1983, The application of salt and water balances to quantify causes of the dryland salinity problem in Victoria. *Proceedings of the Royal Society of Victoria*, 95, 103–11.

Williamson, M., 1996, *Biological invasions*. London: Chapman & Hall.

Wilshire, H. G., 1980, Human causes of accelerated wind erosion in California's deserts. In D. R. Coates and J. D. Vitek (eds), *Geomorphic thresholds*. Stroudsburg: Dowden, Hutchinson & Ross, 415–33.

Wilshire, H. G., Nakata, J. K. and Hallet, B., 1981, Field observations of the December 1977 wind storm, San Joaquin Valley, California. In T. L. Péwé (ed.), *Desert dust: origin, characteristics and effects on man*. Geological Society of America, 233–51.

Wilson, A. T., 1978, Pioneer agriculture explosion and CO_2 levels in the atmosphere. *Nature*, 273, 40–1.

Wilson, E. O., 1988, *Biodiversity*. Washington DC: National Academy Press, 58–70.

—— , 1992, *The diversity of life*. Cambridge, MA: Harvard/Belknap.

Wilson, K. V., 1967, A preliminary study of the effect of urbanization on floods in Jackson, Mississippi. *United States geological survey professional paper*, 575-D, 259–61.

Winkler, E. M., 1970, The importance of air pollution in the corrosion of stone and metals. *Engineering geology*, 4, 327–34.

Winstanley, D., 1973, Rainfall patterns and general atmospheric circulation. *Nature*, 245, 190–4.

Wolman, M. G., 1967, A cycle of sedimentation and erosion in urban river channels. *Geografiska annaler*, 49A, 385–95.

Wolman, M. G. and Schick, A. P., 1967, Effects of construction on fluvial sediment, urban and suburban areas of Maryland. *Water resources research*, 3, 451–64.

Wong, C. S., 1978, Atmospheric input on carbon dioxide from burning wood. *Science*, 200, 197–9.

Woo, M.-K., Lewkowicz, A. G. and Rouse, W. R., 1992, Response of the Canadian permafrost environment to climate change. *Physical Geography*, 13, 287–317.

Wood, C. M., Lee, M., Linker, J. A. and Saunders, P. J. W., 1974, *The geography of pollution, a study of greater Manchester*. Manchester: Manchester University Press.

Woodroffe, C. D., 1990, The impact of sea-level rise on mangrove shorelines. *Progress in Physical Geography*, 14, 483–520.

Woodwell, G. M., 1978, The carbon dioxide question. *Scientific American*, 238, 1, 34–43.

——, 1990, *The earth in transition: patterns and processes of biotic impoverishment*. Cambridge: Cambridge University Press.

——, 1992, The role of forests in climatic change. In N. P. Sharman (ed.), *Managing the world's forests*. Dubuque, Iowa: Kendall/Hunt, 75–91.

Woodwell, G. M., Whittaker, R. H., Reiners, W. A., Likens, G. E., Delwiche, C. C. and Borkin, D. B., 1978, The biota and world carbon budget. *Science*, 199, 141–6.

Wooster, W. S., 1969, The ocean and man. *Scientific American*, 221, 218–23.

World Conservation Monitoring Centre, 1992, *Global biodiversity*. London: Chapman & Hall.

World Resources Institute, 1986, *World resources 1986–7*. New York: Basic Books.

——, 1988, *World resources 1988–9*. New York: Basic Books.

——, 1992, *World resources 1990–91*. New York and Oxford: Oxford University Press.

——, 1996, *World resources 1994–5*. New York and Oxford: Oxford University Press.

——, 1998, *World resources 1996–7*. New York and Oxford: Oxford University Press.

Worthington, E. B. (ed.), 1977, *Arid land irrigation in developing countries: environmental problems and effects*. Oxford: Pergamon.

Wright, H. E., 1974, Landscape development, forest fires and wilderness management. *Science*, 186, 487–95.

Wright, L. W. and Wanstall, P. J., 1977, The vegetation of Mediterranean France: a review. Occasional paper 9, Department of Geography, Queen Mary College, University of London.

Wrigley, E. A., 1965, Changes in the philosophy of geography. In R. J. Chorley and P. Haggett (eds), *Frontiers in geographical teaching*, London: Methuen.

Wymer, J., 1982, *The palaeolithic age*. London: Croom Helm.

Yaalon, D. H. and Yaron, B., 1966, Framework for man-made soil changes – an outline of metapedogenesis. *Soil science*, 102, 272–7.

Yi-fu Tuan, 1971, Man and nature. *Commission on College Geography resource paper*, 10.

Yorke, T. H. and Herb, W. J., 1978, Effects of urbanization on stream-flow and sediment transport in the Rock Creek and Anacostia River basins, Montgomery County, Maryland, 1962–74. *United States Geological survey professional paper*, 1003.

Young, J. E., 1992, *Mining the earth*. Worldwatch Paper 109, 1–53.

Yunus, M. and M. Iqbal (eds), 1996, *Plant response to air pollution*. Chichester: Wiley.

Zabinski, C. and Davis, M. B., 1989, Hard times ahead for Great Lakes forests: a climate threshold model predicts responses to CO_2-induced climate change. In J. B. Smith and D. Tirpak (eds), *The potential effects of global climate change on the United States*. Washington DC: US Environmental Protection Agency, Appendix D, 5-1–5-19.

Zaret, T. M. and Paine, R. T., 1973, Species introduction in a tropical lake. *Science*, 182, 449–55.

Zhang, J., Yan, J. and Zhang, Z. F., 1995, Nationwide river chemistry trends in China: Huanghe and Changyiang. *Ambio*, 24, 274–9.

Zimina, R. P., Polevaya, Z. A. and Yelkin, K. F., 1972, The plowing up of the virgin lands and the Bobac Marmot resources of Central Kazakhstan. *Soviet geography*, 13, 246–56.

Index

Note: Alphabetical arrangement of headings and subheadings is word-by-word, ignoring words such as 'and', 'by', 'in', 'of' and 'for'. Page numbers in *italics* refer to figures, plates and tables. Author references are included only where their work is quoted or discussed in the text. References to particular locations, species, etc. will be found most easily under specific entries rather than by searching under more general headings (e.g. *Aswan Dam* rather than *dams* or *Nile*; *Grimes Graves* rather than *East Anglia* or *Britain*; *prickly pear* rather than *Australia* or *vegetation*).

Aberfan disaster 297, *298*
accelerated geomorphological processes
  see particular processes
acid mine drainage 246–7
acid rain 364–71
  ecological effects 176–7, 245–6,
    367–9
  and weathering 293
Ackerman, W. C. 237
Adams, R. M. 168
aerosols, atmospheric 340–3
  and fog 358
  and plants 93
  see also dust
affluence 15, 16
afforestation and reforestation 60,
  224–5, 308
Africa
  deforestation *55, 58*
  desertification 70, 72, 73, 74, 75, 76
  elephants *150*
  high-altitude (Afromontane) grasslands
    67
  megafauna extinctions 145
  savanna 65–6
  soil erosion 190–1
  Stone Age tools 17
  wildlife habitat losses *155–6*
  see also particular countries, especially
    Egypt; Kenya; South Africa; Sudan;
    Zimbabwe
Africanized honey bee 110, *112*
Afromontane grasslands 67
Agricola, Georgius 129–30
agricultural revolution, population
  growth and 13
agriculture
  animal use, secondary 26–7
  animals benefiting from 118

climatic modification 346–7, 420
deforestation for 80–1, 82, 174, 175,
  188–91, 305, *306*, 345–6
dust storm generation 193–6
ecosystem simplification 29, 430
empires based on *21*
fertilizer use 183–4, 239–44
and global warming *418*
habitat destruction 99, *100–1*, 130–7
hydrological effects 305–7
and lateritization 174–5
mechanization: with animal power
  26–7; effects 98, 177, 192–3
origins and evolution 20–9
pesticide use 121–5, 244
productivity: decline 172–3, 174–5,
  187; increase 182–3, 184
and sedimentation 275–80
selective breeding programmes 103–4
and slope instability 299
soil conservation 199–202, 305
soil drainage *101*, 180–3, 225, 243,
  285, *286*
soil erosion 185–7, 188–96, 202
and soil fertility 41, 58–9, 176,
  183–4, 185, 421
and soil formation 160–1
soil improvement techniques 176, 179
soil structural changes 174–5,
  176–80
specialization 240
water pollution 236, 239–45
see also particular activities, e.g.
  irrigation; ploughing
air pollution
  control 358, 363–4, 366–7, 378
  effects: animals 128–9; forest decline
    97, 98; plants 91–3; soils 176–7;
    weathering 293

long-distance transfer 362–3, *364*
trends 358–60, *364*
types 91, *355*, *358*
urban 332, 355–62
variability 355–8, 374
vehicular *32*, 352
see also acid rain; water pollution *and specific pollutants*
aircraft, water vapour from 342
Aitlan, Lake 116
Alabama, subsidence in 280
Alabaster, J. S. 129
Alaska
engineering problems *287*
fire suppression 45
reindeer decline 113
albedo 331, 332
deliberate modification 377
feedback effects 383
urban areas 348–9
and vegetation 343–5
alder, spread of 80–1
algae
blooms 240, 258–9
planktonic *32*
and water temperature 257
alluvial fans 210–11
alluviation 289, 291
Alps
deforestation 2, 6
glacier fluctuations 393–4
altithermal period 385
aluminium poisoning 246
aluminium smelting 93
Amazonia, deforestation in 345
America
Africanized honey bee 110, *112*
fires and vegetation 48
population decline 15
see also North America; South and Central America; United States
Anderson, D. M. 258
animals
accidental introductions 108–14
biomass turnover 146–7
conservation 120–1, 142, 152–4, 159, 425–6
decline 121–42
dependence on humans 106
disease transmission 147
dispersal 107–14, 118, *119*
domesticated 22, 25, 26, 106–7; naturalization 112–13
dwarfing 148–9
expanding numbers and distribution 115–21
extinctions 142–59; oceanic islands 139, *140*, *141*; Pleistocene 142, 144–9
on feedlots 244–5
feral *51*, 113
and fire 138
game cropping 120–1
genetic impoverishment 154
grazing *see* grazing
habitat changes affecting: agricultural 99, *100–1*, 130–7; estuaries 218;

and extinctions 152–9; mineral extraction 116–17; reservoirs 89; river channelization 215; sensitivity to 151–2
hunting 138–41, 142, 144–9
introduced *51*, 88–9, 108–14, 139, 156; effect on plants 88–9; extinctions caused by 156; feral *51*; islands 88–9, 139
keystone species 67, 104
legislation protecting 426–7
nature reserves 152–4
oceanic islands 88–9, 139–40, *141*
pesticides affecting 121–5, 244
pollution affecting 121–30, 245–6, 249, 256–7, 259
range loss 154–6
savanna 67
secondary utilization 26–7
size: and vulnerability 146–7, 148–9
survival characteristics 150–1
trampling 178
urban 115–16
see also birds; fish; mammals *and particular species*
Antarctica
global warming 393, 406
ozone hole 372, *373*
antelopes 120–1
anthropogenic landforms 261–327
see also particular processes, e.g. erosion; weathering
anthropogenic landscape science 9
anticipatory reaction 416
Appalachians, deforestation in 256
aquatic vegetation
effect of nitrogen on 239–40
invasive 103
aquifers
artificial recharge 233
water abstraction 165–7, 229–33
see also groundwater
Aral Sea, desiccation of 171, *172*, *218*, 227–9
Arctic
oil spills 125
ozone depletion 372
permafrost displacement *395*, *396*
sea ice: and global warming 384, 392–3, 406; removal 376; water transfers affecting 347
Argentina
Pampas *49*, 79–80, 86
arid areas
groundwater abstraction 230, 233
runoff 180
sabkhas 410–11
salinity 161–3
urban groundwater recharge 234–5
see also desertification; deserts; drought; semi-arid areas
Arizona, soil erosion in 197
arroyo trenching 288–91
artificial habitats 37
Ascension Island, deforestation of 2–3
ash tree dieback 98
Ashby, E. 427

Asia
   deforestation   55, 58
   see also Bangladesh; India; Indonesia;
      Japan; Malaysia; Pakistan;
      South-East Asia
asphalt mining   259
Aswan Dam   166, 208–9, 318
Atkinson, B. W.   352
Atlantic Ocean, discharge into
   sediment   320
   waste   272–3
Atlantic-industrial era   21
Atlantropa project   229
atmosphere
   aerosols in see aerosols; dust
   carbon dioxide in see carbon dioxide
   as source of salts   162
   see also air pollution; climate
atolls
   bird extinctions   139–40
   sea level rise   411
attitudes, environmental   426–8
Aubertin, G. M.   251
Australasian fauna   108
Australia
   early humans   12
   experimental catchments   197, 222
   fire: frequency and causes   43; use
      41, 197
   groundwater abstraction   230
   hurricanes   391–2
   introduced species: animals   88, 110,
      118; plants and pests   87, 102–3
   jarrah forests   87
   prickly pear control   102–3
   rabbits   118
   river water chemistry   162
   Salvinia control   103
   savanna   62, 64
   sheep grazing   50–1
   soil erosion   197
   soil salinization   168
   water table changes   164

Badbury Rings   422
badgers   133–4
Bahrain   266
Ball, D. F.   176
Ballantyne, C. K.   299
Bangladesh, flooding in   406–7
Barnston, A. G.   347
Barreis, D. A.   340
barrier islands   321–2, 413
Barry, R. G.   394
Battarbee, R. W.   367
beaches
   coastal protection   311, 317, 321, 322
   formation and replenishment   318
   and sea level rise   414–15
   sediment movement   312–13, 315
   see also coasts
Behre, K.-E.   80
Belfast, lichen growth in   92
Belize
   coastal erosion   320–1
   fire frequency   42
Bendell, J. F.   138

Bennett, C. F.   66
Bering Strait   376
Bidwell, O. W.   160–1
Biglane, K. E.   319
Binford, M. W.   278–9
biodiversity
   hot spots   158, 159
   maintenance   159, 428
   see also species diversity
biological invasions see introductions
biological magnification   121–5, 244
biological pest control   102–3, 118–19
biomass
   human   106
   turnover: and body size   146–7
Bird, E. C. F.   412, 414
birds
   agriculture benefiting   118
   buildings affecting   137
   decline: and insect increase   115
   eggshell thinning   123
   extinctions   139, 140, 141, 143, 148,
      150
   and fire   138
   habitat changes   131, 134–5, 136
   hedgerow removal affecting   131
   islands   139, 140, 141
   lead poisoning   129
   legislative protection   426–7
   and leisure activities   137
   oil pollution   125, 126
   pesticides affecting   123, 124, 125
   scavenging   118–19
   urban   115, 118–19
   woodland, species diversity   134,
      135
   see also particular species
Birks, H. J. B.   52, 149
bison   79, 139, 147–8
Blytt–Sernander model   80
boats, Neolithic   27
Boca Ciega Bay   271
Böckh, A.   226
bog bursts and slides   292
bomb craters   265
Bonan, G. B.   345–6
boreholes, overgrazing at   71, 73
Bormann, F. H.   251, 253
botanic gardens   85
Boulder (Hoover) Dam   204, 210
Boussingault, J. B.   2–3
bracken (Pteridium aquilinum)   51, 91
Braer oil spill   126
Brazel, A. J.   340
Brazil
   coastal erosion   312–13
   gold mining   130
   land reclamation   265
Brazos river   307
breakwaters and piers   312–13, 315–17
Breckland   263, 264, 308
Brighton   317
Britain
   acid rain   364, 369
   agriculture: fertilizer use   241, 242;
      habitat destruction   99, 100–1, 118,
      130–5, 136, 183; and sedimentation

277–8, 279–80; and soil erosion
192–3; *see also* soils, *below*
air pollution 356; trends 358, *359*,
370; and vegetation 91, 92, 93
animals, introduced 117–18, *120*
birds 116, *117*, 130–1, 134–5, 137
bracken encroachment 91
building and construction 198
chalk pits 263, *264*
coasts: erosion 312, 313, *314*,
316–17; pollution 249–50
concrete production 269
conservation and preservation 426–7
constructional landforms *270*
dams and reservoirs 205–6, 255–6
deforestation 52, 81, 253, *254*
Dutch elm disease 86
erosion 299, *422*
excavational landforms 263–5, *269*,
273–5
floods 183, 205–6, 225–6, 284
Global Environment Research Office
11
grasslands 99, *100*, 132–3
grazing 51
groundwater abstraction 167,
229–30, *231*, 233–4
heather burning 83, 196–7
heathland 83, *84, 85*
hedgerow removal 130–1, *132*, 192,
193
lake acidification 367
lead poisoning, wildfowl 129
limestone pavements 266
marl digging 273–5
mining 116–17; drainage waters
246–7; and subsidence 284, *285*;
waste 129, 297, *298*
oil spills 94, 125, *126*
peat 175–6; cutting 265; drainage
183, 285, *286*; erosion 292
photochemical smog 93
plants: decline 99, *100–1*;
introduction and dispersal 85, 90,
322–3; species diversity *153*
rabbits 117–18, 308
ridge and furrow 180–2
rivers: channel modification *213*, 300;
chemical composition 253, *254*;
effluent in 222; regulation by dams
205, 206; *see also* water pollution,
below
salt marshes 322–3, 408, *409*
sand dunes 308, *310*, 321
soils: drainage 180, *181*, 183, 243,
285, *286*, 300; erosion 192–3, 292;
improvement 176; podzolization
175–6; *see also* peat, *above*
subsidence 284, *285*
thunderstorms 352–3, *354–5*
tree dieback 98
urban areas: air pollution 91, 92, 93,
356, 358, *359*, 362; climate 352,
353–5
vegetation: and air pollution 92, 93;
post-glacial changes 80–3; upland
51, 91

water pollution 235, 241, *242*,
247–8; thermal 254–5
wetland losses 183
see also London
British Columbia 276
Broads, origin of 264–5
Broadus, J. 406, *407*
Bronze Age agriculture 277
bronze working 30–1
Brookes, A. 302–3
Brown, J. H. 114
Brunhes, Jean 7
Bruun Rule 414
Bryan, G. W. 123, 125
Bryan, K. 289
Bryson, R. A. 340
Buddemeier, R. W. 411
Budyko, M. I. 345, 385, 387–9, 392,
400
buffalo *88*, 110
Buffon, Count 2
buildings
on clay 183
permafrost areas 288
turbulence creation 353
Burcham, L. T. 50
Burney, A. D. 144
burning *see* fire
bush encroachment 66–7
butterfly habitat losses 133
Butzer, K. W. 289, 291

Cactoblastus moth 102–3
Calabria, landslides in 297–8
California
air pollution *see* Los Angeles
biological pest control 120
birds, suburban 115
coastal erosion *315*
dam leakage 281
dust storms 195–6
fire suppression 45
fish introductions 110, 112, *113*,
142
global warming 398
grasslands 50
mining and sedimentation 275
pesticide use 123
plant introductions and invasions 50,
86
recreational vehicles 137
sea-water incursion 166
see also Los Angeles; San Francisco
Calluna (heather) burning 83, 196–7
Canada
gold-mining 276
industrial pollution 93
permafrost displacement *395, 396*
salt extraction 284
canals
animal dispersal 113–14
irrigation 211–12
Cape Town, pollution in *332*
carbon dioxide, atmospheric
greenhouse effect 331, 333–5
relative importance 339
sources *334, 335*, 420

trends 293, 335, 381–2; *see also*
  global warming
carbon monoxide emissions *360*
Carter, L. J. 186–7
carts, wheeled 27, *28*
Caspian Sea 226–7
catchment experiments 197, 222–3,
  251, *253*, 256
cattle
  domestication 22, 107
  feedlots 245
  soil compaction 178
  see also grazing
cays 320–1, 411
Ceara 312–13
Central America *see* South America *and
  particular countries*
CFCs *334, 337*, 338–9, 372, 374
chalk
  downland losses 99, *100*, 133
  excavation 263, *264*
  for soil improvement 176
Chandler, T. J. 353
Changnon, S. A. 352, 353
channel modification 203–19, 262,
  299–305, 319–20, *321*
  see also dams
channelization 212–15, 319-20,
  321
chaparral 45, 48, 197
Charney, J. 344
chemical fertilizers 183–4, 239–41,
  242, 243
chemical pollution
  agricultural 239–45
  trends 248–51
  urban and industrial 246–8, *249*
  see also acid rain
Chesapeake Bay 275
Cheshire
  marl digging 273–5
  salt extraction 284, *285*
chestnut blight fungus 87
Chicago 230
Childe, V. G. 22
China
  dams 210
  paddy soils 176
  soil erosion 189
  terraces *201*
  water pollution 237
chlorinated hydrocarbons 259
chlorofluorocarbons (CFCs) *334, 337*,
  338–9, 372, 374
Chorley, R. J. 374
CIS (formerly USSR)
  dust storms 196
  game cropping 120–1
  geographers' concern with
    environmental transformation
    8–9
  lake levels 171, *172, 218*, 226–9
  river diversion and water transfers
    215–17, *218*, 227, 347
  virgin lands scheme 135, 137
cities *see* urban areas; urbanization
Clark, G. 21

clay soils
  drainage 225
  engineering problems 183
  improvement 179
Clear Lake 123
clear-cutting experiments 222, 223,
  251, 253
clear-water erosion 210
Clements' Theory of Succession 428–9
Cleopatra's Needle 293–4
cliffs
  erosion and protection 311, *314, 316,
    317*
  and sea level rise 412
climate
  changes *see* climatic change; global
    warming
  feedback effects 331–2, 382–4
  and grassland origins: Pampas 79;
    prairies 78; savanna 63, 65
  modification: by aerosols 340–3;
    deliberate 374–7; by land use
    changes 343–7, 420; by reservoirs
    347–8
  and soil formation 161
  urban 348–55, 378; *see also* air
    pollution
  see also precipitation; temperature;
    wind, *etc.*
climatic change
  analogues for 384
  and arroyo trenching 288–9
  causal factors 328–33
  and desertification 68, 73–5
  glacier response 393–4
  historic 384–5
  and lake levels 226–7
  modelling and prediction 381–9
  and Pleistocene extinctions 142, 144,
    147–8
  and vegetation change 80, 81, 83
  see also global warming
climatic optimum, Holocene 387, *388*,
  396
clouds
  aircraft effects 342
  feedback effects 383
  formation 340
  interception from 346
  seeding experiments 374–5
  urban areas *349*
Clovis peoples 145, 148, 149
coal burning 93, *359*, 366–7
coal mining 266–7, 284
  waste tip instability 297, *298*
coasts
  dune stabilization 307–8, 310–11,
    321–2
  erosion 263, 311–22, *413, 414–15*
  flooding 322, 406–7, 412, *413*
  global warming effects 389–90,
    404–15
  land reclamation 212, *213*, 270, *271,
    272, 273*
  management *417*
  permafrost areas 396
  pollution 258, *259*; oil 125–7

protection 311–13, *314–17*
sediment discharge 318–20
soil salinization 165–7
waste disposal 272–3
see also beaches; deltas; estuaries; salt
marshes; sea level
coconut plantations 320–1
Coe, M. 145–6
Coffey, M. 193, *195*
Collier, C. R. 246
Colorado
cloudiness 342
deforestation 222
earthquakes 324
feedlots *245*
soil erosion 189
Colorado river 206, *208*, 210, 211
common blue butterfly 133
Commonwealth of Independent States *see*
CIS
concrete production 269
conifer plantations, birds in 134, *135*
coniferous forests
decline 96–7
fire protection 45
conservation
animals 120–1, 142, 152–4, 159,
425–6
development 426–7
motivation for 427–8
soils 199–202, 305
construction, soil erosion and 197–9
constructional landforms 262, 269–75
constructive geography 9
continental shelves, sedimentation on
189
contour-strip ploughing 200, *201*
Conway, V. M. 292
Cooke, R. U. *290*, 296
Coones, P. 181–2
Cooper, C. F. 79
copper smelting 30
coral
bleaching 411–12
eutrophication effects 258–9
pollution 127–8
and sea level rise 411–12, 414
stresses 423, *424*
coral atolls, bird extinctions on 139–40
corncockle decline 99, *100*
Cornwall, mining in 129, 279
Corsica, maquis in 77
Costa, J. E. 275
Costa Rica, habitat fragmentation in
157
country breezes 353, *355*
Cow Green reservoir 255–6
Coweeta catchments 223, *253*
craters, bomb 265
Creation, beliefs about 1–2
Crisp, D. T. 255
critical load concept 177
Cronk, Q. C. B. 87–8
crops
evolution 103
food 29
salt tolerance 169, 172–3

selective breeding 104
Culbin Sands 308, 310
cultivated habitats 37
cultivation, domestication distinct from
22
cultural development 15–16
Bronze and Iron Age 30–1
Mesolithic and Neolithic 20–9
Palaeolithic 16–20
urban and industrial 31–6
cultural ecosystems 38

dams
channel modification 303, *304, 305*
number and size 204
and slope instability 297
stream regulation 204–6
see also reservoirs *and particular dams,
e.g.* Aswan
Dansgaard, W. 384
Darby, H. C. 53–4
Darling, F. F. 51
Davis, M. B. 398
DDT 121, *122*, 123, *124*, 125, 244,
252
Dead Sea 229
death rate reduction 14
debris avalanches 191
debris flows 297, *298*, 299
debt-for-nature swaps 61
deflation *see* wind: soil erosion
deforestation *52–61*
antiquity 52, *53*, 55, 80–1
carbon derived from 335, *336*
causes 53, 55, 57; agriculture 80–1,
82, 174, 175, 188–91, 305, *306*,
345–6; gold mining *130*
definition 52
early concern about 2–3, 4–5, 6
effects *59*; on albedo 343–5; on
biodiversity 159; climatic 343–6;
hydrological 2–3, 5, 6, 222–4; on
lake levels 2, 226; lateritization
174–5; soil erosion 188–92; soil
salinization 167–8; on water
quality 251–3, *254*; on water
temperatures 256
by fire 39, 41, 196
islands 144
Mesolithic and Neolithic 80–2
rates and extent 52, *55*–8, 60–1
degraded habitats 37
deltas
erosion 318–20, *321*
global warming effects 406–7, 409
dendrochronology 189
Denevan, M. W. 14–15
Denmark
dune reactivation 308
forest clearance 82
Denson, E. P. 425–6
Denver 324, 342
desertification 67–76
causes 68, 70–1, 72, 73–6
definition 68
and dune reactivation 308–9
dust generation 340–1

irreversibility 71, 73
process 69–70, *71*
rate and extent 68–9, 420
deserts
    albedo 343–4
    dune reactivation and stabilization
        308–9, *310*
    expansion *see* desertification
    global warming effects 389, 397, 398
    rainfall augmentation 348
    resilience and inertia 431
    salinity 161–2
    soil improvement 180
    see also semi-arid areas
Despain, D. G. 46
determinism, environmental 7
developing countries, air pollution in
    358, 360, 370–1
di Castri, F. 108, *109*
Dickinson, R. E. 382
Dieldrin 244
diet, human 19
Dimbleby, G. W. 175
dimethylsulphide 342
discharge *see under* rivers
disease transmission
    accidental: plants 86–7
    animals 147
dispersal
    animals 107–14, 118, *119*
    plants 83–7, 89–91
    see also domestication
dodo 139, *141*
dogs, pollution from 248
domestication
    animals 22, 25, 26, 106–7, 112–13;
        see also pastoralism
    plants 21, 22–4, 29, 83–4
    see also dispersal
dongas *188*
Doornkamp, J. C. 296
Dorset
    coastal erosion 313, *314, 316*
    heathland 83, *85*
Down, C. G. 246–7
drainage 180–3, 243
    habitat destruction *101*
    hydrological effects 225
    for salinity eradication 173
    and subsidence 285, *286*
dredging, estuarine 217–18, *218*, 270,
    271, 272
Dregne, H. E. 68
drought
    and desertification 73, 75
    and global warming 401–2
    and soil erosion 193
dry deposition 364
dumping, landform creation by 269–70,
    271, 272–3
dunes 307–11, 321–2, 403
dung beetles 110
duricrusts 173–4
dust, atmospheric 330–1, 340–2
Dust Bowl 187, 193, *195, 402*, 403
dust storms 171, *172*, 193–6, 340, 403
Dutch elm disease 86

earthquakes 285–6, 323–7
East Africa
    early humans 12; fire use 19; tools
        17
    see also Kenya
East Anglia
    drainage 183, 285, *286*, 300
    dune instability 308
    excavational landforms 263–5
    hedgerow removal *131*
Easter Island 144
economic development, key stages of *21*
economy, global changes in *34*
ecosystems
    diversity 428, 429; *see also* species
        diversity
    interdependence within 4, 7, 431
    invasive plants 87–8
    modification: increasing scale 38
    nutrients 185, 251, 253
    protection and management *417–18*
    resilience and inertia 431
    simplification, agricultural 29, 430
    stability and vulnerability 428–31
    see also particular ecosystems, e.g.
        islands
Eden, M. J. 65
effluent
    in urban runoff 222
eggshell thinning 123
Egypt
    coastal erosion 318
    flooding 406
    irrigation 26
    see also Nile
Ehrlich, P. R. and A. H. 76, 104
El-Kholy, O. A. 69
elasticity, ecosystem 431
electricity generation *see* power stations
elephants 67, *150*
Ellison, J. C. 410
elms 81–2, 86
Elton, C. S. 108
embankments 214–15, 319
Emmanuel, W. R. 397
energy use 33, *34*, 348, 382
environmental change
    causation 15–16; problem of
        determining 431–3
    chronology 15–36, 423, *424–5*
    development of ideas about 1–12
    ecosystem vulnerability to 428–31
    global scale 10–12, 35–6
    modern era 31–6, 420–5
    positive effects 5
    pre-industrial 16–31, 419–20
    predictions 379–81; uncertainties
        380–2, 415–16, 433
    proliferation and complexity 420–5
    reversibility 425–8
    *see also particular human activities and
        environments affected, e.g.*
        agriculture; industrialization; soils;
        urbanization
environmental determinism 7
environmental ethic 427
environmental impact statements 432–3

environmental movement 10, 426–8
environmentalism 7
equifinality 431
equinoxes, precession of 330
erosion 262
    coastal 263, 311–22, *413*, 414–15
    mineral extraction compared with 31
    recreational cause *422*
    rivers 210
    soil *see* soil erosion
    valley-bottom gullies 288–91
    see also sedimentation
estuaries
    dredging and filling 217–18, 270,
        271, 272
    resilience and inertia 431
    sedimentation 275
Ethiopia, early humans in 17
Eucalyptus forest 168, 223
Eurasia, global warming in 390
Eureka model 23
Europe
    acid rain 245, 366–7, 370
    deforestation 53–4
    dune instability 308
    elm decline 81–2
    forest decline 96, 98
    heathland 82–3
    megafauna extinctions 145, 147–8
    sand dunes 308
    sedimentation 277
    soils: drainage *181*; Plaggen 176;
        podzolization 175–6
    Stone Age: settlements 19; tools 17
    tufa decline 294
    urban heat islands *351*, 352
    vegetation change, post-glacial 80–3
    water pollution 237
    see also Britain; Denmark; France;
        Germany, *etc.*
European Network for Research in
    Global Change 11–12
eutrophication 238, 240, 258–9
evaporation 332
    feedback effects 383
    and soil salinization 164
evapotranspiration 344
evolution, extinction as part of 150
excavational landforms 262, 263–9
experimental catchments 197, 222–3,
    251, *253*, 256
extinctions, animal
    future rate 156, 159
    islands 139, *140, 141*
    modern-day 150–9
    prehistoric and historic 142–9

Fairhead, J. *55*
fallow, length of *59*
farming *see* agriculture
faunal realms 107–8
feedbacks, climatic 331–2, 382–4,
    401–2
feedlot runoff 244–5
Fenlands
    channel modification 300
    drainage 183, 285, *286*

feral animals 51, 113
fertility *see under* soils
fertilizers 183–4
    and species diversity 104
    water pollution 239–44
field drainage 180, 243
fire
    and animals 138
    deforestation by 39, 41, 196
    early use 19–20, 41, 419
    frequency and causes 42–3
    in heather management 83, 196–7
    importance and utility 20, 38–9
    natural and anthropogenic compared
        41–3
    pedological effects 185, 196–7
    in savanna maintenance 63, 64, 65,
        66
    smoke generation 341–2
    and soil erosion 196–7
    suppression 44, 45–7
    temperatures 44
    vegetation adaptation 48, 77–9
    vegetation resistance 65
    and vegetation succession 47
fish
    acidification affecting 245–6, 369
    aluminium poisoning 246
    eutrophication effects 240
    hydroelectric turbines affecting 137–8
    introduced 110, 112, *113*, 114, 116,
        142
    and lake level changes 227
    over-exploitation 140
    pesticides affecting 123
    pollution affecting 127, 240, 245–6,
        249, 256–7
    temperature sensitivity 256–7
    turbidity tolerance 129
Fisher, J. 150
fishing 27, 140
flightless rail 139
flint mines 263, *264*
Flohn, H. 383–4, 392
floodplains
    aggradation 276–7
    soil drainage 182
floods
    coastal 322, 406–7, 412, *413*
    control: channel adjustments to 303,
        304; through channel modification
        212–15, 299–300; urban 221
    exacerbated by channel improvements
        215
    magnitude and frequency 225–6;
        below dams 205–6; urban areas
        220–1
    sea level rise causing 406–7, 412, *413*
    and sedimentation 276
    and soil drainage 183
    subsidence causing 284
Florida, coastal pollution in 127
flue gas desulphurization 371
fog
    dispersal 377
    frequency 358, *359*
    interception from 346

food chain, biological magnification in 121–5, 244
forest decline 94–9, 369
forests
  acid rain damage 369
  animal extinctions 152–4
  biodiversity hot spots *158*
  burning *see* fire
  clearance *see* deforestation
  degeneration to maquis 76–8
  fire suppression 44–7
  global warming effects 389, 397, 398, *399*, 400
  interception rates 223, 224
  nutrient cycling 251, 253
  and precipitation 346
  soils 188
  stream-water chemistry 251, 253, *254*
  water budget influence 346
  see also rain forest
fossil-fuel burning *332*, 335, 342–3, 355, 365, 366–7
  future 382
France
  dunes 308
  fire causation *43*
  prehistoric hunters 145, *146*
Frenkel, R. E. 89–90
Friedrich, E. 6
Frink, C. R. 368
frost
  reduction near reservoirs 348
Fuller, J. L. 87–8
future *see* climatic change *and under* environmental change

game cropping 120–1
gases *see* greenhouse gases
Gatun Lake 116
Geertz, C. 58, 59
Geikie, Sir Archibald 379
General Circulation Models 345, 381–2
  in vegetation prediction 396–400
genetic engineering 104–5
geographers, environmental concerns of 3–10
geomorphological processes *see particular landforms, processes and environments, e.g.* erosion; mass movements; weathering
George Lake 367
Georgia 198, 305, *306*
Gerasimov, I. P. 8–9
Germany
  forest decline *95*, *96*, 369
  mining 129–30
GESAMP 259
Ghyben–Herzberg principle 165
Gibbs, J. N. 98
Gilbert, G. K. 275
Gillon, D. 65
glaciation 330, 331
  and human colonization *12*
  and megafauna extinction 142, 144, 147–8
glaciers
  global warming effects 389, 393–4

Glacken, C. J. 1
Gleick, P. H. 203, 400
Glen Canyon Dam 211
Global 2000 Report 159, 379–80
global environmental change 10–12, 35–6
Global Environmental Change (journal) 11
Global Environmental Monitoring System 355
global warming 333, 339, 342–3
  effects 389–90; coasts 389–90, 404–15; forests and trees 389, 397, 398, *399*, 400; hydrological 400–3; ice sheets and glaciers 389, 392–4; permafrost 394–6, 401; precipitation 385, 387–9; vegetation 389, 396–400
  feedback effects 382–4
  importance 378
  responses to 416–18, 433
  sensitivity to 390
  spatial pattern 385, *386*
  uncertainties 415–16
  see also greenhouse gases
globalization
  biological invasions 108, *109*
  environmental change 10–12, 33, 35–6
gold mining *130*, 275, 276, 280, *281*, 326
Goodman, D. 429, 430
Goreau, T. J. 412
Gornitz, V. 345
Gourou, P. 174–5
Graf, W. 197
Grainger, A. *55, 57, 59*, 68
Granoch, Loch 280
grasses
  for dune stabilization 311
  grazing resistance 51
grasslands
  bush encroachment 66–7
  ecosystem simplification 29
  and fire 63, 64, 65, 66, 78–9
  future extent 397, 398
  grazing-induced changes 49–52
  losses to agriculture 29, 80, 99, *100*
  plant invasions *50*
  species diversity 104, 132–3
  see also prairies; savanna
Gray, R. 392
Grayson, D. K. 148, 149
grazing
  and channel form 307
  and desertification 71, *73*
  and grassland maintenance 79
  and runoff 224
  savanna 66–7
  and soil erosion 118
  and soil structure 178
  and species diversity 104
  vegetation changes induced by 49–52, 66–7
  see also cattle; pastoralism
Great Bitter Lake 113, *114*

Great Lakes
  fish introductions 114
  global warming effects 402
  pollution 251, *252*
Great Plains
  prairie development 78–80
  precipitation and drought 346,
    401–2
  wheat cultivation and dust storms
    193–4
grebes, Western 123
Greece 291
Green, F. H. W. 180, *181*
greenhouse effect 331, 333–5, 378
  see also global warming
greenhouse gases 331, 333–9, 381
  reduction 416, 418
  see also carbon dioxide
Greenland ice cap, lead in 362–4
Grimes Graves 263, *264*
Gross, M. G. 272–3
ground subsidence 280–8
groundwater
  abstraction 229–35; and sea-water
    incursion 165–7; and subsidence
    282–4
  dam construction affecting 297
  irrigation use 230, *232*
  nitrate concentrations 241, *242*
  recharge 233–5, 401
  rising levels: and accelerated
    weathering 293; following
    deforestation 168; and irrigation
    163–5, 293
  salinity 162–3, 165–8
Grover, H. D. 50
groynes 312, *314–15, 317*
Guatemala 278–9
Gulf War, smoke from 341–2
gullies *188*
  beneficial 202
  peat 292
  valley-bottom (arroyos) 288–92
gulls 118–19
gypsum subsidence 281

habitats
  artificial: ecological explosions 89
  beneficial changes 116–17
  biodiversity hot spots *158, 159*
  classification 37–8
  fragmentation: animal sensitivity to
    151–2; rain forest 156, *157*; by
    roads 137
  losses: through agricultural changes
    99, *100–1*, 130–7; and species
    extinctions 152–9; wetlands
    183
haggs, peat 292
Haigh, M. J. 261–2
hailstone modification 375, *376*
Hallsands 311, *313*
Happ, S. C. 275–6
hares, introduced 88
Harlan, J. R. 23–4
Hawaii 346
Hawksworth, D. L. 92–3

Hayes, R. L. 412
Haynes, C. V. 148
health
  policy-making *417*
  pollution affecting 238, 239
heat island, urban 348–52, *353*
heather (*Calluna*) burning 83, 196–7
heathlands 82–3, *158*
heavy metal contamination
  atmosphere 362
  from fertilizers 184
  marine organisms *122*, 123
  sea water 249–50
  vegetation 94
  water 238
hedgerow removal 130–1, *132*, 192,
  193
Hellberg, T. 367
Helldén, U. 69
Helliwell, D. R. 90
Helsinki, birds in 115
Henderson-Sellers, A. 333, 345
Hewlett, J. D. 253
High Plains
  drought 401, *402*, 403
  dune reactivation 403
  irrigation 346–7
Hill, A. R. 430
hills
  excavation for land reclamation
    265–6
Hills, T. L. 62–3
Hilo 327
Himalayas, deforestation of 55
Hippophae 310–11
historical geographers
  environmental concerns 9–10
Hole, F. D. 160–1
Hollis, G. E. 221
Holme Fen Post 285, *286*
Holocene climatic optimum 387, *388*,
  396
hominids 17
Honduras 66
Hong Kong
  land reclamation *272, 273*
  landslides 297
Hoover (Boulder) Dam 204, 210
Hopley, D. 411
Houghton, J. T. 381, 385, 392, 406
Hoxne, early humans at 20
Hubbard Brook 251, *253*
Hughes, R. J. 189
Hull, S. K. 98
human–environment relationship
  development of ideas about 1–12
  Mesolithic and Neolithic era 20–9
  modern era 31–6
  Palaeolithic era 19–20
humans
  biomass 106
  communicative skills 20
  cultural and technological development
    15–36
  definition 16–17
  diet 19
  earliest evidence of 12

habitat adaptation through cultural
    change   15; *see also* habitats
    population *see* population
humus   185, *186*, 188
hunter-gatherers   25
    cultural and technological development
    16–20; constraints on   21–2
    population size   13
hunting
    effect on animal populations   138–41
    game cropping   120–1
    and Pleistocene animal extinctions
    142, 144–9
Huntingdonshire   *132*
Huntington, E.   288
hurricanes
    and coastal erosion   320–1
    and global warming   390–2
    modification   376
Husar, R. B. and J. D.   237
Hwang Ho river   214
hydro-isostasy   285–6, 326
hydrocompaction   284–5
hydroelectric turbines, fish affected by
    137–8

ice, sea *see* Arctic
ice cores
    CO_2 concentrations in   331
    pollutants in   362–3, *364*
ice sheets
    global warming response   392–3
icebergs   383
Iceland
    lava flow   327
    whale butchering   *141*
Idso, S. B.   340
Imperata cyclindrica   59
India
    dams   *326*
    tank landscape   211
    teak plantations   174
    water transfers   *216*
Indo-China, bomb craters in   265
Indonesia
    mangrove destruction   59
    shifting agriculture   39
Indus river   319
Indus valley
    irrigation   212, 293; and salinization
    169
industrial organic pollutants   239
industrial pollution
    and acid rain   245–6
    and animals   127, 128–30
    dust and smoke   340
    long-distance transfer   362–4
    vegetation effects   93–4
    water   236
    see also mining
industrialization   32–3
    and population growth   13–14
inertia, ecosystem   431
infiltration capacity
    forest soils   188
    and land use   177, 178–9, 180
infrared radiation   333

Inglewood oilfield   282, 324
Innes, J. L.   97, 98–9, 299
insecticides   121–5, 244
insects
    biological control   120
    dispersal   108, 110, *111–12*
    increase   115
interception   223, 224, 346
Intergovernmental Panel on Climate
    Change (IPCC)   381, 389–90, 392,
    394, 406, 416
International Geosphere–Biosphere
    Programme   11
introductions
    animals   51, 88–9, 108–14, 139, 156
    fish   110, 112, *113*, 114, 116, 142
    on islands   87, 88–9, 139
    pests and diseases   86–7, 99
    plants   83–8, 102–3, 224, 322–3
    predators and parasites   102–3,
    119–20
invasions, biological
    human activities relating to   108, *109*
    successful   114
    see also plants, invasive
IPCC *see* Intergovernmental Panel on
    Climate Change
Iran, salinity in   162
Iraq, salinization in   169
iron
    in mine drainage waters   246–7
    smelting   31
Iron Age agriculture   277
irrigation
    albedo reduction   345
    canals and reservoirs for   111–12
    civilizations based on   *21*
    early evidence   25–6
    groundwater abstraction for   230,
    232, 233
    increase   *163*, 170–1
    precipitation impact   346–7
    and salinization   163–5, 169–71, 173
    toxic trace elements from   239
Isachenko, A. G.   9
islands
    alien plants   87
    animals: extinctions   142, *143*, 144;
    introductions   88–9, 139;
    vulnerability   139–40, *141*
    apparent instability   430
    barrier   321–2, *413*
    ecosystem simplicity   87
    sea-level rise affecting   410, *411*
    species diversity   152, *153*, 430
Israel
    groundwater recharge   233
    pesticide and poison use   125
Italy
    landslides   297–8
    Vaiont Dam   297, *326*
Ives, J. D.   55
Izrael, Y. A.   385, 387–9

jackals   125
Jacks, G. V.   202
Jacobs, J.   23, 116

Jacobsen, T. 168
Jakarta *32*
Jamaica, rats in 120
Japan
  dune stabilization 308, 310
  sea fog 346
  subsidence 284
  urban heat islands *351*
Japanese beetle 108, 110, *111*
Jarman, M. R. 25
Jarvis, P. H. 85
Jebel Aulia Dam 89
Jennings, J. N. 264
Jenny, H. 138
jetties 313, *315, 316*
Jickells, T. D. 257–8
Johannesburg, earth tremors in 326
Johannessen, C. L. 66
Johnston, D. W. 369
Jonglei Canal 299–300
Jordan, River, water from 229

k-selected species 150–1
Kahn, H. 380
Kammerer, J. C. 219
Kansas, dust storms in 403
Kariba, Lake 89, *205*
Kates, R. W. 9–10, 34, 36, 423,
  424–5
Kauppi, P. 397
Kazakhstan, virgin lands in 135, 137
Kellogg, W. W. 393
Kent, M. 94
Kentucky, mining in 246
Kenya
  desertification 74
  fish introductions 116
  Stone Age tools 17
  tea plantations 224
  woodland increase *55*
keystone species 67, 104
Khartoum, wood use in 72
Kinsey, D. W. 411
Knox, J. C. 307
Korea, urban heat islands in *351*
Kovda, V. A. 68
Koyna Dam *326*
Krikjufell eruption 327
Krug, E. C. 368
Krypton-85 342
Kuwait, Gulf War in 341–2

Lafleur, R. A. 319
lake core analysis 42–3, 367
lakes
  acidification 245–6, 367–8, 369
  anthropogenic origins 264–5
  artificial *see* reservoirs
  augmentation proposals 229
  climatic impact 347–8
  deforestation affecting 2, 226
  eutrophication 240
, falling levels 226–9, 402
  fish in 227, 245–6, 369
  pollution 127, 250–1
  salinity 226, 227
  sedimentation 277–80

subsidence as cause 284
  vulnerability to changes 430–1
Lamprey, H. 69, *74*
Lancashire, marl digging in 273–5
land bird extinction 139, *140*
land reclamation 212, *213*
  hills excavated for 265–6
  for urbanization 270, 271, 272, 273
land use *see* vegetation *and particular*
  land uses and activities, e.g.
  agriculture; deforestation; mining;
  urbanization
Landes dunes 308
landforms, anthropogenic 261–327
  see also particular landforms and
  processes, e.g. erosion; weathering
landslides 294–5, 297–8
lapwings 118
large blue butterfly 133
Larson, F. 79
lateritization 173–5
Latin America *see* South and Central
  America
latitude, global warming and 390, 394
lava flow diversion 327
lavaka 291–2
Laysan atoll 88–9, 139
Le Houérou, H. N. 73
Leach, M. *55*
lead, atmospheric 358, *359*, 362–4
lead poisoning 129
Lean, J. *345*
Leeds, air pollution in 91, *92*
legislation, environmental 426–7
leisure *see* recreation and leisure
Leopold, L. B. 289, 303
Lessepsian migration 114
lichens
  pollution sensitivity 91–2
Liddle, M. 421–2
lightning
  fire causation 41–2
  suppression 375
limestone pavements 266
limestone subsidence 280–1
Little Ice Age 289, 328
Llangorse Lake 277–8
Llobregat delta *167*
Loe Pool 279
loess *201*
London
  climate 352–3, *354–5, 356*
  groundwater abstraction 229–30,
  231, 233–4
  ozone concentrations *362*
  water pollution 235, 241, *242*, 249
Long Island, thermal pollution in *255*
Los Angeles
  air pollution 93, 358, *360*, 361–2,
  363
  mass movements 297
  oilfields 282, 324
Lyell, Charles 3

Maat, P. B. 403
Mabbutt, J. A. 69
McLennan, S. M. 190

MacMillan Dam 281
Madagascar, gullies in 291–2
Madras, erosion at 312, *315*
Madurai–Ramanathapuram tank country
    211
Maizels, J. K. 68, 71
Malaysia, shifting agriculture in 39
Maldive Islands *128*
mammals
    biomass turnover 146–7
    extinction *143*, 150
    see also particular species
mammoth extinction 147–8
Manabe, S. 401–2
mangal vulnerability 410
mangrove swamps
    destruction 59–60
    sea level rise affecting 407, *408*,
        409–10
    species diversity *153*
*Man's role in changing the face of the
    earth* (Thomas) 9–10
maquis 76–8
marine pollution 125, 257–60
Marks, P. L. 253
marl digging 273–5
marmots 137
marram grass 311
Marsh, George Perkins 4–5, 115,
    307–8, 419, 425, 433
Marshall, L. G. *145*
marshes
    drainage 182
    erosion 319, *321*
    sea level rise affecting 407–10
    see also mangrove swamps; salt
        marshes
Martin, P. S. 142, 144–5, 147
Maryland, sediment yields and
    sedimentation in 198, 275
mass movements 191, 294–9
Mauritius, bird extinctions in 139, *141*
Mayan civilization 39, 278–9
Mead, Lake 204
Mediterranean region
    fire causation *43*
    maquis 76–8
    valley-bottom cut and fill 289, 291
Mediterranean Sea
    Atlantropa project 229
    fish migration into 113–14
megafauna extinctions, Pleistocene 142,
    144–9
Mellanby, K. 128
Mendel, A. C. 178
Mercer, J. H. 393
Mesolithic forest clearance 80–1
Mesopotamia, salinization in 168–9
Messer, J. B. 256
Messerli, B. 55
metal pollution 238–9, 246–7
metal working 30–1
    see also mining
metapedogenesis 161
methane emissions 331, *334*, 336, *337*,
    338, 384
methyl mercury *122*, *123*, 125

Mexico, Gulf of *271*, 319
Mexico, soil erosion in 202
Meybeck, M. 238
Meyer, W. B. 16
microclimate 377
Middle East
    occupation mounds (tells) 270
    sabkhas 410–11
    salinization 168–9
    see also Egypt; Iran; Israel
Middleton, N. J. 69
Milankovitch Theory 330
Mil'kov, F. N. 9
Milliman, J. D. 189
mining
    and animals 116–17, 129–30
    environmental impacts 31
    landform creation 266–9
    origins and expansion 30–1
    and sedimentation 275, *276*, 279
    seismic effects 326–7
    and subsidence 280, *281*, 284, *285*
    waste 267, 269–70, *271*; channel
        form effects 307; instability 297,
        298; pollution from 246; and
        stream turbidity 129, *130*; and
        vegetation 94
    see also coal mining; oilfields
Mississippi river
    channel modification 214, 299,
        319–20
    delta 319–20, *321*, 409
    pollution 127, 237
    sediment discharge 208, *209*, *210*
Missouri, mining in 268
Missouri river, sediment in 208, *209*, *210*
Mixtec farmers 202
Moa extinction *140*
Mohenjo-Daro *293*
Montreal Protocol 374
More, R. J. 374
Morgan, R. P. C. 192, 199, *200*
Morgan, W. B. 65–6
Morocco, irrigation in *163*
Moss, R. P. 65–6
motorcycles, soil compaction by *178*
mounds 269, *270*
mountains
    deforestation 55
    global warming effects 389
Muhs, D. R. 403
Murdoch, W. W. 429, 430
Murray Valley, groundwater in *164*
Musick, H. B. 50
Myers, N. 56, 57, 159, 185
myxomatosis 308

Nakuru, Lake 116
Namibia
    dune stabilization *309*
    mining *267*
natural gas 366–7
natural habitats and ecosystems 37, 38
nature
    attitudes to 1–2, 426–8
    conservation *see* conservation
    human forces compared with 3–4

nature reserves 152–4
Near East, domestication in 24–5
Nebraska
    dust storms 403
    groundwater abstraction 230, *232*
    hailstones *376*
Negev
    albedo 344
    runoff 180
Neolithic peoples
    agriculture and vegetation clearance
        80–2, 277
    boats 27
    chalk pits 263, *264*
    elm use 81
nerve-gas waste disposal 324
Netherlands, land reclamation in 212,
    213
New Guinea, species diversity in *153*
New Jersey, Japanese beetles in 108,
    110, *111*
New Mexico
    arroyo trenching 289
    salt cedar introduction 224
    shrubland encroachment 50
    subsidence 281
New World
    agricultural evolution 26
    see also Australia; North America;
        South America, *etc.*
New York
    accelerated weathering 293–4
    sewage and waste 248, 272–3
New Zealand
    animal introductions 89
    glaciers and global warming 394
    Moa extinction *140*
    soils 176, 190
    vegetation change *40*, 41, 89
Nigeria, savanna in 65–6
Nihlgård, B. J. 97, *98*
Nile
    dams 89, 166, 208–9, 318
    delta: erosion 318–19; salinization
        166
    embankments 214
    Jonglei Canal diversion 299–300
    sediment load 208–10
nitrates 183–4
    water pollution 238, 239–44, 251,
        253, *254*
nitrogen fertilizers 184
nitrogen oxides *358, 360*, 365, 370,
    371
nitrous oxide 331, *334, 337, 338, 339*
no-regrets policies 416, *417*
no-tillage cultivation 179
nomadic pastoralists 29, 75–6
Norfolk Broads, origin of 264–5
North America
    acid rain 364, *365*
    bison decline 139
    chestnut blight fungus 87
    debris avalanches 191
    deforestation 54, 191
    forest decline 96
    global warming 390, *399*

megafauna extinctions 145, 147, 148,
    149
    prairies 29, 78–9, 80
    sand dunes 308
    urban heat islands *351*, 352
    water pollution 237, 250–1, *252*
    water transfers *216*
    see also Canada; United States
North Carolina
    dune stabilization 321–2
    experimental catchments 223, *253*
North Platte River 303
North Sea 258, 405
North York Moors, heather burning on
    196–7
Norway
    acid rain *366*
    aluminium smelter fumes 93
    lake acidification 369
Nossin, J. J. 297–8
Nriagu, J. O. 238–9
nuclear industries, pollution from 128,
    342
nuclear war, climatic effects of 341
Nurek Dam 325–6
nutrients *see under* soils

oak woodland loss 134
occupation mounds (tells) 270
oceans
    currents 383
    islands *see* islands
    overexploitation 140
    pollution 125, 257–60
    surface temperatures 390–2, 411–12
    see also coasts; sea level changes
oil pollution 125–7, 259–60
    wetlands 94
oil shales 267
oilfields
    environmental impacts 422–3
    fires 341–2
    fluid pressures 324
    seismic activity 327
    subsidence 282, *283*
Oke, T. R. 349, 350
Oklahoma, Dust Bowl in *195*
optimism 380
Opuntia (prickly pear) 102–3
Oregon, global warming in 398
organic compounds, synthetic 239,
    259
organic matter, soil 161, 188, 192, 241
organic soils *see* peat
organic waste *see* sewage
overfishing 140
overgrazing 49–50, 51, 66–7
    and desertification 71, 73
overpumping *see* groundwater:
    abstraction
Oxford
    fog frequency *359*
    ragwort 90
ozone
    and plants 93
    stratospheric depletion 339, 371–4
    urban concentrations 361–2, *363*

Pacific-global era *21*
pack ice *see* sea ice
Pacyna, J. M. 238–9
paddy soils 176
Page, M. J. 190
Pakistan
  accelerated weathering *293*
  irrigation 212; and salinization 169,
    293
  reservoir sedimentation 280
palaeoclimates
  in climatic prediction 384–5, 387
Palaeolithic peoples 16–20
  hunting 142, 144–7, 148–9
Pampas *49*, 79–80
Panama, savanna in 66
Papua New Guinea, soil erosion in *189*
parasites, introduced 102–3, 119–20
Park, R. A. 409
Parkinson, C. L. 393
particulate pollutants 355, 357, 358
  and plants 93
pasque flower decline 99, *100*
pastoralism
  environmental impact 420
  nomadic 29, 75–6
  see also grazing
Patric, J. H. 251
Patten, J. H. C. 181–2
peat 175–6
  cutting (turbary) 265
  drainage 183, 225, 285, *286*
  erosion 292
Pecos River 281
Peierls, B. L. 238
Peking Man 19
Pennines, peat erosion on 292
peregrine falcons *124*
permafrost
  fire suppression effects 45
  global warming degradation 394–6,
    401
  thermokarst 286–8
pessimism 380
pesticides 121–5, 244
pests
  biological control 102–3, 118–19
  introduced 99, 102–3
  see also predators
Peten Lakes 278–9
Peters, R. L. 398
Pethick, J. 409
Petrie, Sir W. F. 31
pH *see* acid rain
Philippines, mangrove destruction in 59
phosphates, water pollution by 238
photochemical smog 93, 360–2
Phytophthora cinnamomi 87
phytoplankton 371
piers and breakwaters 312–13, *315–17*
pigs, introduced 89
Pimental, D. 187
pines
  dune stabilization 310
  interception by 223, 346
  pollution affecting 93
Plaggen soils 176

planktonic algae 342
plantations
  erosion 320–1
  lateritization 174
  runoff 224
plants
  accidental dispersal 86–7
  agricultural improvements affecting
    99, *100–1*
  aquatic 89, 239–40
  collection and eradication 99
  cultivation: distinguished from
    domestication 22
  dependence on humans 83–4
  diseases and pests 86–7, 102–3
  dispersal 83–7, 89–91
  domesticated 21, 22–4, 29;
    naturalization 84–5
  genetic modification 104–5
  global warming response 398
  growth: soil erosion affecting 187
  introduced 83–8, 102–3, 224, 322–3
  invasive 50, 87–8; aquatic 89;
    control 99, 102–3; salt marshes
    322–3; weeds 90–1
  nutrients *see under* soils
  pollution affecting 91–4
  salinity tolerance 171–3
  selective breeding 103–4
  unpalatable: avoidance 51
  see also vegetation *and particular*
    species
Platte river 303, *304*, 307
Pleistocene
  human colonization 12
  megafauna extinctions 142, 144–9
Pliocene climate 387, *388*
plough, invention and diffusion of 26–7
ploughing
  contour-strip 200, *201*
  environmental impact 420
  hydrological effects 243
  and precipitation 346
  ridge-and-furrow patterns 180–2
  and runoff 224
  soil compaction 179
  and soil erosion 192
Pluhowski, E. J. 255
Plynlimon, water quality in 253, *254*
PM10s 360
podzolization 175–6
poisonous plants 51
polar regions
  global warming effects 384, 392–3,
    406
  ozone depletion 371, 372, *373*
polders *213*
policy-making
  for global warming 416–18, 433
pollen analysis 20, 52, 81
pollution
  and animals 121–30, 245–6, 249,
    256–7, 259
  types 34–5
  and vegetation 91–9
  see also air pollution; water pollution
    and particular pollutants

pool and hummock topography 292
Poole Harbour, salt marshes in 323
Poore, M. E. D. 57
population
    decline 14–15
    density: savanna 66; and shifting
        agriculture 421
    distribution 33
    growth 12–15, 142, *143*
    urban: and heat island intensity *351*,
        352; projections 32
Posch, M. 397
Potter, G. L. 344–5
poverty 16, 380
power stations
    and fish deaths 137–8
    flue gas desulphurization 371
    thermal pollution 254–5
prairies 29, 78–9, 80
    dust storms 193–4
pre-industrial civilizations 16–31,
    419–20
precipitation
    acidity *see* acid rain
    and arroyo trenching 289
    and deforestation 344, 345
    enhancement: lakes 348; and land use
        changes 346–7; rainmaking
        experiments 374–5
    feedback effects 383
    and forests 346
    future 385, 387–9, 400, 401
    salinity 162
    urban areas *349*, 352–3
    see also drought
predators, introduced 102–3, 119–20
predictions 379–81
    animal extinctions 159
    climatic *see* climatic change
    uncertainties in 379, 380–2, 405–6,
        415–16, 420, 433
preservation 426
    see also conservation
prickly pear control 102–3
primate habitat losses 154–6
Prince, H. C. 263
Pteridium aquilinum 51, 91
Pyne, S. J. 19–20, 48

Quine, T. A. 192–3

r-selected species 150–1
rabbits 88, 117–18, 308
Rackham, O. 81, 82
radiation *see* greenhouse effect; solar
    radiation
ragwort dispersal 90
railways, plant dispersal along 90
rain forest
    clearance 55–9, 60–1; climatic effects
        344–5; and lateritization 174–5
    habitat fragmentation 156, *157*
    potential resources 57–8
    secondary 61–2
    shifting agriculture 58–9
    species diversity 430
    susceptibility to perturbation 430

rainfall *see* precipitation
rainmaking experiments 374–5
Rangley oilfield 324
Rapp, A. 68, *70*, 75
Ratcliffe, D. A. 116–17
rats 120
Raubwirtschaft 6
ravens 134
Ray, C. 409
reactive adaptation 416
reclamation *see* land reclamation *and
    under* soils
Reclus, Emile 5–6
recombinant DNA 104–5
recreation and leisure
    and conservation 428
    impacts 421–2; on animals 137; on
        vegetation 103
Red Sea, fish dispersal from 113–14
red tides 258
Reed, D. J. 407
Reeves, R. W. *290*
reforestation and afforestation 60,
    224–5, 308
reindeer 113
Renberg, I. 367
reserves, wildlife 152–4
reservoirs
    climatic impact 347–8
    ecological explosions 89
    evaporation 164
    frost reduction 348
    hydro-isostatic effects 285–6, 326
    sea level rise mitigation 404
    sedimentation 187, 206–11, 280, *304*
    seismic effects 324–6
    subsidence and leakage 281
    thermal pollution effects 255–6
    see also dams
resilience, ecosystem 431
resource exploitation, destructive 6–7,
    8–9
Revelle, R. R. 401
Richards, P. W. 60
ridge and furrow 180–2
ringed plover expansion 116, *117*
Rio de Janeiro, building in 265
Rio Grande River 224
Ripley, E. A. 344
rivers
    acidification 369
    bankside vegetation 224, 307
    bypass and diversion channels 215,
        299–300
    canalization: and animal dispersal
        113–14
    channel incision 210
    channel modification 203–19, 262,
        299–305, 319–20, *321*; *see also*
        dams
    channelization 212–15, 319–20, *321*
    clear-water erosion 210
    discharge: and channel morphology
        303; drainage affecting 225;
        regulation by dams 204–6, 211;
        urban areas 303; vegetation
        changes affecting 222–6

embankments   214–15
global warming response   403
incision: and gully formation
   288–92
pollution   127, 235, 237–8, 241, *242*,
   251, 253, *254*; *see also* water
   pollution
salinity   171
sediment load   206–10, 257, 275;
   coastal effects   318–19, *320*
sedimentation   210–11
thermal pollution   254–7
turbidity   257; and fish   129
urban   219–22
water transfers between   215–17, *218*,
   229, 347
see also estuaries; lakes *and particular*
   rivers, e.g. Mississippi; Nile
roads
   and animals   137
   de-icing salt   247–8
   forests   191
   plant dispersal along   89–90
   and slope instability   295, 297
robber economy   6, 7
Robin, G. de Q.   393
Robinson, M.   225
Robinson, P. J.   333
rocks, soil salinity and   162
rodent control   125
Rössing uranium mine   *267*
Romme, W. H.   46
Rozanov, B. G.   170
ruderal habitats   37
runoff
   and global warming   400–1, 403
   and inter-basin water transfers   215,
      217
   and land use   *190*, 224–5
   and soil drainage   225
   and soil structure   180
   urban   220–2
   see also floods
Rybinsk reservoir   348

Sabadell, J. E.   68
sabkhas   410–11, 412
Sahagian, D. L.   404
Sahara
   origins   76
   rate of advance   69
Sahel, drought in   75, 340
saiga antelope   120–1
St Louis, precipitation in   353
Salinas valley   166
salinity
   Dead Sea   229
   groundwater   162–3, 165–8
   lakes   226, 227
   soils, natural   161–3
salinization
   and accelerated weathering   293
   antiquity   168–9
   causes   163–8
   effects   171–3
   extent and severity   168–71, 420
   reduction and control   173

salt
   extraction   284, *285*
   for road de-icing   247–8
salt cedar, introduction of   224, 307
Salt Lake City, cloudiness in   342
salt marshes
   accretion   322–3
   pollution   94, *122*
   and sea level rise   407–10
salt plains (*sabkhas*)   410–11, 412
Salvinia molesta   103
San Diego, pollution in   127
San Francisco
   pine tree drip in   346
   sedimentation   275
sand dunes
   coastal protection   321
   reactivation   308–9, 403
   stabilization   307–11, 321–2
sand fences   *310*, 311, 321
sanitization   48
Saudi Arabia, groundwater abstraction in
   230, 233
Sauer, Carl   7–8, 22–3, 38–9, 45, 185–6
savanna grass (*Imperata*)   59
savanna origins and maintenance   59,
   62–7
Savini, J.   219
Schickendanz, P. T.   347
Schmieder, O.   79, 87
Schneider, S. H.   385
Scolt Head Island   409
Scotland
   acid rain   364, 369
   debris flows   299
   dune stabilization   308, 310
   peat erosion   292
sea birds
   eggshell thinning   123
   islands   139–40
   oil pollution   125, *126*
sea ice
   and global warming   384, 392–3, 406
   removal   376
   water transfers affecting   347
sea level
   and beach formation   318
   and global warming   389–90, 404–15
sea surface temperatures
   and coral bleaching   411–12
   and hurricanes   390–2
sea walls   313, *314, 317*
sea water
   incursion   163, 165–7
   pollution   249–50, 257–60
Sears, P. B.   15
Seattle, effluents in   127
secondary rain forest   61–2
sediment load
   coastal effects   318–19, *320*
   increase   257
   mining-related   129, 275
   reduction below dams   206–10
sedimentation   262
   coral reef damage   127
   lakes   277–80
   rates   189–90, 191, 275–80

reservoirs 187, 206–11, 280, *304*
  and riverbank vegetation 307
  through waste disposal 273
seedbeds, fire effects on 48
seeds
  distribution by animals 49
  germination: and fire 48
Seeswood Pool 279
Seidel, S. 372
seismicity 285–6, 323–7
semi-arid areas
  desertification 75
  global warming effects 401–2, 403
  irrigation 346–7
  reservoir sedimentation 187
  salinity 161–3
semi-natural ecosystems 38
settlements
  nomadic pastoralists 76
  Palaeolithic 19
Severn, thermal pollution of 254
sewage
  compared with urban runoff 248
  coral reef damage 127–8
  water pollution 250–1
sewers, flood runoff affected by 220
Seychelles, mangrove swamps on *408*
shale tips *271*
shear stress 296
shearing resistance 296–7
sheep 50–1, 107
Sheffield, A. T. 190
shellfish, heavy metals in *122*, 123
Sherlock, R. L. 267–8
Shetland Isles, oil spill on *126*
shifting agriculture 39, 41, 58–9, 185,
  251, 421
shingle mining, cliff retreat and 311
ships
  animal introductions 108
  early 27, *28*
shrubland encroachment 50
Siberia, permafrost displacement in *395*
Sierra Nevada mountains 275
Silbury Hill *270*
silver-spotted skipper decline 133
Simmons, I. G. 20
Simon, J. L. 380
Sinai–Negev region 343–4
sink-holes 280, *281*
slash-and-burn *59*, 185, 421
slopes
  instability and failure 262, 294–9
  soil conservation 199–200
  soil erosion 192
smelting 30–1
  and vegetation 93
Smith, C. T. 212
Smith, M. W. 396, 397
Smith, S. V. 411
smog *356*
  photochemical 360–2
smoke
  climatic effects 341–2
  trends *359*
Snelgrove, A. K. 169
snow 332, 400

soil erosion
  causes: agriculture 185–7, 188–96,
    420; fire 196–7; grazing 118;
    salinity 171; urbanization 197–9;
    vegetation clearance 188–92,
    305–7
  effects 187; river channels 305, *306*;
    stream turbidity 129
  peat 292
  rates 187, 188–92, 197–8
  scale and extent 185–7, 420
  usefulness 202
soils
  acidification 175–7, 368–9
  active layer 286–8
  albedo modification 377
  compaction 49, 177–8
  conservation 199–202, 305
  drainage *101*, 173, 180–3, 225, 243,
    285, *286*
  erosion *see* soil erosion
  fertility: improvement 176; under
    shifting agriculture 41, 58–9, 185,
    251, 421; *see also* fertilizers
  and fire 185, 196–7
  formation 160–1, 176
  humus 185, *186*, 188
  infiltration capacity 177, 178–9, 180,
    188
  lateritization 173–5
  modification, classification 160–1
  moisture 402
  nutrients: cycling 251, 253; depletion
    175, 177; and grazing 51; released
    by burning 185
  organic content 161, 188, 192, 241
  ploughing *see* ploughing
  podzolization 175–6
  pollution: effects on vegetation 94
  reclamation 173
  salinity, natural 161–3
  salinization 163–73, 293, 420
  savanna 65
  structure: damage to 177–9, 180; and
    fertilizer use 184; forests 188;
    improvement 176, 179
  tundra 286–8
  see also particular types, e.g. peat
solar radiation 328–31
  dust effects 340
  human energy use compared with 348
  reflection *see* albedo
Somerville, Mary 3–4
South Africa
  air pollution *332*
  fires *43, 45*
  gold mining 280, *281*, 326
South and Central America
  Africanized honey bee 110, *112*
  deforestation *58*
  fire use 41
  fish introductions 116
  savanna 66
  shifting agriculture 39
  sparrow dispersal *119*
  see also Belize; Brazil; Mexico; pampas;
    Venezuela

South Platte River   303, *304*
South Saskatchewan River   206, *207*
South-East Asia
    habitat losses   154–6
    plant domestication   22–3
    see also China; Indonesia; Malaysia
Southampton Water, oil in   94
Soviet geography   8–9
Spain, sea-water incursion in   167
sparrows   118, *119*
Spartina grasses   322–3, 408
species diversity
    birds, urban   115
    coral reefs   127
    and ecosystem vulnerability   428–31
    and farm modernization   *134*
    and fire   48
    grasslands   104, 132–3
    and grazing   49
    introductions affecting   88–9, 116
    islands   152, *153*, 430
    maintenance   159, 428
    rain forest   430
    reduced through selective breeding
        103–4
    secondary rain forest   62
Sphagnum   225
spoil tip reclamation   94
steppe, ploughing of   135, 137
steppe bison   147–8
Stewart, O. C.   78
Stocks, J.   246–7
Stoddart, D. R.   320–1, 410
Stone Age   16–20
    animal extinctions   142, 144–9
    tools   17, *18*, 82, 263
stratospheric ozone depletion   339,
        371–4
streams *see* rivers
strip mining   266, *268*
subnatural ecosystems   38
subsidence   262, 280–8, 405
suburbs, birds in   115
Sudan
    channel modification   299–300
    desertification   69, *70*, 72
Sudd swamps   300
Suez Canal, fish dispersal along   113–14
Suffolk, hedgerow removal in   *131*
sugar beet   193
sulphates
    air pollution   342–3, *364*, 366–7
    critical load   177
    water pollution   237, 246–7
sulphur dioxide emissions   365
    effects: on plants   91–3; on weathering
        293
    and global warming   342–3
    reduction   371
    trends   *359*, 360, *366*, 369–71
    variation   *357*, 358
Sumatra, habitat fragmentation in   *157*
sun, earth's orbit around   330
sunlight
    in photochemical smog formation
        361
sunshine trends   *359*

sunspot cycles   328
sustainable development   61
swamps, river channelization and   215
Swansea, coal-burning in   93
Swaziland
    dambos   *188*
    grassland burning   *39*
Sweden, acid rain in   *366*, 367, 368,
        369
swidden *see* shifting agriculture
Swift, L. W.   256
synanthropes   118
synthetic organic pollutants   239, 259

Tallis, J. H.   292
Tame, River   222
tank landscape   211
tankers, oil   125, *126*, 259–60
Tanzania, stone tools in   17
Tawe valley, floods in   225–6
tea plantations   224
teak plantations   174
technological development   15, 16
    Bronze and Iron Age   30–1
    Mesolithic and Neolithic   20–9
    Palaeolithic   16–20
    urban and industrial   31–6
tells   270
temperature
    and deforestation   343–5
    and dust   340
    feedback effects   382, 383
    urban areas   348–52, 353
    and water vapour   342
    see also global warming
Tennessee Valley, lakes in   89
terns   137
terraces, soil conservation   199–200,
        201, 202
Texas
    precipitation   347
    sediment loads and
        sedimentation   307, 319
    soil erosion and dust storms   *194,
        195*
Thailand, metal working in   30
thallium sulphate poisoning   125
Thames, pollution of   235, 241, *242*,
        249
thermal pollution   254–7
thermokarst   286–8
Thomas, D. S. G.   69
Thomas, R. H.   393
Thomas, W. L.   9
thunderstorms, urban   352–3, *354–5*
Tickell, Sir Crispin   418
tides, sea level changes and   405
tile drains   180, 243
Titus, J. G.   372
Tokyo, subsidence in   284
Tolba, M. K.   69
Tomlinson, T. E.   241
tools, Stone Age   17, *18*, 82, 263
Torrey Canyon oil spill   125
tourism   421–2
    see also recreation and leisure
trampling   49, 51–2, 103, 178

transport
  early developments 27, *28*
  see also railways; roads; tankers;
    vehicles
Transvaal, sink hole in *281*
travertine 294
tree rings
  fire frequency evidence 42–3
  soil erosion evidence 189
trees
  CO$_2$ effects on 400
  in dune stabilization 310
  fire resistance 65
  global warming effects 398, *399*, 400
  on grassland 78, 79
  interception 346
  stresses and dieback *95*, 97–8
  see also deforestation; forests; rain
    forest *and particular species*
Trimble, S. W. 178
Trinidad Pitch Lake 259
tropical cyclones *see* hurricanes
tropical soils
  fertility 41, 58–9, 185, 251, 421
  lateritization 173–5
tropical vegetation *see* rain forest;
    savanna
trout 108, *110*, 246
Trustrum, N. A. 190
tufa decline 294
tundra
  global warming 396, 400–1
  soils 286–8
Tunisia, desertification in 73
turbidity, river 129, 257

UK Global Environment Research Office
    11
ultraviolet radiation 371
uncertainties 379, 380–1, 415–16,
    420
  General Circulation Models 381–2
  ice sheet behaviour 393
  sea level rise 405–6
United Kingdom *see* Britain
United Nations Environment Programme
    (UNEP) 69, 259
  Global Environmental Monitoring
    System 355
United States
  afforestation 60, 224
  air pollution *see* Los Angeles
  animals: introduction and dispersal
    108, 110, *111–12*, 113, 114, 142;
    pesticides and pollution affecting
    123, 127, 245–6; urban 115–16
  arroyo trenching 288–9, *290*
  climate: and land use 345–7;
    predictions 401–2; urban 352,
    353
  cloudiness: and aircraft 342
  coasts: erosion *315*, 319–20, 321–2,
    413; pollution 272–3
  construction: and soil erosion 198
  dams and reservoirs 204, 281;
    sediment load below 206–8, *209*,
    210, 211, 303–5

deforestation: clear-cutting experiments
    222, 223, 251, 253; and climate
    345–6; and sediment yield 191
drought 193, 401–2
Dust Bowl and dust storms 187,
    193–6, *402*, 403
earthquakes 324
experimental catchments 222, 223,
    253, 256
fires: frequency and causes *42*; and
    soil erosion 197; suppression *44*,
    45–7
global environmental change definition
    11
global warming and sea level rise
    398, 401–2, 403, 409, *413*
grasslands and grazing 50
groundwater abstraction 166, 230,
    232, 283–4
hailstone damage *376*
insects, introduced 110, *111–12*
irrigation 230, 232, 346–7
lake acidification 246
mass movements 297
mining 266–7
oilfields 282, 324, 327
permafrost 287
pesticide use 123
plant introduction and dispersal 50,
    86
prairies 29, 78–9, 80
precipitation 345–7, 401–2
rivers: bankside vegetation 224, 307;
    channel form and land use 305–7;
    channel modification *213*, 299,
    319–20; pollution *see* water
    pollution, *below*; sediment loads
    206, 208, *209*, *210*, 303–7, 319,
    320; *see also particular rivers, e.g.*
    Mississippi
road de-icing 248
sand dunes 321–2, 403
sea level rise 409, *413*
sediment discharge 319–20
sedimentation 191, 275–6, 307
soils: conservation 200, *201*; erosion
    186–7, 189, 193–6, 197, 198, *402*,
    403; infiltration capacity *178*
subsidence 280, *281*, 283–4
water pollution: and animals 127;
    feedlots 244–5; trends 237, 251,
    252; urban 248
weathering, urban areas 293–4
wetland losses 135
see also North America *and particular
    states, etc., especially* California;
    Colorado
uranium mining *267*
urban areas
  agricultural origins in 23
  air pollution 355–62; and plants
    91–3; and weathering 293–4
  animals and birds 115–16, 118–19
  antiquity 31–2
  climate 348–55, 378
  groundwater recharge 234–5
  population projections 32

vegetation 70, 72, 91–3
waste 118–19
water pollution 248, *249*
urban heat island 348–52, 353
urbanization 31–2
hydrological effects 219–22, 303
land reclamation for 270, 271, 272, *273*
and mass movements 295
and soil erosion 197–9
and thermal pollution 255
USSR *see* CIS

Vaiont Dam 297, *326*
Valencia, Lake 2, 226
valley-bottom gullies (arroyos) 288–91
van Andel, T. H. 291
Vaudour, J. 294
vegetation
albedo effects 343–5
and animal introductions 88–9
aquatic 89, 103, 239–40
artificial lakes 89
and channel form 307
clearance: and arroyo trenching 289; by fire *see* fire; and coastal erosion 320–1; and infiltration capacity 178–9; permafrost areas 287–8; and salinization 167–8; and soil erosion 188–92, 305–7; *see also* deforestation
and climatic change 80, 81, 83, 396–400
CO_2 effects on 400
and coastal erosion 320–1
deserts 431
dune stabilization 308, 309–11
estuaries 218
and evapotranspiration 344
fire adaptation 48, 77–8, 79
fire resistance 65
global warming effects 389, 396–400
human modification: early evidence 20, 80–1; increasing scale 37–8; *see also particular processes and activities, e.g.* fire
islands 87, 88–9
and megafauna extinction 149
nutrient cycling 251
roads and railways 89–90
succession *47*
trampling 51–2, 103
see also deforestation; forests; plants *and particular types and species*
vehicles
and animals 137
permafrost vegetation damage 288
pollution 32, 352
soil compaction 177, *178*
Venezuela
deforestation 2
lake levels 2, 226
savanna 65
Venice, subsidence in *282*
Viets, F. G. 244
virgin lands scheme 135, 137, 196

Virginia
marshes 409
soil erosion 198
Vita-Finzi, C. 289, 291
Vogl, R. J. 138
volcanic soils, erosion of 202
volcanoes
ash 372
dust 330–1, 340
lava flows 327
Von Humboldt, Alexander 2, 226

Waggoner, P. E. 401
Wagon Wheel Gap 222
Wake Island 139
Waldsterben/Waldschäden (forest death/decline) 94–9
Wales
afforestation 225
air pollution 93
deforestation 253, *254*
floods 183, 225
peat degeneration 292
soil drainage 183
soil erosion 191
woodland birds 134, *135*
Walker, H. J. 320
Walling, D. E. 192–3
warfare, craters caused by 265
Warren, A. 68, 71
Warrick, R. A. 381–2
Warrilow, D. A. 345
Washington, Lake, sewage in 250–1
waste
coastal cities 272–3
dumping 269–70, *271*
mining *see* mining
organic: feedlots 244–5; sewage 127–8, 248, 250–1
urban areas 118–19
water
importance 203
inadequate efforts at controlling 203
inter-basin transfers 215–17, *218*, 227, 229, 347
irretrievable losses *204*
pollution *see* water pollution
salinity 162–3, 165–8, 171
soil *see* drainage
temperature: sea surface 390–2, 411–12; streamwater 254–7
usage: industrial 222; semi-arid areas 233; trends 203, 215, 233
see also groundwater; lakes; rivers; water table
water pollution 235
effects: on fish 127, 240, 245–6, 249, 256–7; on vegetation 94
marine 257–60
oil 125–7, 259–60
reversibility 248–51
scale and extent 237–8
sources 235–6, 238–9; acid rain 245–6; agriculture 236, 239–45; mining 246–7; road de-icing 247–8; sewage 250–1; suspended

sediments 257; urban runoff 248,
249
thermal 254–7
trends 236–8, 241, *242*, 248–51
water quality, deforestation and 251–3,
254
water resources
groundwater 165–7, 229–33
management *417*
water spreading techniques 233
water table
and channel improvement 215
and irrigation 163–5, 293
lowering: and salinization 227; and
subsidence 280, *281*; *see also*
drainage
rising levels 233–4
see also groundwater
water transfers 215–17, *218*
climatic impact 347
for lake augmentation 229
urban areas 222
water vapour 332
from aircraft 342
feedback effects 382
waterlogging, peat formation and
175–6
Watson, A. 273–5
wave energy 409
weather *see* climate
weathering 293–4
and shearing resistance 296
Weaver, J. E. 78
Webb, T. 384, 385
Weber, P. 258–9
weeds 90–1
Welland Canal 114
wells
nitrates in 241
overgrazing around 71, *73*
overpumping 165
waste disposal in 324
Wells, P. V. 78
West Bay, erosion at 313, *316*
West Indies, species diversity on *153*
West Virginia, clear-cutting in 251, 253
Westhoff, V. 38
Wetherald, R. T. 401–2

wetlands
drainage 183, 285
losses 135
oil spills 94
see also mangrove swamps; salt
marshes
Weymouth, coastal defences at *314*
whaling 140, *141*
wheat cultivation, dust storms and 193,
195
Whitaker, J. R. 6–7
Whitney, G. G. 80
Wigley, T. M. L 384, 385
wildfowl, lead poisoning of 129
Wilson, K. V. 221
wind
hurricanes 376
soil erosion 187, 193–6, 200
urban areas *349, 353, 355*
windbreaks 200, 377
Wisconsin
land use changes 307
sedimentation 275
soil conservation 200, *201*
Woeikof, A. 6
Wolman, M. G. 198
wolves 154
Woo, M.-K. 396
wood usage
and desertification 70, *72*
early 17, 19
for iron smelting 31
woodland *see* deforestation; forests
Woodwell, G. M. 384
Wymer, J. 20

Yale, gold-washing at *276*
Yellow River 189
Yellowstone National Park, fires in
45–7
Yorkshire Wolds, forest clearance on
81

Zabinski, C. 398
Zambezi River *205*
Zimbabwe, bush clearance in 67

Index compiled by Ann Kingdom